Jordan Structures in Lie Algebras

Mathematical
Surveys
and
Monographs

Volume 240

Jordan Structures in Lie Algebras

Antonio Fernández López

American Mathematical Society
Providence, Rhode Island

2010 *Mathematics Subject Classification.* Primary 17B05, 17C10; Secondary 17B60, 17B65, 17C65.

For additional information and updates on this book, visit
www.ams.org/bookpages/surv-240

Library of Congress Cataloging-in-Publication Data

Names: López, Antonio Fernández, 1952- author.

Title: Jordan structures in Lie algebras / Antonio Fernández López.

Description: Providence, Rhode Island : American Mathematical Society, [2019] | Series: Mathematical surveys and monographs ; volume 240 | Includes bibliographical references and index.

Identifiers: LCCN 2019010955 | ISBN 9781470450861 (alk. paper)

Subjects: LCSH: Jordan algebras. | Lie algebras.

Classification: LCC QA252.5 .L665 2019 | DDC 512/.482–dc23

LC record available at https://lccn.loc.gov/2019010955

∞ The paper used in this book is acid-free and falls within the guidelines
established to ensure permanence and durability.
Visit the AMS home page at `https://www.ams.org/`

10 9 8 7 6 5 4 3 2 1 24 23 22 21 20 19

To Conchi, my wife, who shared my dreams

Contents

Preface

This is neither a book on Lie algebras nor on Jordan systems. This book is just intended to show how Jordan theory can be applied to Lie algebras. In one sense it is introductory, since very few prerequisites on Lie algebras are presupposed. In fact, Lie readers, especially those accustomed to working with root systems or with the fruitful relationship between Lie groups and Lie algebras, might be disappointed with the fact that most of the Lie algebras considered here are over a ring of scalars where only the invertibility of $2, 3$, and sometimes 5, is assumed, and even in the case of Lie algebras over a field, these are not necessarily finite-dimensional. Yet in this poor soil a germinal form of root system can still grow: the grading induced by a compatible family of idempotents (a notion which is defined in Jordan terms and which connects with the Jacobson–Morozov lemma) resembles in a broad sense the root decomposition associated to a root system. On the other hand, the book is far from being elementary. We allow ourselves to use any Jordan result under the sun if this were required for solving a single Lie question.

Honestly I must say that in spite of the abundance of results that the reader can find on applications of Jordan theory to Lie algebras in the text, the book has some shortcomings, as the lack of a proof, or at least an outline of it, of the structure theorem of simple Lie algebras with a finite grading. Nevertheless, making a virtue of necessity, this important gap might encourage him to study the proof directly from Zelmanov's paper itself.

This book is the fruit of a long collaboration of the author with Esther García and Miguel Gómez Lozano; a great deal of the results exposed in it are due to them. I am also grateful to Alexander Baranov, Jose Brox, Miguel Cabrera, Esperanza Sánchez Campos, and Guillermo Vera De Salas for their valuable suggestions and again to Miguel Gómez Lozano for his help and technical assistance in TEXquestions.

I acknowledge the support of the Spanish Ministerio de Economía y Competitividad [1]

I hope that the reader can enjoy reading the book as much as I did writing it.

Antonio Fernández López
Málaga, Spain
October, 2018

[1] MTM2013-47154-P, MTM2014-52470-P, and MTM2017-84194-P (AEI/FEDER, UE.)

Introduction

The contents of this book are organized into fifteen chapters. Rather than giving a description of the different sections of each chapter we prefer to highlight the more relevant results and notions exposed in them.

The aim of Chapter 1 is to present those notions and basic results that Lie algebras have in common with other algebraic structures as associative and Jordan algebras, so in this chapter we adopt the framework of an algebra that is not necessarily associative. It is shown here that simple algebras are not locally nilpotent.

Although being introductory, Chapter 2 still contains some important, and by no means trivial, results in structure theory, as the existence of a locally nilpotent radical (in the Kurosh sense) in any Lie algebra in which every element is ad-nilpotent.

The study of ad-nilpotent elements of index at most 2 (called *absolute zero divisors*, a term borrowed from Jordan theory, or *sandwiches*, as A. I. Kostrikin called them) is the goal of Chapter 3. A Lie algebra is said to be *nondegenerate* if it has no nonzero absolute zero divisors. Among other relevant results gathered in this chapter we mention a theorem, due to A. I. Kostrikin and E. Zelmanov, proving that every 2-torsion free Lie algebra generated by absolute zero divisors is locally nilpotent.

Ad-nilpotent elements of index at most 3 (called *Jordan elements*) are a key tool in the applications of Jordan techniques to Lie algebras. In Chapter 4 we study the properties of Jordan elements in a nondegenerate Lie algebra, their connection with abelian inner ideals, and conditions guaranteeing their existence, as the celebrated Kostrikin Descent Lemma.

A Jordan element x of a Lie algebra L is called *von Neumann regular* if $x \in \mathrm{ad}_x^2 L$. There is a good reason for adopting this terminology: they play in Lie algebras a role similar to that played by von Neumann regular elements in associative and Jordan algebras. In Chapter 5 it is shown that any von Neumann regular element e in a Lie algebra L over a ring of scalars in which $2, 3$ and 5 are invertible can be extended to a $\mathfrak{sl}_2$-triple (e, f, h), where f is also von Neumann regular (the pair (e, f) is then called an *idempotent* of L), and gives rise to a 5-grading on L (Jacobson–Morozov Lemma), which resembles the Peirce decomposition associated to an idempotent of a Jordan pair. We define the socle of a Lie algebra L as the sum of its minimal abelian inner ideals and prove that when L is nondegenerate, the socle is a direct sum of ideals, each of which is a simple nondegenerate Lie algebra coinciding with its socle. We also characterize nondegenerate Lie algebras with an essential socle as an essential subdirect product of strongly prime Lie algebras with nonzero socle.

A distinguished family within the class of von Neumann regular elements is that of the so-called extremal elements: an element e of a Lie algebra L over a field $\mathbb{F}$ that is not an absolute zero divisor is *extremal* if $\mathrm{ad}_e^2 L = \mathbb{F}e$. Over an algebraically closed field of characteristic $p > 5$, classical Lie algebras (the modular version of the complex finite-dimensional simple Lie algebras) are characterized by the property of being generated by extremal elements. Extremal elements are studied in Chapter 6, where it is given a classification-free proof of the fact that a simple finite-dimensional Lie algebra over a field $\mathbb{F}$ of characteristic 0 or $p > 3$ that contains an extremal element is either generated by extremal elements or $\mathrm{char}(\mathbb{F}) = 5$ and L is the 5-dimensional Witt algebra over $\mathbb{F}$.

Following a paper by E. García and M. Gómez, we prove in Chapter 7 that any 6-torsion free Lie algebra L is *strongly prime*; that is, L is prime and nondegenerate if and only if $[x, [y, L]] \neq 0$ for any nonzero elements $x, y \in L$. Similar characterizations are known for Jordan systems.

Chapter 8 is the core of the book since it is in this chapter where a first Lie–Jordan connection is introduced by associating a Jordan algebra L_a with a Jordan element a of a Lie algebra L (over a ring of scalars in which 6 is invertible). Most properties of the Lie algebra are inherited by their Jordan algebras, and some of the notable elements studied in the previous chapters, as von Neumann regular or extremal elements, are characterized in terms of their associated Jordan algebra. As an application of this Lie–Jordan connection, we provide a classification-free proof of the fact that every simple finitary Lie algebra over an algebraically closed field of characteristic 0 is spanned by its extremal elements and give the proof of Zelmanov's solution of the Kurosh–Lie problem for the particular case of a Lie algebra over a field of characteristic 0 (in its general version for Lie algebras over an arbitrary field, this result of E. Zelmanov has important implications in group theory, extending significatively the positive solution of the Restricted Burnside Problem and his work on compact torsion groups).

In Chapter 9 we study the Kostrikin radical of a Lie algebra L, namely, the smallest ideal I of L such that the Lie algebra L/I is nondegenerate. By a strongly prime ideal of a Lie algebra L we mean an ideal P of L such that the algebra L/P is strongly prime. The Kostrikin radical of a Lie algebra is contained in the intersection of its strongly prime ideals, but while for associative algebras the Baer radical is equal to the intersection of the prime ideals, an analogous result does not hold for Lie algebras. Conditions guaranteeing the coincidence of the Kostrikin radical of a Lie algebra with the intersection of its strongly prime ideals are provided in this chapter.

A well-known theorem due to E. Zelmanov proves that any Lie PI-algebra with algebraic adjoint representation over a field of characteristic 0 is locally finite. In particular, Lie algebras with algebraic adjoint representation of bounded degree are locally finite. Using the transference of properties of a Lie algebra to its Jordan algebras, we give in Chapter 10 an independent proof of the second Zelmanov's theorem quoted above. We also prove that any strongly prime Lie algebra with algebraic adjoint representation of bounded degree over a field of characteristic 0 is simple and finite-dimensional over its centroid, which is an algebraic extension of the base field.

Mimicking the Jordan pair case, in Chapter 11, we associate a Jordan pair to every abelian inner ideal B of a Lie algebra L. This Jordan pair, called the *subquotient of L with respect to B*, inherits regularity conditions from L and shares with it the inner ideal structure within B. For instance, the subquotient is a nondegenerate Artinian Jordan pair whenever L is nondegenerate and B has finite length, namely, if there is a bound for the lengths of proper chains of inner ideals of L contained in B.

In Chapter 12 we introduce the notion of *complemented inner ideal*, similar to that introduced by O. Loos and E. Neher for inner ideals of a Jordan pair, and prove that any abelian inner ideal B of finite length of a nondegenerate Lie algebra L gives rise to a finite $\mathbb{Z}$-grading $L = L_{-n} \oplus \cdots \oplus L_n$ with $B = L_n$, whence B is a complemented inner ideal. In particular, simple nondegenerate Artinian Lie algebras are complemented. The converse is also true: A Lie algebra is *complemented*, that is, every inner ideal is complemented, if and only if it is a direct sum of simple nondegenerate Artinian Lie algebras. As an instance of how the Jordan information contained in the subquotient is relevant, we show that if B is an abelian inner ideal of a simple Lie algebra L such that the subquotient V of L with respect to B is a simple exceptional Jordan pair, then L is the Tits–Kantor–Koecher algebra of V and therefore a Lie algebra of type E_6 or E_7.

In Chapter 13 we study the inner ideal structure of the following types of Lie algebras: Lie algebras coming from a centrally closed prime ring of characteristic not 2 or 3, Lie algebras of skew-symmetric elements of a centrally closed prime ring R with involution of characteristic 0 or greater than 5 ($\mathrm{char}(R) \neq 2, 3$ if R is simple), and exceptional Lie algebras over an algebraically closed field of characteristic 0.

As an application of Zelmanov's theorem for simple Lie algebras having a finite $\mathbb{Z}$-grading, and the fact that any nonzero von Neumann regular element in a Lie algebra gives rise to a 5-grading, we give in Chapter 14 a "Jordan proof" of Baranov's structure theorem of simple finitary Lie algebras over an algebraically closed field of characteristic 0, extend this result to strongly prime Lie algebras containing extremal elements, and give a new proof of the characterization, due to A. A. Baranov and J. Rowley, of simple locally finite Lie algebras of diagonal type.

Finally, in Chapter 15 we deal with Banach–Lie algebras. Using the algebraic information obtained in the previous chapter, we show that strongly prime Banach–Lie algebras containing extremal elements have the same algebraic-topological structure as that of the classical Banach–Lie algebras of compact operators on Hilbert spaces. We also prove the uniqueness of the complete norm topology in *primitive* (a notion of primitivity for Lie algebras is introduced in this monograph) Banach–Lie algebras, the automatic continuity of derivations of nondegenerate Banach–Lie algebras with essential socle, and the fact that nondegenerate Banach–Lie algebras spanned by extremal elements are finite-dimensional.

CHAPTER 1

Nonassociative Algebras

In this first chapter we are concerned with those aspects of the theory of Lie algebras which are commonly found in other algebraic structures, as associative rings and Jordan algebras. For this reason we adopt the general framework of an arbitrary (not necessarily associative) algebra A. We also take the opportunity to standardize notation and terminology. Basic definitions and notation are given in Section 1.1. In Section 1.2 we introduce the multiplication algebra $\mathcal{M}(A)$ and the centroid $\Gamma(A)$ of A, and study the behavior of $\Gamma(A)$ when A is semiprime, prime, or simple. In Section 1.3 we study the extended centroid $\mathcal{C}(A)$ of a semiprime algebra A. The notions of nilpotency and local nilpotency for nonassociative algebras are introduced in Section 1.4, where it is proved that a simple algebra cannot be locally nilpotent. In Section 1.5, we take a glance at the right and the symmetric Martindale algebras of quotients of a semiprime ring R and prove that the center of any of these algebras of quotients coincide with the extended centroid of R as defined in Section 1.3. As an example of a nonassociative algebra, and because it plays a relevant role in the construction of the exceptional Lie algebras, we introduce in Section 1.6 the split Cayley algebra (or vector algebra) as realized by Zorn as 2×2-matrices.

1.1. Definitions and notation

Nonassociative information can be found in the books [**Sch66, ZSSS82**].
We put $X \subset Y$, equivalently, $Y \supset X$, to express that the set X is a subset of Y, without excluding the case that $X = Y$ (symbols $\subseteq$ and $\supseteq$ never be used in any part of the book). We also warn the reader that the term *identity* will be used in a broad sense, so it should not be confused with the formal concept of identity.

- We write 1_X to denote the identity map on the set X.
- Given a linear map φ, $\ker(\varphi)$ will denote its kernel and $\operatorname{im}(\varphi)$ its image, and by $\varphi(x)$, φx, $(x)\varphi$, $x\varphi$, we mean the image of the element x under φ.
- Given a real number α, we denote by $[\alpha]$ the integer part of α.
- By a *ring of scalars* we mean a commutative (associative) ring with 1. Unless otherwise specified, Φ will denote an arbitrary ring of scalar and $\mathbb{F}$ will stand for a field. We write Φ^* to denote the *multiplicative group* of invertible elements of Φ, so $\mathbb{F}^* = \mathbb{F} \setminus \{0\}$. As usual, $\operatorname{char}(\mathbb{F})$ denotes the characteristic of $\mathbb{F}$.
- Let M be a Φ-module. For any subset $M \subset X$, denote by $\operatorname{span}_\Phi(X)$ the Φ-submodule of M spanned by X.
- We write $\Phi[\lambda]$, $\mathbb{F}[\xi]$ or $\mathbb{F}[x]$ to denote the ring of polynomials in one variable.
- By an *algebra over* Φ (in short, a Φ-algebra, or simply an algebra when no confusion can arise) we understand a unitary Φ-module A with a bilinear operation

which we will usually denote by juxtaposition. Thus no associativity condition is assumed, neither is it supposed the existence of a unit element in A. A *unital algebra* is one that has a *unit element* 1: $1x = x = x1$ for all $x \in A$. Any nonunital algebra A can be imbedded in its *unital hull* $\hat{A} = \Phi 1 + A$ with product extending that of A and having 1 as unit element. As usual, $[x, y] = xy - yx$ is the commutator of x, y.

• Given $a \in A$, l_a (resp. r_a) denote the left (resp. right) multiplication by a.

• An associative algebra will be usually denoted by R. By a *ring* we mean an associative $\mathbb{Z}$-algebra. We denote by $\mathrm{M}_n(R)$ the associative algebra of n-by-n matrices with entries in an associative algebra R, and by $Z = Z(R)$ the *center* of R.

• Given a Φ-module M, $\mathrm{End}_\Phi(M)$ stands for the associative Φ-algebra of endomorphisms of M.

• We denote by $[ij]$ the entry matrix with 1 in the (i, j) position.

• Let A be a Φ-algebra and let X, Y be nonempty subsets of A. We write XY to denote the Φ-submodule of A spanned by the set $\{xy : x \in X, y \in Y\}$. If $Y = X$ we will simply write X^2 instead of XX.

• A *subalgebra* S of A, denoted by $S \leq A$, is a Φ-submodule S of A such that $S^2 \subset A$. Given $X \subset A$, we write $\langle X \rangle_A$ to denote the subalgebra of A generated by X. If $X = \{x\}$ we write $\langle x \rangle_A$ by $\langle \{x\} \rangle_A$.

• A (two-sided) *ideal* I of A, denoted by $I \lhd A$, is a Φ-submodule I of A such that $AI + IA \subset I$. Clearly every ideal is a subalgebra. By abuse of notation, we write 0 to denote the ideal $\{0\}$. Note that every algebra is an ideal of its unital hull. Given $X \subset A$, we write $\mathrm{id}_A(X)$ to denote the ideal of A generated by X, and put $I * J$ for the ideal generated by IJ. If $X = \{x\}$, we write $\mathrm{id}_A(x)$ in place of $\mathrm{id}_A(\{x\})$. If R is an associative algebra, then the product IJ of two ideals I, J of R is an ideal, but this does not remain true for arbitrary algebras.

• If I is an ideal of A, then the Φ-module A/I becomes a Φ-algebra by defining $(x + I)(y + I) := xy + I$, for all $x, y \in A$, called the *quotient of A with respect to the ideal I*. Usually we will write $\bar{x}$ to denote the coset $x + I$.

• The notions of *homomorphism* and *isomorphism* between algebras are defined in the obvious way. We write $A \cong B$ if A is isomorphic to B and denote by $\mathrm{Aut}(A)$ the group of automorphisms of A.

• An algebra A is said to be *semiprime* if for any ideal I of A, $I^2 = 0$ implies $I = 0$; *prime* if $A \neq 0$ and for I, J ideals of A, $IJ = 0$ implies $I = 0$ or $J = 0$; and *simple* if $A^2 \neq 0$, and 0 and A are the only ideals of A.

Derivations. A Φ-linear map $d : A \to A$ such that $d(xy) = d(x)y + xd(y)$, $x, y \in A$, is called a *derivation*, , i.e. d satisfies the equivalent conditions:

$$l_{d(x)} = [d, l_x], \quad r_{d(y)} = [d, r_y], \text{ for all } x, y \in A. \tag{1.1}$$

Consequently, $l_{d^m(x)} = [d, l_{d^{m-1}}(x)]$ and $r_{d^m(x)} = [d, r_{d^{m-1}}(x)]$, for all $x \in A$. A simple induced argument shows that this implies

$$l_{d^m(x)} = \sum_{k=0}^{m} (-1)^k \binom{m}{k} d^{m-k} l_x d^k, \quad r_{d^m(x)} = \sum_{k=0}^{m} (-1)^k \binom{m}{k} d^{m-k} r_x d^k. \tag{1.2}$$

The nth-power of any derivation d of A satisfies the *Leibniz rule*:

$$d^n(xy) = \sum_{k=0}^{n} \binom{n}{k} d^{n-k}(x)d^k(y). \tag{1.3}$$

For $\alpha, \beta \in \Phi$, we have:

$$(d - (\alpha+\beta)1_A)^n(xy) = \sum_{k=0}^{n} \binom{n}{k} (d - \alpha 1_A)^{n-k}(x)(d - \beta 1_A)^k(y), \tag{1.4}$$

which can be proved by using induction and basic properties of the binomial coefficients.

LEMMA 1.1. *Let d be a derivation of A such that $d^n = 0$ for some $n \geq 2$. If A is n-torsion free, then $(d^{n-1}(A))^2 = 0$.*

PROOF. It follows from $d^n(xd^{n-2}(y)) = 0$ by using (1.3) together with the assumption that A is n-torsion free. □

Nilpotent Derivations and Automorphisms. Let A be a Φ-algebra and let d be a *nilpotent derivation* of A, i.e. $d^m = 0$ for some $m \geq 1$. Then the usual exponential power series makes sense and the formula

$$\exp(d) = 1_A + d + \frac{1}{2}d^2 + \cdots + \frac{1}{(m-1)!}d^{m-1}$$

defines a Φ-linear map of A whenever $(m-1)! \in \Phi^*$. But to guarantee that $\exp(d)$ is an automorphism further conditions on Φ are required.

LEMMA 1.2. *Let A be a Φ-algebra and let d be a derivation of A such that $d^m = 0$ for some $m > 1$. If $(2m-2)! \in \Phi^*$, then $\exp(d) \in \mathrm{Aut}(A)$.*

PROOF. By the Leibniz rule, $d^k(xy) = 0$, $m \leq k \leq 2m-2$ and $x, y \in A$. Then the proof follows as in the case that Φ is a field of characteristic 0 (see [**Hum72**, 2.3]). □

We can sometimes relax these restrictions on the ring of scalars Φ. For instance, if $m = 2$ then only 2-torsion free is required, and if $m = 3$, then 3-torsion free plus invertibility of 2 are enough to guarantee that $\exp(d)$ is an automorphism.

COROLLARY 1.3. *Let A be an algebra over a field $\mathbb{F}$ of characteristic $p > 2$, and let d be a derivation of A. If $d^m = 0$ for some $1 < m \leq (p+1)/2$, then $\exp(d) \in \mathrm{Aut}(A)$.*

PROOF. Since $1 < m \leq (p+1)/2$, $2m-2 < p$, so $(2m-2)! \in \mathbb{F}^*$ and Lemma 1.2 applies. □

Gradings and pregradings. Let A be a Φ-algebra and let G be a group with e as neutral element. A *G-grading* on A is a family $\{A_g\}_{g\in G}$ of Φ-submodules of A such that $A = \bigoplus_{g\in G} A_g$ and $A_gA_h \subset A_{gh}$ for all $g, h \in G$. A G-grading $A = \bigoplus_{g\in G} A_g$ is said to be *finite* if the set $G^* := \{g \in G : A_g \neq 0\}$, called the *support of the grading*, is finite. A grading is said to be *nontrivial* if its support contains an element different from e. By a *graded algebra* we understand an algebra equipped with a nontrivial grading.

Along with gradings we consider the less restrictive situation where the sum $A = \sum_{g\in G} A_g$ is not necessarily direct but the inclusions $A_g A_h \subset A_{gh}$ for all $g, h \in G$ still hold. In this case we say that the family $\{A_g\}_{g\in G}$ is a *G-pregrading* of A and that A is a *pregraded algebra*. Finite pregradings and nontrivial pregradings are defined analogously. One of the advantages of pregradings is that for any ideal I of an algebra A, any pregrading of A induces a pregrading of A/I. This is not so for gradings.

A subalgebra S of a pregraded algebra $A = \sum_{g\in G} A_g$ is said to be *graded* provided that $S = \sum_{g\in G}(S \cap A_g)$. For any ideal I of a graded algebra A, the algebra A/I with the pregrading inherited from A is graded if and only if the ideal I is graded (as a subalgebra).

We are mainly interested in $\mathbb{Z}$-gradings, namely, in the case that G is the additive group of the integer numbers. A finite $\mathbb{Z}$-grading of an algebra A with support contained in $\{0, \pm 1, \dots, \pm n\}$ but not in $\{0, \pm 1, \dots, \pm(n-1)\}$ will be called a $(2n+1)$-*grading*. In this case A is called a $(2n+1)$-*graded algebra*. Similar notions are defined for $\mathbb{Z}$-pregradings.

The reader is referred to [**EK13**] for a comprehensive study of gradings in associative, Lie, and Jordan algebras.

EXAMPLE 1.4. Let A be an algebra over an algebraically closed field $\mathbb{F}$ of characteristic 0 and let d be an *algebraic derivation* of A, i.e. such that $p(d) = 0$ for some nonzero polynomial $p(\xi) \in \mathbb{F}[\xi]$. As a consequence of (1.4), d yields a finite $(\mathbb{F}, +)$-grading on A given by

$$A_\lambda := \{x \in A : (d - \lambda 1_A)^m x = 0 \text{ for some } m \geq 1\}$$

since $A_\lambda = 0$ if $\lambda \in \mathbb{F}$ is not an eigenvalue of d. Note that if d is not nilpotent, then the grading is nontrivial.

Annihilators. Let $\mathcal{I}_A$ be the complete lattice of all ideals of A. For any $I \in \mathcal{I}_A$, let $\mathrm{Ann}_A(I)$ (or simply $\mathrm{Ann}(I)$ when there is no risk of confusion) denote the *annihilator* of I in A, i.e. the largest ideal J of A such that $IJ + JI = 0$. Note that an algebra A is prime if and only if for any nonzero ideal I of A, $\mathrm{Ann}(I) = 0$.

PROPOSITION 1.5. *Let I, J, and J_λ, $\lambda \in \Lambda$, be ideals of A. We have:*

(1) $I \subset J \Rightarrow \mathrm{Ann}(J) \subset \mathrm{Ann}(I)$.
(2) $I \subset \mathrm{Ann}(\mathrm{Ann}(I))$.
(3) $\mathrm{Ann}(I) = \mathrm{Ann}(\mathrm{Ann}(\mathrm{Ann}(I)))$.
(4) $\mathrm{Ann}(\sum J_\lambda) = \bigcap \mathrm{Ann}(J_\lambda)$.

Moreover, if A is semiprime, then

(5) $I \cap J = 0 \Leftrightarrow I \subset \mathrm{Ann}(J)$.
(6) $A = \mathrm{Ann}(\mathrm{Ann}(I \oplus \mathrm{Ann}(I)))$.
(7) $\mathrm{Ann}(I * J) = \mathrm{Ann}(I \cap J)$.

PROOF. Statements (1) and (2) are clear and prove that the map $I \mapsto \mathrm{Ann}(I)$ is a symmetric Galois connection on the complete lattice $\mathcal{I}_A$, with closure operator (3) $I \mapsto \mathrm{Ann}(\mathrm{Ann}(I))$; (4) and (5) are also clear and prove that for a semiprime algebra A the annihilator is a pseudo-complementation on $\mathcal{I}_A$; (6) is a direct consequence of (4) and (5). Therefore we only need to prove (7):

$$I * J \subset I \cap J \Rightarrow \mathrm{Ann}(I \cap J) \subset \mathrm{Ann}(I * J).$$

Conversely,

$$(\operatorname{Ann}(I*J)\cap I\cap J)^2\subset \operatorname{Ann}(I*J)\cap(I*J)=0\Rightarrow \operatorname{Ann}(I*J)\cap I\cap J=0,$$

so $\operatorname{Ann}(I*J)\subset\operatorname{Ann}(I\cap J)$. □

Essential ideals. Let A be a Φ-algebra. An ideal E of A is said to be *essential* if it has nonzero intersection with any nonzero ideal of A. Denote by $\mathcal{E}_A$ the filter of the essential ideals of A.

PROPOSITION 1.6. *Let I,E be ideals of an algebra A with $E\in\mathcal{E}_A$. We have:*

(1) *If $\operatorname{Ann}(I)=0$, then I is essential; the converse is true if I is semiprime as an algebra.*
(2) *If the algebra E is semiprime (resp. prime), then A is semiprime (resp. prime).*

PROOF. (1) $\operatorname{Ann}(I)=0$ implies that I is essential since $IJ+JI\subset I\cap J$ for all $J\lhd A$. If I is semiprime, then $I\cap\operatorname{Ann}(I)=0$, so $\operatorname{Ann}(I)=0$ whenever I is essential.

(2) Suppose that E is an essential ideal of A which is semiprime (resp. prime) as an algebra and let I,J be nonzero ideals of A. Then $(I\cap E)^2\neq 0$ (resp. $(I\cap E)(J\cap E)\neq 0$), so $I^2\neq 0$ (resp. $IJ\neq 0$). Thus A is semiprime (resp. prime). □

LEMMA 1.7. *Let A be semiprime and let I,E,F be ideals of A, with $E,F\in\mathcal{E}_A$. Then:*

(1) *$E*F$ and $I\oplus\operatorname{Ann}(I)$ are essential ideals.*
(2) *$\operatorname{Ann}(I\cap E)=\operatorname{Ann}(I)$.*
(3) *There exists $J\in\mathcal{I}_A$ such that $J\subset E$, $J\cap I=0$ and $I+J\in\mathcal{E}_A$.*

PROOF. (1) By Proposition 1.5(7), $\operatorname{Ann}(E*F)=\operatorname{Ann}(E\cap F)$. Hence $E*F$ is essential by Proposition 1.6(1) since $E\cap F$ is essential. On the other hand, by Proposition 1.5(4-5), $\operatorname{Ann}(I+\operatorname{Ann}(I))=\operatorname{Ann}(I)\cap\operatorname{Ann}(\operatorname{Ann}(I))=0$, which proves that $I+\operatorname{Ann}(I)$ is essential.

(2) $I\cap E\subset I$ implies $\operatorname{Ann}(I)\subset\operatorname{Ann}(I\cap E)$. Conversely, since A is semiprime, $I\cap E\cap\operatorname{Ann}(I\cap E)=0$. Hence $I\cap\operatorname{Ann}(I\cap E)=0$ since E is essential. Thus $\operatorname{Ann}(I\cap E)\subset\operatorname{Ann}(I)$.

(3) Put $J:=E\cap\operatorname{Ann}(I)$. Clearly, $J\subset E$ and $J\cap I\subset\operatorname{Ann}(I)\cap I=0$. Furthermore,

$$E\cap\operatorname{Ann}(I+J)=(E\cap\operatorname{Ann}(I))\cap\operatorname{Ann}(J)=J\cap\operatorname{Ann}(J)=0,$$

so $\operatorname{Ann}(I+J)=0$ and hence $I+J$ is essential by Proposition 1.6(1). □

PROPOSITION 1.8. *Let A be semiprime and let M be an ideal of A. Then the quotient $\overline{A}:=A/\operatorname{Ann}(M)$ is a semiprime algebra. Moreover, if the algebra M is prime, then $\overline{A}$ is prime.*

PROOF. Let $\overline{I}:=I/\operatorname{Ann}(M)$ be an ideal of $\overline{A}$ such that $I*I\subset\operatorname{Ann}(M)$. By Proposition 1.5, $M\subset\operatorname{Ann}(\operatorname{Ann}(M))\subset\operatorname{Ann}(I*I)=\operatorname{Ann}(I)$. Hence we have $I\subset\operatorname{Ann}(\operatorname{Ann}(I))\subset\operatorname{Ann}(M)$, i.e. $\overline{I}=0$, which proves that $\overline{A}$ is semiprime.

If the algebra M is prime, then $\overline{M}:=(M\oplus\operatorname{Ann}(M))/\operatorname{Ann}(M)\cong M$ is an ideal of $\overline{A}$ which is prime as an algebra. Hence, by Lemma 1.6(2), it is enough to show that $\overline{M}$ is an essential ideal of $\overline{A}$. Let $I \supset \mathrm{Ann}(M)$ be an ideal of A such that $I \cap (M \oplus \mathrm{Ann}(M)) \subset \mathrm{Ann}(M)$. By the Modular Law, $(I \cap M) \oplus \mathrm{Ann}(M) \subset \mathrm{Ann}(M)$ and hence $I \cap M \subset \mathrm{Ann}(M) \cap M = 0$, so $I = \mathrm{Ann}(M)$, i.e. $\overline{I} = 0$. This proves that $\overline{M}$ is essential, as desired. □

Essential subdirect products. A *subdirect product* of a (nonempty) family $\{A_\lambda\}_{\lambda\in\Lambda}$ of nonzero algebras is any subalgebra A of the full direct product $\prod A_\lambda$ such that the canonical projections $\pi_\lambda : A \to A_\lambda$ are onto. A subdirect product is said to be *essential* if it contains an essential ideal of the full direct product.

LEMMA 1.9. *Let $\{A_\lambda\}_{\lambda\in\Lambda}$ be a family of prime algebras and let $\{I_\lambda\}_{\lambda\in\Lambda}$ be a family of nonzero ideals $I_\lambda \subset A_\lambda$. Then $\bigoplus I_\lambda$ is an essential ideal of the full direct product $\prod A_\lambda$. As a consequence, any subdirect product of the family $\{A_\lambda\}_{\lambda\in\Lambda}$ containing the ideal $\bigoplus I_\lambda$ is essential.*

PROOF. $\mathrm{Ann}_{\prod A_\lambda}(\bigoplus I_\lambda) = \prod \mathrm{Ann}_{A_\lambda}(I_\lambda) = 0$, since for each index λ, A_λ is a prime algebra and I_λ is a nonzero ideal of A_λ and hence $\mathrm{Ann}_{A_\lambda}(I_\lambda) = 0$. This proves, by Proposition 1.6(1), that $\bigoplus I_\lambda$ is an essential ideal of the full direct product $\prod A_\lambda$. □

The reader is referred to [**CCFL12**, Proposition 2.2] for related results on essential subdirect products.

1.2. Multiplication algebra and centroid

It is useful to regard any Φ-algebra A as a left-module over its multiplication algebra. Let $\mathrm{End}_\Phi(A)$ be the associative Φ-algebra of the endomorphisms of the Φ-module A. The *multiplication ideal* $\mathcal{M}_0(A)$ is the subalgebra of $\mathrm{End}_\Phi(A)$ generated by the left and the right multiplications l_a, r_b for all $a, b \in A$, while the *multiplication algebra* is defined by $\mathcal{M}(A) := \Phi 1_A + \mathcal{M}_0(A)$. Then A is a left $\mathcal{M}(A)$-module for the evaluation action with the $\mathcal{M}(A)$-submodules being the ideals of A.

The *centroid* $\Gamma(A)$ of a Φ-algebra A is the centralizer of $\mathcal{M}(A)$ in $\mathrm{End}_\Phi(A)$. Note that the map $\alpha \mapsto \alpha 1_A$ defines a ring homomorphism of Φ into $\Gamma(A)$. Note also that

$$\Gamma(A) = \mathrm{C}_{\mathrm{End}_\Phi(A)}\, \mathcal{M}_0(A) = \mathrm{C}_{\mathrm{End}(A,+)}\, \mathcal{M}(A)$$

since $T \in \mathrm{End}(A,+)$ is Φ-linear if and only if it centralizes the set of all $\alpha 1_A$, $\alpha \in \Phi$.

PROPOSITION 1.10. *Let A be a Φ-algebra. We have:*

(1) *If $A^2 = A$ or $aA = 0 \Rightarrow a = 0$, then $\Gamma(A)$ is a commutative ring with unit element and A becomes a faithful $\Gamma(A)$-module.*
(2) *If A is semiprime, then $\Gamma(A)$ is a reduced ring.*
(3) *If A is prime, then $\Gamma(A)$ is an integral domain.*
(4) *If A is simple, then $\Gamma(A)$ is a field.*

PROOF. (1) (2) and (3) can be easily verified, and (4) follows from (1) together with Schur's lemma. □

As a consequence of the preceding proposition we obtain that the centroid of a semiprime algebra is independent of the ring of scalars. Given an arbitrary algebra A over Φ, denote by $A_{\mathbb{Z}}$ the underlying nonassociative ring of A. It is clear that any ideal of A is an ideal of $A_{\mathbb{Z}}$, and conversely, the linear span $(I)_\Phi$ of an ideal I of $A_{\mathbb{Z}}$ is an ideal of A.

PROPOSITION 1.11. *Let A be a Φ-algebra. We have:*

(1) *A is semiprime (resp. prime) if and only if so is $A_{\mathbb{Z}}$.*
(2) *If A is semiprime, then $\Gamma(A) = \Gamma(A_{\mathbb{Z}})$ as rings.*

PROOF. (1) Clearly, if $A_{\mathbb{Z}}$ is semiprime (resp. prime), then so is A. Suppose conversely that A is semiprime (a similar argument works for the prime case) and let $I \triangleleft A_{\mathbb{Z}}$ such that $I^2 = 0$, then $(I)_{\Phi}^2 = 0$ and hence $I \subset (I)_{\Phi} = 0$.

(2) It is clear that $\Gamma(A) \subset \Gamma(A_{\mathbb{Z}})$. Suppose now that A is semiprime and let $\gamma \in \Gamma(A_{\mathbb{Z}})$. For every $\alpha \in \Phi$ the scalar map $\alpha 1_A \in \Gamma(A_{\mathbb{Z}})$. Since $\Gamma(A_{\mathbb{Z}})$ is commutative, γ commutates with all $\alpha 1_A$. This proves that γ is Φ-linear, so it belongs to the centroid of A. □

LEMMA 1.12. *Let R be a semiprime ring. Then the map $z \mapsto l_z$ defines an imbedding of $Z(R)$ into $\Gamma(R)$. If R is unital, this imbedding is onto and therefore $Z(R) \cong \Gamma(R)$.*

PROOF. Straightforward. □

LEMMA 1.13. *Let R be a simple ring. Then either R is unital (in this case $Z(R) \cong \Gamma(R)$), or $Z(R) = 0$.*

PROOF. Suppose that $Z(R) \neq 0$ and let z be a nonzero central element of R. Then $0 \neq l_{z^2} \in \Gamma(R)$, and since this is a field, l_{z^2} is invertible. Let $\gamma \in \Gamma(R)$ be its inverse. Then for any $x \in R$ we have $x = \gamma(z^2 x) = \gamma(z)zx$, which proves that $\gamma(z)z$ is the unit element of R. □

1.3. Extended centroid and central closure

We study in this section the extended centroid and central closure of a semiprime Φ-algebra A. By a *partially defined centralizer* (in short, p.d.c.) of A we mean a pair (f, I) where I is an ideal of A and $f : I \to A$ is a homomorphism of left $\mathcal{M}(A)$-modules. An *essentially defined centralizer* (in short, e.d.c.) of A is a p.d.c. (f, I) where the ideal I is essential. (According to these definitions the elements of the centroid of A are just the *totally defined centralizers*.) Note that every p.d.c. (f, I) can be extended to an e.d.c. (f', E) taking $E = I \oplus \mathrm{Ann}(I)$ and defining $f'(x + y) = f(x)$ for all $x \in I$, $y \in \mathrm{Ann}(I)$.

In the set of all e.d.c. of A we define an equivalence relation by $(f, I) \cong (g, J)$ if there exists an $E \in \mathcal{E}_A$ such that $E \subset I \cap J$ and $f = g$ on E. We denote by $[f, I]$ the equivalence class of (f, I).

LEMMA 1.14. *Let A be semiprime and let (f, I), (g, J) be e.d.c. of A. We have:*

(1) *$f^{-1}(E) := \{x \in I : f(x) \in E\} \in \mathcal{E}_A$ for every $E \in \mathcal{E}_A$.*
(2) *If $f(E) = 0$ for some $E \in \mathcal{E}_A$, $E \subset I$, then $f(I) = 0$.*
(3) *If $(f, I) \cong (g, J)$, then $f(I) = 0$ if and only if $g(J) = 0$.*

PROOF. (1) By module theory we know that $f^{-1}(E)$ is an ideal. So it suffices to prove that $f^{-1}(E) \cap U \neq 0$ for every nonzero ideal U of A. Since $I \in \mathcal{E}_A$, $I \cap U \neq 0$. If $f(I \cap U) = 0$, then $0 \neq I \cap U \subset f^{-1}(E) \cap U$, so we may assume that $f(I \cap U)$ is nonzero. Then $f(I \cap U) \cap E \neq 0$. Hence $f^{-1}(E) \cap U \neq 0$.

(2) $f(I)E = f(IE) \subset f(E) = 0$ implies $f(I) \subset \mathrm{Ann}(E) = 0$, since $f(I)$ is an *ideal* and E is essential.

(3) $(f, I) \cong (g, J)$ implies that there exists an $E \in \mathcal{E}_A$ such that $(f-g)(E) = 0$. Hence $f(I) = 0$ if and only if $g(J) = 0$ by (2). □

THEOREM 1.15. *Let A be semiprime. Then the set of the equivalence classes of e.d.c. of A becomes an unital ring by defining:*

$$[f, I] + [g, J] = [f + g, I \cap J], \quad [f, I][g, J] = [fg, I * J].$$

PROOF. We only need to check that the operations are independent of the representatives, but this follows from Lemma 1.14. □

DEFINITION 1.16. The unital ring constructed above is denoted by $\mathcal{C}(A)$ and called the *extended centroid* of A. Note that $\Gamma(A)$ can be regarded as a subring of $\mathcal{C}(A)$ via the map $\gamma \mapsto [\gamma, A]$.

THEOREM 1.17. *Let A be semiprime. Then $\mathcal{C}(A)$ is commutative and von Neumann regular. If A is prime, then $\mathcal{C}(A)$ is a field.*

PROOF. Let (f, I) and (g, J) be two e.d.c. of A. We have by Lemma 1.7 that $I * J = \mathcal{M}(A)(IJ)$ is an essential ideal of A. Furthermore, for all $x \in I, y \in J$ and $T \in \mathcal{M}(A)$, we have

$$(fg - gf)(T(xy)) = T(f(xg(y))) - T(g(f(x)y)) = T(f(x)g(y)) - T(f(x)g(y)) = 0.$$

This proves that $\mathcal{C}(A)$ is commutative. Von Neumann regularity follows as in the case of the ring of homomorphisms of a semisimple module. Let $[f, E] \in \mathcal{C}(A)$. As $\ker(f)$ is an ideal of A and E is an essential ideal, we have by Lemma 1.7(3) that there exists and ideal I of A contained in E such that $I \cap \ker(f) = 0$ and $I \oplus \ker(f)$ is an essential ideal of A. Now, since $f(I)$ is an ideal of A and $I \oplus \ker(f)$ is an essential ideal, there exists an ideal J of A contained in $I \oplus \ker(f)$ such that $J \cap f(I) = 0$ and $J \oplus f(I)$ is an essential ideal of A. As $I \cap \ker(f) = 0$, for any $a \in f(I)$, $a = f(x)$ for a *unique* $x \in I$. Define $g : f(I) \oplus J \to A$ by $g(f(x) + y) = x$ for all $x \in I$ and $y \in J$. It is clear that $(g, f(I) \oplus J)$ is an e.d.c. of A and that $[f, E] = [f, I \oplus \ker(f)] = [fgf, I \oplus \ker(f)]$. This proves that $\mathcal{C}(A)$ is von Neumann regular.

Suppose finally that A is prime and let $[f, E] \in \mathcal{C}(A)$ be nonzero. Then $\mathrm{im}(f)$ is an essential ideal and $\ker(f) = 0$. So we can define $g : \mathrm{im}(f) \to A$ by $g(f(x)) = x$ for all $x \in E$. Clearly, $(g, \mathrm{im}(f))$ is an e.d.c. of A and $[g, \mathrm{im}(f)] = [f, E]^{-1}$, which proves that $\mathcal{C}(A)$ is a field. □

Let A be a semiprime Φ-algebra. In the set of all essential defined centralizes of A, we define a partial order relation by $(f, I) \leq (g, J)$ if $I \subset J$ and $g = f$ on I. By Zorn's lemma, we have:

LEMMA 1.18. *Every equivalence class of e.d.c. of A contains a maximal one.*

We have shown in Theorem 1.11(2) the independence of the centroid of a semiprime algebra with respect to the ring of scalars. This is also true for the extended centroid as we see below. Recall that we denote by $A_{\mathbb{Z}}$ the underlying nonassociative ring of a Φ-algebra A and that $A_{\mathbb{Z}}$ is semiprime whenever A is semiprime (Lemma 1.11(1)).

PROPOSITION 1.19. *Let A be a semiprime Φ-algebra. Then $\mathcal{C}(A) = \mathcal{C}(A_{\mathbb{Z}})$.*

PROOF. $\mathcal{C}(A)$ is clearly contained in $\mathcal{C}(A_{\mathbb{Z}})$, so only the reverse inclusion need be checked. Since $\Gamma(A_{\mathbb{Z}}) \subset \mathcal{C}(A_{\mathbb{Z}})$ and $\mathcal{C}(A_{\mathbb{Z}})$ is commutative, to prove that $\mathcal{C}(A_{\mathbb{Z}})$ is contained in $\mathcal{C}(A)$ it suffices to show that for every maximal e.d.c. (f, M) of the ring $A_{\mathbb{Z}}$, M is invariant under $\Gamma(A_{\mathbb{Z}})$, and therefore an ideal of A.

Given $\gamma \in \Gamma(A)$, put $P := M + \gamma(M)$. Then P is an essential ideal of $A_{\mathbb{Z}}$. Now we see that f can be extended to an e.d.c. (g, P) of $A_{\mathbb{Z}}$. Define

$$g(m_1 + \gamma m_2) := f(m_1) + \gamma f(m_2)$$

for all $m_1, m_2 \in M$. If $m_1 + \gamma m_2 = n_1 + \gamma n_2$, then $m_1 - n_1 = \gamma(n_2 - m_2)$. Since $\Gamma(A_{\mathbb{Z}}) \subset \mathcal{C}(A_{\mathbb{Z}})$ and $\mathcal{C}(A_{\mathbb{Z}})$ is commutative (Theorem 1.17),

$$[\gamma f, M] = [f\gamma, \gamma^{-1}(M)].$$

Hence

$$f(m_1 - n_1) = f\gamma(n_2 - m_2) = \gamma f(n_2 - m_2),$$

so $f(m_1) + \gamma f(m_2) = f(n_1) + \gamma f(n_2)$, which proves that $g : P \to A_{\mathbb{Z}}$ is well-defined. Then (g, P) is an e.d.c. of $A_{\mathbb{Z}}$ with $(f, M) \leq (g, P)$. Since (f, M) is maximal, $(f, M) = (g, P)$, so $\gamma(M) \subset M$ for all $\gamma \in \Gamma(A_{\mathbb{Z}})$, as desired. □

Suppose now that A is prime. We know that $\Gamma(A)$ is a commutative integral domain (Theorem 1.10(3)) and $\mathcal{C}(A)$ is a field (Theorem 1.17). Let $\mathbb{F}$ be the field of fractions of $\Gamma(A)$. Then $\mathcal{C}(A)$ is a field extension of $\mathbb{F}$, via the imbedding $\gamma \mapsto [\gamma, A]$ of $\Gamma(A)$ into $\mathcal{C}(A)$.

PROPOSITION 1.20. *Let A be prime. If for any nonzero ideal I of A there exists a nonzero $\gamma \in \Gamma(A)$ such that $\gamma(A) \subset I$, then $\mathcal{C}(A)$ coincides with the field of fractions of $\Gamma(A)$.*

PROOF. Let $[f, I] \in \mathcal{C}(A)$. Then there exists a nonzero element γ in $\Gamma(A)$ such that $\gamma(A) \subset I$. Hence $f\gamma \in \Gamma(A)$. □

DEFINITION 1.21. A semiprime (prime) algebra A is said to be *centrally closed* if $\Gamma(A) = \mathcal{C}(A)$. Note that every simple algebra is a centrally closed prime algebra.

LEMMA 1.22. *Let R be a centrally closed prime ring. Then either R is unital (and therefore $Z(R) \cong \Gamma(R) \cong \mathcal{C}(R)$) or $Z(R) = 0$.*

PROOF. It follows as in Lemma 1.13. □

LEMMA 1.23. *If R is prime and M is a minimal ideal M, then $\mathcal{C}(R) = \Gamma(M)$.*

PROOF. Note that $M = MM$ and that for any nonzero ideal I of R, we have $M \subset I$. Given any e.d.c. (f, I) of R, we have $f(M) = f(M)M \subset M$ and hence f_M can be realized as a centralizer in M. This proves that $\mathcal{C}(R) \subset \Gamma(M)$. Conversely, for any $\gamma \in \Gamma(M)$, (γ, M) is an e.d.c. of R: $M = MM$ and for all $x, y \in M$ and $a \in R$ we have $\gamma(a(xy)) = \gamma((ax)y) = ax\gamma(y) = a\gamma(xy)$. Similarly we get $\gamma((xy)a) = \gamma(xy)a$. □

An *involution* of a Φ-algebra A is a linear map $* : A \to A$ such that $(ab)^* = b^*a^*$ and $a^{**} = a$ for all $a, b \in A$. By a *ring involution* of an algebra A we understand an involution $*$ of A where linearity is replaced by just *additivity*: $(a + b)^* = a^* + b^*$.

Let A be prime. Any ring involution $*$ of A induces an involution (also denoted by $*$) in its extended centroid defined by $q^* := [f^*, I^*]$ for every $q = [f, I] \in \mathcal{C}(A)$, where $f^*(x) := (f(x^*))^*$ for every $x \in I^*$.

DEFINITION 1.24. A ring involution $*$ of a prime algebra A is said to be of the *first kind* if the involution induced in $\mathcal{C}(A)$ is the identity; otherwise, $*$ is said to be of the *second kind.*

Baxter and Martindale proved in [**BM79**] that any semiprime algebra A, with extended centroid $\mathcal{C}$, can be imbedded in an $\mathcal{C}$-algebra $\tilde{A}$ (called the *central closure*) satisfying the following properties:

(CC1) $\tilde{A}$ is semiprime.
(CC2) $\tilde{A}$ is a faithful $\mathcal{C}$-module generated by A.
(CC3) $\Gamma(\tilde{A}) = \mathcal{C}$.

Moreover, if A is prime, then $\tilde{A}$ is prime as well. An axiomatic characterization of the central closure of a semiprime algebra is given in [**CCR12**, Theorem 2.4].

1.4. Nilpotency and local nilpotency

A nonempty subset S of an algebra A is said to be *nilpotent* if there exists a natural number n such that for every $m \geq n$, any product of m elements of S, no matter how associated, is zero. If S is nilpotent, then the smallest such n is called the *index of nilpotency* (*index* in short) of S. Note that if S is multiplicatively closed (for instance, $S = A$) and if any product of n elements of S, no matter how associated, is zero, then the same is true for any $m \geq n$. It is clear what we will mean by a *nilpotent ideal* or a *nilpotent algebra.* A nonempty subset $S \subset A$ is *locally nilpotent* if any finite subset of X generates a nilpotent subalgebra of A.

PROPOSITION 1.25. *Simple algebras are not locally nilpotent.*

PROOF. Let A be simple and let a be a nonzero element in A. Then $\mathcal{M}_0(A)a$ is a nonzero ideal of A and hence there exists $f \in \mathcal{M}_0(A)$ such that $f(a) = a$. Fix $b_1, \ldots, b_r \in A$ such that f lies in the subalgebra of $\mathcal{M}_0(A)$ generated by the set $\{l_{b_i}, r_{b_i} : 1 \leq i \leq r\}$. If A were locally nilpotent, then the subalgebra generated by the set $\{a, b_1, \ldots, b_r\}$ would be nilpotent. But $a = f^n(a)$ for any positive integer n, a contradiction. □

1.5. Martindale algebras of quotients

Following [**BMM96**], we introduce in this section the Martindale (right and symmetric) algebras of quotients of a semiprime associative algebra.

Martindale right algebra of quotients. In what follows, R will denote a semiprime associative Φ-algebra and $\mathcal{E}$ the filter of the essential ideals of R. By Lemma 1.7, if $I, J \in \mathcal{E}$, then $IJ = I * J \in \mathcal{E}$. Given $I \in \mathcal{E}$, by $f : I_R \mapsto R_R$ we mean a homomorphism of right R-modules which is also Φ-linear. Consider the set

$$\{(f, I) : I \in \mathcal{E},\ f : I_R \to R_R\}$$

and define $(f, I) \cong (g, J)$ if there exists an essential ideal E such that $E \subset I \cap J$ and $f = g$ on E. Let $[f, I]$ denote the equivalence class of (f, I). The set of all equivalence classes is endowed with a structure of associative Φ-algebra by defining:

$$[f, I] + [g, J] = [f + g, I \cap J], \quad [f, I][g, J] = [fg, JI], \quad \alpha[f, I] = [\alpha f, I], \alpha \in \Phi.$$

The algebra thus constructed (denoted by $Q_r(R)$ and called the *Martindale right algebra of quotients* of R) is characterized by the following properties:

(RQ1) R is a subalgebra of $Q_r(R)$ via the monomorphism $a \mapsto [l_a, R]$.
(RQ2) For every $q \in Q_r(R)$, there exists $I \in \mathcal{E}$ such that $qI \subset R$.
(RQ3) For every $q \in Q_r(R)$ and $I \in \mathcal{E}$, $qI = 0$ if and only if $q = 0$.
(RQ4) For every $I \in \mathcal{E}$ and $f : I_R \to R_R$, there exists $q \in Q_r(R)$ such that $f(x) = qx$ for all $x \in I$.

PROPOSITION 1.26. *Let R be semiprime and let Q be a subalgebra of $Q_r(R)$ containing R. Then Q is semiprime. Moreover, if R is prime, so is Q.*

PROOF. Let $q, w \in Q$ be such that $qQw = 0$ and $q \neq 0$. By (RQ2) and (RQ3), there exists an essential ideal I of R such that $qI + wI \subset R$ and $qI \neq 0$. Take $0 \neq x \in qI$. Then $xR(wI) = 0$. Hence it easily follows that Q is prime if R is prime. Taking $w = q$ we prove that Q is semiprime. □

Two important examples of subalgebras of $Q_r(R)$ are the Martindale symmetric algebra of quotients and the central closure of R.

Martindale symmetric algebra of quotients. Given a semiprime associative Φ-algebra R, set

$$Q_s(R) = \{q \in Q_r(R) : Iq \subset R \text{ for some } I \in \mathcal{E}\}.$$

One can easily check that $Q_s(R)$ is a subalgebra of $Q_r(R)$, called the *Martindale symmetric algebra of quotients* of R. As noted by D. S. Passman, $Q_s(R)$ may also be characterized by the following four properties:

(SQ1) R is a subalgebra of $Q_s(R)$ via the monomorphism $a \mapsto [l_a, R]$.
(SQ2) For every $q \in Q_s(R)$, there exists $I \in \mathcal{E}$ such that $qI + Iq \subset R$.
(SQ3) For every $q \in Q_s(R)$ and $I \in \mathcal{E}$, $qI = 0$ (or $Iq = 0$) implies $q = 0$.
(SQ4) Given $I \in \mathcal{E}$, $f : I_R \to R_R$ and $g :_R I \to_R R$ such that $xf(y) = g(x)y$ for all $x, y \in I$, there exists $q \in Q_s(R)$ such that $f(x) = qx$ and $xq = g(x)$ for all $x \in I$.

PROPOSITION 1.27. *Let R be semiprime and let $\mathcal{C} := \mathcal{C}(R)$ be the extended centroid of R. Then $\mathcal{C} = Z(Q_r(R)) = Z(Q_s(R))$ and $\tilde{R} = \mathcal{C}R \leq Q_s(R)$.*

PROOF. Let (f, I) be an e.d.c. of R and put $q := [f, I]$. Regarding R as a subalgebra of $Q_r(R)$, for any $x \in I$ we have

$$xq = [l_x, R][f, I] = [l_x f, IR] = [l_{f(x)}, IR] = [l_{f(x)}, R] = f(x) \in R$$

which proves that $\mathcal{C}$ is actually contained in $Q_s(R)$. Now let $[g, J] \in Q_r(A)$. For any $x, y \in I$ and $z \in J$, we have

$$fg(xzy) = f(g(xz)y) = g(xz)f(y) = g(xzf(y)) = gf(xzy).$$

Hence

$$[f, I][g, J] = [f, I][g, IJ] = [fg, IJI] = [gf, IJI] = [g, JI][f, I] = [g, J][f, I],$$

which proves that $[f, I] \in Z(Q_r(R))$. Therefore $\mathcal{C} \subset Z(Q)$ for any subalgebra Q of $Q_r(R)$ containing R. Conversely, let Q be as above and let $[g, J] \in Z(Q)$. For any $a \in R$ we have

$$[gl_a, J] = [gl_a, RJ] = [g, J][l_a, R] = [l_a, R][g, J] = [l_a g, JR] = [l_a g, J].$$

Hence, for any $x \in J$, $g(ax) = ag(x)$. So (g, J) is an e.d.c. of R. This proves that $Z(Q) \subset \mathcal{C}$. The assertion relative to the central closure is clear since $R \leq Q_s(R)$ and $\tilde{R}$ is generated by R as a $\mathcal{C}$-module. □

1.6. The split Cayley algebra

Following [**Jac79**], let V be a 3-dimensional vector space over a field $\mathbb{F}$ of characteristic 0. Take a basis i, j, k and the symmetric bilinear form $\langle\, , \rangle$ over $\mathbb{F}$ such that i, j, k are orthogonal and unit vectors. We endow V with a structure of algebra by defining:

$$i \times i = j \times j = k \times k = 0, \quad i \times j = k, \quad j \times k = i, \quad k \times i = j. \tag{1.5}$$

Denote by $\mathcal{C}$ the set of 2×2-matrices of the form

$$\begin{pmatrix} \alpha & a \\ b & \beta \end{pmatrix} \quad \alpha, \beta \in \mathbb{F}, \quad a, b \in V.$$

Addition and multiplication by elements of $\mathbb{F}$ are as usual, so that $\mathcal{C}$ is an 8-dimensional vector space. An algebra product is defined in $\mathcal{C}$ by

$$\begin{pmatrix} \alpha & a \\ b & \beta \end{pmatrix} \begin{pmatrix} \gamma & c \\ d & \delta \end{pmatrix} = \begin{pmatrix} \alpha\gamma - \langle a, d\rangle & \alpha c + \delta a + b \times d \\ \gamma b + \beta d + a \times c & \beta\delta - \langle b, c\rangle \end{pmatrix}.$$

The algebra $\mathcal{C}$ thus obtained is called the *split Cayley algebra* . This algebra has unit element, the unit matrix 1, and it is not associative but satisfies a weakening of the associative las called the *alternative law*:

$$x^2 y = x(xy), \quad yx^2 = (yx)x. \tag{1.6}$$

The Cayley algebra $\mathcal{C}$ has an involution $x \mapsto \overline{x}$ such that $\overline{x} = -x$ if x is an element of trace 0, $\alpha + \beta = 0$. Denote by $\mathrm{Her}_3(\mathcal{C})$ the space of 3×3 Hermitian Cayley matrices relative to this anti-automorphism, i.e. $\mathrm{Her}_3(\mathcal{C})$ consists of the matrices

$$X = \begin{pmatrix} \xi_1 & x_3 & \overline{x}_2 \\ \overline{x}_3 & \xi_2 & x_1 \\ x_2 & \overline{x}_1 & \xi_3 \end{pmatrix}, \quad \xi_i \in \mathbb{F}, \quad x_i \in \mathcal{C}.$$

If $X, Y \in \mathrm{Her}_3(\mathcal{C})$, then

$$X \bullet Y = \frac{1}{2}(XY + YX) \in \mathrm{Her}_3(\mathcal{C}),$$

where XY is the usual matrix product. The multiplication in $\mathrm{Her}_3(\mathcal{C})$ satisfies

$$X \bullet Y = Y \bullet X, \quad (X^2 \bullet Y) \bullet X = X^2 \bullet (Y \bullet X). \tag{1.7}$$

These identities are the defining properties of a class of algebras called Jordan algebras, which will be studied in Chapter 8.

1.7. Exercises

EXERCISE 1.28. (Modular Law) Let I, J, K be ideals of an algebra A with $J \subset I$. Show: $I \cap (J + K) = J + I \cap K$.

EXERCISE 1.29. Let d be a derivation of an algebra A and let $[A, A]$ be the span of the commutators of A. Show that $d([A, A]) \subset [A, A]$.

EXERCISE 1.30. (Brešar) Let A be a nonassociative algebra over a field $\mathbb{F}$ such that the subspace aA is infinite-dimensional for every nonzero element $a \in A$. If d is a derivation of A such that d^n has finite rank for some $n \geq 1$, then $d^{2n-1} = 0$.

EXERCISE 1.31. Let d be a derivation of an algebra A over a field $\mathbb{F}$. Let $\mathbb{F}[\epsilon] = \mathbb{F} + \mathbb{F}\epsilon$, $\epsilon^2 = 0$, be the algebra of *dual numbers* and set $\overline{A} = \mathbb{F}[\epsilon] \otimes_{\mathbb{F}} A$. Show that the map $1_{\overline{A}} + \epsilon d$ is an automorphism of $\overline{A}$.

EXERCISE 1.32. Let $A = A_{-n} \oplus \cdots \oplus A_n$ be a $(2n+1)$-grading of an algebra A over a field $\mathbb{F}$. Show that the linear map $d : A \to A$ sending x_i to ix_i for all $x_i \in A_i$, $-n \leq i \leq n$ is a derivation of A. Conversely, show that if d is a derivation of an algebra A over a field $\mathbb{F}$ and A_α, A_β, $\alpha, \beta \in \mathbb{F}$, are eigenspaces of d, then $A_\alpha A_\beta \subset A_{\alpha+\beta}$.

EXERCISE 1.33. Let R be an associative algebra. Show that R is semiprime (resp. prime) if and only if $aRa = 0$ (resp. $aRb = 0$) implies $a = 0$ (resp. $a = 0$ or $b = 0$), $a, b \in R$.

EXERCISE 1.34. Let E be an essential ideal of a semiprime algebra A. Show that $\mathrm{Ann}(I) = \mathrm{Ann}(I \cap E)$, $I \lhd A$.

EXERCISE 1.35. Show that any subdirect product of a family of semiprime algebras is a semiprime algebra.

• A nonzero ideal U of an algebra A is called *uniform* if for any nonzero ideals I, J of A contained in U, $I \cap J \neq 0$.

EXERCISE 1.36. This exercise is part of [**FLGR98**, Proposition 3.1]. Let A be semiprime. Show:

(1) For $0 \neq U \lhd A$, the following conditions are equivalent:
   (i) U is uniform;
   (ii) $\mathrm{Ann}(U)$ is maximal among all $\mathrm{Ann}(I)$, $0 \neq I \lhd A$;
   (iii) the algebra $A/\mathrm{Ann}(U)$ is prime.
(2) For each uniform ideal U, there exists a unique maximal uniform ideal M containing U, $M = \mathrm{Ann}(\mathrm{Ann}(U))$.
(3) The sum of all maximal uniform ideals of A is direct.

• A symmetric bilinear form $\langle\,,\rangle$ defined on an algebra A over a field $\mathbb{F}$ is called *associative* or *invariant* if $\langle xy, z\rangle = \langle x, yz\rangle$ for all $x, y, z \in A$. Note that if I is an ideal of A, then $I^\perp := \{x \in L : \langle x, I\rangle = 0\}$ is an ideal of A. In particular, $\mathrm{Rad}(\langle\,,\rangle) := A^\perp \lhd A$.

EXERCISE 1.37. (Dieudonné) Let A be a semiprime algebra equipped with a nondegenerate associative symmetric bilinear form $\langle\,,\rangle$ over a field $\mathbb{F}$. Show:

(a) For any $I \triangleleft A$, $I \cap I^{\perp} = 0$, i.e. the restriction of $\langle\,,\rangle$ to I is nondegenerate.
(b) $I^{\perp} = \mathrm{Ann}(I)$.
(c) A is uniquely expresable as a direct sum of simple ideals.

EXERCISE 1.38. Let R be a unital associative algebra with a complete system orthogonal idempotents $\mathcal{E} = \{e_0, e_1, \ldots, e_n\}$. Show that $R = \bigoplus_{-n}^{n} R_i$, where

$$R_i = \sum_{p-q=i} e_p R e_q, 1 = -n, \ldots, n,$$

is a $\mathbb{Z}$-grading of R.

EXERCISE 1.39. Show that in any alternative algebra the linear map

$$D_{a,b} = [l_a, l_b] + [l_a, r_b] + [r_a, r_b]$$

is a derivation.

EXERCISE 1.40. Show that the split Cayley algebra $\mathcal{C}$ is alternative.

EXERCISE 1.41. In the notation of Section 1.6, let $T : V \to V$ be a linear map of trace 0 and denote by T^* its adjoint relative to the bilinear form $\langle\,,\rangle$, i.e. $\langle aT, b\rangle = \langle a, bT^*\rangle$ for all $a, b \in V$. Show that the map

$$\begin{pmatrix} \alpha & a \\ b & \beta \end{pmatrix} \to \begin{pmatrix} 0 & aT \\ -bT^* & 0 \end{pmatrix}$$

is a derivation of the split Cayley algebra $\mathcal{C}$.

EXERCISE 1.42. Let $\mathcal{C}_0$ be the subspace of $\mathcal{C}$ consisting of the elements of trace 0, i.e. $\alpha + \beta = 0$. Show that $\mathcal{C}_0$ coincides with the subspace spanned by the commutators of $\mathcal{C}$.

CHAPTER 2

General Facts on Lie Algebras

Although this is an introductory chapter on Lie algebras, it still contains some important and by no means trivial results in structure theory. The material is organized as follows. In Section 2.1 we give basic definitions and first examples of Lie algebras and fix the notation. Classical examples of infinite-dimensional Lie algebras are presented in Section 2.2 following a geometric approach based on pairs of dual vector spaces, and regarding these linear Lie algebras as those defined by simple rings with minimal one-sided ideals. Introduced by G. Benkart and J. Faulkner, inner ideals in Lie algebras are the analogues of the one-sided ideals in associative algebras and the inner ideals in Jordan systems. In Section 2.3 we give the first steps on an inner ideal theory for Lie algebras and prove a structure theorem for minimal inner ideals. A theorem of inheritance of primeness by ideals is the goal of Section 2.4. Solubility and nilpotency are studied in Section 2.5. In Section 2.6 it is shown the existence of a locally nilpotent radical (in the Kurosh sense) in any Lie algebra L such that for every $x \in L$ the adjoint operator ad_x is nilpotent. A similar questions is considered in Section 2.7, where a radical (largest locally nilpotent strong ideal) is introduced for Lie algebras with a finite pregrading, and in Section 2.8, where the existence of locally finite radical is proved for Lie algebras over an arbitrary field such that the adjoint operator of every element is algebraic.

2.1. Definitions and examples

Throughout this section, Φ will denote an arbitrary ring of scalars. A *Lie algebra* is a Φ-algebra L whose product, denoted by a bracket $[.,.]$, satisfies the following conditions:

(L1) $[x, x] = 0$

(L2) $[x, [y, z]] + [y, [z, x]] + [z, [x, y]] = 0$

for all $x, y, z \in L$. Note that (L1) implies anticommutativity: (L1') $[x, y] = -[y, x]$ for all $x, y \in L$, and that if L is 2-torsion free, then (L1') and (L1) are equivalent. Note also that in presence of (L1'), condition (L2), called the *Jacobi identity*, can be rephrased by saying that the *adjoint operator* $\mathrm{ad}_a : x \mapsto [a, x]$ (also denoted by $\mathrm{ad}(a)$) is a derivation of L, called the *inner derivation* determined by a . We will often use upper case letters to denote adjoint operators; that is, we will write X instead of ad_x for $x \in L$.

EXAMPLE 2.1. Let R be an associative Φ-algebra. Then the commutator $[x, y] = xy - yx$ is anticommutative and satisfies the Jacobi identity. Thus the Φ-module R with the new product given by the commutator is a Lie Φ-algebra, denoted by R^-, with adjoint operators $\mathrm{ad}_x = l_x - r_x$. In particular, any Φ-module X gives rise to the Lie algebra $\mathfrak{gl}(X) := \mathrm{End}_\Phi(X)^-$.

Other Lie algebras associated to R are:

$$\overline{R} = R^-/Z, \quad R' = [R, R], \quad \overline{R}' = R'/R' \cap Z.$$

Note that $\overline{R}' \cong (R' + Z)/Z = [\overline{R}, \overline{R}]$ can be regarded as an ideal of $\overline{R}$.

REMARK 2.2. It is proved in [**AAJZ15**, Theorem 1] that if R is a finitely generated associative algebra with an idempotent e such that $ReR = R(1-e)R = R$, then the Lie algebra R' is finitely generated.

EXAMPLE 2.3. If R has an involution $*$, then the set $K = \mathrm{Skew}(R, *)$ of the skew-symmetric elements of R is a subalgebra of R^- and therefore a Lie algebra. Other Lie algebras of skew symmetric elements are:

$$\overline{K} = K/K \cap Z, \quad K' = [K, K], \text{ and } \overline{K}' = K'/K' \cap Z.$$

REMARK 2.4. A. N. Grishkov and I. P. Shestakov prove in [**GS05**] that a Lie algebra L is isomorphic to a Lie algebra of type $\mathrm{Skew}(R, *)$ for a certain associative algebra with involution $(R, *)$ if and only if L admits a trilinear product $\{x, y, z\}$ such that, with respect to this triple product and the original Lie multiplication, L forms a so-called *Lie–Jordan algebra*; that is, the following identities hold:

(1) $\{x, y, z\} = \{z, y, x\}$,
(2) $[[x, y], z] = \{x, y, z\} - \{y, x, z\}$,
(3) $[\{x, y, z\}, t] = \{[x, t], y, z\} + \{x, [y, t], z\} + \{x, y, [z, t]\}$,
(4) $\{\{x, y, z\}, t, v\} = \{\{x, t, v\}, y, z\} - \{x, \{y, v, t\}, z\} + \{x, y, \{z, t, v\}\}$.

The next two theorems due to I. N. Herstein and W. E. Baxter are included for future reference.

THEOREM 2.5. [**Her69**, Theorem 1.12] *Let R be a simple noncommutative ring of characteristic not 2. Then $\overline{R}'$ is a simple Lie $\mathbb{Z}$-algebra.*

THEOREM 2.6. [**Her69**, Theorem 2.15] *Let R be a simple ring of characteristic not 2 with involution $*$. If $Z(R) = 0$ or $\dim_{Z(R)} R > 16$, then the Lie algebra $\overline{K}'$ is simple.*

REMARK 2.7. Let R and K be as above. Suppose further that R contains a nonzero idempotent e such that $ee^* = e^*e = 0$. It is proved in [**AAJZ15**, Theorem 3] that if R is finitely generated, then the Lie algebra $\overline{K}'$ is finitely generated as well.

The above examples include the simple finite-dimensional Lie algebras of types A, B, C and D (see [**Hum72**]) with R being a finite-dimensional simple associative algebra over an algebraically closed field of characteristic 0. Examples of infinite-dimensional simple Lie algebras (as those of the finitary algebras introduced by A. A. Baranov [**Bar99**]) occur when R is a simple non-Artinian ring with minimal one-sided ideals.

EXAMPLE 2.8. Let A be a not necessarily associative algebra. It is easy to check that the commutator $[d_1, d_2] = d_1 d_2 - d_2 d_1$ of two derivations of A is again a derivation. Thus the Φ-module $\mathrm{Der}(A)$ of all derivations of A is a subalgebra of the Lie algebra $\mathfrak{gl}(A)$.

EXAMPLE 2.9. Those readers interested in classical Lie theory, i.e. representations, Cartan subalgebras, root systemas, and Dynkin diagrams, are encouraged to check (following Jacobson's exposition in [**Jac79**]) that the algebra of derivations $\mathrm{Der}(\mathcal{C})$ of the split Cayley algebra is a split simple Lie algebra of type G_2.

An interesting example of infinite-dimensional $\mathbb{Z}$-graded Lie algebra is the Lie algebra of derivations of the $\mathbb{F}$-algebra $\mathbb{F}[x]$ of polynomials in a variable x.

EXAMPLE 2.10. Let $\mathbb{F}$ be a field of characteristic 0. It is easily checked:

(1) $\mathrm{Der}(\mathbb{F}[x])$ is isomorphic to the Lie $\mathbb{F}$-algebra L defined on $\mathbb{F}[x]$ by the product $[p,q] := pq' - qp'$, with p', q' denoting the derivatives of these polynomials.
(2) $L = \bigoplus_{n\geq -1} L_n$, $L_n = \mathbb{F}x^{n+1}$, is a $\mathbb{Z}$-grading.
(3) $\mathrm{Der}(\mathbb{F}[x])$ is simple.

Witt algebras. There is a finite-dimensional counterpart of the Lie algebra $\mathrm{Der}(\mathbb{F}[x])$ in positive characteristic, the so-called p-dimensional Witt algebra.

EXAMPLE 2.11. Given a prime number p, the p-dimensional Witt algebra $W(p)$ is defined as follows: Let $\mathbb{F}$ be a field of characteristic p, take a p-dimensional vector space over $\mathbb{F}$ with basis $\{x_0, x_1, \dots, x_{p-1}\}$, and define a product by

$$[x_i, x_j] = (j-i)x_{i+j-1},$$

with the convention that $x_k = 0$ whenever $k \notin \{0, 1, \dots, p-1\}$. It is readily seen that $W(p)$ is a simple Lie algebra when $p > 2$, and that $W(3) \cong \mathfrak{sl}_2(\mathbb{F})$: take $e = -x_0$, $f = x_2$ and $h = -2x_1$.

REMARK 2.12. The p-dimensional simple Lie algebra $W(p)$ was found by Witt, although he never published this example or its generalizations. This was the first instance of a simple Lie algebra with nonzero absolute zero divisors, which will be studied in the next chapter.

Lie algebra defined by a group. For this subsection we follow the exposition of [**Kos90**, 7.1]. Let G be a group (with product denoted by juxtaposition), let $(x,y) := x^{-1}y^{-1}xy$ be the *commutator* of the two elements $x, y \in G$, and set $x^y = y^{-1}xy$ to denote the conjugate of x by y. The following identities involving commutators and conjugate are easily checked.

(i) $x^{-1}x^y = (x,y)$,
(ii) $(xy, z) = (x,z)^y(y,z)$, $\quad (x, yz) = (x,z)(x,y)^z$.

Moreover, we have the following identity discovered by P. Hall.

(iii) $((x,y),z^x)((z,x),y^z)((y,z),x^y) = 1$.

A *central series* of a group G is a decreasing chain of normal subgroups

$$G = H_1 \supset H_2 \supset H_3 \supset \cdots$$

such that $[H_i, H_j] \subset H_{i+j}$ for any positive integers i, j. The factor groups H_i/H_{i+1} are abelian, since $[H_i, H_i] \subset H_{2i} \subset H_{i+1}$. We write these factor groups additively and set $\bar{x}_i$ to denote the coset mod H_{i+1} of the element $x_i \in H$. With these notations, $\overline{x_i y_i} = \bar{x}_i + \bar{y}_i$, $x_i, y_i \in H_i$. Then the abelian group

$$L = L(G, \{H_i\}) := \bigoplus_{i \geq 1} H_i/H_{i+1},$$

is a $\mathbb{Z}$-graded Lie algebra with the elements of $L_i = H_i/H_{i+1}$ being homogeneous of degree i.

For $x_i \in H_i$, $x_j \in H_j$, set $[\bar{x}_i, \bar{x}_j] = \overline{(x_i, x_j)}$. It is not difficult to check that this bracket is well defined, and it is clear that $[L_i, L_j] \subset L_{i+j}$. Then extend the definition of the bracket by linearity to all pairs of elements $\bar{x} = \sum x_i$, $\bar{y} = \sum y_i$. It follows from (i):

(iv) $\overline{x_i^{y_j}} = \bar{x}_i$ for any $x_i \in H_i$, $y_j \in H_j$.

Now bilinearity of the bracket follows from (ii) and (iv), and the Jacobi identity is a consequence of (iii) and (iv).

REMARK 2.13. The most important example of central series of a group G is the *lower central series*, with terms $\xi_1(G) = G$, $\xi_{i+1}(G) = (G, \xi_i(G))$. The reader is referred to Kostrikin's book [**Kos90**] for an application of the Lie algebra $L(G, \{\xi_i(G)\})$ to the restricted Burnside problem for prime power exponent.

Centralizers, annihilators and center. The *centralizer* of a subset X of a Lie algebra L is the set $\mathrm{C}_L(X) = \{a \in L : [a, X] = 0\}$. Using the Jacobi identity, we get:

(i) If $X \leq L$, then $\mathrm{C}_L(X) \leq L$.
(ii) If $I \triangleleft L$, then $\mathrm{C}_L(I) \triangleleft L$, and hence
(iii) $\mathrm{Ann}_L(I) = \mathrm{C}_L(I)$, for any ideal I of L.

In particular, $\mathrm{Ann}_L(L) = \mathrm{C}_L(L)$ is the *center* of L, denoted by $Z(L)$.

Commutators. Let $M = \{x_1, \ldots, x_n\}$ be a nonempty subset of a Lie algebra L. The elements of the closure of M with respect to $[.,.]$ are called *commutators* in M. The *length* of a commutator ρ in M, denoted by $\mathrm{length}(\rho)$, is defined inductively:

$$\mathrm{length}(a_1) = 1, \ \mathrm{length}([\rho_1, \rho_2]) = \mathrm{length}(\rho_1) + \mathrm{length}(\rho_2),$$

so the length of the commutator $[[x_1, x_2], [x_3, x_1]]$ is 4. Two commutators ρ_1, ρ_2 in $\{x_1, \ldots, x_n\}$ have the same *composition* if each letter x_i occur in ρ_1 and ρ_2 the same number of times. Let $k \in \mathbb{Z}$ with $k \geq 1$. Denote by $\mathrm{Seq}_k(M)$ the set of all sequences $(x_{i_1}, x_{i_2}, \ldots, x_{i_k})$ of M of length k and put $\mathrm{Seq}(M) := \bigcup_{k \geq 1} \mathrm{Seq}_k(M)$. Given $(x_{i_1}, x_{i_2}, \ldots, x_{i_k}) \in \mathrm{Seq}(M)$, we define the *left commutator* $[x_{i_k}, \ldots, x_{i_1}]$ recursively as follows:

$$[x_{i_1}] = x_{i_1} \text{ and } [x_{i_k}, x_{i_{k-1}}, \ldots, x_{i_1}] = [x_{i_k}, [x_{i_{k-1}}, \ldots, x_{i_1}]] \text{ if } k > 1.$$

Forgetting the brackets, to any commutator ρ in M of length k it is associated a sequence $(x_{i_1}, x_{i_2}, \ldots, x_{i_k}) \in \mathrm{Seq}_k(M)$. Then, using the Jacobi identity, we can write

$$\rho = \sum_{\sigma \in \mathfrak{S}_k} \alpha_\sigma [x_{i_{\sigma(k)}}, x_{i_{\sigma(k-1)}}, \ldots, x_{i_{\sigma(1)}}],$$

where $\mathfrak{S}_k$ denotes the group of the permutations of the set $\{1, 2, \ldots, k\}$ and for all $\sigma \in \mathfrak{S}_k$, $\alpha_\sigma \in \{-1, 0, 1\}$.

The adjoint representation. A *representation* of a Lie Φ-algebra L is a homomorphism $\Psi : L \to \mathfrak{gl}(X)$, where X is a Φ-module. It follows from Example 2.8 that for any Lie algebra L, the embedding of $\mathrm{Der}(L)$ into $\mathfrak{gl}(L)$ is a representation of the Lie algebra $\mathrm{Der}(L)$. Another relevant example of representations is the adjoint representation of a Lie algebra.

PROPOSITION 2.14. *Let M be a Lie Φ-algebra. We have:*

(1) *The map $x \mapsto \mathrm{ad}_x$ defines a homomorphism $\mathrm{ad} : M \to \mathfrak{gl}(M)$, whose kernel is the center of M and whose image $\mathrm{ad}(M)$ is the ideal of all inner derivations of M; in fact, for any $d \in \mathrm{Der}(M)$ and $a \in M$, we have $[d, \mathrm{ad}_a] = \mathrm{ad}_{d(a)}$.*

(2) *If $Z(M) = 0$ and L is a subalgebra of $\mathrm{Der}(M)$ containing $\mathrm{ad}(M)$, then $\mathrm{ad}(M) \cong M$ is an essential ideal of L. Hence, if M is semiprime (resp. prime), then L is semiprime (resp. prime).*

PROOF. (1) Clearly the map ad is Φ-linear , and for all $a, b, x \in M$, we have

$$\mathrm{ad}_{[a,b]}(x) = [[a, b], x] = [a, [b, x]] - [b, [a, x]] = [\mathrm{ad}_a, \mathrm{ad}_b](x),$$

which proves that ad is a homomorphism of Lie algebras, with $\ker(\mathrm{ad}) = Z(M)$ and $\mathrm{ad}(M) \subset \mathrm{Der}(M)$. Now let $a \in M$ and $d \in \mathrm{Der}(M)$. For any $x \in L$, we have

$$[d, \mathrm{ad}_a](x) = d([a, x]) - [a, d(x)] = [d(a), x] + [a, d(x)] - [a, d(x)] = \mathrm{ad}_{d(a)}(x),$$

which proves that $\mathrm{ad}(M)$ is an ideal of $\mathrm{Der}(M)$.

(2) Let $d \in \mathrm{Der}(M)$. By (1), $[d, \mathrm{ad}(M)] = 0$ implies $d(M) \subset Z(M)$. Thus, if M has trivial center (for instance, if M is semiprime), then $\mathrm{Ann}_L(\mathrm{ad}(M)) = 0$, which proves, by Proposition 1.6, that $\mathrm{ad}(M)$ is an essential ideal of L, and that if M is semiprime (resp. prime) then L is semiprime (resp. prime). □

• Given a Lie Φ-algebra L, denote by $\mathrm{Ad}(L)$ the subalgebra of $\mathrm{End}_\Phi(L)$ generated by $\mathrm{ad}(L)$, namely, according to the general definition (see Section 1.2), $\mathrm{Ad}(L)$ is the multiplication ideal $\mathcal{M}_0(L)$. The *adjoint representation* $x \mapsto X(= \mathrm{ad}_x)$ from L to $\mathfrak{gl}(L)$ transforms identities of the Lie algebra L into identities of the associative algebra $\mathrm{Ad}(L)$, which allows us to find new identities by means of simplifications. For example, if $a \in L$ is such that $A^2 = 0$, then, for every $x \in L$, $\mathrm{ad}_{A^2 x}$ is also zero and therefore

$$0 = \mathrm{ad}_{A^2 x} = \mathrm{ad}_{[a,[a,x]]} = [A, [A, X]]$$

for every $x \in$ L. But

$$[A, [A, X]] = A^2 X - 2AXA + XA^2 = -2AXA$$

because $A^2 = 0$, and therefore $2AXA = 0$. We use this trick (due to A. I. Kostrikin) without further mention from now on.

• We can also consider the adjoint representation of the Lie algebra $\mathrm{ad}(L)$. To distinguish visually one representation from another, we will keep writing ad to denote the adjoint representation of L, and will write $\mathbf{ad} : \mathrm{ad}(L) \to \mathfrak{gl}(\mathrm{ad}(L))$. According to this notation, for any $a, x \in L$, we have $[A, [A, X]] = \mathbf{ad}_A^2(X)$. In fact,

$$ad_{A^n x} = \mathbf{ad}_A^n(X) = (l_A - r_A)^n(X) = \sum_{k=0}^{n} (-1)^k \binom{n}{k} A^{n-k} X A^k. \tag{2.1}$$

This formula, a particular case of (1.2), will be frequently used in what follows. Note that $\mathbf{ad}_A^n = 0$ in $\mathrm{End}_\Phi(\mathrm{ad}(L))$ does not mean that $\mathbf{ad}_A^n = 0$ in $\mathrm{End}_\Phi(\mathrm{Ad}(L))$; for example, it does not necessarily mean that $\mathbf{ad}_A^n(XY) = 0$. But there are at least two facts we can assume:

- $\mathbf{ad}_A^n([X, d]) = 0$ for every $d \in \mathrm{Der}(L)$,
- $\mathbf{ad}_A^n(XY) = \sum_{i=1}^{n-1} \binom{n}{i} \mathbf{ad}_A^{n-i}(X)\mathbf{ad}_A^i(Y)$,

since $\mathrm{ad}(L)$ is an ideal of $\mathrm{Der}(L)$ and $\mathbf{ad}_A$ is an associative derivation.

DEFINITION 2.15. By *an ad-nilpotent* (resp. *ad-algebraic*) *element* of a Lie algebra L over Φ (resp. over a field $\mathbb{F}$) we mean an element $x \in L$ such that the adjoint operator ad_x is nilpotent (resp. algebraic over $\mathbb{F}$). An ad-nilpotent element of index at most 2 is called an *absolute zero divisor*.

REMARK 2.16. G. Benkart and I. M. Isaacs showed in [**BI77**] the existence of nonzero ad-nilpotent elements in every nonzero finite-dimensional Lie algebra over an algebraically closed field of arbitrary characteristic.

LEMMA 2.17. *Let $a \in L$ be such that $A^n = 0$ for some $n \geq 2$. We have:*

(1) *If L is n-torsion free, then $A^{n-1}(L)$ is an abelian subalgebra of L.*
(2) *If $n > 2$ and L is $(2n-3)!$-torsion free, then*

$$[A^{n-2}x, A^{n-1}y] + [A^{n-1}x, A^{n-2}y] = 0$$

for all $x, y \in L$. (If $n = 3$, only 3-torsion free is necessary.)

PROOF. (1) It follows from Lemma 1.1.

(2) Since $n > 2$, $2n - 3 \geq n$, so $A^{2n-3} = 0$. Then using the Leibniz rule we get for all $x, y \in L$,

$$0 = A^{2n-3}[x, y] = \sum_{k=0}^{2n-3} \binom{2n-3}{k} [A^{2n-3-k}x, A^k y].$$

Since $A^k = 0$ for $k \geq n$ and $A^{2n-3-k} = 0$ for $2n - 3 - k \geq n$, we actually have $[A^{2n-3-k}x, A^k y] = 0$ up possibly to $k = n - 2$ and $k = n - 1$. Hence the equation above reduces to

$$0 = \binom{2n-3}{n-1}[A^{n-2}x, A^{n-1}y] + \binom{2n-3}{n-2}[A^{n-1}x, A^{n-2}y],$$

with $\binom{2n-3}{n-1} = \binom{2n-3}{n-2}$ since $n-1$ and $n-2$ are complementary. Using that L is $(2n-3)!$-torsion free, we get $[A^{n-2}x, A^{n-1}y] = -[A^{n-1}x, A^{n-2}y]$. (If $n = 3$, $\binom{2n-3}{n-2} = 3$.) □

The Vandermonde argument. Using the well known Vandermonde determinant, we provide a sufficient condition for a submodule V of a Lie algebra L invariant under automorphism to be an ideal of L. We begin with some technical definitions.

An automorphism of the form $\exp(\mathrm{ad}_x)$, where ad_x is nilpotent, is called an *inner automorphism*. The subgroup of $\mathrm{Aut}(L)$ generated by the inner automorphisms of L, denoted by $\mathrm{Int}(L)$, is a normal subgroup of $\mathrm{Aut}(L)$:

$$f \exp(\mathrm{ad}_x) f^{-1} = \exp(\mathrm{ad}_{fx})$$

for every $f \in \mathrm{Aut}(L)$.

DEFINITION 2.18. Let L be a Lie Φ-algebra. By a *distinguished generating set of* L we understand a generating set X of ad-nilpotent elements of L such that:

(i) $n_x! \in \Phi^*$, where n_x denotes the index of ad_x for any $x \in X$, and
(ii) $\exp(i\,\mathrm{ad}_x) \in \mathrm{Aut}(L)$ for $x \in X$ and any integer i such that $1 \leq i \leq n_x$.

EXAMPLES 2.19. (1) Let L be a Lie algebra over a field $\mathbb{F}$ of characteristic 0 or $p \geq 5$, and let X be a generating subset of L consisting of ad-nilpotent element of index less or equal to 3. Then X is a distinguished generating set of L. Indeed, condition (i) is clear since $n_x \leq 3$, and condition (ii) follows from Corollary 1.3 since $p \geq 5$.

(2) Let $L = L_{-n} \oplus \cdots \oplus L_n$ be a simple Lie algebra with a nontrivial finite $\mathbb{Z}$-grading over a field $\mathbb{F}$ of characteristic $p \geq 4n+1$. Then the set $X := \bigcup_{k=1}^{m} L_{\pm k}$, consisting of ad-nilpotent elements of index $m \leq 2n+1$, is a distinguished generating set of L.

Condition (i) of Definition 2.18 is clear since $m \leq 2n+1 < p$, so it suffices to show that $\exp(\mathrm{ad}_x) \in \mathrm{Aut}(L)$ for all $x \in X$. But this follows from Lemma 1.2: $2m - 2 \leq 2(2n+1) - 2 = 4n$, so $(2m-2)!$ is invertible in $\mathbb{F}$, as desired.

(3) By Lemma 1.2, if $x \in L$ is ad-nilpotent of index $n_x \leq 5$, invertibility of $2, 3, 5$ and 7 is required to get that $\exp(\mathrm{ad}_x) \in \mathrm{Aut}(L)$, but when L has a nontrivial 5-grading, invertibility of $2, 3$ and 5 is enough. So, for $n = 2$ in (2), we only need $p > 5$.

DEFINITION 2.20. Given a distinguished generating set X of a Lie Φ-algebra L, we denote by $\mathrm{Elem}(X)$ the subgroup of $\mathrm{Aut}(L)$ generated by $\exp(i\,\mathrm{ad}_x)$, $x \in X$, $1 \leq i \leq n_x$. Clearly, $\mathrm{Elem}(X)$ is a subgroup of $\mathrm{Int}(L)$ which we will call the *elementary group* of X.

LEMMA 2.21. *Let L be a Lie Φ-algebra with a distinguished generating set X. Then every Φ-submodule V of L which is invariant under* $\mathrm{Elem}(X)$ *is an ideal of* L.

PROOF. Fix $x \in X$ and $v \in V$. Since L is generated by X, it suffices to show that $[x, v] \in V$. As V is invariant under $\mathrm{Elem}(X)$, $v_i := \exp(i\,\mathrm{ad}_x)v \in V$ for all $1 \leq i \leq n = n_x$. Consider now the following system of linear equations with unknowns v, $\mathrm{ad}_x v$, $\frac{1}{2!}\mathrm{ad}_x^2 v, \ldots, \frac{1}{(n-1)!}\mathrm{ad}_x^{n-1} v$:

$$\sum_{k=0}^{n-1} i^k \frac{\mathrm{ad}_x^k v}{k!} = v_i, \quad i = 1, \ldots, n.$$

The matrix for this linear system is a Vandermonde matrix whose determinant is invertible in Φ since $i - j \in \Phi^*$ for $1 \leq i < j \leq n$ by (i) in Definition 2.18. Therefore the matrix is invertible and we conclude that each one of the unknown belongs to V, including $\mathrm{ad}_x v = [x, v]$. $\square$

2.2. Linear Lie algebras

In this section we study a class of Lie algebras coming from prime rings with minimal one-sided ideals (which are here presented following Jacobson's approach [**Jac64**], but avoiding the topological point of view). Among these Lie algebras we will find those called finitary, introduced and studied by A. A. Baranov.

Pairs of dual vector spaces. The key notion in this approach to prime rings with minimal one-sided ideals is that of a *pair of dual vector spaces* $(X, Y, \langle\,,\rangle)$, where X is a left vector space, Y a right vector space (both over a division ring Δ), and $\langle\,,\rangle$ a nondegenerate bilinear map from $X \times Y$ to Δ. Any left vector space X over a division ring Δ gives rise to the *canonical pair* $(X, X^*, \langle\,,\rangle)$, where X^* is the *dual* of X and $\langle x, \varphi\rangle = x\varphi$, for all $x \in X$, $\varphi \in X^*$.

DEFINITION 2.22. Let $\mathcal{P} = (X, Y, \langle\,,\rangle)$ and $\mathcal{P}_0 = (X_0, Y_0, \langle\,,\rangle_0)$ be two pairs of dual vector spaces over the same division ring Δ. Then $\mathcal{P}_0$ is said to be a *subpair* of $\mathcal{P}$ if $X_0 \leq X$, $Y_0 \leq Y$, and $\langle\,,\rangle_0$ is the restriction of $\langle\,,\rangle$ to $X_0 \times Y_0$.

DEFINITION 2.23. A subpair $\mathcal{P}_0 = (X_0, Y_0)$ of a pair $\mathcal{P} = (X, Y, \langle\,,\rangle)$ is said to be a *direct subpair* if $X = X_0 \oplus Y_0^\perp$ and $Y = Y_0 \oplus X_0^\perp$, where

$$Y_0^\perp = \{x \in X : \langle x, Y_0\rangle = 0\} \text{ and } X_0^\perp = \{y \in Y : \langle X_0, y\rangle = 0\}.$$

PROPOSITION 2.24. *Let $\mathcal{P} = (X, Y, \langle\,,\rangle)$ be a pair of dual vector spaces.*

(1) *If $V \leq X$ and $W \leq Y$ are finite-dimensional, then (V, W) can be imbedded in a finite-dimensional subpair of $\mathcal{P}$.*
(2) *Every finite-dimensional subpair of $\mathcal{P}$ is direct.*
(3) *If $\mathcal{P}_0 = (X_0, Y_0)$ is a direct subpair of $\mathcal{P}$, then $\mathcal{P}_0^\perp := (Y_0^\perp, X_0^\perp)$ is also direct subpair of $\mathcal{P}$ and $\mathcal{P} = \mathcal{P}_0 \oplus \mathcal{P}_0^\perp$.*

PROOF. (1) See [**Jac64**, IV.Lemma 15.1].

(2) Let $\mathcal{P}_0 = (X_0, Y_0)$ be a finite-dimensional subpair of $\mathcal{P}$, and take dual bases $\{x_1, \ldots, x_n\} \subset X_0$ and $\{y_1, \ldots, x_n\} \subset Y_0$, $\langle x_i, y_j\rangle = \delta_{ij}$. Given $x \in X$, $x - \sum\langle x, y_i\rangle x_i \in Y_0^\perp$, so $X = X_0 + Y_0^\perp$. Now let $x_0 = \sum \alpha_i x_i \in Y_0^\perp$. Then $0 = \langle x_0, y_j\rangle = \alpha_j$ for each $1 \leq j \leq n$, so $X = X_0 \oplus Y_0^\perp$. That $Y = Y_0 \oplus X_0^\perp$ follows by symmetry.

(3) It suffices to show that $X_0 = X_0^{\perp\perp}$ (the proof of $Y_0 = Y_0^{\perp\perp}$ is similar). From $Y = Y_0 \oplus X_0^\perp$ we get

$$0 = Y^\perp = (Y_0 \oplus X_0^\perp)^\perp = Y_0^\perp \cap X_0^{\perp\perp},$$

and from $X = X_0 \oplus Y_0^\perp$, using the Modular Law (because $X_0 \subset X_0^{\perp\perp}$), we get

$$X_0^{\perp\perp} = X_0^{\perp\perp} \cap X = X_0^{\perp\perp} \cap (X_0 \oplus Y_0^\perp) = X_0 \oplus (Y_0^\perp \cap X_0^{\perp\perp}) = X_0.$$

□

• Let $(X, Y, \langle\,,\rangle)$ be a pair of dual vector spaces. Two sequences of vectors $\{x_1, \ldots, x_n\} \subset X$ and $\{y_1, \ldots, y_n\} \subset Y$ are said to be *bi-orthogonal* if $\langle x_i, y_j\rangle = \delta_{ij}$. In this case, they span a subpair of dual vector spaces of $(X, Y, \langle\,,\rangle)$.

Adjointable linear maps. Let $(X, Y, \langle\,,\rangle)$ be a pair of dual vector spaces over a division ring Δ. A linear map $a : X \to X$ is *adjointable* or *continuous* if there exists a linear map $a^\# : Y \to Y$, necessarily unique and called the *adjoin of* a, such that $\langle xa, y\rangle = \langle x, a^\# y\rangle$ for all $x \in X$, $y \in Y$. Note that we write the maps of a left vector space on the right (thus composing them from left to right), and the maps of a right vector space on the left (thus composing them from right to left), so by our conventions if a, b are adjointable so is ab with (careful!) $(ab)^\# = a^\# b^\#$. We denote by $\mathcal{L}_Y(X)$ the ring of adjointable linear maps of X, and by $\mathcal{F}_Y(X)$ the ideal of those maps having finite rank. One can see that such rings are algebras over a ring of scalars Φ when Δ is a Φ-algebra. We will simply write $\mathcal{L}(X)$, $\mathcal{F}(X)$ instead of $\mathcal{L}_{X^*}(X)$, $\mathcal{F}_{X^*}(X)$, with respect to the canonical pair (X, X^*).

For $x \in X$, $y \in Y$, write y^*x to denote the linear map defined by

$$(x')y^*x = \langle x', y\rangle x$$

for all $x' \in X$. Then $y^*x \in \mathcal{F}_Y(X)$, with adjoint $(y^*x)^\# y' = y\langle x, y'\rangle$ for all $y' \in Y$.

Given $V \leq X$ and $W \leq Y$ we denote by W^*V the subgroup of the abelian group $(\mathcal{F}_Y(X), +)$ generated by the set $\{w^*v : w \in W, v \in V\}$.

The following lemma, whose proof is left to the reader, shows how to handle these linear maps.

LEMMA 2.25. *Let $(X, Y, \langle\,,\rangle)$ be a pair of dual vector spaces. We have:*

(1) *Every $a \in \mathcal{F}_Y(X)$ can be expressed as $a = \sum y_i^* x_i$, where both subsets of vectors $\{y_i\} \subset Y$ and $\{x_i\} \subset X$ are linearly independent.*
(2) *$y^*\alpha x = (y\alpha)^* x$ for all $\alpha \in \Delta$, $x \in X$, $y \in Y$.*
(3) *$\mathcal{F}_Y(X) = Y^*X$ is isomorphic as abelian group to $Y \otimes_\Delta X$.*
(4) *$a(y^*x)b = (a^\# y)^*(xb)$, for all $a \in \mathcal{L}_Y(X)$, $b \in \mathrm{End}_\Delta(X)$.*
(5) *$(y_1^*x_1)(y_2^*x_2) = y_1^*\langle x_1, y_2\rangle x_2$, for all $x_1, x_2 \in X$, $y_1, y_2 \in Y$.*

Prime rings with nonzero socle. For a semiprime ring R, the sum of its minimal right ideals coincides with the sum of its minimal left ideals. This ideal, called *socle* of R, will be denoted by $\mathrm{Soc}(R)$. The reader is referred to the books [**BMM96**] and [**Jac64**] for an account of socle theory of semiprime rings, and encouraged to prove some results proposed in the exercises by himself.

THEOREM 2.26. [**Jac64**, IV.9, Structure theorem] *Up to isomorphism, a ring R is prime with minimal one-sided ideals if and only if there exists a pair $(X, Y, \langle\,,\rangle)$ of dual vector spaces over a division ring Δ such that*

$$\mathcal{F}_Y(X) \lhd R \leq \mathcal{L}_Y(X).$$

In this case, $\mathrm{Soc}(R) = \mathcal{F}_Y(X)$, R is left and right primitive, the division ring Δ is uniquely determined by R up to isomorphism, and $\mathcal{C}(R) \cong Z(\Delta)$.

Trace of a finite rank linear map. Following [**FLGGL06**], we give in this subsection a characterization of the trace for finite-rank linear maps $a \in \mathcal{F}_Y(X)$. This characterization is intrinsic (in the sense that it avoids imbedding into finite matrices), elementary (since it can be easily computed in terms of any representation of a as a finite sum of rank-one operators), and consistent with the usual notion of trace of a square matrix over a commutative ring. However, unlike the commutative case (endomorphisms of free modules of finite rank over a commutative unital ring), the trace of a finite-rank linear map on a left vector space over a division ring Δ is not an element of Δ, but a residue class module $[\Delta, \Delta]$. For $a = \sum_i y_i^* x_i \in \mathcal{F}_Y(X)$, set

$$\mathrm{tr}(a) := \overline{\sum_i \langle x_i, y_i\rangle} = \sum_i \langle x_i, y_i\rangle + [\Delta, \Delta].$$

It is shown in [**FLGGL06**, Lemma 1.4] that the map $\operatorname{tr} : \mathcal{F}_Y(X) \to \frac{\Delta}{[\Delta,\Delta]}$ is well defined, $Z(\Delta)$-linear, and onto. And if $a = \sum_{i,j} {y_i}^* \alpha_{ij} x_j$, with $\langle x_i, y_j \rangle = \delta_{ij}$, then

$$\operatorname{tr}(a) = \sum_i \alpha_{ii} + [\Delta, \Delta],$$

where $\sum_i \alpha_{ii}$ coincides with the usual trace of the matrix $(\alpha_{ij}) \in \mathrm{M}_n(\Delta)$, with n being the rank of a.

The special linear algebra $\mathfrak{fsl}_Y(X)$. Associated to a pair of dual vector spaces $(X, Y, \langle\,,\rangle)$ over a division ring Δ of characteristic not 2, we have the following *linear Lie algebras*:

(1) The *general linear algebra* $\mathfrak{gl}_Y(X) = \mathcal{L}_Y(X)^-$.
(2) The *general linear algebra of finite rank operators* $\mathfrak{fgl}_Y(X) = \mathcal{F}_Y(X)^-$.
(3) The *special linear algebra* $\mathfrak{fsl}_Y(X) = [\mathcal{F}_Y(X), \mathcal{F}_Y(X)]$.

Set $L := \mathfrak{fsl}_Y(X)$. If $\dim_\Delta X > 1$, then the Lie algebra $L/L \cap Z(L)$ is simple by Theorem 2.5. If X is actually infinite-dimensional over Δ, then $L \cap Z(L) = 0$ and $\mathfrak{fsl}_Y(X)$ is itself a simple Lie algebra. The same is true if Δ is a finite-dimensional central division ring of characteristic 0.

As could be expected, the elements of $\mathfrak{fsl}_Y(X)$ are precisely the traceless linear maps of $\mathfrak{fgl}_Y(X)$.

THEOREM 2.27. [**FLGGL06**, Theorem 1.7] *Let $(X, Y, \langle\,,\rangle)$ be a pair of dual vector spaces over Δ. The map $a \mapsto \operatorname{tr}(a)$ is a homomorphism of $\mathfrak{fgl}_Y(X)$ onto $\Delta^-/[\Delta, \Delta]$, with kernel equal to $\mathfrak{fsl}_Y(X)$.*

Involutions and inner products. Let Δ be a division ring with an involution $\alpha \mapsto \overline{\alpha}$ and let X be a left vector space over Δ. Recall that a *Hermitian* (resp. *skew-Hermitian*) form on X is a map $(x, y) \mapsto \langle x, y\rangle$ of $X \times X \to \Delta$ satisfying the following conditions:

(i) It is linear in the first variable,
(ii) $\langle y, x\rangle = \epsilon\overline{\langle x, y\rangle}$, $\epsilon = \pm 1$, for all $x, y \in X$.

In the particular case that the involution of Δ is the identity (and therefore, Δ is a field), the Hermitian (resp. skew-Hermitian) form is *symmetric* (resp. *alternate*).

Every left vector space X with a nondegenerate Hermitian or skew-Hermitian form $\langle\,,\rangle$ becomes a pair of dual vector spaces $(X, \overline{X}, \langle\,,\rangle)$, where $\overline{X}$ coincides with X as an abelian group, but it is regarded as a right vector space over Δ by defining $x \cdot \alpha = \overline{\alpha}x$, for all $\alpha \in \Delta$, $x \in X$. Set $\mathcal{L}_X(X) := \mathcal{L}_{\overline{X}}(X)$ and $\mathcal{F}_X(X) := \mathcal{F}_{\overline{X}}(X)$ and observe that the adjoint operator $a \mapsto a^\#$ is actually a ring involution on $\mathcal{L}_X(X)$, which will be denoted by $a \mapsto a^*$.

The following lemma, whose proof is left to the reader, shows how finite rank linear maps behave with respect to the adjoint involution.

LEMMA 2.28. *Let X be a left vector space with a nondegenerate Hermitian (resp. skew-Hermitian) form $\langle\,,\rangle$ over a divison ring with involution $(\Delta, -)$ of characteristic not 2. For any $\alpha \in \Delta$, $x, y \in X$, we have:*

(1) $(\alpha x)^* y = x^* \overline{\alpha} y$ *and* $(x^*y)^* = y^*x$ *(resp.* $(x^*y)^* = -y^*x$*).*
(2) *The operator defined by* $[x, y] = x^*y - \epsilon y^*x$ *belongs to* $\operatorname{Skew}(\mathcal{F}_X(X), *)$ *and it will be called a skew-trace.*

If V, W are subspaces of X, we write $[V, W]$ to denote the set of all finite sums of the skew-traces $[v, w]$, for all $v \in V$, $w \in W$. With this notation, we have:

(3) $\mathrm{Skew}(\mathcal{F}_X(X), *) = [X, X]$.
(4) *If $\langle\,,\rangle$ is skew-Hermitian, then $x^*x = [(1/2)x, x]$ is a skew-trace for any $x \in X$. In fact, for any skew-trace $[x, y]$ we have:*

$$[x, y] = (x+y)^*(x+y) - x^*x - y^*y.$$

(5) *If $\langle\,,\rangle$ is Hermitian, then $(\alpha x)^*x$ is a skew-trace if and only if $\overline{\alpha} = -\alpha$.*

DEFINITION 2.29. Let R be a prime ring with nonzero socle. An involution $*$ of R is said to be of *transpose type* if R contains a *division idempotent* e, i.e. eRe is a division ring, such that $e^* = e$. Otherwise $*$ is said to be of *symplectic type.*

PROPOSITION 2.30. *Let X be a left vector space with a nondegenerate Hermitian (resp. alternate) form $\langle\,,\rangle$ over a division ring with involution $(\Delta, -)$ of characteristic not 2. Then the adjoint involution of the prime ring $\mathcal{L}_X(X)$ is of transpose (resp. symplectic) type.*

PROOF. If $\langle\,,\rangle$ is Hermitian, we can choose an anisotropic vector $x \in X$. Then $e = x^*\langle x, x\rangle^{-1}x \in \mathcal{F}_X(X)$ is a self-adjoint division idempotent, so the adjoint involution $*$ is of transpose type.

If $\langle\,,\rangle$ is alternate, then any $x \in X$ is isotropic, and since any division idempotent e of $\mathcal{L}_X(X)$ is of rank one (say $e = y^*x$), we have

$$ee^* = (y^*x)(y^*x)^* = -y^*\langle x, x\rangle y = 0,$$

which proves that the adjoint involution $*$ is of symplectic type. □

THEOREM 2.31. [**BMM96**, Theorem 4.6.8] *The $*$-subrings R of $\mathcal{L}_X(X)$ containing $\mathcal{F}_X(X)$, relative to a nondegenerate Hermitian or alternate form, are precisely the prime rings with involution and nonzero socle. Transpose involutions correspond to Hermitian forms and symplectic involutions to alternate forms. Furthermore, R is simple if and only if $R = \mathcal{F}_X(X)$.*

The finitary orthogonal algebra $\mathfrak{fo}(X, \langle\,,\rangle)$. Let X be a vector space over a field $\mathbb{F}$ of characteristic not 2 and $\dim_{\mathbb{F}} X > 2$, which is endowed with a nondegenerate symmetric bilinear form $\langle\,,\rangle$. Associated to the inner space $(X, \langle\,,\rangle)$ we have the following Lie algebras over the field $\mathbb{F}$ (see [**Bar99**]):

(i) The *ortogonal algebra* $\mathfrak{o}(X, \langle\,,\rangle) = \mathrm{Skew}(\mathcal{L}_X(X), *)$.
(ii) The *finitary orthogonal algebra* $\mathfrak{fo}(X, \langle\,,\rangle) = \mathrm{Skew}(\mathcal{F}_X(X), *)$.

The finitary symplectic algebra $\mathfrak{fsp}(X, \langle\,,\rangle)$. Let X be a vector space over a field $\mathbb{F}$ of characteristic not 2 and $\dim_{\mathbb{F}} X > 2$, which is endowed with a nondegenerate alternate bilinear form $\langle\,,\rangle$. Associated to the inner space $(X, \langle\,,\rangle)$ we have the following Lie algebras over the field $\mathbb{F}$ (see [**Bar99**]):

(i) The *symplectic algebra* $\mathfrak{sp}(X, \langle\,,\rangle) = \mathrm{Skew}(\mathcal{L}_X(X), *)$.
(ii) The *finitary symplectic algebra* $\mathfrak{fsp}(X, \langle\,,\rangle) = \mathrm{Skew}(\mathcal{F}_X(X), *)$.

REMARK 2.32. In this remark we give some indications to prove the equality

$$\mathrm{Skew}(\mathcal{F}_X(X), *) = [\mathrm{Skew}(\mathcal{F}_X(X), *), \mathrm{Skew}(\mathcal{F}_X(X), *)]$$

when X is a vector space of dimension greater than 2 endowed with a nondegenerate (symmetric or alternate) bilinear form $\langle , \rangle$ over a field $\mathbb{F}$ of characteristic not 2. For brevity we write R to denote the ring $\mathcal{F}_X(X)$ and K for the Lie algebra $\mathrm{Skew}(\mathcal{F}_X(X), *)$.

Suppose that the form is alternate. It suffices to show that for any vector $x \in X$, the linear map $a = x^*x$ belongs to $[K, K]$. Since $a^2 = 0$, $[[a, b], a] = 2aba$ for any $b \in K$. Use now the von Neumann regularity of a in R to get $b \in K$ such that $[[a, b], a] = a$.

Assume then that the bilinear form is symmetric. We must prove that every skew-trace $a = [x, y] = x^*y - y^*x$, where $x, y \in X$ are linearly independent, belongs to $[K, K]$. We divide the proof into three cases according to the dimension of the radical of the 2-dimensional subspace $V = \mathbb{F}x \oplus \mathbb{F}y$.

(1) $\mathrm{Rad}(V) = 0$. Since $\dim_{\mathbb{F}} X > 2$, there exist $x_1, x_2, x_3 \in X$ such that $\{x_1, x_2\}$ is a basis of V, $\langle x_i, x_i\rangle \neq 0$ for $i = 1, 2, 3$, and $\langle x_i, x_j\rangle = 0$ for $i \neq j$. As $a = \alpha[x_1, x_2]$ for some $\alpha \in \mathbb{F}$, we may assume $a = [x_1, x_2]$. Take $b = [x_1, x_3]$, $c = [x_2, x_3]$, and verify that $a = \beta[b, c]$ for some scalar β. So $a \in [K, K]$.

(2) $\dim \mathrm{Rad}(V) = 1$. We may assume $\langle x, x\rangle \neq 0$ and $\langle x, y\rangle = \langle y, y\rangle = 0$. Let $z \in X$ be such that $\langle z, x\rangle = 0$ and $\langle z, y\rangle = 1$. Setting $b = [y, z]$, we get $a = [b, a] \in [K, K]$.

(3) $\dim \mathrm{Rad}(V) = 2$. Then $a^2 = 0$ and we can proceed as in the alternate case. This completes the proof.

2.3. Inner ideals of Lie algebras

Inner ideals of Lie algebras are the analogues of the one-sided ideals in associative algebras and of the inner ideals in Jordan algebras. Since their introduction by J. R. Faulkner [**Fau73**] and G. Benkart [**Ben77**], inner ideals have proved to be useful for classifying both finite-dimensional and infinite-dimensional simple Lie algebras. Most of the contents of this section are taken from [**Ben77**].

DEFINITION 2.33. Let L be a Lie algebra over a ring of scalars Φ. An *inner ideal* of L is a Φ-submodule B of L such that $[[B, L], B] \subset B$. An *abelian inner ideal* is an inner ideal B which is also an abelian subalgebra, i.e. $[B, B] = 0$.

EXAMPLE 2.34. Let a be an absolute zero divisor of a Lie algebra L over a ring of scalars Φ. Then Φa is an abelian inner ideal of L.

LEMMA 2.35. *Suppose that* $2 \in \Phi^*$ *and let* B *be an abelian subalgebra of* L. *Then* B *is an abelian inner ideal of* L *if and only if* $\mathrm{ad}_b^2 L \subset B$ *for every* $b \in B$.

PROOF. From $[B, B] = 0$, we get $[b, [c, a]] = [c, [b, a]]$, for any $b, c \in B$ and $a \in L$. Hence

$$2[b, [c, a]] = [b + c, [b + c, a]] - [b, [b, a]] - [c, [c, a]].$$

As $\frac{1}{2} \in \Phi$, B is inner if and only if $\mathrm{ad}_b^2 L \subset B$ for every $b \in B$. □

PROPOSITION 2.36. *Let* I *be an ideal of a Lie algebra* L. *Then any ideal* J *of* I *is an inner ideal of* L.

PROOF. $[J, [J, L]] \subset [J, I] \subset J$. □

As a consequence of the grading properties, we have.

EXAMPLE 2.37. Let $L = L_{-n} \oplus \cdots \oplus L_n$ be a $(2n+1)$-graded Lie algebra. Then L_n and L_{-n} are abelian inner ideals of L.

PROPOSITION 2.38. *Let A be an associative algebra over Φ, with $\frac{1}{2} \in \Phi$.*

(1) *If $\mathcal{L}$ is a left ideal and $\mathcal{R}$ is a right ideal of A such that $\mathcal{L}\mathcal{R} = 0$, then $\mathcal{R}\mathcal{L}$ and $\mathcal{R} \cap \mathcal{L}$ are abelian inner ideals of the Lie algebra A^-.*
(2) *Let $x \in A$ be such that $x^2 = 0$. Then xAx is an abelian inner ideal of A^-.*

PROOF. Note that any of the Φ-submodules (say B), $\mathcal{R}\mathcal{L}$, $\mathcal{R} \cap \mathcal{L}$, xAx of A, satisfies $BB = 0$. Hence, for any $b, c \in B$ and $a \in A$, we have $[[b, a], c] = bac + cab$. Now it is easy to check that B is an inner ideal in each one of the three cases. □

Basic structure theory of inner ideals. Throughout this subsection, L will denote a Lie algebra over an arbitrary ring of scalar Φ. For Φ-submodules $S_1, S_2, \ldots, S_n$ of L, we denote by $S_1 S_2 \cdots S_n$ the linear span of $[S_1, S_2, \ldots, S_n]$, where $[S_1] = S_1$ and $[S_1, S_2, \ldots, S_n] = [S_1, [S_2, \ldots, S_n]]$ and set $S^n = SS^{n-1}$ for $n > 1$. Using induction on n it can be proved that if S, T are Φ-submodules of L, then for any $n \geq 1$, we have:

$$S^n T \subset \overbrace{S \cdots S}^{n} T. \tag{2.2}$$

LEMMA 2.39. *Let B be an inner ideal of L. We have:*

(1) *BBL is an inner ideal and a subalgebra of L.*
(2) *$[LB, B^n] \subset B^n$ for $n \geq 1$.*
(3) *$LB^n \subset B^{n-1}$ for $n \geq 2$.*
(4) *If $B^2 = B$, then B is an ideal of L.*
(5) *For any Φ-submodule V of B with $VVB \subset V$, V^3 is an inner ideal of L.*
(6) *For every Φ-submodule V of B, $V^2 V^2 L \subset VVB$.*

PROOF. (1) Since $BBL \subset B$, we have $[BBL, [BBL, L]] \subset [B, [B, L]] = BBL$ and $[BBL, BBL] \subset [B, BBL] = [B, [B, BL]] \subset BBL$. The proof of (2) follows from induction on n: For $n = 1$ is clear since B is an inner ideal of L, and for $n > 1$, $[LB, B^n] = [LB, BB^{n-1}] \subset [LB, B]B^{n-1} + B[LB, B^{n-1}] \subset BB^{n-1} = B^n$. The proof of (3) follows from induction on n and (2): $LB^2 \subset [L, B]B + BLB \subset B$, and for $n > 2$, $LB^n = LBB^{n-1} \subset [L, B]B^{n-1} + BLB^{n-1} \subset B^{n-1} + BB^{n-2} = B^{n-1}$. Part (4) is a consequence of (3) with $n = 2$. Using (2.2) we prove (5)

$$V^3 V^3 L \subset V^3 V(VVL) \subset V^3 VB \subset VV(VVB) \subset V^3.$$

A similar argument gives (6). □

The above lemma shows that for each inner ideal B of L there is a descending chain of inner ideals given by $B_0 = B$, $B_1 = BBL$, $B_2 = B_1 B_1 L, \ldots$ Moreover, (5) applied to $V = B$ shows that if B is an inner ideal, so is B^3. Still another way of manufacturing inner ideals is to begin with an arbitrary Φ-submodule M of L and to define: $T(M) := \{t \in L : [M, [t, L]] \subset M\}$.

LEMMA 2.40. *For any Φ-submodule M of L, $T(M)$ is an inner ideal and a subalgebra of L which contains M if M is an inner ideal of L.*

PROOF. That $T(M)$ contains M whenever M is an inner ideal follows from the definition of $T(M)$. To show that $T(M)$ is itself an inner ideal, let $s, t \in T(M)$, $x, y \in L$, $a \in M$. Then

$$\begin{aligned}[a,[[t,[s,x]],y]] &= [a,[t,[[s,x],y]]] + [a,[[t,y],[s,x]]] \\ &\equiv [a,[[t,y],[s,x]]] \mod M \\ &\equiv [[a,[t,y]],[s,x]] + [[t,y],[a,[s,x]]] \mod M \\ &\equiv 0 \mod M.\end{aligned}$$

Thus $T(M)$ is an inner ideal of L. Moreover, for $s, t \in T(M)$, $x \in L$ and $a \in M$, we see that

$$[a,[[t,s],x]] = [a,[[t,x],s]] + [a,[t,[s,x]]] \subset M.$$

Hence $T(M)$ is a subalgebra of L. □

The inner ideal $T(M)$ defined in this form often enable us to reduce our consideration to inner ideals which are also subalgebras for which more can be said:

LEMMA 2.41. *Let T be both an inner ideal and a subalgebra of L. Then for all $n \geq 1$, T^n is an inner ideal and $\bigcap_{n=1}^{\infty} T^n$ an ideal of L.*

PROOF. It is a consequence of Lemma 2.39(3). □

Minimal inner ideals. An inner ideal B of a Lie algebra L is *minimal* if it is nonzero and it does not properly contain any nonzero inner ideal of L.

EXAMPLE 2.42. Let L be a Lie algebra over a field $\mathbb{F}$ and let $a \in L$ be a nonzero absolute zero divisor. Then $\mathbb{F}a$ is a minimal inner ideal of L.

THEOREM 2.43. *Let B be a minimal inner ideal of L. Then either:*

(1) *$B = \Phi b$ where b is a nonzero absolute zero divisor of L and Φ is a field,*
(2) *$B = BBL$ and $B^2 = 0$, or*
(3) *B is an ideal of L which is simple as a Lie algebra and every proper inner ideal of B is abelian.*

PROOF. If B contains a nonzero absolute zero divisor b of L, the minimality of B gives $B = \Phi b$ and Φ is a field. So we can assume that B does not contain any nonzero absolute zero divisor of L. As BBL is an inner ideal of L (Lemma 2.39(1)) contained in B, either $BBL = B$ or $BBL = 0$. But the last possibility can be excluded because B has no nonzero absolute zero divisors. So $B = BBL$ and by Lemma 2.39(1) B is also a subalgebra of L. By Lemma 2.39(5), B^3 is an inner ideal of L, so there are two possibilities: $B^3 = 0$ or $B^3 = B$. In the first case, Lemma 2.39(6) applied to B gives $B^2B^2L \subset B^3 = 0$, namely, every element of B^2 is an absolute zero divisor of L. As B is a subalgebra which does not contain nonzero absolute zero divisors of L, $B^2 = 0$ and we are in case (2). We may suppose then $BBL = B$ and also $B^3 = B$. Because B is a subalgebra, we have $B \supset B^2 \supset B^3 = B$, so $B = B^2$. According to Lemma 2.39(4), B is an ideal of L. Since by Proposition 2.36 any ideal of B is an inner ideal of L, it follows that B is a simple algebra. Furthermore, for any proper inner ideal V of B, we have by (2.2) that

$$V^2V^2L \subset V^2VVL \subset V^2VB \subset VVVB \subset V^2,$$

showing that V^2 is an inner ideal of L contained in B. If $V^2 = B$ then $V^2 \supset V$ and $V \supset V^3 \supset V^2 = B$ contrary to the assumption. So $[V,V] = V^2 = 0$ for any proper inner ideal V of B, and we are in case (3). □

2.4. Inheritance of primeness by ideals

Let L be a Lie algebra over an arbitrary ring of scalars Φ. Following [**Gar03**], we prove that ideals which are semiprime algebras inherits primeness from L.

By the Jacobi identity, if I and J are ideals of a Lie algebra L, then $[I, J]$ is an ideal of L. In particular, given an ideal I, we have the sequence of ideals $\{I^n\}_{n\geq 1}$ defined as follows: $I^1 = I$ and $I^n = [I, I^{n-1}]$ for $n > 1$.

DEFINITION 2.44. Given an ideal I of a Lie algebra L the *eventual annihilator* of I in L is the set defined by $\mathrm{Evann}_L(I) = \bigcup_{n\geq 1} \mathrm{Ann}_L(I^n)$.

LEMMA 2.45. *Let L be a Lie algebra, let I be an ideal of L, and let J be an ideal of I. Then the eventual annihilator of J in I is an ideal of L.*

PROOF. Using the Jacobi identity it is proved inductively that $[J^{n+1}, L] \subset J^n$ for all positive integer n. Now take $x \in \mathrm{Evann}_I(J)$. Then $x \in \mathrm{Ann}_I(J^n)$ for some $n \geq 1$. We claim that $[x, L] \subset \mathrm{Ann}_I(J^{n+1})$, which would complete the proof. Indeed, by the Jacobi identity, $[J^{n+1}, [x, L]] \subset [[J^{n+1}, x], L] + [x, [J^{n+1}, L]] = 0$, since $J^{n+1} + [J^{n+1}, L] \subset J^n$ and $x \in \mathrm{Ann}_I(J^n)$. □

PROPOSITION 2.46. *Let L be prime and let I be an ideal of L which is a semiprime algebra. Then I is a prime algebra.*

PROOF. We will prove that for every nonzero ideal J of I, $\mathrm{Ann}_I(J) = 0$. We will first show that

$$J \cap \mathrm{Evann}_I(J) = 0. \tag{2.3}$$

Indeed, let $S_n := J \cap \mathrm{Ann}_I(J^n)$, which is an ideal of I since J^n is an ideal of I, so both J and $\mathrm{Ann}_I(J^n)$ are ideals of I. Moreover, $S_n^{n+1} = [S_n, S_n^n] \subset [\mathrm{Ann}_I(J^n), J^n] = 0$. Therefore S_n is a nilpotent ideal of I and hence $S_n = 0$ since I is a semiprime algebra. Then

$$J \cap \mathrm{Evann}_I(J) = J \cap (\cup_{n\geq 1} \mathrm{Ann}_I(J^n)) = \cup_{n\geq 1}(J \cap \mathrm{Ann}_I(J^n)) = \cup_{n\geq 1} S_n = 0,$$

proving (2.3).

Now let $\mathrm{id}_L(\mathrm{Ann}_I(J))$ be the ideal of L generated by $\mathrm{Ann}_I(J)$. Obviously, $\mathrm{Ann}_I(J) \subset \cup_{n\geq 1} \mathrm{Ann}_I(J^n) = \mathrm{Evann}_I(J)$ and since $\mathrm{Evann}_I(J)$ is an ideal of L by Lemma 2.45, $\mathrm{id}_L(\mathrm{Ann}_I(J)) \subset \mathrm{Evann}_I(J)$. Therefore, by (2.3),

$$J \cap \mathrm{id}_L(\mathrm{Ann}_I(J)) \subset J \cap \mathrm{Evann}_I(J) = 0. \tag{2.4}$$

We claim that $\mathrm{id}_L(\mathrm{Ann}_I(J)) = 0$. For if not, $I_0 := I \cap \mathrm{id}_L(\mathrm{Ann}_I(J))$ would be a nonzero ideal of L and hence $\mathrm{Ann}_L(I_0) = 0$, using the primeness of L twice. Then $0 \neq J \not\subset \mathrm{Ann}_L(I_0)$ and $0 \neq [J, I_0] \subset J \cap I_0 \subset J \cap \mathrm{id}_L(\mathrm{Ann}_I(J)) = 0$ by (2.4), which is a contradiction. □

2.5. Solvability and nilpotency

Let L be a Lie algebra. For any ideal I of L, we have the *derived series*

$$I = I^{[0]} \supset I^{[1]} \supset \cdots \supset I^{[k]} \supset \cdots,$$

where $I^{[k]} = [I^{[k-1]}, I^{[k-1]}]$ for $k \geq 1$. An ideal I of a Lie algebra L is *solvable* if $I^{[n]} = 0$ for some natural number n. Note that a Lie algebra is semiprime (see Section 1.1) if and only if it does not contain nonzero solvable ideals.

LEMMA 2.47. *Let L be a Lie algebra. We have:*

(1) *If L is solvable, then so are all subalgebras and homomorphic images of L.*
(2) *If I is a solvable ideal of L such that L/I is solvable, then L itself is solvable.*
(3) *If I and J are solvable ideals of L, then so is $I+J$.*

PROOF. (1) If K is a subalgebra of L, then $K^{[n]} \subset L^{[n]}$. Similarly, if $f : L \to M$ is an epimorphism, then $f(L^{[n]}) = M^{[n]}$.

(2) If $I^{[n]} = 0$ and $L^{[m]} \subset I$ for some positive integers n, m, then $L^{[n+m]} = (L^{[m]})^{[n]} \subset I^{[n]} = 0$, so L is solvable.

(3) Using the canonical isomorphism $(I+J)/I \cong J/(I \cap J)$, it follows from (1) and (2) that $I+J$ is solvable. □

THEOREM 2.48. *Let L be a Lie algebra satisfying the ascending chain condition on ideals (for instance, L is finite-dimensional over a field). Then L contains a largest solvable ideal, called the* radical *of L and denoted by* $\operatorname{Rad}(L)$. *Furthermore, $L/\operatorname{Rad}(L)$ is semiprime.*

PROOF. Let M be a maximal solvable ideal of L, whose existence is guaranteed by the ascending chain condition. For any other solvable ideal I of L we have by Lemma 2.47(iii) that $M+I$ is solvable, so $I \subset M$ by maximality of M and therefore $M = \operatorname{Rad}(L)$. We also have by Lemma 2.47(ii) that $L/\operatorname{Rad}(L)$ does not contain nonzero solvable ideals, in other words, $L/\operatorname{Rad}(L)$ is a semiprime Lie algebra. □

Nilpotency. In Section 1.4 we considered a notion of nilpotency for any subset of an arbitrary algebra A. Here we show that that notion agrees with the usual one for Lie algebras.

Let I be an ideal of a Lie algebra L. Recall that the powers I^n were defined recursively by setting $I^1 = 1$ and $I^k = [I, I^{k-1}]$ for $k > 1$.

LEMMA 2.49. *Let I be an ideal of a Lie algebra. Then I is nilpotent in the general nonassociative sense if there exists a natural n such that $I^n = 0$.*

PROOF. Let ρ be a commutator defined by a sequence of elements of I. As was already pointed out, ρ can be expressed as a linear combination of left commutators of the same length and equal composition as ρ, so I is nilpotent in the general nonassociative sense whenever $I^n = 0$ for some $n \geq 1$. □

REMARK 2.50. Nilpotent Lie algebras are clearly solvable, $L^{[n]} \subset L^{2^n}$ for $n \geq 1$, but the converse is not true. The 2-dimensional nonabelian Lie algebra $L = \mathbb{F}a + \mathbb{F}b$, $[a,b] = b$, is soluble, $L^{[1]} = \mathbb{F}b$, $L^{[2]} = 0$, but not nilpotent, $L^n = \mathbb{F}b$ for $n \geq 2$. Nevertheless, as a consequence of Lie's theorem, if L is a finite-dimensional solvable Lie algebra over an algebraically closed field of characteristic 0, then its derived algebra $[L, L]$ is nilpotent [**Hum72**, Corollary C of II.4.1]. Note also that this example also shows that the nilpotent version of Lemma 2.47(2) does not hold. However, 2.47(3) does remain true for nilpotent ideals.

PROPOSITION 2.51. *Let I and J be nilpotent ideals of a Lie algebra L. Then the ideal $I+J$ is also nilpotent.*

PROOF. Let $I^n = 0$ and $J^m = 0$ for some integers $n, m \geq 1$. We claim that $(I+J)^{n+m} = 0$. It is enough to see that any left commutator

$$\rho = [x_{kr_k}, \ldots, x_{k1}, y_{ks_k}, \ldots, y_{k1}, \ldots, x_{1r_1}, \ldots, x_{11}, y_{1s_1}, \ldots, y_{11}],$$

where $x_{ij} \in I$ and $y_{lh} \in J$, vanishes whenever $\text{length}(\rho) \geq n+m$. Since ρ belongs to $I^{r_1+\cdots+r_k} \cap J^{s_1+\cdots+s_k}$, $\rho = 0$ if $\text{length}(\rho) = r_1+\cdots+r_k+s_1+\cdots+s_k \geq n+m$. □

COROLLARY 2.52. *Every Lie algebra satisfying the ascending chain condition on ideals contains a largest nilpotent ideal.*

PROOF. Let M be a maximal nilpotent ideal of L and let I be a nilpotent ideal of L. By Proposition 2.51, $M+I \supset M$ is nilpotent, so $M+I = M$, i.e. $I \subset M$. □

The Engel–Jacobson theorem. Let R be an associative algebra over a field $\mathbb{F}$. A subset $S \subset R$ is called a *weakly closed set* if there exists a map $\gamma : S \times S \to \mathbb{F}$ such that $st + \gamma(s,t)ts \in S$ for arbitrary $s, t \in S$. It is clear that any subalgebra L of the Lie algebra R^- is a weakly closed subset of the associative algebra R. Note also that any multiplicative semigroup S of R is a weakly closed subset (taking $\gamma(s,t) = 0$ for any $s, t \in S$).

THEOREM 2.53. [**Jac79**, II.2.Theorem 1] *Let S be a weakly closed subset of the associative algebra R of linear transformations of a finite-dimensional vector space over a field. Assume that every $s \in S$ is a nilpotent element of the associative algebra R. Then the subalgebra of R generated by S is nilpotent.*

Now let L be a Lie algebra over a field $\mathbb{F}$. A subset $S \subset L$ is called a *Lie set* if $[a,b] \in S$ for any $a, b \in S$. For a subset $X \subset L$, the Lie set of L generated by X is denoted by $S\langle X \rangle$.

It is clear that if S is a Lie subset of a Lie algebra L, then the set $\{\text{ad}_s : s \in S\}$ is a weakly closed subset of the associative algebra $\text{Ad}(L)$. Hence, it follows from Theorem 2.53.

THEOREM 2.54. (Engel-Jacobson Theorem) *Let L be a finite-dimensional Lie algebra and let $\{a_1, \ldots, a_m\} \subset L$ be a generating set of L. If every $s \in S\langle a_1, \ldots, a_m \rangle$ is ad-nilpotent, then the Lie algebra L is nilpotent.*

2.6. The locally nilpotent radical

Let L be a Lie algebra over an arbitrary ring of scalars Φ. Following [**ZS73**], we will prove in this section that L contains a largest locally nilpotent ideal $\text{LocN}(L)$. However, this is not yet an efficient radical, since the fundamental Kurosh property of a radical $\mathbf{r}$, i.e. $\mathbf{r}(L/\mathbf{r}(L)) = 0$, is not always satisfied (the 2-dimensional nonabelian Lie algebra in Remark 2.50 shows that $\text{LocN}(L)$ is not a radical in the Kurosh sense). Nevertheless, if L is nil, i.e. every $x \in L$ is ad-nilpotent, then $\text{LocN}(L)$ is a radical in the Kurosh sense.

The locally nilpotent radical of an associative algebra. To illustrate this question we consider first the case of an associative algebra R over an arbitrary ring of scalars Φ. In this case, $\text{LocN}(R)$ is a radical in the Kurosh sense without any further condition.

PROPOSITION 2.55. *Every associative algebra R contains a largest locally nilpotent ideal* $\text{LocN}(R)$. *The algebra $R/\text{LocN}(R)$ does not contain nonzero locally nilpotent ideals.*

PROOF. (1) First we prove by induction that if R is generated by a finite set X, then for any integer $n \geq 1$ the subalgebra R^n is generated by the *finite set* $Y_n := \bigcup_{n \leq k \leq 2n-1} X^k$. The result is true for $n = 1$, since $Y_1 = X$. Let $n \geq 1$ and suppose that R^n is generated by Y_n. We must prove that R^{n+1} is generated by Y_{n+1}. It is clear that R^{n+1} is Φ-spanned by products of the form $a_1 \cdots a_m b$ ($m \geq 1$), where $a_i \in X^{k_i}$ with $n \leq k_i \leq 2n - 1$, and $b \in X^l$ with $l \geq 1$, and therefore these products have length $k_1 + \cdots + k_m + l \geq n + 1$. By the Euclidean algorithm, $k_1+\cdots+k_m+l = q(n+1)+r$ for suitable integers $q \geq 1$ and $0 \leq r \leq n$. Therefore, $k_1+\cdots+k_m+l = (q-1)(n+1)+n+1+r$ with $n+1 \leq n+1+r \leq 2n+1 = 2(n+1)-1$.

(2) Now we show that the class of locally nilpotent algebras is closed with respect to extensions. Suppose that R contains a locally nilpotent ideal I such that the quotient R/I of R by I is locally nilpotent. We will prove that R is locally nilpotent. Without loss of generality we may assume that R is finitely generated. Then there exists an integer $n \geq 1$ such that $R^n \subset I$. As we just have seen, the algebra R^n is also finitely generated, and therefore nilpotent of index (say m). So $R^{nm} = (R^n)^m = 0$.

(3) The assertion (2) immediately implies that the sum of two locally nilpotent ideals I, J of an algebra R is a locally nilpotent ideal, since $(I + J)/I \cong J/(I \cap J)$. By an obvious induction we conclude that the sum of any finite number of locally nilpotent ideals is a locally nilpotent ideal, which implies that the sum of any (infinite) family of locally nilpotent ideals of R is a locally nilpotent ideal.

Let $\operatorname{LocN}(R)$ denote the sum of all locally nilpotent ideals of R. We have seen that $\operatorname{LocN}(R)$ is a locally nilpotent ideal of R which contains all other locally nilpotent ideals of R.

(4) Finally we show that the quotient $R/\operatorname{LocN}(R)$ of R by $\operatorname{LocN}(R)$ does not contain any nonzero locally nilpotent ideal. Let $\bar{I} = I/\operatorname{LocN}(R)$ be a locally nilpotent ideal of $R/\operatorname{LocN}(R)$, I being an ideal of R containing $\operatorname{LocN}(R)$. By (2), the ideal I is locally nilpotent, which implies that I is contained in $\operatorname{LocN}(R)$; that is, $\bar{I} = 0$. This proves the proposition. □

As an application of Preposition 2.55, we prove the (ordinary) Kurosh–Levitzly problem for associative algebras.

THEOREM 2.56. (Kaplansky) *Let R be an associative algebra over a field $\mathbb{F}$ such that for some $n > 1$, $a^n = 0$ for every element $a \in R$. If $\mathbb{F}$ has at least n elements, then R is locally nilpotent.*

PROOF. By factoring R by its locally nilpotent radical, we may suppose that it is semiprime. Then it is enough to show that for any element $a \in R$, $a^{n-1}Ra^{n-1} = 0$. Given $a, b \in R$ and $\alpha \in \mathbb{F}$, we have

$$0 = (a + \alpha b)^n = F_0(a, b) + \alpha F_1(a, b) + \cdots + \alpha^n F_n(a, b),$$

where each $F_i(a, b)$ is homogeneous of degree i at b and $n-i$ at a, $F_0(a, b) = a^n = 0$, and $F_n(a, b) = b^n = 0$. Then

$$\alpha F_1(a, b) + \cdots + \alpha^{n-1} F_{n-1}(a, b) = 0.$$

Choose $n - 1$ distinct nonzero scalars $\alpha_i \in \mathbb{F}$, $1 \leq i \leq n - 1$, and consider the equation

$$(\alpha_{ij})F(a, b) = 0,$$

where (α_{ij}) denote the matrix with entries $\alpha_{ij} = \alpha_i^j$, $i, j \in \{1, 2, \dots, n-1\}$, and $F(a,b)$ the column vector formed by the $F_i(a,b)$. Since the determinant of the matrix (α_{ij}) is $\alpha_1 \cdots \alpha_{n-1} \prod_{i<j}(\alpha_i - \alpha_j) \neq 0$, we get $F_i(a,b) = 0,\ 1 \leq i \leq n$. In particular,

$$F_1(a,b) = a^{n-1}b + a^{n-2}ba + \cdots + ba^{n-1} = 0,$$

so

$$a^{n-1}ba^{n-1} = (-a^{n-2}ba - \cdots - ba^{n-1})a^{n-1} = (-a^{n-2}b - \cdots - ba^{n-2})a^n = 0,$$

as required. □

REMARK 2.57. In [**Lev53**], J. Levitzky would drop the restriction on the field. Moreover, as will be commented in Chapter 8, just assuming that R is nil, the identity $X^n = 0$ can be replaced by any other nontrivial polynomial identity.

The largest locally nilpotent ideal of a Lie algebra. Return to the case of a Lie algebra L. Given a nonempty subset Y of L, we denote by $\langle Y \rangle$ the subalgebra of L generated by Y. Let X and E be nonempty finite subsets of L. By $K(X,E)$ we mean the ideal of the algebra $\langle X \cup E \rangle$ generated by X. So every element of $K(X,E)$ is a linear combination of commutators ρ in $X \cup E$ containing some $x \in X$ in its expression, i.e. $\mathrm{occ}_X(\rho) \geq 1$ with $\mathrm{occ}(\rho)$ denoting the number of occurrences of elements of X in ρ. For each $i \geq 1$, let $K^{(i)}$ be the set of commutators ρ in $X \cup E$ such that $1 \leq \mathrm{occ}_X(\rho) \leq i$. It is clear that $K(X,E)$ is the direct limit of the sequence of subalgebras $\langle K^{(1)} \rangle \leq \langle K^{(2)} \rangle \leq \cdots$. Actually this sequence stabilizes at $\langle K^{(1)} \rangle$.

LEMMA 2.58. $K(X,E) = \langle K^{(1)} \rangle$.

PROOF. We can write any $a \in K(X,E)$ as a linear combination of left commutators ρ in $X \cup E$ with $\mathrm{occ}_X(\rho) \geq 1$. So it suffices to prove that such a left commutator ρ belongs to $\langle K^{(1)} \rangle$. Suppose that $\mathrm{occ}_X(\rho) \geq 2$, i.e.

$$\rho = [u_r, \dots, u_1, x_1, v_s, \dots, v_1, x_2, w_t, \dots, w_1],$$

where $u_i, v_j, w_k \in \langle X \cup E \rangle$ and $x_1, x_2 \in X$. Setting $\tau = [v_s, \dots, v_1, x_2, w_t, \dots, w_1]$ in $K(X,E)$, we have $\rho = [u_r, \dots, u_1, x_1, \tau]$, with $\mathrm{occ}_X(\tau) < \mathrm{occ}_X(\rho)$, Thus, using the Jacobi identity repeatedly, we can represent ρ as a linear combination of commutators $[\rho_i, \tau_i]$ where $\rho_i, \tau_i \in K(X,E)$ and $\max\{\mathrm{occ}_X(\rho_i), \mathrm{occ}_X(\tau_i)\} < \mathrm{occ}_X(\rho)$. Induction on $\mathrm{occ}_X(\rho)$ completes the proof. □

LEMMA 2.59. *Let X and E be nonempty finite subsets of L. Suppose that E is contained in a locally nilpotent ideal I of L. Then the algebra $K(X,E)$ is finitely generated.*

PROOF. Since the ideal I is locally nilpotent, the subalgebra $\langle E \cup [X,E] \rangle$ of I is nilpotent and therefore finitely generated as a Φ-module. Let S be a generating set of the Φ-module $\langle E \cup [X,E] \rangle$. Then $[X,S]$ is a generating set of the Φ-module $K([X,E],E)$. To prove the lemma it is sufficient to show that $K^{(1)} \subset \langle X \cup [X,S] \rangle$, since in this case it follows from Lemma 2.58 that

$$K(X,E) = \langle K^{(1)} \rangle \subset \langle X \cup [X,S] \rangle \subset K(X,E)$$

and hence $K(X,E) = \langle X \cup [X,S] \rangle$ is finitely generated.

Let $a \in K^{(1)}$ be written (by the Jacobi identity) as a linear combination of left commutators $\rho = [e_{i_r}, \ldots, e_{i_1}, x, f_{j_s}, \ldots, f_{j_1}]$, where the e_{i_h} and the f_{j_k} belong to E and $x \in X$. If $r + s = 0$, then $\rho = x \in X \subset \langle X \cup [X, S]\rangle$. If $r + s > 0$, then $\rho \in K([X, E], E) = ([X, S])_\Phi \subset \langle X \cup [X, S]\rangle$. This completes the proof. $\square$

LEMMA 2.60. *Let I and J be locally nilpotent ideals of L. Then $I + J$ is locally nilpotent.*

PROOF. Let $X \subset I$ and $Y \subset J$ be finite subsets. We must prove that the subalgebra $\langle X \cup Y\rangle$ is nilpotent. By Lemma 2.59, the algebras $K(X, Y) \subset I$ and $K(Y, X) \subset J$ are finitely generated and therefore nilpotent. As both $K(X, Y)$ and $K(Y, X)$ are ideals of $\langle X \cup Y\rangle$, it follows from Proposition 2.51 that

$$\langle X \cup Y\rangle = K(X, Y) + K(Y, X)$$

is nilpotent. This proves that $I + J$ is locally nilpotent. $\square$

PROPOSITION 2.61. *L contains a largest locally nilpotent ideal* $\mathrm{LocN}(L)$.

PROOF. Let M be a maximal locally nilpotent ideal of L, whose existence is guaranteed by Zorn's lemma, and let I be any locally nilpotent ideal of L. By Lemma 2.60 the ideal $M + I$ is locally nilpotent, so $I \subset M$ by the local nilpotent maximality of M. $\square$

The locally nilpotent radical of a nil Lie algebra. Following the exposition given in [**Zel92b**], we prove in this subsection that if L is *nil*, i.e. every $x \in L$ is ad-nilpotent, then $\mathrm{LocN}(L)$ is a radical in the Kurosh sense.

LEMMA 2.62. *Let $\mathcal{A}\langle X\rangle$ be the free associative algebra on a finite set of free generators $X = \{x_1, \ldots, x_m\}$ and let $n \geq 1$. Then for any n-sequence $(x_{i_1}, \ldots, x_{i_n})$ of X,*

$$x_{i_1} \cdots x_{i_n} = \sum \rho_1^{t_1} \cdots \rho_s^{t_s},$$

where $\rho_1, \ldots, \rho_s$ are commutators in the elements of the sequence (x_{i_j}), the exponents t_i are greater than or equal to 1, every summand on the right hand side has the same composition as $x_{i_1} \cdots x_{i_n}$, and all commutators $\rho_1, \ldots, \rho_s$ are distinct.

PROOF. If all the factors x_{i_j} are distinct, the lemma is proved (single elements are commutators of length one). So suppose that there exists a natural number $1 \leq k \leq n$ such that the first k factors $x_{i_1}, \ldots, x_{i_k}$ appear just one time in the product and there exists $k + 1 < l \leq n$ such that $x_{i_{k+1}} = x_{i_l}$. We can move the factor x_{i_l} to the $k+2$-position by transposing adjacent factors, what yields summands involving commutators. We illustrate this process with an example:

$$\begin{aligned} x_1x_2x_3x_4x_2 &= x_1x_2x_3[x_4, x_2] + x_1x_2x_3x_2x_4 \\ &= x_1x_2x_3[x_4, x_2] + x_1x_2[x_3, x_2]x_4 + x_1x_2^2x_3x_4. \end{aligned}$$

Making transpositions as many times as necessary, we can write

$$x_{i_1} \cdots x_{i_n} = x_{i_1} \cdots x_{i_k} x_{i_{k+1}}^t x_{k+1+t} \cdots x_{i_n} + x_{i_1} \cdots x_{i_k} \left(\sum \rho_1^{t_1} \cdots \rho_r^{t_r}\right),$$

where in the first summand on the right hand side the element $x_{i_{k+1}}$ appears just t-times. Now we repeat this process for the product $x_{k+1+t} \cdots x_{i_n}$, and each one of the summands $\rho_1^{t_1} \cdots \rho_r^{t_r}$, working in the second case with commutators as we did with elements. Induction on the number of factors allows us to complete the proof. $\square$

For m, k natural numbers, denote by $r_{m,k}$ the number of distinct commutators of length less than k in the set $X = \{x_1, \dots, x_m\}$ of the free associative algebra $\mathcal{A}\langle X\rangle$.

LEMMA 2.63. *Let R be an associative algebra and let $S = \{a_1, \dots, a_m\}$ be a nonempty subset of elements of R. Suppose that every commutator in S of length less than k is nilpotent of index at most M, and let $N = r_{m,k}Mk$. Then for any N-sequence $(a_{i_1}, \dots, a_{i_N})$ of S, we have $a_{i_1} \cdots a_{i_N} = \sum_j v_j \rho_j w_j$, where v_j, w_j are associative monomials in S (possibly of degree 0), ρ_j are commutators in S of length k, and each summand on the right hand side has the same composition as $a_{i_1} \cdots a_{i_N}$.*

PROOF. By Lemma 2.62, for any N-sequence (x_{i_j}) of X,

$$x_{i_1} \cdots x_{i_N} = \sum \rho_1^{t_1} \cdots \rho_s^{t_s},$$

where $\rho_1, \dots, \rho_s$ are commutators in the elements of the sequence (x_{i_j}), the exponents t_i are greater than or equal to 1, every summand on the right hand side has the same composition as $x_{i_1} \cdots x_{i_N}$, and all commutators $\rho_1, \dots, \rho_s$ are distinct. Let $\varphi : \mathcal{A}\langle X\rangle \to R$ be the homomorphism sending x_i to a_i, $1 \le i \le m$. Then $a_{i_1} \cdots a_{i_N} = \sum \varphi(\rho_1)^{t_1} \cdots \varphi(\rho_s)^{t_s}$. If the claim of the lemma is false, then in at least one summand $\sigma = \varphi(\rho_1)^{t_1} \cdots \varphi(\rho_s)^{t_s}$, we have $\text{length}(\varphi(\rho_i)) = \text{length}(\rho_i) < k$ for all $1 \le i \le s$. Since $s \le r_{m,k}$, this implies that

$$\text{length}(\sigma) = \sum_{i=1}^{s} t_i \,\text{length}(\rho_i) < r_{m,k}Mk,$$

a contradiction. This proves the lemma. □

DEFINITION 2.64. A Lie algebra L is said to be *nil* if every $x \in L$ is ad-nilpotent.

PROPOSITION 2.65. *Let L be a nil Lie algebra and let I be an ideal of L such that both I and L/I are locally nilpotent algebras. Then L is locally nilpotent.*

PROOF. Without loss of generality we may suppose that L is finitely generated. Then L/I is also finitely generated and hence there exists an integer $k \ge 1$ such that $L^k \subset I$. We will show that the subalgebra L^k of the ideal I is finitely generated and therefore nilpotent. Let $\{a_1, \dots, a_m\}$ be a generating set of L. Consider all commutators in $\{a_i\}$ of length less than k and let M be the maximum of their indexes of nilpotency. We claim that the algebra L^k is generated by all commutators ρ in $\{a_i\}$ whose lengths are subject to the inequality

$$k \le \text{length}(\rho) < k + kMm^k.$$

Let $N = kMm^k$. Consider a commutator $\rho = [a_{i_t}, \cdots, a_{i_1}]$ of length $t \ge k+N$, and let us represent it as $\rho = \text{ad}(a_{i_t}) \cdots \text{ad}(a_{i_{t-N+1}})[a_{i_{t-N}}, \dots, a_{i_1}]$. Applying Lemma 2.63 to the product $\text{ad}(a_{i_t}) \cdots \text{ad}(a_{i_{t-N+1}})$ (with $R = \text{Ad}(L)$), we get

$$\rho = \sum_j [\rho_j', \rho_j''],$$

where $k \le \text{length}(\rho_j'), \text{length}(\rho_j'') < \text{length}(\rho)$. A recursive argument proves that $L^k \subset I$ is finitely generated and therefore nilpotent.

Let $d \geq 1$ be an integer such that $(L^k)^d = 0$. Then the algebra L is nilpotent of index at most $N(d-1)+k$. Indeed, any commutator $\rho = [a_{i_{N(d-1)+k}}, \dots, a_{i_1}]$ can be written as

$$\rho = \operatorname{ad}(a_{i_{N(d-1)+k}}) \cdots \operatorname{ad}(a_{i_{k+1}})[a_{i_k}, \dots, a_{i_1}].$$

Applying Lemma 2.63 to each one of the $d-1$ segments

$$\operatorname{ad}(a_{i_{N(d-1)+k}}) \cdots \operatorname{ad}(a_{i_{N(d-2)+k+1}}), \ \dots, \ \operatorname{ad}(a_{i_{N+k}}) \cdots \operatorname{ad}(a_{i_{k+1}}),$$

we get the inclusion $\rho \in \operatorname{ad}(L^k)^{d-1} L^k = (L^k)^d$, which proves the proposition. □

PROPOSITION 2.66. (Kostrikin (1956), Plotkin (1958)) *Every nil Lie algebra L has a locally nilpotent ideal* $\operatorname{LocN}(L)$ *which contain all other locally nilpotent ideals of L. The algebra $L/\operatorname{LocN}(L)$ does not contain any nonzero locally nilpotent ideal.*

PROOF. Let $\operatorname{LocN}(L)$ be the largest locally nilpotent ideal of L (Proposition 2.61). It follows from Proposition 2.65 that $L/\operatorname{LocN}(L)$ does not contain any nonzero locally nilpotent ideal. □

2.7. A locally nilpotent radical for graded Lie algebras

Let $L = \sum_k L_k$ be a Lie algebra with a finite $\mathbb{Z}$-pregrading. Set $L^* = \bigcup_{k \neq 0} L_k$ and called an ideal I of L *strong* if it is generated by a subset of L^*. It is clear that any strong ideal is a graded ideal. Following [**Zel83a**], we prove in this section that L has a largest locally nilpotent strong ideal which is radical in the Kurosh sense.

Strong ideals and pregradings. Let $L = \sum_{k=-n}^{n} L_k$ be a Lie algebra with a $(2n+1)$-pregrading. Denote by $G = \operatorname{Gr}(L)$ the Φ-submodule of L defined by $G = \sum_{i=-n}^{n} G_i$, where $G_{-n} = 0$ and $G_k = L_k \cap (\sum_{i<k} L_i)$ for any $-n < k \leq n$. It is easy to see that $\operatorname{Gr}(L)$ is an ideal called the *graduating ideal* of the pregraded Lie algebra L.

LEMMA 2.67. *Let L be a Lie Φ-algebra with a $(2n+1)$-pregrading. If L is generated by L^*, then* $\operatorname{Gr}(L)$ *is a nilpotent strong ideal of L.*

PROOF. $G = \operatorname{Gr}(L)$ is a graded Φ-submodule of L satisfying $[G_k, L_j] \subset G_{k+j}$, $-n \leq k, j \leq n$, which proves that G is an ideal of L. Moreover, since L is generated by L^*, G is in fact a strong ideal. We claim that G is nilpotent of index $m \leq 2n+1$, i.e. for any $(2n+1)$-sequence $(g_1, \dots, g_{2n+1})$ of elements of G, the left commutator $[g_1, \dots, g_{2n+1}]$ vanishes. Let $g_i \in G_{s_i}$ for $1 \leq i \leq 2n+1$ and $-n \leq s_i \leq n$. Since $G_{s_i} \subset L_{s_i}$ and $L_r = 0$ for $r > n$, we may assume $s_1 + \cdots + s_{2n+1} \leq n$. On the other hand, for $s_i > -n$ (recall that $G_{-n} = 0$), $g_i \in G_{s_i} \subset \sum_{j<s_i} L_j$, so we can write $g_i = \sum_{j_i=-n}^{s_i-1} g_{j_i}$. Then

$$g = \sum_{j_1, \dots, j_{2n+1}} [g_{j_1}, \dots, g_{j_{2n+1}}] = 0$$

since

$$\begin{aligned} j_1 + \cdots + j_{2n+1} &\leq (s_1 - 1) + \cdots + (s_{2n+1} - 1) \\ &= s_1 + \cdots s_{2n+1} - (2n+1) \leq n - (2n+1) = -n - 1 < n. \end{aligned}$$

Therefore G is nilpotent of index at most $2n+1$. □

PROPOSITION 2.68. *Let L be a Lie Φ-algebra with a $(2n+1)$-pregrading. If L is generated by L^* and has no nonzero nilpotent strongly ideal, then:*

(a) *the sum $L = \sum_{k=-n}^{n} L_k$ is direct (so L is actually a graded Lie algebra), and*

(b) *for any ideal I of L, either $I \subset L_0 \cap Z(L)$ or $I \cap L^* \neq 0$.*

PROOF. (a) By Lemma 2.67, $G = \mathrm{Gr}(L) = 0$. Let $\sum_{i=-n}^{n} x_i = 0$, where for each index i, $x_i \in L_i$. Suppose $x \neq 0$ and let k be the largest integer such that $x_k \neq 0$. Then $x_k \in G_k$, a contradiction. Therefore the pregrading is actually a grading.

(b) Let I be an ideal of L. If I is contained in L_0, then for any $k \neq 0$, $[I, L_k] = 0$, which implies $L \subset Z(L)$, since L is generated by L^*. Suppose that I is not contained in L_0 and that $I \cap L^* = 0$. Then I does not contain any nonzero graded ideal and therefore the pregrading induced by L in L/I is not a grading. Hence, by (a), $\mathrm{Gr}(L/I)$ is a nonzero nilpotent ideal. Let J be its pre-image in L. We have $J^m \subset I$ for some integer $m \geq 1$. Since J^m is a graded ideal, it must be zero, a contradiction, so $I \cap L^* \neq 0$, as desired. $\square$

COROLLARY 2.69. *If L is semiprime, then any finite $\mathbb{Z}$-pregrading of L is actually a grading.*

Lie subalgebras generated by nonzero homogeneous elements. Let $L = \sum_{k=-n}^{n} L_k$ be a Lie algebra with a $(2n+1)$-pregrading over a field $\mathbb{F}$ of characteristic 0 or $p > 2n+1$, and let B be the subalgebra of L generated by a nonempty finite subset $\{a_1, \ldots, a_m\}$ of L^*. Then $B = \sum_{k=-n}^{n} B_k$ is a graded subalgebra of L, $B_k \subset B \cap L_k$, with $B_0 = \sum_{k\neq 0}[B_k, B_{-k}]$. Denote by B_+ and B_- the vector subspaces $\sum_{k>0} B_k$ and $\sum_{k<0} B_k$ respectively. Given a nonempty subset $X \subset L$, set:

(i) $\mathrm{ad}(X) = \{\mathrm{ad}_x : x \in X\} \subset \mathrm{Ad}(L)$,

(ii) $\mathrm{Ad}(X)$ the associative subalgebra of $\mathrm{Ad}(L)$ generated by $\mathrm{ad}(X)$, and

(iii) $\widehat{\mathrm{Ad}(X)} = \mathbb{F}1_L + \mathrm{Ad}(X)$.

Note that $\mathrm{Ad}(B) = \sum_{j=1}^{\infty} \mathrm{ad}(B)^j$, where for each $j \geq 1$, $\mathrm{ad}(B)^j$ is the subspace of $\mathrm{Ad}(B)$ generated by the products $\mathrm{ad}_{b_1} \cdots \mathrm{ad}_{b_j}$, with $b_i \in B$ for $1 \leq i \leq j$. For brevity's sake, for any $j \geq 1$, we will write $\mathrm{ad}(B)^{1\leq i\leq j}$ to denote the sum $\sum_{i=1}^{j} \mathrm{ad}(B)^i$, and $\mathrm{ad}(B)^{0\leq i\leq j} = \mathbb{F}1_L + \mathrm{ad}(B)^{1\leq i\leq j}$.

PROPOSITION 2.70. *There exists a function $f(n,m)$ such that*

$$\mathrm{Ad}(B) = \sum_{i=1}^{f(n,m)} \mathrm{ad}(B)^i.$$

PROOF. (1) By the grading properties, we have $\mathrm{ad}(B_+)^{2n+1} = \mathrm{ad}(B_-)^{2n+1} = 0$.

(2) Now let $a \in B_k$ and let $(x_1, \ldots, x_{2n+1})$ be a sequence of elements of B_{-k}, $k \neq 0$, $-n \leq k \leq n$. For any $1 \leq r \leq 2n+1$, we have

$$\mathrm{ad}_a^r \,\mathrm{ad}_{x_1} \cdots \mathrm{ad}_{x_r} \equiv r!\, \mathrm{ad}_{[a,x_1]} \cdots \mathrm{ad}_{[a,x_r]}$$
$$\mod \mathrm{ad}(B)^{0\leq i\leq 2r-1}\, \mathrm{ad}\left(\sum_{i>0} B_{ik}\right) + \mathrm{ad}(B)^{1\leq i\leq r-1}.$$

We will use the upper case notation $A = \mathrm{ad}_a$, $X_i = \mathrm{ad}_{x_i}$, and omit the module of the congruence to simplify the proof. For $r = 1$ is trivial, $AX_1 \equiv [A, X_1]$. Suppose then that $r > 1$. We have

$$A^r X_1 \cdots X_r \equiv \sum_{i=1}^{r} A^{r-1} X_1 \cdots [A, X_i] \cdots X_r,$$

since the missing summand, $A^{r-1}X_1 \cdots X_r A$, belongs to $\mathrm{ad}(B)^{1\le i\le 2r-1}\,\mathrm{ad}(B_k)$. Similarly, each one of the summands on the right hand side of the above congruence is in turn congruent with a sum of $r-1$ terms of the form

$$A^{r-2} X_1 \cdots [A, X_i] \cdots [A, X_j] \cdots X_r,$$

where $1 \le i < j \le r$. Two summands are omitted because they are congruent with 0, that ending at A and the one of the form $A^{r-2}X_1 \cdots [A, A, X_i] \cdots X_r$. In fact, it is not difficult to verify that any product involving a factor which is a commutator with variables A and X_i, $1 \le i \le r$, where the number of occurrences of A is greater than the number of occurrences of X_i, is congruent with 0. Proceeding in this way and collecting all the summands obtained in each step, we obtain the required congruence $A^k X_1 \cdots X_r \equiv r![A, X_1] \cdots [A, X_r]$.

(3) $\mathrm{ad}_{[a,x_1]} \cdots \mathrm{ad}_{[a,x_{2n+1}]} \in \mathrm{ad}(B)^{0\le i\le 2n}\,\mathrm{ad}\left(\sum_{i>0} B_{ik}\right) + \mathrm{ad}(B)^{1\le i\le 2n}$.

It follows from (2), since $\mathrm{ad}_a^{2n+1} = 0$ and $\mathrm{char}(\mathbb{F}) = 0$ or $p > 2n+1$.

(4) For any finite sequence $(b_1, \ldots, b_s)$, $s > 1$, of elements of B and any permutation σ of $\{1, \ldots s\}$, we have

$$\prod_{i=1}^{s} \mathrm{ad}_{b_{\sigma(i)}} \equiv \prod_{i=1}^{s} \mathrm{ad}_{b_i} \mod \mathrm{ad}(B)^{1\le i\le s-1}.$$

Note that any transposition of two adjacent factors yields a congruence modulo $\mathrm{ad}(B)^{s-1}$.

(5) Hence, independent of the sign of r, we get

$$\prod_{i=1}^{2n+1} \mathrm{ad}_{[a,x_1]} \cdots \mathrm{ad}_{[a,x_{2n+1}]} \in \mathrm{ad}(B)^{0\le i\le 2n}\,\mathrm{ad}(B_+) + \mathrm{ad}(B)^{1\le i\le 2n}.$$

If $r > 0$, it follows from (3). Suppose then $r < 0$. We have by (4) that, modulo $\mathrm{ad}(B)^{1\le i\le 2n}$, the operator $(2n+1)!\prod_{i=1}^{2n+1} \mathrm{ad}_{[a,x_1]} \cdots \mathrm{ad}_{[a,x_{2n+1}]}$ is a symmetric function and therefore congruent with

$$\mathrm{ad}^{2n+1}_{[a,x_1+\cdots+x_{2n+1}]} - \sum_{i=1}^{2n+1} \mathrm{ad}^{2n+1}_{[a,x_1+\cdots+\hat{x}_i+\cdots+x_{2n+1}]} + \cdots + \sum_{i=1}^{2n} \mathrm{ad}^{2n+1}_{[a,x_i]},$$

where each summand lies in $\mathrm{ad}(B)^{0\le i\le 2n}\,\mathrm{ad}(\sum_{i<0} B_{ik}) + \mathrm{ad}(B)^{1\le i\le 2n}$ by (3).

(6) $\mathrm{ad}(B_0)^{2nm+1} \subset \mathrm{ad}(B)^{0\le i\le 2nm}\,\mathrm{ad}(B_+) + \mathrm{ad}(B)^{1\le i\le 2nm}$.

Consider the operator

$$T = \prod_{i=1}^{2nm+1} \mathrm{ad}_{[x_i,y_i]}, \text{ where } x_i \in B_k,\ y_i \in B_{-k},\ 1 \le i \le 2nm+1.$$

Since B is generated by the set $\{a_1, \ldots, a_m\}$, we may assume (using the Jacobi identity) that for each $1 \le i \le 2nm+1$, $x_i \in \{a_1, \ldots, a_m\}$. There exist $2n+1$

indexes $1 \le i_1 < \cdots < i_{2n+1} \le 2nm+1$ such that $x_{i_1} = \cdots = x_{2n+1} = a \in \{a_1, \ldots, a_m\}$. Using the same argument as in (4), we get

$$T \equiv \mathrm{ad}_{[x_{2nm+1}, y_{2nm+1}]} \cdots \mathrm{ad}_{[x_{2n+1}, y_{2n+1}]} \cdots \mathrm{ad}_{[x_i, y_i]} \cdots \mathrm{ad}_{[x_1, y_1]} \quad \mathrm{mod}\ \mathrm{ad}(B)^{1 \le i \le 2nm}.$$

Now we have by (2)

$$\begin{aligned}\mathrm{ad}_{[x_{2nm+1}, y_{2nm+1}]} \cdots \mathrm{ad}_{[x_1, y_1]} &\in \mathrm{ad}(B)^{2m(n-1)} \mathrm{ad}_{[a, y_{2n+1}]} \cdots \mathrm{ad}_{[a, y_i]} \cdots \mathrm{ad}_{[a, y_1]} \\ &\subset \mathrm{ad}(B)^{0 \le i \le 2nm} \mathrm{ad}(B_+) + \mathrm{ad}(B)^{2n(m-1)} \mathrm{ad}(B)^{1 \le i \le 2n} \\ &\subset \mathrm{ad}(B)^{0 \le i \le 2nm} \mathrm{ad}(B_+) + \mathrm{ad}(B)^{1 \le i \le 2nm},\end{aligned}$$

which completes the proof of (6).

(7) $\mathrm{ad}(B)^{2n(m+1)+1} \subset \mathrm{ad}(B)^{0 \le i \le 2n(m+1)} \mathrm{ad}(B_+) + \mathrm{ad}(B)^{1 \le i \le 2n(m+1)}$.

Let $\prod_{i=1}^{2n(m+1)+1} \mathrm{ad}(x_i)$ an operator of the left hand side of (7). Without loss of generality we may suppose that for each index i, $x_i \in B_- \cup B_0 \cup B_+$. Now, by (4), if some $x_i \in B_+$, we can move this factor to the end proving (7). Thus we may assume that none of the x_i belongs to B_+. Then there exists $0 \le t \le 2n(m+1)+1$ such that $x_i \in B_-$ for $i \le t$ and $x_i \in B_0$ for $t < i \le 2n(m+1)+1$. If $t \le 2n$, it follows from (6) that

$$\begin{aligned}\prod_{i=1}^{2n(m+1)+1} \mathrm{ad}(x_i) &\in \mathrm{ad}(B)^{2n} \mathrm{ad}(B_0)^{2nm+1} + \mathrm{ad}(B)^{1 \le i \le 2n(m+1)} \\ &\subset \mathrm{ad}(B)^{0 \le i \le 2n(m+1)} \mathrm{ad}(B_+) + \mathrm{ad}(B)^{1 \le i \le 2n(m+1)}\end{aligned}$$

which proves (7) in this case. Suppose then that $t > 2n$. Then (7) also holds since $\mathrm{ad}(B_-)^{2n+1} = 0$ by (1).

(8) Set $N = 2n(m+1)+1$. We will prove by induction on k that

$$\mathrm{ad}(B)^{kN} \subset \mathrm{ad}(B)^{0 \le i \le kN-k} \mathrm{ad}(B_+)^k + \mathrm{ad}(B)^{1 \le i \le kN-1}, \quad 1 \le k \le 2n$$

The case $k = 1$ is (7). Suppose that the inclusion is true for k and prove it for $k+1$. Using (6) and (7) we get

$$\begin{aligned}\mathrm{ad}(B)^{(k+1)N} &\subset \mathrm{ad}(B)^N \left(\mathrm{ad}(B)^{0 \le i \le kN-k} \mathrm{ad}(B_+)^k + \mathrm{ad}(B)^{1 \le i \le kN-1}\right) \\ &\subset \mathrm{ad}(B)^{0 \le i \le kN-k} \mathrm{ad}(B)^N \mathrm{ad}(B_+)^k + \mathrm{ad}(B)^{1 \le i \le (k+1)N-1} \\ &\subset \mathrm{ad}(B)^{0 \le i \le (k+1)N-k-1} \mathrm{ad}(B_+)^{k+1} + \mathrm{ad}(B)^{1 \le i \le (k+1)N-1}.\end{aligned}$$

Taking $k = 2n+1$ in (8), it follows from (1) that

$$\mathrm{ad}(B)^{(2n+1)N} \subset \mathrm{ad}(B)^{1 \le i \le (2n+1)N-1} = \sum_{i=1}^{(2n+1)N-1} \mathrm{ad}(B)^i,$$

so $f(n,m) = (2n+1)(2n(m+1)+1) - 1$ is the required function. ☐

LEMMA 2.71. *For any natural number j, there exists a function $f_j(n,m)$ such that*

$$\mathrm{ad}(B)^{f_j(n,m)} \subset \sum_{k \ne 0} \mathrm{ad}(B^{[j]} \cap L_k) \widehat{\mathrm{Ad}(B)}.$$

Hence, if the Lie algebra B is solvable, then the associative algebra Ad(B) *is nilpotent.*

PROOF. See [**Zel83a**, Lemma 12 and Corollaries 1, 2]. ☐

A locally nilpotent strong radical. Recall that we are considering a Lie algebra $L = \sum_{k=-n}^{n} L_k$ with a $(2n+1)$-pregrading over a field $\mathbb{F}$ of characteristic 0 or $p > 2n+1$, and that an ideal I of L is said to be strong if it is generated by a subset of $L^* = \bigcup_{k\neq 0} L_k$.

LEMMA 2.72. *Let I be a strong ideal of L such that $I \cap L^*$ is a locally nilpotent set. Then I is locally nilpotent.*

PROOF. (1) Given $0 \neq k$, $-n \leq k \leq n$, and $m \geq 1$, let $x_i \in I \cap L_k$, $1 \leq i \leq m$, and $a \in L_{-k}$. Denote by X the set of commutators in $\{a, x_i : 1 \leq i \leq m\}$ having degree at least 1 in x_i, length at most $f_1(n, m+1)$, and lying in L^*. By hypothesis, X is a nilpotent set, and hence, by Lemma 2.71, $\mathrm{ad}(X)$ is a nilpotent subset of the associative algebra $\mathrm{Ad}(L)$. Suppose $\mathrm{ad}(X)^s = 0$. Then $\{\mathrm{ad}_a, \mathrm{ad}_{x_i} : 1 \leq i \leq m\}$ is nilpotent of index at most $f_1(n, m+1)s$.

(2) Let $x_i \in I \cap L_k$, $1 \leq i \leq m$, as in (1), and let $a_i \in L_{-k}$, $1 \leq i \leq m$. We will prove by induction on m that the set $\{\mathrm{ad}_{[a_i, x_i]} : 1 \leq i \leq m\}$ is nilpotent. For $m = 1$ this has been proved in (1). Assume that the subalgebra U of $\mathrm{Ad}(L)$ generated by the set $\{\mathrm{ad}_{[a_i, x_i]} : 1 \leq i \leq m-1\}$, is nilpotent, of index, say t, and therefore finite-dimensional. Let $\{\mathfrak{u}_1, \ldots, \mathfrak{u}_r\}$ be a basis of U, $\hat{U} = \mathbb{F}1_l + U$, and $X = \{x_i, \mathfrak{u}_j x_i : 1 \leq i \leq m,\ 1 \leq j \leq r\}$. Using the Jacobi identity we obtain

$$\prod_{i=1}^{m} \mathrm{ad}_{[a_1, x_1]} \cdots \mathrm{ad}_{[a_m, x_m]} \in (\mathrm{ad}([Ua_n, X]) + \mathrm{ad}([a_n, X]))\,\hat{U}.$$

By (1), $\{\mathrm{ad}_{a_n}, \mathrm{ad}_{x_i}, \mathrm{ad}_{\mathfrak{u}_j x_i}, \mathrm{ad}_{\mathfrak{u}_j a_n} : 1 \leq i \leq m,\ 1 \leq j \leq r\}$ is a nilpotent set of index s. Then $\{\mathrm{ad}_{[a_i, x_i]} : 1 \leq i \leq m\}$ is nilpotent of index at most mt. We have thus proved that the set $\sum_{k\neq 0} \mathrm{ad}([L_{-k}, I_k])$ is locally nilpotent.

(3) Let $x_1, \ldots x_m \in I \cap L^*$, and $y_1, \ldots y_r \in I_0 = \sum_{k\neq 0}[L_{-k}, I_k]$. We prove now that the associative algebra $R := \mathrm{Ad}(\{x_1, \ldots, x_m, y_1, \ldots, y_r\})$ is nilpotent (and therefore that I is locally nilpotent). By (2), $V := \mathrm{Ad}(\{y_1, \ldots, y_r\})$ is a finite-dimensional nilpotent algebra, of index, say q. Let $\{\mathfrak{v}_1, \ldots, \mathfrak{v}_s\}$ be a basis of V, and denote by $\tilde{X}$ the finite set $\{x_i, \mathfrak{v}_j x_i : 1 \leq i \leq m,\ 1 \leq j \leq s\}$. Then $R^q \subset \mathrm{ad}(\tilde{X})\hat{R}$.

(4) By induction on i we get $R^{iq} \subset \mathrm{ad}(\tilde{X})^i \hat{R}$. For $i = 1$ it has just seen. Now for $i > i$ we have

$$R^{iq} = R^{(i-1)q} R^q \subset \mathrm{ad}(\tilde{X})^{i-1} \hat{R} R^q \subset \mathrm{ad}(\tilde{X})^{i-1} R^q \subset \mathrm{ad}(\tilde{X})^i \hat{R}.$$

Since $\tilde{X}$ contained in $I \cap L^*$, it is nilpotent, so, by Lemma 2.71, $\mathrm{ad}(\tilde{X})$ is nilpotent, of index, say t. Then it follows from (4) that $R^{tq} \subset \mathrm{ad}(\tilde{X})^t \hat{R} = 0$, which proves the lemma. □

LEMMA 2.73. *Let L be a Lie algebra with a nontrivial finite pregrading over an arbitrary ring of scalars Φ, Then L has a largest locally nilpotent strong ideal.*

PROOF. Let M be a maximal locally nilpotent strong ideal of L, whose existence is guaranteed by Zorn's lemma. For any locally nilpotent strong ideal I of L, we have by Lemma 2.60 that the strong ideal $M + I$ is locally nilpotent, so $I \subset M$. □

We denote the largest locally nilpotent strong ideal of L by $\mathrm{LocSN}(L)$

LEMMA 2.74. (1) *The Lie algebra $\overline{L} = L/\mathrm{LocSN}(L)$ contains no nonzero locally nilpotent strong ideals.* (2) *Any locally nilpotent ideal of $\overline{L}$ lies in $\overline{L}_0 \cap Z(\overline{L})$.*

PROOF. (1) Any strong ideal of $\overline{L}$ is of the form $\overline{I}$ for a unique strong ideal I of L containing $\mathrm{LocSN}(L)$. Suppose that $\overline{I}$ is locally nilpotent. We must show that I is locally nilpotent. By Lemma 2.72, it suffices to verify that $I \cap L^*$ is a locally nilpotent set. Let $a_1, \dots, a_m \in I \cap L^*$ and let A be the subalgebra of I generated by these elements a_i. Since $\overline{I}$ is locally nilpotent, there exists a natural number r such that $A^{[r]} \subset \mathrm{LocSN}(L)$. We finish the proof of (1) by proving that the associative algebra $\mathrm{Ad}(A)$ is nilpotent. Denote by B the subalgebra generated by the commutators $\rho \in A^{[r]} \cap L^*$ of length at most $f_r(n,m)$ in $a_1, \dots, a_m$. Then B is a finitely generated subalgebra of $\mathrm{LocSN}(L)$ and therefore nilpotent. Hence, by Lemma 2.71, $\mathrm{Ad}(B)$ is nilpotent, of index, say s. Furthermore, using Lemma 2.71, now applied to the subalgebra A, we get $\mathrm{ad}(A)^{f_r(n,m)} \subset \mathrm{ad}(B)\widehat{\mathrm{Ad}(A)}$ and hence, by induction, we get

$$\mathrm{ad}(A)^{sf_r(n,m)} \subset \mathrm{ad}(B)^{s-1}\widehat{\mathrm{Ad}(A)}\,\mathrm{ad}(A)^{f_r(n,m)} \subset \mathrm{ad}(B)^{s-1}\,\mathrm{ad}(B)\widehat{\mathrm{Ad}(A)} = 0,$$

which implies by Proposition 2.70 that $\mathrm{Ad}(A)$ is nilpotent, proving (1).

(2) Let $\overline{J}$ be a nonzero locally nilpotent ideal of $\overline{L}$. Since, by (1), $\overline{L}$ has no nonzero locally nilpotent strong ideal, $\overline{J} \cap \overline{L}^* = 0$. Hence, by Proposition 2.68, $\overline{J} \subset Z(\overline{L}) \cap \overline{L}_0$. □

PROPOSITION 2.75. *Let $\widetilde{\mathrm{LocSN}}(L)$ be the pre-image of $Z(\overline{L})$ under the homomorphism map from L onto $\overline{L}$, i.e. $Z(\overline{L}) = \widetilde{\mathrm{LocSN}}(L)/\mathrm{LocSN}(L)$. We have:*

1. *(1) Any locally nilpotent ideal of L lies in $\widetilde{\mathrm{LocSN}}(L)$.*
2. *(2) $[\widetilde{\mathrm{LocSN}}(L), L] \subset \mathrm{LocSN}(L)$.*
3. *(3) $L/\widetilde{\mathrm{LocSN}}(L)$ contains no nonzero locally nilpotent ideals.*

PROOF. (1) Let I be a locally nilpotent ideal of L. Then $\overline{I}$ is a locally nilpotent ideal of $\overline{L}$ and hence it is contained in $Z(\overline{L})$ by Lemma 2.74, equivalently, I is contained in $\widetilde{\mathrm{LocSN}}(L)$.

(2) It follows from (1).

(3) Let $I/\widetilde{\mathrm{LocSN}}(L)$ be a locally nilpotent of $L/\widetilde{\mathrm{LocSN}}(L)$. In virtue of the isomorphism $\overline{I}/Z(\overline{L}) \cong I/\widetilde{\mathrm{LocSN}}(L)$, this implies that $\overline{I}/Z(\overline{L})$ is a locally nilpotent ideal of $\overline{L}/Z(\overline{L})$, which in turn implies that $\overline{I}$ is a locally nilpotent ideal of $\overline{L}$. By Lemma 2.74, $\overline{I} \subset Z(\overline{L})$ and hence, by the above isomorphism, $I/\widetilde{\mathrm{LocSN}}(L) = 0$. □

2.8. The locally finite radical

An arbitrary $\mathbb{F}$-algebra A is *locally finite* if any finitely generated subalgebra of A is finite-dimensional. Following [**ZS73**], we will prove in this section that any Lie algebra L over an arbitrary field $\mathbb{F}$ contains a largest locally finite ideal $\mathrm{LocF}(L)$. But as in the case of the locally nilpotent radical, this is not yet an efficient radical. Nevertheless, if L is *algebraic*, namely, every element $x \in L$ is ad-algebraic, then $\mathrm{LocF}(L)$ is a radical in the Kurosh sense.

The largest locally finite ideal of a Lie algebra. Let L be a Lie algebra over a field $\mathbb{F}$. Following the notation of Section 2.6, if X and E are nonempty finite subsets of L, we denote by $K(X, E)$ the ideal of the algebra $\langle X \cup E \rangle$ generated by X; that is, the linear span of all commutators ρ in $X \cup E$ containing some element $x \in X$ in their expressions.

LEMMA 2.76. *Suppose that E is contained in a locally finite ideal I of L. Then the algebra $K(X, E)$ is finitely generated.*

PROOF. The proof is similar to that of Lemma 2.59. As I is locally finite, the subalgebra $\langle E \cup [X, E]\rangle$ of I is finite-dimensional and hence so is the ideal $K([X, E], E)$ of $\langle E \cup [X, E]\rangle$. Let B be a basis of $K([X, E], E)$. To prove the lemma it is sufficient to show that $K^{(1)} \subset \langle X \cup B\rangle$, since in this case it follows from Lemma 2.58 that

$$K(X, E) = \langle K^{(1)}\rangle \subset \langle X \cup B\rangle \subset K(X, E)$$

and hence $K(X, E) = \langle X \cup B\rangle$ is finitely generated.

Let $w \in K^{(1)}$ be written (by the Jacobi identity) as a linear combination of left commutators $\rho = [e_{i_r}, \ldots, e_{i_1}, x, f_{j_s}, \ldots, f_{j_1}]$, where the e_{i_h} and the f_{j_k} belong to E and $x \in X$. If $r + s = 0$, then $\rho = x \in X \subset \langle X \cup B\rangle$. If $r + s > 0$, then $\rho \in K([X, E], E) = \mathrm{span}_{\mathbb{F}}(B) \subset \langle X \cup B\rangle$. This completes the proof. □

LEMMA 2.77. *Let I, J be ideals of L. If I, J are locally finite, then $I + J$ is locally finite.*

PROOF. It suffices to prove that the subalgebra $\langle X \cup Y\rangle$, where $X \subset I$ and $Y \subset J$ are finite subsets, is finite-dimensional. By Lemma 2.76 the subalgebras $K(X, Y) = \mathrm{id}_{\langle X\cup Y\rangle}(X) \subset I$ and $K(Y, X) = \mathrm{id}_{\langle X\cup Y\rangle}(Y) \subset J$ are finitely generated and therefore finite-dimensional. Hence $\langle X \cup Y\rangle = K(X, Y) + K(Y, X)$ is finite-dimensional. □

PROPOSITION 2.78. *L contains a largest locally finite ideal* $\mathrm{LocF}(L)$.

PROOF. Let M be a maximal locally finite ideal of L, whose existence is guaranteed by Zorn's lemma, and let I be any locally finite ideal of L. It follows from Lemma 2.77 that $M + I$ is locally finite, so $I \subset M$ by the local finite maximality of M. □

The locally finite radical of an algebraic Lie algebra. Let L be a Lie algebra over an arbitrary field $\mathbb{F}$. Following [**ZS73**], we prove that if every $x \in L$ is ad-algebraic, then the largest locally finite ideal $\mathrm{LocF}(L)$ of L is a radical in the Kurosh sense.

Let $E = \{e_1, \ldots, e_n\}$ be a finite set of ad-algebraic elements of L; that is, for each $1 \le i \le n$, there exist $m_i \in \mathbb{N}$ and $\lambda_{i_0}, \lambda_{i_1}, \ldots, \lambda_{i_{m_i-1}} \in \mathbb{F}$ such that

$$\mathrm{ad}_{e_i}^{m_i} = \lambda_{i_0} 1_L + \lambda_{i_1} \mathrm{ad}_{e_1} + \cdots + \lambda_{i_{m_i-1}} \mathrm{ad}_{e_i}^{m_i-1}, \tag{2.5}$$

and suppose that for any couple of indexes $i, j \in \{1, \ldots, n\}$,

$$[e_i, e_j] = \sum_{k=1}^{n} \alpha_{ijk} e_k + x_{ij}, \quad \alpha_{ijk} \in \mathbb{F},\ x_{ij} \in L. \tag{2.6}$$

LEMMA 2.79. *Let $E = \{e_1, \ldots, e_n\}$ be as above and denote by X the set of the elements x_{ij} in (2.6). Then the algebra $K(X, E)$ is finitely generated.*

PROOF. Set $m = \max\{m_1, \ldots, m_n\}$. We will prove that the algebra $K(X, E)$ is generated by the set

$$Y := X \cup \left(\bigcup_{1\le r\le n(m-1)} \{[e_{i_1}, \ldots, e_{i_r}, x] : (e_{i_1}, \ldots, e_{i_r}) \in \mathrm{Seq}_r(E),\ x \in X\} \right).$$

Denote by S the subalgebra of L generated by Y. Since $X \subset S \subset K(X,E) \triangleleft \langle X \cup E \rangle$, all we need to prove is that S is an ideal of $\langle X \cup E \rangle$, which in virtue of the Jacobi identity reduces to verify that $[X,Y] \subset S$, which is clear, and that $[E,Y] \subset S$, which will be seen now. The inclusion $[E,X] \subset S$ follows from the definition of S, so we only need to verify that for any $e \in E$, any $(e_{i_1},\dots,e_{i_r}) \in \mathrm{Seq}_r(E)$, with $1 \le r \le n(m-1)$, and $x \in X$, $[e,e_{i_1},\dots,e_{i_r},x] \in S$. If $r < n(m-1)$ there is nothing to prove, so we may assume that $r = n(m-1)$ and prove that for any sequence $(e_{i_1},\dots,e_{n(m-1)+1})$ of elements of E and any $x \in X$, the left commutator $c = [e_{i_1},\dots,e_{n(m-1)+1},x]$ belongs to S. This will be achieved in one 2-step process that involves permutations of the sequence $(e_{i_1},\dots,e_{n(m-1)+1})$ and uses the relations between the e_i and the equations of their adjoint operators, ad_{e_i}, to reduce the lengths of the left commutators. Given $\sigma \in \mathfrak{S}_{n(m-1)+1}$, denote by σc the left commutator $[e_{i_{\sigma(1)}},\dots,e_{i_{\sigma(n(m-1)+1)}},x]$. We claim that $c-\sigma c \in S$. Without loss of generality we may assume that σ is a transposition, say $\sigma = (1,2)$. Denoting by upper case letters the adjoint operators and using the relations (2.6) we get

$$\begin{aligned} c-\sigma c =& E_{i_1}E_{i_2}E_{i_3}\cdots E_{i_{n(m-1)+1}}(x) - E_{i_2}E_{i_1}E_{i_3}\cdots E_{i_{n(m-1)+1}}(x) \\ =& [E_{i_1},E_{i_2}]E_{i_3}\cdots E_{i_{n(m-1)+1}}(x) = \mathrm{ad}_{[e_{i_1},e_{i_2}]}E_{i_3}\cdots E_{i_{n(m-1)+1}}(x) \\ =& \sum_{k=1}^{n} \alpha_{i_1 i_2 k} E_{i_k}E_{i_3}\cdots E_{i_{n(m-1)+1}}(x) + X_{i_1 i_2}E_{i_3}\cdots E_{i_{n(m-1)+1}}(x) \\ =& \sum_{k=1}^{n} \alpha_{i_1 i_2 k}[e_{i_k},e_{i_3},\dots,e_{i_{n(m-1)+1}},x] + [x_{i_1 i_2},[e_{i_3},\dots,e_{i_{n(m-1)+1}},x]] \in S, \end{aligned}$$

since the length of the sequences involving the e_{i_j} is less than or equal to $n(m-1)$.

We complete the proof by proving that $\sigma c \in S$ for some $\sigma \in \mathfrak{S}_{n(m-1)+1}$. Write $c = E_{i_1}\cdots E_{i_{n(m-1)+1}}(x)$ as before. As the set E contains only n elements, one of the adjoint operators, (say $E_{i_j} = E_1$) occurs at least m times in the expression of c, namely, there exists a subsequence $1 \le j_1 \le \cdots \le j_m \le n(m-1)+1$ such that $E_{i_{j_k}} = E_1$ for $1 \le k \le m$. Let $\sigma \in \mathfrak{S}_{n(m-1)+1}$ be a permutation such that $\sigma(k) = j_k$ for $1 \le k \le m$. Then σc has the form $E_1^m E_{i_{\sigma(m+1)}}\cdots E_{i_{\sigma(n(m-1)+1)}}(x)$. Since $m_1 \le m$, we can write, by (2.5), E_1^m as a linear combination of the operators E_1^k, $0 \le k \le m_1 - 1$. In any of the summands of this linear combination the number of the E_1 involved is less than or equal to $n(m-1)$. This proves that $\sigma c \in S$, as desired. □

DEFINITION 2.80. A Lie algebra L over a field $\mathbb{F}$ is said to be *algebraic* if every $x \in L$ is ad-algebraic.

PROPOSITION 2.81. *Let I be an ideal of L such that both I and L/I are locally finite. If L is algebraic, then L is locally finite.*

PROOF. Let $E = \{e_1,\dots,e_n\}$ be a finite subset of elements of L and let $\bar e_1,\dots,\bar e_n$ be their images in $\overline{L} = L/I$. As $\overline{L}$ is locally finite, the set $\overline{E} = \{\bar e_1,\dots,\bar e_n\}$ generates a finite-dimensional subalgebra of $\overline{L}$. By enlarging E if necessary, we may assume that $\langle \overline{E} \rangle$ is spanned by $\overline{E}$ itself. Then, for each $i,j \in \{1,\dots,n\}$, we have

$$[e_i,e_j] = \sum_{k=1}^{n} \alpha_{ijk}e_k + x_{ij}, \quad \alpha_{ijk} \in \mathbb{F},\ x_{ij} \in I. \tag{2.7}$$

Denote by X the set of all x_{ij}. It follows from (2.7) that $X \subset \langle E \rangle$, and from Lemma 2.79 that the algebra $K(X, E)$ is finitely generated. As $X \subset I$, the local finiteness of I implies that $K(X, E)$ is finite-dimensional. Since, again by (2.7), the Lie algebra $\langle E \rangle / K(X, E)$ is finite-dimensional, we get that $\langle E \rangle$ is finite-dimensional. This proves that L is locally finite. □

COROLLARY 2.82. *If L is algebraic, then* $\mathrm{LocF}(L/\mathrm{LocF}(L)) = 0$.

PROOF. Let I be an ideal of L such that $\mathrm{LocF}(L) \subset I$ and the Lie algebra $\overline{I} = I/\mathrm{LocF}(L)$ is locally finite. As $\mathrm{LocF}(L)$ is a locally finite ideal of I, we have by Proposition 2.81 that I is locally finite. Hence $\overline{I} = 0$ by the local finite maximality of $\mathrm{LocF}(L)$. □

COROLLARY 2.83. *If L is soluble and algebraic, then L is locally finite.*

PROOF. We may assume $L \neq 0$. Then $L^{(n)} = [L^{(n-1)}, L^{(n-1)}] = 0$ with $L^{(n-1)} \neq 0$ for some $n > 1$. We will prove by induction on k that $L^{(n-k)}$ is locally finite for $1 \leq k \leq n-1$ and hence that $L = L^{(n-(n-1))}$ is locally finite. For $k = 1$ this is clear since $L^{(n-1)}$ is in fact abelian. Suppose then that this is true for $1 \leq k < n-1$. Then $L^{(n-(k+1))}/L^{(n-k)}$ is abelian (therefore locally finite), and $L^{(n-k)}$ is locally finite by the induction hypothesis. As L is algebraic, it follows from Proposition 2.81 that $L^{(n-(k+1))}$ is locally finite. □

REMARK 2.84. As proved in Proposition 1.25, a simple algebra cannot be locally nilpotent. Similarly, by [**BS94**, Corollary 2], a simple locally finite Lie algebra L over a field (without restriction of the characteristic) cannot be *locally solvable*, i.e. not every finitely generated subalgebra of L is a solvable algebra.

2.9. Exercises

EXERCISE 2.85. Let I and J be ideals of a Lie algebra L. Show that $[I, J]$ is an ideal of L.

EXERCISE 2.86. Let $a \in R$. Show that the adjoint operator ad_a is a derivation of the associative algebra R.

EXERCISE 2.87. Let L be a prime Lie algebra and let $Q_m(L)$ denote its maximal algebra of quotients [**SM04**, Defintion 3.7]. Suppose that L contains a minimal ideal M. Show that $Q_m(L)$ is isomorphic to $\mathrm{Der}(M)$.

EXERCISE 2.88. Show that $\mathrm{Der}(\mathbb{F}[x_1, x_2, \ldots, x_n])$, where $\mathbb{F}$ be a field of characteristic 0, is a free $\mathbb{F}[x_1, x_2, \ldots, x_n]$-module of rank n with basis $\frac{\partial}{\partial x_1}, \frac{\partial}{\partial x_2}, \ldots, \frac{\partial}{\partial x_n}$, and Lie product

$$\left[p\frac{\partial}{\partial x_i}, q\frac{\partial}{\partial x_j}\right] = p\frac{\partial q}{\partial x_i}\frac{\partial}{\partial x_j} - q\frac{\partial p}{\partial x_j}\frac{\partial}{\partial x_i}$$

EXERCISE 2.89. In relation with the Lie algebra $L(G, \{H_i\})$ defined by a group G and a central series $\{H_i\}$. Show that $\bar{x}_i = \bar{y}_i$ implies $[\bar{x}_i, \bar{x}_j] = [\bar{y}_i, \bar{x}_j]$ for $x_i, y_i \in H_i,\ x_j \in H_j$.

EXERCISE 2.90. (Mackey) Let $(X, Y, \langle\, ,\rangle)$ be an infinite-dimensional pair of dual vector spaces over a division ring. Show that there exist infinite sequences of vectors $\{x_n\}_{n=1}^{\infty} \subset X$ and $\{y_n\}_{n=1}^{\infty} \subset Y$ such that $\langle x_n, y_m \rangle = \delta_{nm}$ for all $n, m \geq 1$.

EXERCISE 2.91. Let $(X, Y, \langle\,,\rangle)$ be a pair of dual vector spaces over a division algebra Δ such that X and Y are countably infinite-dimensional. Show that $\mathcal{F}_Y(X)$ is isomorphic to the algebra $\mathrm{M}_\infty(\Delta)$ of infinite matrices with a finite number of nonzero entries over Δ.

EXERCISE 2.92. Let $(X, Y, \langle\,,\rangle)$ be a pair of dual vector spaces, let $a \in \mathcal{F}_Y(X)$ be nilpotent of index n, and take $x \in X$ such that $xa^{n-1} \neq 0$. Show:

(1) There exists $y \in Y$ such that $\langle xa^{n-1}, y\rangle = 1$ and $\langle xa^k, y\rangle = 0$ for any integer k such that $0 \leq k < n-1$.
(2) $\{x, xa, \ldots, xa^{n-1}\}$ and $\{(a^{n-1})^\# y, \ldots, a^\# y, y\}$ are bi-orthogonal.
(3) $X = \mathrm{span}(x, xa, \ldots, xa^{n-1}) \oplus \{(a^{n-1})^\# y, \ldots, a^\# y, y\}^\perp$.
(4) Both direct summands in (3) are invariant under a. [1]

EXERCISE 2.93. Let $(X, Y, \langle\,,\rangle)$ be a pair of dual vector spaces. Show that the ring $\mathcal{F}_Y(X)$ is von Neumann regular.

EXERCISE 2.94. Let $(X, Y, \langle\,,\rangle)$ be a pair of dual vector spaces. Show:

(1) Any left ideal of the ring $\mathcal{F}_Y(X)$ is of the form Y^*V, where $V \leq X$, and any right ideal is of the form W^*X, where $W \leq Y$.
(2) The correspondences $V \mapsto Y^*V$ and $W \mapsto W^*X$ are lattice isomorphisms.
(3) $\mathrm{rann}(Y^*V) = (V^\perp)^*X$ and $\mathrm{lann}(W^*X) = Y^*W^\perp$, where rann and lann respectively denote the right and left annihilators.
(4) $\mathcal{F}_Y(X)$ is a simple ring.

EXERCISE 2.95. Let R be a semiprime ring and let $a \in \mathrm{Soc}(R)$. Show that $Ra = Re$, where e is an idempotent.

EXERCISE 2.96. Let R be a semiprime ring and let e be a nonzero idempotent in R. Show that Re is a minimal left ideal if and only if eRe is a division ring if and only if eR is a minimal right ideal.

EXERCISE 2.97. Let R be a prime ring and let $e \in R$ be a division idempotent of R. Set $X = eR$, $Y = Re$, $\Delta = eRe$, and $\langle x, y\rangle = xy$ for all $x \in X$, $y \in Y$. Show:

(1) $(X, Y, \langle\,,\rangle)$ is a pair of dual vector spaces over the division ring Δ.
(2) For any $a \in R$, the map $r_a : X \to X$, given by $xr_a = xa$ for all $x \in X$, belongs to $\mathcal{L}_Y(X)$, with adjoint $r_a^\# = l_a$, where $l_a : Y \to Y$ is defined by $l_a y = ay$ for all $y \in Y$.
(3) Via the map $a \mapsto r_a$, the ring R is a subring of $\mathcal{L}_Y(X)$, with

$$\mathrm{Soc}(R) = \mathcal{F}_Y(X) \text{ and } \mathcal{C}(R) \cong Z(\Delta).$$

EXERCISE 2.98. Let $\mathbb{F}$ be a field of characteristic 0, and let L be the realization of the Lie algebra $\mathrm{Der}(\mathbb{F}[x])$ given in Example 2.10. Setting $B_n = \sum_{k\geq n} L_k$, $n \geq 1$. Show:

(1) B_n is an inner ideal of L.
(2) L has no nonzero abelian inner ideals.

EXERCISE 2.99. The reader is referred to Section 2.2 for notation.

(1) Let $(X, Y, \langle\,,\rangle)$ be a pair of dual vector spaces over a division ring Δ, and let $V \leq X$, $W \leq Y$ be vector subspaces such that $\langle V, W\rangle = 0$. Show that W^*V is an abelian inner ideal of the general linear algebra $\mathfrak{gl}_Y(X)$ contained in $\mathfrak{fsl}_Y(X)$.

[1] This exercise provides a proof of the canonical Jordan form for nilpotent matrices.

(2) Let X be a left vector space with a nondegenerate Hermitian (resp. skew-Hermitian) form $\langle\,,\rangle$ over $(\Delta, -)$ of characteristic not 2, and let $V \leq X$ be a totally isotropic subspace. Show that $[V, V]$ is an abelian inner ideal of $\mathrm{Skew}(\mathcal{L}_X(X), *)$ contained in $\mathrm{Skew}(\mathcal{F}_X(X), *)$, with $*$ being the adjoint involution.

(3) Let X be a vector space over a field $\mathbb{F}$ of characteristic not 2, $\dim_{\mathbb{F}} X > 2$, which is endowed with a nondegenerate symmetric bilinear form $\langle\,,\rangle$, and let $H = \mathbb{F}x \oplus \mathbb{F}y$ be a *hyperbolic plane*, i.e. $x, y \in X$ are isotropic vectors satisfying $\langle x, y\rangle = 1$. Then $X = H \oplus H^\perp$, where

$$H^\perp = \{z \in X : \langle z, H\rangle = 0\}.$$

Show that $[x, H^\perp]$ is an abelian inner ideal of the orthogonal algebra $\mathfrak{o}(X, \langle\,,\rangle)$ contained in $\mathfrak{fo}(X, \langle\,,\rangle)$.

EXERCISE 2.100. With the notation of Exercise 2.99(3) and denoting by R the ring $\mathcal{F}_X(X) = X^*X$, show that the derivation ad_b, where $b = y^*x - y * x$, induces a 5-grading in R with $R_{-2} = \mathbb{F}x^*x$, $R_{-1} = x^*H^\perp + (H^\perp)^*x$, $R_0 = \mathbb{F}x^*y + (H^\perp)^*H^\perp + \mathbb{F}y^*x$, $R_1 = y^*H^\perp + (H^\perp)^*y$, and $R_2 = \mathbb{F}y^*y$. Moreover, setting $K = \mathrm{Skew}(R, *) = \mathfrak{fo}(X, \langle\,,\rangle)$, we have $K_2 = K_{-2} = 0$, $K_{-1} = [y, H^\perp]$, $K_0 = \mathbb{F}d + [H^\perp, H^\perp]$, and $K_1 = [x, H^\perp]$.

EXERCISE 2.101. Let $\{\alpha_1, \alpha_2, \alpha_1 + \alpha_2, 2\alpha_1 + \alpha_2, 3\alpha_1 + \alpha_2, 3\alpha_1 + 2\alpha_2\}$ be the system of positive roots of the exceptional Lie algebra $L = G_2$, expressed on the basis of fundamental roots. Show $L_{3\alpha_1+2\alpha_2}$ and $L_{3\alpha_1+2\alpha_2} + L_{3\alpha_1+\alpha_2}$ are abelian inner ideals.

EXERCISE 2.102. Show that if $L/Z(L)$ is nilpotent, so is L.

EXERCISE 2.103. Show that a Lie algebra L is nilpotent if and only if the associative algebra $\mathrm{Ad}(L)$ is nilpotent.

EXERCISE 2.104. Let R be a finite-dimensional associative algebra such that every $x \in R$ is nilpotent. Show that R is nilpotent.

EXERCISE 2.105. Let L be a solvable finite-dimensional Lie algebra over an algebraically closed field of characteristic 0. Using Lie's theorem show that its derived algebra $[L, L]$ is nilpotent.

EXERCISE 2.106. Show that $A[A, A, X_1]X_2X_3 \equiv 0$ (Proposition 2.70).

CHAPTER 3

Absolute Zero Divisors

Ad-nilpotent elements of index less than or equal to 2 are called absolute zero divisors. As conjectured by A. I. Kostrikin, a simple finite-dimensional Lie algebra over a field of characteristic $p > 5$ is *classical* (a modular version of a complex finite-dimensional simple Lie algebra) if and only if it is nondegenerate, i.e. it does not contains nonzero absolute zero divisors On the other hand, sandwiches (as A. I. Kostrikin called them) played a fundamental role in Zelmanov's solution (via Lie algebras) of the Restricted Burnside Problem. Lie algebras considered in this chapter are over a ring of scalars Φ, with some restrictions on the torsion when required. In Section 3.1 we give the basic identities involving absolute zero divisors. In Section 3.2 we prove a fundamental result (due to A. I. Kostrikin and E. Zelmanov) settling that every 2-torsion free Lie algebra generated by absolute zero divisors is locally nilpotent. Using this result, A. N. Grishkov proved the local nilpotency of the ideal generated by the absolute zero divisors of a Lie algebra over a field of characteristic 0 (Section 3.3). In Section 3.4 we characterize nondegeneracy in finite-dimensional Lie algebras over a field of characteristic 0 in terms of the Killing form, and prove (using Grishkov's theorem) the nondegeneracy of every simple Lie algebra over a field of characteristic 0. In positive characteristic, some conditions are given which guarantee the nondegeneracy of a simple Lie algebra. Revisiting a classical Herstein's result, we show in Section 3.5 that if R is a semiprime 2-torsion free ring, then the Lie algebra $R^-/Z(R)$ is nondegenerate. In Section 3.6 we consider an analogous question for the Lie algebra of the skew-symmetric elements of a ring with involution.

3.1. Identities involving absolute zero divisors

The notational convention of writing adjoint operators by their corresponding capital letters ($X = \mathrm{ad}_x$, $Y = \mathrm{ad}_y$) will be used frequently in this section. Recall that by an absolute zero divisor of a Lie algebra L we understand an ad-nilpotent element $a \in L$ of index at most 2.

EXAMPLE 3.1. Let R be an associative algebra and let $a \in R$ be an absolute zero divisor of R, i.e. $a^2 = 0$ and $aRa = 0$. Then $a + z$ in an absolute zero divisor of the Lie algebra R^- for every central element $z \in R$.

EXAMPLE 3.2. Let $\mathbb{F}$ be a field of characteristic $p > 3$. Then the standard basis vector x_{p-1} is an absolute zero divisor of the Witt algebra $W(p)$ over $\mathbb{F}$ (see Example 2.11).

LEMMA 3.3. *Let L be a 2-torsion free Lie algebra and let $a, x, y, z \in L$ with a being an absolute zero divisor. We have:*

(1) $AXA = 0$.
(2) $AXYA = AYXA$.

(3) $\mathrm{ad}^2_{[a,x]} = -AX^2A$.
(4) $\mathrm{ad}^3_{[a,x]} = 0$.
(5) $AXYAz = AXZAy = AYZAx$.
(6) *If L is also 3-torsion free, then* $AX^2AX^2A = 0$

PROOF. (1) $A^2 = 0$ implies

$$0 = \mathrm{ad}_{A^2x} = [A, [A, X]] = \mathbf{ad}^2_A X = (l_A - r_A)^2 X = -2AXA,$$

so $AXA = 0$ since L is 2-torsion free.

(2) By (1), $0 = A[X, Y]A = AXYA - AYXA$.

(3) $\mathrm{ad}^2_{Ax} = [A, X]^2 = (AX - XA)^2 = -AX^2A$, since $A^2 = 0 = AXA$.

(4) $\mathrm{ad}^3_{Ax} = \mathrm{ad}^2_{Ax}\,\mathrm{ad}_{Ax} = -AX^2A(AX - XA) = 0$, by (1) and (3).

(5) Using (L1), (L2) and (1), we get

$$(AX)YAz = (AX)AYz + (AX)ZAy = (AXA)Yz + AXZAy = AXZAy.$$

Hence, by (2), $AXYAz = AYXAz = AYZAx$.

(6) By (1),

$$\begin{aligned} 0 = A\mathbf{ad}^4_X(A)A &= A(l_X - r_X)^4(A)A \\ &= A\sum_{k=0}^{4}(-1)^k\binom{4}{k}X^kAX^{4-k}A = 6AX^2AX^2A, \end{aligned}$$

since the remaining summands involve either AXA or A^2 as a factor. As L is 6-torsion free, $AX^2AX^2A = 0$. □

Some consequences.

PROPOSITION 3.4. *Let L be a 2-torsion free Lie algebra. Then the absolute zero divisors of L span a subalgebra.*

PROOF. Let a, b be absolute zero divisors of L. By Lemma 3.3(3),

$$\mathrm{ad}^2_{[a,b]} = -AB^2A = 0,$$

so $[a, b]$ is an absolute zero divisor. □

REMARK 3.5. While the span of the absolute zero divisors of an associative algebra (Example 3.1) is an ideal, this is no true for Lie algebras, as follows from Example 3.2 and the fact that the Witt algebra $W(p)$ is simple (Example 2.11).

PROPOSITION 3.6. *Let L be a 2-torsion free Lie algebra and let $a \in L$ be an absolute zero divisor. Then $[a, L]$ is an abelian inner ideal of L.*

PROOF. By Lemma 2.17(1), $[a, L]$ is an abelian subalgebra. Now let $x, y \in L$. Eliminating the term involving A^2 and using Lemma 3.3(1), we get

$$\mathrm{ad}_{Ax}\,\mathrm{ad}_{Ay} = [A, X][A, Y] = -AXYA.$$

Then, for any $z \in L$, $[Ax, [Ay, z]] = AXYAz \subset [a, L]$, so $[a, L]$ is an abelian inner ideal. □

3.2. A theorem on sandwich algebras

The contents of this section are taken from the paper [**ZK90**]. Our aim is to prove that a 2-torsion free Lie algebra generated by a finite number of absolute zero divisors is nilpotent. The major part of the work required for the proof takes place in an associative context, via the adjoint representation (note that a Lie algebra L is nilpotent if and only if so is the associative algebra $\mathrm{Ad}(L)$). Although, as will be seen now, the proof is quite long and technical, in the case of an algebra generated by three absolute zero divisors it is straightforward:

Let $S = \{a, b, c\}$ be a generating set of L consisting of absolute zero divisors. Then L is nilpotent of index at most 4. It suffices to verify that any left commutator $\rho = [x_1, x_2, x_3, x_4]$, with $x_i \in S$ for $1 \leq i \leq 4$, vanishes. But this is clear in all cases. For instance, $\rho = [a, c, b, a] = -ACAb = 0$.

Sandwichers. Making an exception in our notational convention, we will assume in this section that A is a 2-torsion free associative Φ-algebra and L is a subalgebra of the Lie algebra A^- generating A. For any integer $k \geq 1$, let L^k denote the Φ-submodule of A defined recursively as follows: $L^1 = L$ and $L^{k+1} = LL^k$. According to this notation, $A = \sum_{k=1}^{\infty} L^k$. The following inclusions are easily verified for any positive integer n and any $x \in L$.

(I1) $[L, L^n] \subset L^n$.
(I2) $xL^n \subset [x, L^n] + L^n x$.
(I3) $L^n x \subset \mathrm{ad}_L^n x + \sum_{j=1}^{n-1} L^{n-j} x L^j + xL^n$.

Definition 3.7. Let $k \geq 1$ be an integer. An element $x \in A$ is called a *sandwicher of strength* k (*in* A *with respect to* L) if $xL^j x = 0$ for all $0 \leq j \leq k$, with $xL^0 x = x^2$. We denote by S_k the set of elements of L that are sandwichers of strength k. It is clear that $S_1 \supset S_2 \supset \cdots$. Note that any $x \in S_1$ (and therefore, any $x \in S_k$, $k \geq 1$) is an absolute zero divisor of L.

Lemma 3.8. *Let $k \geq 1$ be an integer and let x, y, z be elements of L such that $x \in S_k$, $y \in S_1$, and $z \in S_2$. Then:*

(i) *For any $0 \leq j \leq k$ we have*

$$xyL^j xy = xyL^j yx = yxL^j yx = yxL^j xy = 0.$$

Hence $[S_k, S_1] \subset S_k$.

(ii) *For any $0 \leq j \leq k+1$ we have*

$$xzL^j xz = xzL^j zx = zxL^j zx = zxL^j xz = 0.$$

Hence $[S_k, S_2] \subset S_{k+1}$.

Proof. (i) It clearly holds for $j = 0$. Let $1 \leq j \leq k$.

(1) Using (I1) and (I2) we obtain $xyL^j xy \subset (x[y, L^j]x)y + xL^j(yxy) = 0$.

(2) By (I1), for any $a \in L^j$ we have $2xyayx = x[[y,a],y]x \in xL^j x = 0$. As A is 2-torsion free this proves that $xyL^j yx = 0$.

(3) Clearly, $y(xL^j x)y = 0$.

(4) $yxL^j yx = 0$ follows from (1) by symmetry. This completes the proof of (i).

(ii) By (i) we only need to check the formulas for $j = k+1$. Note that the case $k = 1$ also follows from (i). Suppose then that $k > 1$.

(5) Using (I3) we obtain

$$zxL^{k+1}xz \subset z(x\,\mathrm{ad}_L^{k+1}\,x)z + z(xL^kx)Lz + \cdots + z(xLx)L^kz + zx^2L^{k+1}z = 0,$$

since $z \in S_2$ and $x \in S_k$.

(6) By (I3),

$$\begin{aligned} zxL^{k+1}zx \subset & zx(L^{k-1}[L,[L,z]])x + zxL^kzLx + zxL^{k-1}zL^2x \\ \subset & zxL^kzLx + zxL^{k-1}zL^2x, \end{aligned}$$

since $L^{k-1}[L,[L,z]] \subset L^k$ and $x \in S_k$. Now let $i \geq 2$ and $j \geq 1$ be integers such that $i+j = k+1$ (the case that $i = 1, j = k$ is clear: $zxLzL^kx = 0$ since $zxLz = 0$ because $z \in S_2$). We have

$$\begin{aligned} zxL^izL^jx \subset & z(x[L,[L^{i-1},z]]L^jx) + (zxLz)L^{i-1+j}x + zxL^{i-1}zL^{j+1}x \\ & + (zxz)L^{i+j}x = zxL^{i-1}zL^{j+1}x, \end{aligned}$$

since $[L,[L^{i-1},z]]L^j \subset L^k$ (and therefore $x[L,[L^{i-1},z]]L^jx = 0$) and $z \in S_2$. Repeating this argument we get $zxL^izL^jx \subset zxL^{i-1}zL^{j+1}x \subset \cdots \subset (zxLz)L^kx = 0$, as desired.

(7) $xzL^{k+1}xz = 0$ follows from (6) by symmetry.

(8) $xz[L^{k-1},[L,[L,z]]]x \subset xzL^{k-1}x \subset xL^kx = 0$. Hence, by (I3),

$$\begin{aligned} xzL^{k+1}zx \subset & xzL^kzLx + xzL^{k-1}zL^2x + x(zL^2z)L^{k-1}x + x(zLz)L^kx \\ & + xz^2L^{k+1}x = xzL^kzLx + xzL^{k-1}zL^2x. \end{aligned}$$

Let $i+j = k+1$ with $i > 2$. As before, $xz[L^{i-2},[L,[L,z]]]L^jx \subset xL^kx = 0$. Hence

$$\begin{aligned} xzL^izL^jx \subset & xzL^{i-1}zL^{j+1}x + xzL^{i-2}zL^{j+2}x + x(zL^2z)L^{k-1}x + x(zLz)L^kx \\ & + xz^2L^{k+1}x \subset xzL^{i-1}zL^{j+1}x + xzL^{i-2}zL^{j+2}x. \end{aligned}$$

Repeating this process we get $xzL^izL^jx = 0$. This proves that $xzL^{k+1}zx = 0$. □

LEMMA 3.9. *Let $x_1, \ldots, x_n$ $(n \geq 1)$ be elements of S_1 such that $[x_i, x_j] = 0$ for all $1 \leq i,j \leq n$. Then $x_1 \cdots x_n L^j x_1 \cdots x_n = 0$ for all $0 \leq j \leq n$.*

PROOF. By induction on n. For $n = 1$ the assertion is clear. Let $n > 1$ and let $x_1, \ldots, x_n$ be elements of S_1 such that $[x_i, x_j] = 0$ for all $1 \leq i,j \leq n$. Before proceeding with the induction, we note the following identity for any $x, a_1, a_2 \in A$.

$$xa_1a_2 = [[x,a_1],a_2] + a_1xa_2 + a_2xa_1 - a_2a_1x \tag{3.1}$$

and observe that in the second and third summands on the right hand side of the equality, the element x has moved forward a position, while in the fourth summand x has gained two positions.

Suppose the lemma is true for any integer $1 \leq k \leq n$. Let $x_1, \ldots, x_{n+1} \in S_1$ with $[x_i, x_j] = 0$, $1 \leq i,j \leq n+1$. We must prove that $x_1 \cdots x_{n+1} L^j x_1 \cdots x_{n+1} = 0$ for any $0 \leq j \leq n+1$. It is clear by the induction hypothesis and the commutativity of the x_i that it suffices to verify the equality for $j = n+1$. Let $a_1, \ldots, a_{n+1} \in L$.

Using (3.1) we have

$$\begin{aligned} x_1 \cdots x_n x_{n+1} a_1 \cdots a_{n+1} x_1 \cdots x_{n+1} =& x_1 \cdots x_n [[x_{n+1}, a_1], a_2] a_3 \cdots a_{n+1} x_1 \cdots x_n x_{n+1} \\ &+ x_1 \cdots x_n a_1 x_{n+1} a_2 a_3 \cdots a_{n+1} x_1 \cdots x_n x_{n+1} \\ &+ x_1 \cdots x_n a_2 x_{n+1} a_1 a_3 \cdots a_{n+1} x_1 \cdots x_n x_{n+1} \\ &- x_1 \cdots x_n a_2 a_1 x_{n+1} a_3 \cdots a_{n+1} x_1 \cdots x_n x_{n+1}, \end{aligned}$$

where the first summand on the right hand side vanishes by the induction hypothesis. Should in any of the last three summands, (say the second one) happened that x_{n+1} is followed by at least two a_i, namely, we had $\cdots a_1 x_{n+1} a_2 a_3 \cdots$, then applying again (3.1), we would split this second summand (and likewise the third and the forth ones) into four summands

$$\begin{aligned} x_1 \cdots x_n a_1 x_{n+1} a_2 \cdots a_{n+1} x_1 \cdots x_{n+1} =& x_1 \cdots x_n a_1 [[x_{n+1}, a_2], a_3] \cdots a_{n+1} x_1 \cdots x_{n+1} \\ &+ x_1 \cdots x_n a_1 a_2 x_{n+1} a_3 \cdots a_{n+1} x_1 \cdots x_n x_{n+1} \\ &+ x_1 \cdots x_n a_1 a_3 x_{n+1} a_2 \cdots a_{n+1} x_1 \cdots x_n x_{n+1} \\ &- x_1 \cdots x_n a_1 a_3 a_2 x_{n+1} \cdots a_{n+1} x_1 \cdots x_n x_{n+1}, \end{aligned}$$

where the first summand on the right hand side is zero by the induction hypothesis, and in the second, the element x_{n+1} has moved forward a position (two positions in the third and forth summands). Reiterating this process as many times as necessary, we reach to summands of the form

$$\pm x_1 \cdots x_n a_{\sigma(1)} \cdots a_{\sigma(n)} x_{n+1} a_{\sigma(n+1)} x_1 \cdots x_n x_{n+1}$$

for a suitable permutation $\sigma \in \mathfrak{S}_{n+1}$, and summands of the form

$$\pm x_1 \cdots x_n a_{\tau(1)} \cdots a_{\tau(n)} a_{\tau(n+1)} x_{n+1} x_1 \cdots x_n x_{n+1}$$

for suitable permutation $\tau \in \mathfrak{S}_{n+1}$. In both cases, these summands vanish. □

LEMMA 3.10. *Let $x \in L$ and $y_1, y_2, y_3, y_4 \in S_1$. Put $z = [y_4, y_3, y_2, y_1, x]$ and suppose that $[y_{\sigma(4)}, y_{\sigma(3)}, y_{\sigma(2)}, y_{\sigma(1)}, x] = z$ for every $\sigma \in \mathfrak{S}_4$. Then $z \in S_2$.*

PROOF. It follows from the assumption on z that for all $i \in \{1, 2, 3, 4\}$,

$$z = [y_i, z_i] \text{ for some } z_i \in L. \tag{3.2}$$

We now proceed in steps. In the first two steps, (3.2) is translated in two properties. Then we use these properties to obtain some identities, which will be used to prove the assertion of the lemma.

(1) $zy_i = 0$ and $y_i z = 0$ for all $i \in \{1, 2, 3, 4\}$.
By(3.2), $zy_i = [y_i, z_i]y_i = y_i z_i y_i - z_i y_i^2 = 0$. Similarly $y_i z = 0$.

(2) For $i \in \{1, 2, 3, 4\}$ and $u \in L$, we have $zuy_i = -y_i uz$.
Using (3.2) and the equality $y_i[z_i, u]y_i = 0$, we get

$$zuy_i = [y_i, z_i]uy_i = y_i z_i u y_i = y_i u z_i y_i = y_i u[z_i, y_i] = -y_i uz.$$

Now let $i, j, k \in \{1, 2, 3.4\}$. We have:

(3) $zLy_i Ly_j y_k = 0$.
By (1) and (2), $(zLy_i)Ly_j y_k = y_i L(zLy_j)y_k = y_i Ly_j L(zy_k) = 0$.

(4) $zL^2 y_i y_j y_k = 0$.
Using (I3), (1), (2), and (3), we get

$$zL^2 y_i y_j y_k \subset z[L, [L, y_i]]y_j y_k + zLy_i Ly_j y_k + zy_1 L^2 y_j y_k = 0.$$

Now let $i, j, k, l \in \{1, 2, 3, 4\}$. We have:

(5) $zL^2 y_i L y_j y_k y_l = 0$.

By (I3), (1), (2), and (4), we get

$$zL^2 y_i L y_j y_k y_l \subset z[L, [L, y_i]] L y_j y_k y_l + zL y_i L^2 y_j y_k y_l + z y_i L^3 y_j y_k y_l = 0.$$

(6) $zL^3 y_i y_j y_k y_l = 0$.

By (I3), (1), (2), (4), and (5), we obtain

$$\begin{aligned} zL^3 y_i y_j y_k y_l \subset & z[L, [L, [L, y_i]]] y_j y_k y_l + zL^2 y_i L y_j y_k y_l \\ & + zL y_i L^2 y_j y_k y_l + z y_i L^3 y_j y_k y_l = 0. \end{aligned}$$

(7) $zL^2 y_i y_j L y_k y_l = 0$.

Using (I3), (1), (2) and (3), we get

$$zL^2 y_i y_j L y_k y_l \subset z[L, [L, y_i]] y_j L y_k y_l + zL y_i L y_j L y_k y_l + z y_i L^2 y_j L y_k y_l = 0.$$

Put $x_1 = [y_1, x]$, $x_2 = [y_2, x_1]$ and $x_3 = [y_3, x_2]$, and observe that, according with the hypothesis, $z = [y_4, x_3]$.

(8) $z^2 = 0$.

It follows from (1) and (2):

$$z^2 = z[x_3, y_4] = z x_3 y_4 = z[x_2, y_3] y_4 = (z x_2 y_3) y_4 = -y_3 x_2 (z y_4) = 0.$$

(9) $zLz = 0$.

Let $u \in L$. As in (8), using (1) repeatedly we obtain

$$\begin{aligned} zuz = zu[y_4, x_3] &= zuy_4x_3 - zux_3y_4 = zuy_4y_3x_2 - zuy_4x_2y_3 - zuy_3x_2y_4 \\ &+ zux_2y_3y_4 = zuy_4y_3y_2x_1 - zuy_4y_3x_1y_2 - zuy_4y_2x_1y_3 + zuy_4x_1y_2y_3 \\ &- zuy_3y_2x_1y_4 + zuy_3x_1y_2y_4 + zuy_2x_1y_3y_4 - zux_1y_2y_3y_4 = 0 \end{aligned}$$

(10) $zL^2 z = 0$. The proof is similar to that of (9):

$$\begin{aligned} zL^2 z = zL^2[y_4, y_3, y_2, y_1, x] &\subset zL^3 y_1 y_2 y_3 y_4 + \sum zL^2 y_i L y_j y_k y_l \\ &+ \sum zL^2 y_i y_j L y_k y_l + \sum (zL^2 y_i y_j y_k) L y_l + \sum (zL^2 y_i y_j y_k) y_l L = 0 \end{aligned}$$

by (4), (5), (6), and (7).

From steps (8), (9), and (10) we conclude that $z \in S_2$, as desired. □

LEMMA 3.11. *For $z \in S_1$ and $x, y \in L$, we have $[[z, x], y]z = z[[z, x], y]$.*

PROOF. Straightforward using that $z[x, y]z = 0$. □

From here up to and including Theorem 3.16, we assume that the Lie algebra L is generated by a finite set X of sandwichers of strength 1. Observe that X also generates the associative algebra A.

LEMMA 3.12. *If $|X| = n$, then $A = \sum_{k=1}^{2n} L^k$.*

PROOF. Let $X = \{x_1, x_1, \ldots, x_n\}$. For each integer $i \in \{1, \ldots, n\}$, let L_i denote the Φ-submodule $\Phi x_i + [L, x_i]$ of L. It is easy to check: (1) $L_i^3 = 0$, and (2) $L = \sum_{i=1}^{n} L_i$. By assumption, $A = \sum_{k=1}^{\infty} L^k$. Set $B = \sum_{k=1}^{2n} L^k$. Suppose that B is strictly contained in A. Then there exists a natural number $m > 2n$ and a m-sequence $(y_1, \ldots, y_m)$ of elements of L such that the product $y_1 \cdots y_m$ does not belong to B and m is the minimum integer with this property. We claim that for any permutation σ of the set $\{1, \ldots, m\}$, the product $y_{\sigma(1)} \cdots y_{\sigma(m)}$ does not belong to B either. We only need to consider the case of a transposition of two consecutive factors $\tau = (j, j+1)$, $1 \le j < m$. Let $y = y_1 \cdots y_m$. Then $y = y' + y''$, where $y' = y_1 \cdots y_{j-1} y_{j+1} y_j \cdots y_m$ and $y'' = y_1 \cdots y_{j-1}[y_j, y_{j+1}] \cdots y_m$. Since y'' is a product of a $m-1$-sequence of elements of L, $y'' \in B$ by assumption, and hence y' does not belong to B, as claimed. Using (2), we can split y into a sum of products of m factors each of which belongs to $\bigcup_{i-1}^{m} L_i$. At least one of these summands, say $\xi = z_{1,i_1} \cdots z_{j,i_j} \cdots z_{m,i_m}$, where $z_{j,i_j} \in L_{i_j}$, $1 \le i_j \le n$, does not belong to B. Since $m > 2n$, there exists $1 \le k \le n$ such that $z_{j,i_j} \in L_k$ for at least three indices (j, i_j), $1 \le j \le m$. By our claim, we can bring the corresponding three factors together, which leads to a contradiction by (1). □

LEMMA 3.13. *If $|X| = n$, then every sandwicher of strength $2n$ generates an ideal with zero multiplication.*

PROOF. Let $x \in A$ be a sandwicher of strength $2n$ and let I be the ideal of A generated by x, i.e. $I = \Phi x + xA + Ax + AxA$. Then $I^2 = 0$ since, by Lemma 3.12, $xAx = \sum_{k=1}^{2n} xL^k x = 0$. □

LEMMA 3.14. $S_2 \subset \operatorname{LocN}(A)$.

PROOF. We may assume that $\operatorname{LocN}(A) = 0$, and therefore that A is semiprime. Put $n = |X|$. By Lemma 3.13, any sandwicher of strength $2n$ generates an ideal of A with zero multiplication. Since A is semiprime, this implies that A does not contain nonzero sandwichers of strength $2n$. Suppose that $S_2 \neq 0$. As

$$S_2 \supset S_3 \supset \cdots \supset S_{2n-1} \supset S_{2n} = 0,$$

we have $S_i \neq 0$ and $S_{i+1} = 0$ for some $i \in \{2, \ldots, 2n-1\}$. Using Lemma 3.8(ii) we see that $[S_i, S_i] \subset [S_i, S_2] \subset S_{i+1} = 0$. Hence, by Lemma 3.9, any element of $S_i^{2n} \subset S_1^{2n}$ is a sandwicher of strength $2n$, so $S_i^{2n} = 0$. Using now Lemma 3.8(i) we get $S_i S_1 \subset S_1 S_i + [S_i, S_1] \subset S_1 S_i + S_i$. But, *by assumption*, S_1 generates A, so $S_i A \subset (AS_i)_\Phi + (S_i)_\Phi$. This means that $I = A(S_i)_\Phi + (S_i)_\Phi$ is the (two-sided) ideal of A generated by S_i. An easy induction proves that $I^k = A(S_i^k)_\Phi + (S_i^k)_\Phi$ for all integer $k \ge 1$. Thus $I^{2n} = 0$ and $S_i \subset I = 0$, a contradiction. □

The next lemma is needed for the proof of Theorem 3.16. There we will obtain the stronger result that the second possibility in the lemma leads to a contradiction. We denote by U the set of nonzero commutators in X.

LEMMA 3.15. *Either A is nilpotent or there exists an element $x \in X$ and a nonempty finite subset X_1 of $[U, U, x]$ with $xX_1^k \neq 0$ for all $k \ge 1$.*

PROOF. If A is nilpotent, we have finished. Suppose that A is not nilpotent. Then X is not contained in $\operatorname{LocN}(A)$, so we can change A to $A/\operatorname{LocN}(A)$ and assume that $\operatorname{LocN}(A) = 0$, and therefore that A is semiprime. We use induction on the cardinal $|X|$ of the set X. If $|X| \le 3$, then A is nilpotent as observed at the beginning of the section. So we may asume $|X| \ge 4$ and $0 \notin X$. Take $x \in X$ and put $X' := X \setminus \{x\}$. Let L' (resp. A') be the subalgebra of L (resp. A) generated by X'. If A' is not nilpotent then the induction hypothesis gives an element $x' \in X' \subset X$

and a nonempty finite subset $X_1' \subset [U', U', x'] \subset [U, U, x']$, where U' denotes the set of nonzero commutators in X', such that $0 \neq x'(X_1')^k \subset x'X_1^k$ for any natural number k. Assume therefore that A' is nilpotent. Then U' is finite. Now we proceed in steps.

(1) Let $u \in U$ be such that $[u, X'] = 0$. Then $[u, x] \neq 0$.

Otherwise $[u, X] = 0$ and hence, since X generates A, $u \in Z(A)$. This leads to a contradiction because A is semiprime and $u^2 = 0$ by Lemma 3.8(i).

(2) Let $u \in U$ be as in (1). Then there exist $v_1, v_2 \in U'$ such that

$$u_1 := [v_2, v_1, u, x] \neq 0 \text{ and } [L', u_1] = 0.$$

By (1), we have that $[u, x] \neq 0$. Since A' is nilpotent, there exist a largest natural number r such that $[y_r, \ldots, y_1, u, x] \neq 0$ for some $y_1, \ldots, y_r \in X'$. Let s be the smallest nonnegative integer such that $[v_s, \ldots, v_1, u, x] \neq 0$, where for each index $1 \leq i \leq s$, $v_i \in U'$, with $\sum_{i=1}^s \operatorname{length}(v_i) = r$. Put $u_1 = [v_s, \ldots, v_1, u, x]$. Using the Jacobi identity, we can express u_1 as a linear combination of left commutators of the form $[z_r, \ldots, z_1, u, x]$ with $z_i \in X'$ for $1 \leq i \leq r$. Hence, by maximality of r, X' is contained in the centralizer of u_1 in L', so $[L', u_1] = 0$. Set $v_0 = u$. We claim:

$$u_1 = [v_{\sigma(s)}, \ldots, v_{\sigma(1)}, v_{\sigma(0)}, x], \tag{3.3}$$

for any permutation σ of the set $\{0, 1, \ldots, s\}$. Since $[L', u] = 0$ we have that $[v_i, u] = 0$ for all v_i, $1 \leq i \leq s$. This allows us to move, say $u = v_{\sigma(i)}$, to the penultimate position, i.e. we may suppose that $v_{\sigma(0)} = u$. So

$$[v_{\sigma(s)}, \ldots, v_{\sigma(0)}, x] = [v_{\tau(s)}, \ldots, v_{\tau(i)}, v_{\tau(i-1)}, \ldots, v_{\tau(1)}, u, x]$$

for some $\tau \in \mathfrak{S}_s$. Using the identity

$$\operatorname{ad}_{v_{\tau(i)}} \operatorname{ad}_{v_{\tau(i-1)}} - \operatorname{ad}_{v_{\tau(i-i)}} \operatorname{ad}_{v_{\tau(i)}} = \operatorname{ad}_{[v_{\tau(i)}, v_{\tau(i-1)}]},$$

we can interchange the consecutive terms $v_{\tau(i)}$ and $v_{\tau(i-1)}$ obtaining

$$[v_{\tau(s)}, \ldots, v_{\tau(i)}, v_{\tau(i-1)}, \ldots, v_{\tau(1)}, u, x] \equiv [v_{\tau(s)}, \ldots, v_{\tau(i-1)}, v_{\tau(i)}, \ldots, v_{\tau(1)}, u, x]$$

module the term $[v_{\tau(s)}, \ldots, [v_{\tau(i)}, v_{\tau(i-1)}], \ldots, v_{\tau(1)}, u, x]$. Repeating this process we can express $[v_{\tau(s)}, \ldots, v_{\tau(1)}, u, x] - u_1$ as a linear combinations of left commutators of the form $[w_{i_{s-1}}, \ldots w_{i_1}, u, x]$, with $w_{i_j} \in U'$ for all the indexes i_j. If $[v_{\tau(s)}, \ldots, v_{\tau(1)}, u, x] - u_1 \neq 0$, then at least one of these left commutators is nonzero, which contradicts the minimality of s. So $[v_{\sigma(s)}, \ldots, v_{\sigma(0)}, x] = u_1$, as claimed. If $s = 1$, then it follows from Lemma 3.11 that $u_1 x = [v_1, [u, x]]x = x[v_1, [u, x]] = xu_1$, a contradiction. If $s > 2$, then we have by (3.3) and Lemma 3.10 that $u_1 \in S_2$, a contradiction since, by Lemma 3.14, $S_2 \subset \operatorname{LocN}(A) = 0$ by assumption. Therefore, $u_1 = [v_2, v_1, u, x]$, which completes the proof of (2).

(3) There exists $u_0 \in U'$ such that $[L', u_0] = 0$.

Since A' is nilpotent and U' is a finite set, there exists $u_0 \in U'$ such that $[U', u_0] = 0$. As U' spans L', this implies that $[L', u_0] = 0$.

(4) There exist sequences $u_0, u_1, \ldots, u_k, \ldots$ in U' and $v_1, \ldots, v_k, \ldots$ in U such that $[L', u_k] = 0$ for all $k \geq 0$, and $u_k = [v_{2k}, v_{2k-1}, u_{k-1}, x]$ for all $k \geq 1$.

This is proved by a straightforward induction using (2) and starting with the element u_0 given by (3).

(5) Now we show that the element x and the set $X_1 := [U', U', x]$ have the properties required in the lemma. By (4), for each $k \geq 1$, $[u_{2k}, x]$ is a linear combination of associative monomials of length $6k + 2$ in the variables $v_0 = u_0, v_1, \ldots, v_{4k}$, each of

which with occurrence 1, and x with occurrence $2k+1$. Since by (1) $[u_{2k}, x] \neq 0$, at least one of these monomials, say $\zeta = z_1 \cdots z_{6k+2}$, is nonzero. Then every interval between two occurrences of x in ζ must have length greater than or equal to 2. Because there are $2k+1$ occurrences of x in ζ, there are $2k$ of these intervals in ζ. A counting argument proves that at most an interval has length 3 and that there is no interval of length greater than 3. So there are at least k consecutive intervals of length 2 in ζ. This means that for some $l \in \{0, 1, \ldots, 3k+1\}$, we have $z_{l+3i+1} = x$ for all $0 \leq i \leq k$, and $z_{l+3i+2}, z_{l+3i+3} \in U'$ for all $0 \leq i \leq k-1$. Then we get

$$x[[x, z_{l+3}], z_{l+2}] \cdots [[x, z_{l+3k}], z_{l+3k-1}] = z_{l+1} \cdots z_{l+3k} z_{l+3k+1} \neq 0,$$

which proves that for any positive integer k, $xX_1^k \neq 0$, as desired. □

The idea of the proof of the next theorem is to construct, under the assumption that A is not nilpotent and using Lemma 3.9, nonzero sandwichers of arbitrary strength, which leads to a contradiction by Lemma 3.13.

THEOREM 3.16. *A is nilpotent.*

PROOF. Suppose that A is not nilpotent. Then (since A is finitely generated) $\mathrm{LocN}(A)$ is a proper ideal of A, and we may also assume by Proposition 2.55 that $\mathrm{LocN}(A) = 0$, and therefore that A is semiprime. We show in two steps that this assumption leads to a contradiction. Denote by U the set of nonzero commutators in X.

(1) For any integer $n \geq 1$, there exist $x_1, \ldots, x_n \in U$ and a nonempty finite subset X_1 of U satisfying:

(i) $[x_i, x_j] = 0$ for all $i, j \in \{1, \ldots, n\}$.
(ii) $[x_i, y] = 0$ for all $y \in X_1$ and all $i \in \{1, \ldots, n\}$.
(iii) $x_1 \cdots x_n X_1^k \neq 0$ for all $k \geq 1$.

We prove this by induction on n. If $n = 1$ the assertion follows from Lemmas 3.15 and 3.11. Assume that $n > 1$. Let $x_1 \cdots x_{n-1}$ and X_0 be as given by the induction hypothesis. Let U_0 be the set of nonzero commutators in X_0, and let A_0 (resp. L_0) be the subalgebra of A (resp. of L) generated by X_0. Put $I_0 = \{y \in A_0 : x_1 \cdots x_{n-1} y = 0\}$. Clearly I_0 is a right ideal of A_0, and since $[x_i, X_0] = 0$ for all $1 \leq i \leq n-1$ by (ii), I_0 is in fact a (two-sided) ideal of A_0. Let $y \mapsto \bar{y}$ the natural homomorphism of A_0 onto $\overline{A}_0 = A_0/I_0$. It follows from (iii) that $\overline{A}_0$ is not nilpotent ($A_0^l \subset I_0$ for some $l \geq 1$ would imply $x_1 \cdots x_{n-1} X_0^l \subset x_1 \cdots x_{n-1} I_0 = 0$, a contradiction). This fact allows us to apply Lemma 3.15 to the pair $(\overline{A}_0, \overline{L}_0)$. Then there exist an element $x_n \in X_0$ and a nonempty finite subset X_1 of $[[x_n, U_0], U_0]$ such that, for all $k \geq 1$, $x_n X_1^k$ is not contained in I_0; that is, $x_1 \cdots x_{n-1} x_n X_1^k \neq 0$. So $x_1, \ldots, x_n$ and X_1 satisfy (iii). Since $x_n \in X_0$, it follows from (ii) for the case $n-1$, that $[x_i, x_n] = 0$ for any $1 \leq i \leq n-1$, i.e. the set $\{x_1, \ldots, x_{n-1}, x_n\}$ satisfies (i); and by the Jacobi identity, it also satisfies (ii).

(2) Put $m = |X|$. Applying (1) with $n = 2m$, we find $x_1, \ldots, x_{2m} \in U \subset S_1$ such that $[x_i, x_j] = 0$ for all $i, j \in \{1, \ldots, 2m\}$ and $x_1 \cdots x_{2m} \neq 0$. By Lemmas 3.9 and 3.13, the nonzero element $x_1 \cdots x_{2m}$ generates an ideal I such that $I^2 = 0$, a contradiction since A is semiprime. □

THEOREM 3.17. *Let L be a 2-torsion free Lie algebra. If L is generated by a finite number of absolute zero divisors, then L is nilpotent.*

PROOF. Suppose that L is generated by a finite set $X = \{x_1, \dots, x_n\} \subset L$ of absolute zero divisors, i.e. $\mathrm{ad}_{x_i}^2 L = 0$ for all $1 \le i \le n$. The proof follows immediately from the previous theorem via the adjoint representation. Replace L by $\mathrm{ad}(L)$, take $A := \mathrm{Ad}(L)$ and observe that, by Lemma 3.3(1), $\mathrm{ad}_{x_i} \in S_1$ for every $1 \le i \le n$. By Theorem 3.16 A is nilpotent, and therefore so is L. □

COROLLARY 3.18. *Let L be a 2-torsion free Lie algebra. Then the subalgebra generated by its absolute zero divisors is locally nilpotent.*

REMARKS 3.19. (1) Nilpotency of a Lie algebra generated by a finite number of absolute zero divisors (Corollary 3.18) was proved by A. I. Kostrinkin [**Kos90**, Theorem 5.3.1] for Lie algebras over a field of characteristic $p > 5$. Later, E. Zelmanov and A. I. Kostrikin [**ZK90**] proved this theorem for 2-torsion free Lie algebras, restriction which can be avoided by a suitable modification of the definition of absolute zero divisor. A different proof based on a theorem of Furstenberg for dynamical systems can be found in [**Zel92b**]. Unfortunately, none of these proofs is constructive. A constructive proof would give a formula or an upper bound for the nilpotency class of the *free sandwich algebra* on d generators (see Commentary 5.5.3 of [**Kos90**]).

(2) The reader is referred to the recent paper by J. Anquela, T. Cortés and E. Zelmanov [**ACZ16**] where the local nilpotency of the McCrimmon radical of a completely arbitrary Jordan system is derived from Kostrinkin–Zelmanov's theorem.

3.3. Absolute zero divisors generate a locally nilpotent ideal

If L is actually a Lie algebra over a field of characteristic 0, then even the ideal generated by the absolute zero divisors is locally nilpotent. The proof of this result (due to A. N. Grishkov) we give here is taken verbatim from [**Kos90**, Theorem 5.4.2].

Given a nonempty set X and a field $\mathbb{F}$, denote by $\mathfrak{L}\langle X\rangle = \mathfrak{L}_{\mathbb{F}}\langle X\rangle$ the free Lie $\mathbb{F}$-algebra on X (see [**Jac79**] for definition). For each $r > 0$, let $\mathfrak{L}_r\langle X\rangle$ be the linear span of the set of commutators in X of length r. Then $\mathfrak{L}\langle X\rangle$ is the direct sum of the subspaces $\mathfrak{L}_r\langle X\rangle$ and this decomposition is a $\mathbb{Z}$-grading on $\mathfrak{L}\langle X\rangle$.

THEOREM 3.20. *Let L be a Lie algebra over a field $\mathbb{F}$ of characteristic 0. Then the ideal $\mathrm{K}_1(L)$ generated by the absolute zero divisors of L is locally nilpotent.*

PROOF. Take a copy X of a generating set of L containing the set of the absolute zero divisors of L, denoted by C in the copy X. Set $\mathfrak{L} := \mathfrak{L}(X)$ and let $\mathfrak{I}$ be the ideal of $\mathfrak{L}$ generated by the set $\bigcup_{c\in C}[c,[c,\mathfrak{L}]]$. Then C is a set of absolute zero divisors of $\mathfrak{L}/\mathfrak{I}$ and the natural homomorphism of this Lie algebra onto L maps $\mathrm{id}_{\mathfrak{L}/\mathcal{J}}(C)$ onto $\mathrm{K}_1(L)$. Given any $a \in \mathfrak{L}$, denote its coset $a + \mathfrak{I}$ by the same letter a. We can also assume that $L = \mathfrak{L}/\mathfrak{I}$. It is enough to show that $\mathrm{id}_L(C)$ is a locally nilpotent ideal of L. As $\mathfrak{I}$ is an homogeneous ideal, $L = \bigoplus_{i=1}^{\infty} L_i$ with L_i being the subspace spanned by the commutators of length i in X. There is a topology of L associated with the $\mathbb{Z}$-grading on L just defined for which the ideals $I_k = \bigoplus_{i\ge k} L_i$ form a basis of neighborhoods of zero. Consider the completion of this topology, namely, the algebra $\hat{L}$ of formal power series $\sum_{i=1}^{\infty} u_i$, $u_i \in L_i$, with the usual operations of addition and multiplication. Let $\mathcal{S}$ be the subalgebra of $\hat{L}$ generated by its absolute cero divisors and let $\overline{\mathcal{S}}$ denote its closure in $\hat{L}$. We claim that $[L, \overline{\mathcal{S}}] \subset \overline{\mathcal{S}}$. Let $x \in L$ and $a \in \overline{\mathcal{S}}$. Then, for each $m \geq 1$, there exists an $a_m \in \mathcal{S}$ such that $a - a_m \in \hat{I}_m$, with $\hat{I}_m = \{\sum_{i=m}^{\infty} u_i : u_i \in L_i\}$. Note that both $\mathcal{S}$ and the $\hat{I}_m$ are invariant under automorphisms of $\hat{L}$, and that (since $\operatorname{char}(\mathbb{F}) = 0$), $\exp(\lambda \operatorname{ad}_x) \in \operatorname{Aut}(\hat{L})$ for every $\lambda \in \mathbb{F}$. So $\exp(\lambda \operatorname{ad}_x)a - \exp(\lambda \operatorname{ad}_x)a_m \in \hat{I}_m$. Hence

$$a + \lambda[x, a] + \frac{\lambda^2}{2!}[x, [x, a]] + \cdots \in \mathcal{S} \mod \hat{I}_m$$

for each $\lambda \in \mathbb{F}$. A Vandermonde argument (see Lemma 2.21) proves that $[x, a] \in \mathcal{S} \mod \hat{I}_m$, so $[x, a] \in \overline{\mathcal{S}}$ as claimed. Since $C \subset \mathcal{S}$, it follows that all homogeneous elements whose expression involve one of the letters c must lie in $\overline{\mathcal{S}}$. Since every element $u \in \operatorname{id}_L(C)$ is contained in the subalgebra generated by its homogeneous components, which also lie in $\operatorname{id}_L(C)$, to establish the local nilpotency of $\operatorname{id}_L(C)$ it suffices to show that the Lie algebra generated by a finite set $\{v_1, \ldots, v_d\}$ of homogeneous elements of lengths $l_1, \ldots, l_d$ in $\operatorname{id}_L(C)$ is nilpotent. Let s be a natural number such that $s > \max\{l_1, \ldots, l_d\}$. Since $v_1, \ldots, v_d \in \overline{\mathcal{S}}$, there exist elements $a_1, \ldots, a_d \in \mathcal{S}$ such that $w_i = v_i - a_i \in \hat{I}_s$. By Theorem 3.17, the subalgebra generated by $a_1, \ldots, a_d$ is nilpotent, of index, say $q - 1$. Let $[v_{i_1}, \cdots, v_{i_q}]$ be a left commutator of length q in $\{v_1, \ldots, v_d\}$. Then

$$[v_{i_1}, \cdots, v_{i_q}] = [a_{i_1} + w_{i_1}, \cdots, a_{i_q} + w_{i_q}] = [a_{i_1}, \cdots, a_{i_q}] + w = w,$$

where $w \in \hat{I}_{r+1}$, with $r = l_{i_1} + \cdots + l_{i_q}$. Since $\hat{I}_{r+1} \cap L_r = 0$, we have that $[v_{i_1}, \cdots, v_{i_q}] = 0$, which proves that the subalgebra of L generated by $v_1, \ldots, v_d$ is nilpotent. □

REMARK 3.21. Suppose that every element of L is ad-nilpotent. Then it follows from Proposition 2.65 and transfinite induction that the Kostrikin radical of L (see Chapter 9 for the definition) is locally nilpotent.

3.4. Nondegenerate Lie algebras

A Lie algebra L is said to be *nondegenerate* if it has no nonzero absolute zero divisors, i.e. $\operatorname{ad}_x^2 L = 0$ implies $x = 0$. By a *strongly prime Lie algebra* we mean a Lie algebra which is both prime and nondegenerate. Nondegeneracy implies semiprimeness: if L is a nondegenerate Lie algebra and I is an ideal of L such that $[I, I] = 0$, then any $y \in I$ satisfies $\operatorname{ad}_y^2 L \subset [I, I] = 0$, thus $y = 0$, and therefore semisimplicity for finite-dimensional Lie algebras (see [**Hum72**]). The converse is also true for finite-dimensional Lie algebras over a field of characteristic 0.

PROPOSITION 3.22. *Suppose that* $\mathbb{F}$ *is a field of characteristic not* 2 *and* L *is a finite-dimensional Lie algebra over* $\mathbb{F}$. *If its Killing form* $\kappa(x, y) := \operatorname{tr}(\operatorname{ad}_x \operatorname{ad}_y)$, $x, y \in L$, *is nondegenerate, then* L *is nondegenerate.*

PROOF. Let $x \in L$ be such that $[x, [x, L]] = 0$. Then for any $y \in I$ we have $0 = \operatorname{ad}_{[x,[x,y]]} = X^2Y - 2XYX + YX^2 = -2XYX$. Since $\frac{1}{2} \in \mathbb{F}$, $(XY)^2 = 0$. Hence $\kappa(x, y) = \operatorname{tr}(XY) = 0$ for $y \in Y$, which implies $x = 0$ by nondegeneracy of the Killing form. □

COROLLARY 3.23. *Let* L *be a finite-dimensional Lie algebra over a field of characteristic* 0. *Then the following conditions are equivalent:*

(i) L *is semiprime,*
(ii) L *is semisimple,*

(iii) *the Killing form is nondegenerate,*
(iv) *L is nondegenerate.*

PROOF. A Lie algebra is semiprime if and only if it does not contain nonzero solvable ideals, so (i) implies (ii); (ii) $\Leftrightarrow$ (iii) is the well known semisimplicity criterion (see [**Hum72**, 5.1. Theorem]); (iii) $\Rightarrow$ (iv) follows from Proposition 3.22; and (iv) $\Rightarrow$ (i) is clear. □

Over a field of characteristic 0, simple, not necessarily finite-dimensional, Lie algebras are automatically nondegenerate.

COROLLARY 3.24. *Every simple Lie algebra L over a field of characteristic* 0 *is nondegenerate.*

PROOF. By Theorem 3.20, $\mathrm{K}_1(L)$ is a locally nilpotent ideal of L. Since L is simple, $\mathrm{K}_1(L) \neq 0$ would be imply that $L = \mathrm{K}_1(L)$ is locally nilpotent, but a simple algebra cannot be locally nilpotent (Proposition 1.25). □

The positive characteristic case. Unlike the case of Lie algebras of characteristic 0, in positive characteristic there exist simple Lie algebras having nonzero absolute zero divisors, as the Witt algebra $W(p)$ over a field of characteristic $p > 3$. Thus some conditions are required to guarantee nondegeneracy in simple Lie algebras of positive characteristic. The following result combines Kostrinkin–Zelmanov's theorem with the Vandermonde argument to get some criteria of nondegeneracy.

PROPOSITION 3.25. *Let L be a simple Lie algebra over a field of characteristic $p \geq 5$. If L is generated by a set X such that $\mathrm{ad}_x^3 L = 0$ for all $x \in X$, then L is nondegenerate.*

PROOF. By Example 2.19(1), X is a distinguished generating set and since the subspace V of L generated by the absolute zero divisors of L is invariant under $\mathrm{Aut}(L)$, it follows from Lemma 2.21 that V is an ideal of L. As V is locally nilpotent by Corollary 3.18, it must be equal to zero by Proposition 1.25 since L is simple. □

COROLLARY 3.26. *Let L be a simple Lie algebra over a field of characteristic $p \geq 5$. If L has a nontrivial short $\mathbb{Z}$-grading $L = L_{-1} \oplus L_0 \oplus L_1$, then L is nondegenerate.*

PROOF. As the grading is nontrivial, L is generated by $L_1 \cup L_{-1}$, which is a set of ad-nilpotent elements of index at most 3, so Proposition 3.25 applies. □

This corollary is in fact a particular case of the following more general result.

PROPOSITION 3.27. *Let L be a simple Lie algebra with a nontrivial finite $\mathbb{Z}$-grading $L = L_{-n} \oplus \cdots \oplus L_n$ over a field $\mathbb{F}$ of characteristic $p \geq 4n + 1$. Then L is nondegenerate.*

PROOF. As proved in Example 2.19(2), $X = \bigcup_{k=1}^{n} L_{\pm k}$ is a distinguished generating set of L. Now let V be the subspace of L generated by its absolute zero divisors. Clearly V is invariant under $\mathrm{Aut}(L)$, so V is an ideal of L by Lemma 2.21. Since V is locally nilpotent by Corollary 3.18, it must be equal to 0 by Proposition 1.25. □

If $n = 2$, then only characteristic greater than 5 is required.

PROPOSITION 3.28. *Let* $L = L_{-2} \oplus L_{-1} \oplus L_0 \oplus L_1 \oplus L_2$ *be a simple* 5*-graded Lie algebra over a field of characteristic greater than* 5*. Then* L *is nondegenerate.*

PROOF. It follows as in the previous case using Example 2.19(3) instead of Example 2.19(2). ☐

PROPOSITION 3.29. *Let* L *be a Lie algebra with a nontrivial finite* $\mathbb{Z}$*-grading* $L = L_{-n} \oplus \cdots \oplus L_n$ *over a field* $\mathbb{F}$ *of characteristic* 0 *or* $p \geq 4n+1$*. Then* $\mathrm{K}(L) \subset \widetilde{\mathrm{LocSN}}(L)$*, where* $\mathrm{K}(L)$ *denotes the Kostrikin radical of* L*.*

PROOF. We follow the proof given in [**Zel83a**, Proposition 4]. Let K_1 be the linear span of the absolute zero divisors of L. Clearly, K_1 is invariant under the automorphisms of L. Hence, as in the proof of Proposition 3.27, it is an ideal of L, so $K_1 = \mathrm{K}_1(L)$. By Theorem 3.18, $\mathrm{K}_1(L)$ is locally nilpotent, which implies by Proposition 2.75(i) that it is contained in $\widetilde{\mathrm{LocSN}}(L)$. To prove the inclusion $\mathrm{K}(L) \subset \widetilde{\mathrm{LocSN}}(L)$ we use transfinite induction. Suppose that $\mathrm{K}_\beta(L) \subset \widetilde{\mathrm{LocSN}}(L)$ for every $\beta < \alpha$. If α is a limit ordinal, then $\mathrm{K}_\alpha(L) = \bigcup_{\beta<\alpha} \mathrm{K}_\beta(L) \subset \widetilde{\mathrm{LocSN}}(L)$. Otherwise, and again by Theorem 3.18, $(\mathrm{K}_\alpha(L) + \widetilde{\mathrm{LocSN}}(L))/\widetilde{\mathrm{LocSN}}(L)$ is a locally nilpotent ideal of $L/\widetilde{\mathrm{LocSN}}(L)$. Then, by Proposition 2.75(iii), $\mathrm{K}_\alpha(L) \subset \widetilde{\mathrm{LocSN}}(L)$. This completes the proof. ☐

DEFINITION 3.30. A Lie algebra is said to be *Artinian* if it satisfies the descending chain condition on inner ideals.

PROPOSITION 3.31. [**Ben77**, Lemma 1.13] *Let* L *be a nondegenerate simple Artinian Lie algebra. Then every proper inner ideal* V *of* L *is abelian.*

PROOF. The reader is referred to the proof of Lemma 2.39 for notation. Assume initially that V is also a subalgebra. By Lemma 2.41, $B = \bigcap_{n=1}^{\infty} V^n$ is an ideal of L which must be then 0 because B is proper. And by Lemma 2.41 again, the subalgebras $V = V^1 \supset V^2 \supset \dots$ form a descending chain on inner ideals. Therefore there is an integer m such that $V^m = \bigcap_{n=1}^{\infty} V^n = 0$. If $m = 2$ then $[V, V] = V^2 = 0$, the desired conclusion. If $m > 2$, we have by Lemma 2.39(2)

$$V^{m-1}V^{m-1}L \subset V^{m-1}V^{m-2} \subset V^{m-1}V = V^m = 0.$$

Then every element of V^{m-1} is an absolute zero divisor of L and therefore $V^{m-1} = 0$ by nondegeneracy of L. An induction argument shows that $V^2 = 0$. Now if $V \neq L$ is an arbitrary inner ideal, then, by Lemma 2.40, $T(V)$ is an inner ideal and a subalgebra of L which contains V, so $T(V) = L$ or $T(V)^2 = 0$. If $T(V) = L$, then $VLL \subset V$, but since L is simple, $L = L^2$ and so $VL \subset V$. This tells us that V is an ideal of L, and therefore $V = 0$ since it is proper by assumption. Thus $V^2 \subset T(V)^2 = 0$. In both case, V is abelian. ☐

REMARK 3.32. The above proposition, applied to finite-dimensional simple Lie algebras over a field of characteristic 0 or $p > 5$, shows that proper inner ideals of these algebras are necessarily abelian [**Ben77**]. On the other hand, the Artinian condition is not superfluous. Consider the Lie algebra $L = \mathrm{Der}(\mathbb{F}[x])$ over a field $\mathbb{F}$ of characteristic 0. As stated in Exercise 2.98, this algebra is simple, non-Artinian, and no nonzero inner ideal of L is abelian.

3.5. Absolute zero divisors in the Lie algebra of a ring

As a minor part of its investigation on the Lie structure of associative rings, I. N. Herstein proves in [**Her70**, Lemma 1] that if R is a semiprime 2-torsion free ring, then the Lie algebra $\overline{R} = R^-/Z$, where $Z = Z(R)$ denotes the center of R, is semiprime. In fact, $\overline{R}$ is nondegenerate (resp. strongly prime) if R is semiprime (resp. prime).

The following two results, standard applications of the fact that the adjoint operator ad_a, $a \in R$, is a derivation of R, will be used in what follows.

LEMMA 3.33. *Let R be a semiprime ring, let I be an essential ideal of R, and let δ be a derivation of R. If $\delta(I) = 0$ then $\delta = 0$. If I is commutative then so is R.*

PROOF. $\delta(R)I \subset \delta(RI) + R\delta(I) = 0$ implies $\delta(R) = 0$ since I is essential and R is semiprime. Suppose now that I is commutative. Then, for any $x \in I$, $\mathrm{ad}_x I = 0$. Since $x \in I$ is arbitrary, this actually shows that $[I, a] = 0$ for every $a \in R$, which again implies $[a, R] = 0$, i.e. $a \in Z$, so R is commutative. □

LEMMA 3.34. *Let R be a 2-torsion free ring, let I be an ideal of R, and let $a \in R$. We have:*

(1) *If R is semiprime and $\mathrm{ad}_a^2 I = 0$, then $[a, I] = 0$.*
(2) *If R is prime and $I \neq 0$, then $[a, I] = 0$ implies $a \in Z(R)$.*

PROOF. (1) For any $x, y \in I$ we have by the Leibniz rule

$$0 = \mathrm{ad}_a^2(xy) = 2\,\mathrm{ad}_a(x)\,\mathrm{ad}_a(y).$$

So $\mathrm{ad}_a(x)\,\mathrm{ad}_a(y) = 0$, since R is 2-torsion free. Hence

$$\mathrm{ad}_a(x)y\,\mathrm{ad}_a(x) = \mathrm{ad}_a(xy)\,\mathrm{ad}_a(x) = 0,$$

which implies $\mathrm{ad}_a I = 0$, since R is semiprime.

(2) It follows from Lemma 3.33, since any nonzero ideal of a prime ring is essential. □

PROPOSITION 3.35. *Let R be a semiprime ring. We have:*

(1) *Let $a \in R$ such that $\mathrm{ad}_a^n R \subset Z(R)$. Then $\mathrm{ad}_a^n R = 0$ for $n \geq 1$, and $n(\mathrm{ad}_a^{n-1} R)(\mathrm{ad}_a^{n-1} R) = 0$ for $n \geq 2$,*
(2) *If R is 2-torsion free, then the Lie algebra $\overline{R} = R^-/Z(R)$ is nondegenerate and $Z([R, R]) = Z(R) \cap [R, R]$.*
(3) *If R is prime and 2-torsion free, then $\overline{R}$ is strongly prime.*

PROOF. We may assume that R is not commutative. Otherwise $R = Z(R)$ and the result is trivial.

(1) Let n be a positive integer. Since ad_a is a derivation of the ring R, we have by the Leibniz rule that $\mathrm{ad}_a^n(x)a = \mathrm{ad}_a^n(xa) \in Z(R)$ for all $x \in R$. Hence, $0 = [\mathrm{ad}_a^n(x)a, y] = \mathrm{ad}_a^n(x)[a, y]$ for all $y \in R$. Taking $y = \mathrm{ad}_a^{n-1}(x)$, we obtain $\mathrm{ad}_a^n(x)^2 = 0$, which implies $\mathrm{ad}_a^n R = 0$, since $\mathrm{ad}_a^n R \subset Z(R)$ and semiprime rings do not contain nonzero nilpotent central elements. If $n \geq 2$, we have again by the Leibniz rule that $0 = \mathrm{ad}_a^n(x\,\mathrm{ad}_a^{n-2}(y)) = n\,\mathrm{ad}_a^{n-1}(x)\,\mathrm{ad}_a^{n-1}(y)$, as desired.

(2) Let $a \in R$ be such that $\mathrm{ad}_a^2 R \subset Z(R)$. Then $\mathrm{ad}_a^2 R = 0$ by (1) and hence, by Lemma 3.34, $a \in Z(R)$, so $\overline{R}$ is nondegenerate. Now, if $a \in Z([R, R])$ then $\mathrm{ad}_a^2 R = 0$ and hence $a \in Z(R)$ by we have just proved. Thus $Z([R, R]) \subset Z(R) \cap [R, R]$. The reverse inclusion is obvious.

(3) Let J be a nonzero ideal of R. We claim that $[R, J]$ is not contained in $Z(R)$. Otherwise we would have by nondegeneracy of $\overline{R}$ that $J \subset Z(R)$ and hence that R would by commutative by Lemma 3.34, which is a contradiction since R is not commutative by assumption. Now let $I_1/Z(R), I_2/Z(R)$ be nonzero ideals of $\overline{R}$, i.e. for each $k = 1, 2$, I_k is an ideal of R^- containing $Z(R)$ strictly. By [**Her70**, Theorem 5], for each $k = 1, 2$ there exists an ideal J_k of R such that $0 \neq [R, J_k] \subset I_k$. As R is prime, $J_1 \cap J_2 \neq 0$, so $[R, J_1 \cap J_2]$ is not contained in $Z(R)$. Hence

$$\overline{0} \neq \overline{[R, J_1 \cap J_2]} \subset \overline{[R, J_1]} \cap \overline{[R, J_2]} \subset \overline{I}_1 \cap \overline{I}_2,$$

where $x \mapsto \overline{x}$ denotes the natural homomorphism of R^- onto $\overline{R}$. As $\overline{R}$ is nondegenerate, and therefore semiprime, $\overline{I}_1 \cap \overline{I}_2 \neq \overline{0}$ implies $[\overline{I}_1, \overline{I}_2] \neq \overline{0}$, which proves that $\overline{R}$ is prime. □

3.6. Absolute zero divisors in Lie algebras of skew-symmetric elements

Following Chapter 1, given a prime ring R, i.e. an associative $\mathbb{Z}$-algebra, we denote by $\mathcal{C}$ the extended centroid of R. Recall that $\mathcal{C}$ is a field containing the centroid Γ of R and the central closure $\tilde{R} = \mathcal{C}R$ of R is a prime associative algebra over $\mathcal{C}$. We also recall that any involution $*$ of R induces an involution in $\mathcal{C}$, also denoted by $*$, and therefore it can be extended to an involution of $\mathcal{C}R$. Recall that an involution $*$ of R is said to be of the first kind if it acts as the identity on $\mathcal{C}$; otherwise $*$ is said to be of the second kind. We put $K = \text{Skew}(R, *)$ and $S = \text{Sym}(R, *)$, so, if 2 is invertible in Γ, then $R = S \oplus K$.

We will show in this section that if R is a prime ring of characteristic different from 2 with involution $*$, then the Lie algebra $K/Z(K)$ is strongly prime when $*$ is of the second kind, or when $*$ is of the first kind and R is not an order in a simple ring of dimension less than 16 over its center.

R prime with involution of the second kind.

PROPOSITION 3.36. *Let R be a prime ring of characteristic not 2 with involution $*$ of the second kind. Then $K/Z(K)$ is strongly prime and $Z(K) = K \cap Z(R)$.*

PROOF. Without loss of generality we may assume that R is not commutative. Put $\overline{K} = K/K \cap Z(R)$ and let $t \in K$ be such that $\bar{t}$ is an absolute zero divisor of $\overline{K}$, i.e. $\text{ad}_t^2 K \subset Z(R)$. Consider first the case that $\text{ad}_t^2 K = 0$. Put $\mathcal{C}_k = \text{Skew}(\mathcal{C}, *)$ and $\mathcal{C}_s = \text{Sym}(\mathcal{C}, *)$. Since $*$ is of the second kind, we may take $\alpha \in \mathcal{C}_k$, $\alpha \neq 0$. Then $\mathcal{C}_k = \alpha\mathcal{C}_s$ and $\mathcal{C}_s = \alpha\mathcal{C}_k$. Let I be a nonzero $*$-ideal of R such that $\alpha I \subset R$, $I = I_s \oplus I_k$ where $I_s = \text{Sym}(I, *)$ and $I_k = \text{Skew}(I, *)$. Then $\mathcal{C}I$ is a nonzero ideal of the *prime ring* $\mathcal{C}R$. Clearly, $\text{ad}_t^2(\mathcal{C}I_k) = \mathcal{C}\,\text{ad}_t^2 I_k = 0$, and since $\alpha I_s \subset K$, we also have

$$\begin{aligned}\text{ad}_t^2(\mathcal{C}I_s) &= \text{ad}_t^2(\alpha\mathcal{C}_k I_s) + \text{ad}_t^2(\alpha\mathcal{C}_s I_s) = \mathcal{C}_k\,\text{ad}_t^2(\alpha I_s)\\ &\quad + \mathcal{C}_s\,\text{ad}_t^2(\alpha I_s) \subset \mathcal{C}_k\,\text{ad}_t^2 K + \mathcal{C}_s\,\text{ad}_t^2 K = 0,\end{aligned}$$

so $\text{ad}_t^2(\mathcal{C}I) = 0$. Hence, by Lemma 3.34, $t \in Z(R)$, i.e. $\bar{t} = 0$. Therefore, we may assume that there exists an $x \in K$ such that $z := \text{ad}_t^2 x$ is a nonzero central element of R. Then

$$z\,\text{ad}_t^2 R = z\,\text{ad}_t^2(S + K) = \text{ad}_t^2(zS) + z\,\text{ad}_t^2(K) \subset \text{ad}_t^2 K + zZ(R) \subset Z(R).$$

Hence, for any $b \in \mathrm{ad}_t^2 R$ and any $x \in R$, we have $zR(bx - xb) = R((zb)x - x(zb)) = 0$, which implies that $b \in Z(R)$ since R is prime and $z \neq 0$. We have thus proved that $\mathrm{ad}_t^2 R \subset Z(R)$. Then, by Lemma 3.34(1), $t \in Z(R)$, i.e. the Lie algebra $\overline{K}$ is nondegenerate. In particular, every absolute zero divisor of K is contained in $K \cap Z(R)$. Therefore, $Z(K) \subset K \cap Z(R)$, the reverse inclusion being obvious. Let us finally see that $\overline{K}$ is prime. Since $\overline{K}$ is nondegenerate, it suffices to prove that any nonzero ideals $\overline{I}_1$, $\overline{I}_2$ of $\overline{K}$ has nonzero intersection.

Let $\overline{I} = I/K \cap Z(R)$ be a nonzero ideal of $\overline{K}$, i.e, I is an ideal of K which strictly contains $K \cap Z(R)$. By [**Eri72**, Theorem 1(a)], there exists a nonzero $*$-ideal J of R such that $[\mathrm{Skew}(J,*), K] \subset I$. We claim that for any nonzero $*$-ideal J of R, $[\mathrm{Skew}(J,*), K]$ is not contained in $Z(R)$.

Suppose on the contrary that $[\mathrm{Skew}(J,*), K] \subset Z(R)$. By nondegeneracy of $\overline{K}$, $\mathrm{Skew}(J,*) \subset Z(R)$. If $\mathrm{Skew}(J,*) = 0$, then $y^* = y$ for all $y \in J$, and hence, for all $a, b \in R$, we have

$$b^*a^*y = (yab)^* = yab = (ya)^*b = a^*(yb)^* = a^*b^*y,$$

which implies that R is commutative, since R is prime and J is a nonzero ideal, but the commutativity of R was discarded at the beginning of the proof. Suppose then that $0 \neq \mathrm{Skew}(J,*) \subset Z(R)$. Then $\mathrm{Skew}(J,*)$ contains a nonzero central element z. Clearly, $z\,\mathrm{Sym}(J,*) \subset \mathrm{Skew}(J,*) \subset Z(R)$, and hence

$$zJ \subset z\,\mathrm{Sym}(J,*) + z\,\mathrm{Skew}(J,*) \subset Z(R),$$

so zJ is a nonzero commutative ideal of R, which, by Lemma 3.33, forces the commutativity of R again. Therefore, $[\mathrm{Skew}(J,*), K]$ is not contained in $Z(R)$, as claimed.

Now let $\overline{I}_1 = I_1/K \cap Z(R)$ and $\overline{I}_2 = I_2/K \cap Z(R)$ be nonzero ideals of $\overline{K}$. By the result cited above, there exist two nonzero $*$-ideals J_1, J_2 of R such that $[\mathrm{Skew}(J_1,*), K)] \subset I_1$ and $[\mathrm{Skew}(J_2,*), K] \subset I_2$. Since R is prime, $J_1 \cap J_2$ is a nonzero $*$-ideal of R, and hence, by we have just proved, $[\mathrm{Skew}(J_1 \cap J_2,*), K]$ is not contained in $Z(R)$. This clearly implies that $\overline{I}_1 \cap \overline{I}_2$ is nonzero, so $\overline{K}$ is prime. □

R prime with involution of the first kind.

LEMMA 3.37. (Levitzki's lemma) *Let J be a nonzero right ideal of a ring R. Suppose that given any $a \in J$, $a^n = 0$ for some fixed integer n. Then R is not semiprime.*

PROOF. See [**Her69**, Lemma 1.1]. □

LEMMA 3.38. *Let R be a 2-torsion free semiprime ring with involution. If t is an absolute zero divisor of K, then $t^2 \in Z(R)$. Moreover, if $t^2 = 0$ then $t = 0$.*

PROOF. For every $s \in S$, $ts + st \in K$. Therefore,

$$[t,[t^2,s]] = [t, t^2s - st^2] = [t,[t, ts + st]] = 0.$$

Furthermore, for every $k \in K$,

$$[t,[t^2,k]] = [t, t[t,k]] + [t,[t,k]t] = t[t,[t,k]] + [t,[t,k]]t = 0.$$

Thus, for every $r \in R$, $[t,[t^2,r]] = 0$, hence, $[t^2,[t^2,r]] = t[t,[t^2,r]] + [t,[t^2,r]]t = 0$ which proves that t^2 is an absolute zero divisor of R^-. Then, $t^2 \in Z(R)$ by Lemma 3.34(1) since R is semiprime.

Suppose now that $t^2 = 0$. Then $2tKt = [t, [t, K]] = 0$, which implies $tKt = 0$ since R is 2-torsion free. Hence, for any $s \in S$, $(ts)^3 = (t(sts)t)s \in (tKt)s = 0$. Thus every element of the right ideal tR has cube zero. Since R is semiprime, this implies by Levitzki's lemma that $t = 0$. □

PROPOSITION 3.39. *Let R be a prime ring of characteristic not 2 with involution $*$ of the first kind. If R is not an order in a simple algebra Q of dimension less than 16 over its center, then K is strongly prime and $Z(K) = K \cap Z(R) = 0$.*

PROOF. Let t be an absolute zero divisor of K. Then, for every $k \in K$, $\mathrm{ad}_t^2 k = 0$ gives $0 = (kt - tk)t$, and hence $(kt - tk)Rt^2 = 0$. Since R is prime, this implies that $t \in Z(K)$. By [**Eri72**, Theorem 2], the subring $\langle K \rangle$ of R generated by K contains a nonzero $*$-ideal I of R. Then $t \in Z(K)$ implies $[t, I] = 0$, which gives $t \in Z(R)$ by Lemma 3.34. But this leads to a contradiction since $t \in K$ and $*$ is of the first kind. Therefore, $t^2 = 0$ and hence $t = 0$ by Lemma 3.38, which proves that K is nondegenerate. Primeness of K follows as in the proof of Proposition 3.36, using [**Eri72**, Corollary] instead of [**Eri72**, Theorem 1(a)] to prove that for any nonzero ideal I of K there exists a nonzero $*$-ideal J of R such that $[\mathrm{Skew}(J, *), K] \subset I$. □

We complete the information of the structure of K given in the previous proposition by studying the remaining cases.

PROPOSITION 3.40. *Let R be a prime ring of characteristic not 2 with involution $*$ of the first kind.*

(1) *If R is an order in a simple algebra Q of dimension 9 or 16 over its center, then K is nondegenerate.*
(2) *If R is an order in a simple algebra Q of dimension 4 over its center, then K is either strongly prime or abelian.*

PROOF. (1) Arguing as in the proof of Proposition 3.39 (using [**Her69**, Theorem 2.2] instead of [**Eri72**, Theorem 2]), we prove that the Lie algebra $\mathrm{Skew}(Q, *)$ is nondegenerate. Since $\mathrm{Skew}(Q, *) = Z(R)^{-1}K$, this implies that K is nondegenerate as well.

(2) Suppose that R is an order in a simple algebra Q of dimension 4 over its center Z. Then $\overline{Q} := \overline{Z} \otimes_Z Q$, where $\overline{Z}$ is the algebraic closure of Z, is the algebra of 2 by 2 matrices over $\overline{Z}$ with involution transpose or symplectic. In the first case $\mathrm{Skew}(\overline{Q}, *)$ is abelian and in the second case it is simple. Hence K is either abelian or strongly prime. □

3.7. Exercises

EXERCISE 3.41. (Brešar) Let A be a 2-torsion free nonassociative Φ-algebra and let d be a derivation of A such that $d^2 = 0$. Show that the Φ-submodule $D = d(A)$ satisfies $D^2 = 0$ and $(DA)D \subset D$. Notice that Proposition 3.6 is a particular case of this result.

EXERCISE 3.42. (Golubkov) Show that a Lie algebra generated by four absolute zero divisors $\{a, b, c, d\}$ is nilpotent of index at most 6. (It is enough to check that the left commutators $[c, b, d, c, b, a]$ and $[a, b, d, c, b, a]$ vanish.)

EXERCISE 3.43. What about a Lie algebra generated by five absolute zero divisors?

EXERCISE 3.44. In Lemma 3.12, take $n = 3$, $y = y_1 y_2 y_3 y'_1 y'_2 y'_3 y''_1$, where the factors with subindex i belong to L_i, $1 \leq i \leq 3$, and make the computations necessary to show that $y \in B$.

EXERCISE 3.45. In the step 5 of Lemma 3.15, write the expression of the commutator $[u_4, x]$ as a sum of (nonzero) associative monomials.

EXERCISE 3.46. Show that any subdirect product of nondegenerate Lie algebras is a nondegenerate Lie algebra.

EXERCISE 3.47. Let L be a finite-dimensional Lie algebra over a field $\mathbb{F}$. Show that its Killing form is associative.

EXERCISE 3.48. Let L be a finite-dimensional Lie algebra over a field $\mathbb{F}$. of characteristic not 2 such that its Killing form is nondegenerate. Show that L is uniquely expresable as a direct sum of simple ideals.

CHAPTER 4

Jordan Elements

Throughout this chapter, L will be a Lie algebra over a ring of scalars Φ (some restrictions on the torsion of Φ will be required). By a Jordan element of L we mean an element $a \in L$ such that $\mathrm{ad}_a^3 L = 0$. Although without given them this name, Jordan elements were used by G. Benkart in the characterization of those finite-dimensional simple Lie algebras over a field of characteristic $p > 5$ which are classical. The reason why we choose this terminology is twofold. On the one hand, the quadratic operator $A^2 = \mathrm{ad}_a^2$ satisfies the identity $\mathrm{ad}^2_{A^2x} = A^2X^2A^2$, $x \in L$, which resembles the fundamental Jordan identity $U_{U_ax} = U_aU_xU_a$, and on the other hand, as will be seen in Chapter 8, it is possible to attach a Jordan algebra L_a to any Jordan element a of a Lie algebra L (becoming in this way an important Jordan tool in the solution of the Kurosh Problem). We list in Section 4.1 the main identities involving Jordan elements, among them the Jordan identity referred to above. In Section 4.2 we study the relationship between Jordan elements and abelian inner ideals. Section 4.3 is a fairly detailed development of the main consequences of the Jordan identity in nondegenerate Lie algebras, proving among other results a quadratic characterization of the annihilator of an ideal. Minimal abelian inner ideals in nondegenerate Lie algebras are studied in Section 4.4. The Kostrikin Descent Lemma and the refinement of this result given by M. Gómez and E. García are the aims of Section 4.5. In Section 4.6 we determine the structure of the Jordan elements of the Lie algebra R^- of a semiprime ring R. Finally, in Section 4.7, we study the Jordan elements of the Lie algebra $K = \mathrm{Skew}(R, *)$, where R is a centrally closed prime ring with involution.

4.1. Identities involving Jordan elements

Except for slight modifications, the results of this section as well as those of the next one are taken from [**Ben77**].

DEFINITION 4.1. By a *Jordan element* of a Lie algebra L we mean an ad-nilpotent element $a \in L$ of index at most 3.

EXAMPLE 4.2. Let R be a ring and let $a, z \in R$ be such that $a^2 = 0$ and $z \in Z(R)$. Then $a + z$ is a Jordan element of the Lie algebra R^-:

$$\mathrm{ad}^3_{a+z} R = [a, aRa] = 0.$$

EXAMPLE 4.3. Let $L = L_{-n} \oplus \cdots \oplus L_n$ be a Lie algebra with a finite $\mathbb{Z}$-grading. Then any $x \in L_n \cup L_{-n}$ is Jordan element of L.

LEMMA 4.4. *Let L be a 3-torsion free Lie algebra and let a be a Jordan element of L. For any $x, y \in L$, we have:*

(1) $A^2XA = AXA^2$,
(2) $A^2XA^2 = 0$,
(3) $\mathrm{ad}^2_{A^2x} = A^2X^2A^2$,
(4) A^2x *is a Jordan element,*
(5) $A^2X^2AXA^2 = A^2XAX^2A^2$,
(6) $[A^2x, Ay] = [A^2y, Ax] = -A^2XAy$,
(7) $[A^2x, A^2y] = 0$,
(8) $\mathrm{ad}_{A^2x}\mathrm{ad}_{A^2y} = A^2XYA^2 = A^2YXA^2$,

where capital letters denote the adjoint operators with respect to those elements: $X = \mathrm{ad}_x$, $A = \mathrm{ad}_a$.

PROOF. Because $A^3x = 0$ for all $x \in L$,

$$0 = [A, [A, [A, X]]] = \mathbf{ad}^3_A(X) = (l_A - r_A)^3X = -3A^2XA + 3AXA^2 = 0,$$

which proves (1) since L is 3-torsion free. Multiplying (1) on the right by A gives (2). Similarly,

$$\mathrm{ad}^2_{A^2x} = (A^2X - 2AXA + XA^2)^2 = A^2X^2A^2,$$

where we have used (1) and (2) to eliminate the remaining terms, so proving (3). Item (4) follows from (2) and (3):

$$\mathrm{ad}^3_{A^2x} = \mathrm{ad}_{A^2x}\mathrm{ad}^2_{A^2x} = (A^2X - 2AXA + XA^2)A^2X^2A^2 = 0.$$

To prove (5), write $B := \mathrm{ad}_{[x,[x,[x,a]]]} = \mathbf{ad}^3_X(A) = (l_X - r_X)^3A$. Using (2) we get

$$\begin{aligned} 0 = A^2BA^2 = A^2(l_X - r_X)^3(A)A^2 &= A^2(X^3A - 3X^2AX + 3XAX^2 - AX^3)A^2 \\ &= 3A^2XAX^2A^2 - 3A^2X^2AXA^2, \end{aligned}$$

which implies $A^2X^2AXA^2 = A^2XAX^2A^2$, since L is 3-torsion free. (6) and (7) are particular cases of (2) and (1) of Lemma 2.17 (the last equality in (6) follows from the Jacobi identity). By (1) and (2), together with $A^3 = 0$, we have

$$\mathrm{ad}_{A^2x}\mathrm{ad}_{A^2y} = (A^2X - 2AXA + XA^2)(A^2Y - 2AYA + YA^2) = A^2XYA^2,$$

while the equality $A^2XYA^2 = A^2YXA^2$ is a direct consequence of (2). This proves (8) and completes the proof. □

4.2. Jordan elements and abelian inner ideals

Let L be a Lie algebra over a ring of scalars Φ. Recall that a Φ-submodule B of L is an inner ideal if $[B[B, L]] \subset B$, and an abelian inner ideal is an inner ideal which is also an abelian subalgebra.

LEMMA 4.5. *Let B be an abelian inner ideal of L. Then every $b \in B$ is a Jordan element of L.*

PROOF. $\mathrm{ad}^3_b L = \mathrm{ad}_b(\mathrm{ad}^2_b L) \subset [b, B] = 0$. □

Note that Example 4.3 is a particular case of this lemma since both L_n and L_{-n} are abelian inner ideals. The converse is also true if L is 3-torsion free. In fact, we have the following more general result.

PROPOSITION 4.6. *Let L be a 3-torsion free Lie algebra and let $a \in L$ be a Jordan element. If V is an inner ideal of L, then $\mathrm{ad}_a^2 V$ is an abelian inner ideal. In particular, $[a]_L := \mathrm{ad}_a^2 L$* (the principal inner ideal determined by a) *and $(a)_L := \Phi a + [a]_L$ are abelian inner ideals.*

PROOF. Let V be an inner ideal of L. It follows from 4.4(7) that $\mathrm{ad}_a^2 V$ is an abelian subalgebra. Thus it suffices to show that $\mathrm{ad}_x \mathrm{ad}_y L \subset \mathrm{ad}_a^2 V$ for $x = \mathrm{ad}_a^2 b$ and $y = \mathrm{ad}_a^2 c$, $b, c \in V$. By 4.4(8)we have

$$\mathrm{ad}_x \mathrm{ad}_y = \mathrm{ad}_{A^2 b} \mathrm{ad}_{A^2 c} = A^2 BCA^2.$$

Consequently,

$$\mathrm{ad}_x \mathrm{ad}_y L = \mathrm{ad}_a^2 \mathrm{ad}_b \mathrm{ad}_c \mathrm{ad}_a^2 L \subset \mathrm{ad}_a^2 \mathrm{ad}_b \mathrm{ad}_c L \subset \mathrm{ad}_a^2[V, [V, L]] \subset \mathrm{ad}_a^2 V,$$

which proves that $\mathrm{ad}_a^2 V$ is an abelian inner ideal. In particular, $[a] = \mathrm{ad}_a^2 L$ is an abelian inner ideal, and the same is true for $(a) = \Phi a + [a]$:

$$[a, [a]] = \mathrm{ad}_a^3 L = 0 \quad \text{and} \quad [(a), [(a), L]] = [[a], [[a], L]] \subset [a] \subset (a).$$

□

4.3. Jordan elements in nondegenerate Lie algebras

Recall that a Lie algebra L is nondegenerate if $\mathrm{ad}_x^2 L = 0$ implies $x = 0$.

PROPOSITION 4.7. *Let I be an ideal of a 3-torsion free Lie algebra L. If L is nondegenerate, then I is a nondegenerate Lie algebra.*

PROOF. Let a be an absolute zero divisor of I. Then a is a Jordan element of L. Indeed, $\mathrm{ad}_a^3 L = \mathrm{ad}_a^2[a, L] \subset \mathrm{ad}_a^2 I = 0$. So, by 4.4(3), for every $x \in L$, we have

$$\mathrm{ad}_{A^2 x}^2 L = \mathrm{ad}_a^2 \mathrm{ad}_x^2 \mathrm{ad}_a^2 L \subset \mathrm{ad}_a^2 I = 0.$$

Applying the nondegeneracy of L twice, we get $a = 0$. □

REMARK 4.8. Inheritance of nondegeneracy by ideals was obtained in [**Zel83a**, Lemma 4] as a consequence of the inheritance of the Kostrikin radical, which will be studied later.

Recall that a Lie algebra is said to be strongly prime if it is prime and nondegenerate.

COROLLARY 4.9. *Let I be an ideal of a 3-torsion free Lie algebra L. If L is strongly prime, then I is a strongly prime Lie algebra.*

PROOF. By Proposition 4.7, I is nondegenerate and therefore semiprime, so we can apply Proposition 2.46 to get that I is also prime. □

The annihilator of an ideal I of a semiprime associative algebra R can be described in quadratic terms: $xIx = 0 \Rightarrow (xI)^2 = 0 \Rightarrow xI = 0 \Rightarrow x \in \mathrm{Ann}_R(I)$. A similar result holds for nondegenerate Lie algebras.

PROPOSITION 4.10. *Let I be an ideal of a 6-torsion free Lie algebra L. If I is a nondegenerate Lie algebra, then $\mathrm{Ann}_L(I) = \{a \in L : \mathrm{ad}_a^2 I = 0\}$.*

PROOF. Let $a \in L$. Clearly $a \in \operatorname{Ann}_L(I)$ implies $\operatorname{ad}_a^2 I = 0$. Suppose conversely that $\operatorname{ad}_a^2 I = 0$. Since I is an ideal of L, the uppercase notation for adjoint representations is well defined when their arguments are restricted to I. We will assume that all the implicit arguments are in the ideal I. According with this convention, $A^2 = 0$ and the identities of Lemma 3.3 remain valid. Given $x \in I$, set $b = [a, x]$. By 3.3(4), b is a Jordan element of I, with $B^2 = -AX^2A$ by 3.3(3). Then, by 4.4(3),

$$\operatorname{ad}_{B^2x}^2 = B^2X^2B^2 = AX^2AX^2AX^2A = 0$$

since $AX^2AX^2A = 0$ by 3.3(6). Hence $B^2x = 0$ by nondegeneracy of I. Note that x was fixed and related to b, so we are not finished yet. Linearizing

$$B^2x = -AX^2Ax = 0, \ x \in I,$$

we get for any $y \in I$ and $n \in \mathbb{Z}$,

$$0 = A(X + nY)^2A(x + ny) = 3nAX^2Ay + 3n^2AY^2Ax,$$

where we have eliminated the terms AX^2Ax and AY^2Ay and used the identities (2) and (5) of Lemma 3.3. Take $n = 1$ and $n = 2$ to get the system of linear equations

$$3AX^2Ay + 3AY^2Ax = 0, \quad 6AX^2Ay + 12AY^2Ax = 0,$$

which yields the solution $6AX^2Ay = 0$. Since L is 6-torsion free, this implies

$$\operatorname{ad}_{Ax}^2 y = B^2y = -AX^2Ay = 0$$

for all $y \in I$, and hence that $Ax = 0$ because I is nondegenerate. Since $x \in I$ is arbitrary, this proves that $a \in \operatorname{Ann}_L(I)$. [1] □

COROLLARY 4.11. *Let I be an ideal of a 6-torsion free Lie algebra L. If I is nondegenerate, then the Lie algebra $L/\operatorname{Ann}_L(I)$ is nondegenerate.*

PROOF. Denote by $x \mapsto \overline{x}$ the homomorphism of L onto $\overline{L} = L/\operatorname{Ann}_L(I)$, and let $\overline{a}$ be an absolute zero divisor of $\overline{L}$. Then $\operatorname{ad}_a^2 I \subset I \cap \operatorname{Ann}_L(I) = 0$. Hence $a \in \operatorname{Ann}_L(I)$ by Proposition 4.10, i.e. $\overline{a} = 0$. □

PROPOSITION 4.12. *Let E be an essential ideal of a 6-torsion free Lie algebra L. If E is nondegenerate, then L is nondegenerate.*

PROOF. Let a be an absolute zero divisor of L. Then $\operatorname{ad}_a^2 E = 0$, and hence, Proposition by 4.10, $a \in \operatorname{Ann}_L(E)$. But $\operatorname{Ann}_L(E) = 0$ by Proposition 1.6(1). □

PROPOSITION 4.13. [**FLGGL09**, Proposition 1.6] *Let L be a 6-torsion free Lie algebra and let I be an ideal of L. If I is nondegenerate, then every Jordan element of I is a Jordan element of L.*

PROOF. Let a be a Jordan element of I. Then $\operatorname{ad}_a^4 L = \operatorname{ad}_a^3[a, L] \subset \operatorname{ad}_a^3 I = 0$. So for any $x \in L$, $y \in I$ we have

$$\begin{aligned} 0 &= \operatorname{ad}_{A^4x} y = (A^4X - 4A^3XA + 6A^2XA^2 - 4AXA^3 + XA^4)y \\ &= -4A^3XAy + 6A^2XA^2y - 4AXA^3y = 6A^2XA^2y. \end{aligned} \tag{4.1}$$

[1] A different proof of Proposition 4.10 is given in [**Zel83a**, Corollary 1, page 543].

By 4.4(4), $A^2[a,x]$ is a Jordan element of I. Hence we have by (4.1)

$$\begin{aligned}\operatorname{ad}^2_{A^2[a,x]} y &= \operatorname{ad}^2_a \operatorname{ad}^2_{[a,x]} \operatorname{ad}^2_a y = (A^2(AX - XA)^2 A^2)y \\ &= (-A^2XA^2XA^2 + A^2XAXA^3)y = -A^2XA^2XA^2y = 0.\end{aligned}$$

Then, by Proposition 4.10, $A^3x = A^2[a,x] \in I \cap \operatorname{Ann}(I) = 0$, so a is a Jordan element of L. □

PROPOSITION 4.14. *Let L be a 6-torsion free Lie algebra and let E be an essential ideal of L. If E is nondegenerate, then L contains nonzero Jordan elements if and only if so does E.*

PROOF. By Proposition 4.13, every Jordan element of E is a Jordan element of L. Let $a \in L$ be a nonzero Jordan element of L. Since E is an essential ideal, $\operatorname{Ann}_L(E) = 0$ and hence, by Proposition 4.10, $\operatorname{ad}^2_a E \neq 0$. Then it follows from Lemma 4.5 and Proposition 4.6 that E contains a nonzero Jordan element. □

LEMMA 4.15. *Let E be an essential ideal of a Lie algebra L and let $\alpha \in \Phi$. If E is α-torsion free, then so is L.*

PROOF. Let $\operatorname{Ann}_L(\alpha) = \{x \in L : \alpha x = 0\}$. Then $\operatorname{Ann}_L(\alpha)$ is an ideal of L, and since E is essential and α-torsion free, $\operatorname{Ann}_L(\alpha) = 0$, which proves that L is α-torsion free. □

COROLLARY 4.16. *Let L be a a 6-torsion free Lie algebra. If L is nondegenerate, then the Lie algebra* $\operatorname{Der}(L)$ *is also nondegenerate.*

PROOF. As L is nondegenerate, $Z(L) = 0$. Hence, by Proposition 2.14, L is isomorphic to the essential ideal $\operatorname{ad}(L)$ of $\operatorname{Der}(L)$. Since $\operatorname{Der}(L)$ is also 6-torsion free by the above lemma, it is nondegenerate by Proposition 4.12. □

PROPOSITION 4.17. *Let L be a Lie algebra over a ring of scalars Φ such that $6 \in \Phi^*$ and let $z \in L$ be an absolute zero divisor. We have:*

(1) *For any Jordan element $a \in L$, $[a,z] = w_1 + w_2 + w_3$ is a sum of absolute zero divisors.*
(2) *The span C_1 of the absolute zero divisors of L is an inner ideal.*

PROOF. (1) By Lemma 1.2, $\exp(A) \in \operatorname{Aut}(L)$, and hence

$$\exp(A)z = z + Az + \frac{1}{2}A^2z$$

is an absolute zero divisor. Furthermore, by 4.4(3)

$$\operatorname{ad}^2_{A^2z} = A^2Z^2A^2 = 0,$$

so A^2z is also an absolute zero divisor. Thus $[a,z] = \exp(A)z - z - \frac{1}{2}A^2z$ is a sum of three absolute zero divisors.

(2) It suffices to show that for every $x \in L$ and z_1, z_2 absolute zero divisors, $[[x,z_1],z_2] \in C_1$. Since $[x,z_1]$ is a Jordan element by 3.3(4)), we have by (1) that $[[x,z_1],z_2] \in C_1$. □

REMARK 4.18. Suppose that L is a simple Lie algebra over a field $\mathbb{F}$ of characteristic $p > 3$. It follows from (1) of the above proposition that if L is generated by Jordan elements, then the linear span of its absolute zero divisors is an ideal of L. This result had been already obtained in Proposition 3.25 using the Vandermonde argument.

4.4. Minimal abelian inner ideals

Let B be a minimal abelian inner ideal of a Lie algebra L. Since every inner ideal contained in B is abelian, it is cleat that B is a minimal inner ideal of L. Conversely, every minimal inner ideal of L which is also abelian is a minimal abelian inner ideal of L. So we can interchange the adjectives abelian and minimal without changing the meaning.

LEMMA 4.19. *Let L be a 3-torsion free nondegenerate Lie algebra and let B be an abelian subalgebra of L. Then B is a minimal abelian inner ideal of L if and only if $B = \operatorname{ad}_b^2 L$ for every nonzero element $b \in B$.*

PROOF. If B is a minimal abelian inner ideal of L, then it follows from Lemma 4.5 and Proposition 4.6 that $B = \operatorname{ad}_b^2 L$ for every $0 \neq b \in B$. The converse is trivial. □

COROLLARY 4.20. *Suppose that L is 3-torsion free and nondegenerate. Let B be a minimal abelian inner ideal of L and let $a \in L$ be a Jordan element. Then either $\operatorname{ad}_a^2 B = 0$ or it is a minimal abelian inner ideal.*

PROOF. By Proposition 4.6, $\operatorname{ad}_a^2 B$ is an abelian inner ideal of L. Since L is nondegenerate, for any nonzero element $\operatorname{ad}_a^2 b \in \operatorname{ad}_a^2 B$, we have (again by 4.6) that $\operatorname{ad}_{A^2 b}^2 L$ is a nonzero abelian inner ideal of L. Then using 4.4(3) and the minimality of B, we get

$$\operatorname{ad}_{A^2 b}^2 L = \operatorname{ad}_a^2(\operatorname{ad}_b^2 \operatorname{ad}_a^2 L) = \operatorname{ad}_a^2 B,$$

which proves that $\operatorname{ad}_a^2 B$ is minimal by Lemma 4.19. □

COROLLARY 4.21. *Let L be a 6-torsion free Lie algebra, let I be an ideal of L, and let B be an abelian subalgebra of I. If I is nondegenerate, then B is a minimal inner ideal of L if and only if it is a minimal inner ideal of I.*

PROOF. Let B be an minimal abelian inner ideal of I. It follows from Proposition 4.13 that every $b \in B$ is a Jordan element of L. Hence, for every nonzero $b \in B$, $B = \operatorname{ad}_b^2 I$ is an abelian inner ideal of L by Proposition 4.6. Since B is minimal as an inner ideal of I, it is also minimal as an inner ideal of L. Conversely, suppose that B is a minimal abelian inner ideal of L contained in I. As I is nondegenerate, it follows from Proposition 4.6 that $\operatorname{ad}_b^2 I$ is a nonzero inner ideal of L for any nonzero element $b \in B$,. Hence $B = \operatorname{ad}_b^2 I$ by minimality of B, which proves that B is a minimal abelian inner ideal of I. □

4.5. On the existence of Jordan elements

In this section we study conditions guaranteeing the existence of nonzero Jordan elements in Lie algebras. We keep using capital letter to denote adjoint operators.

The Kostrikin Descent Lemma.

LEMMA 4.22. [**Zel92b**, III.2.10] *Let L be a Lie algebra over a field $\mathbb{F}$ of characteristic 0 or $p \geq 5$ and let $a \in L$ be such that $\operatorname{ad}_a^n = 0$ for $4 \leq n \leq p-1$. Then for every $x \in L$ we have $\operatorname{ad}_{A^{n-1}x}^{n-1} = 0$.*

PROOF. From $A^n = 0$ we get, for any $x \in L$, $\mathbf{ad}_A^n X = \operatorname{ad}_{A^n x} = 0$, i.e.

$$-nA^{n-1}XA + \binom{n}{2}A^{n-2}XA^2 + \cdots + (-1)^{n-1}nAXA^{n-1} = 0. \tag{4.2}$$

Multiplying (4.2) on the right by A^{n-2} we get

$$A^{n-1}XA^{n-1}=0, \tag{4.3}$$

since n is invertible in $\mathbb{F}$. Now, multiplying (4.2), first on the right and then on the left by A^{n-3}, we get the system of linear equations

$$-nA^{n-1}XA^{n-2}+\binom{n}{2}A^{n-2}XA^{n-1}=0$$

$$(-1)^{n-2}\binom{n}{2}A^{n-1}XA^{n-2}+(-1)^{n-1}nA^{n-2}XA^{n-1}=0,$$

whose determinant is

$$\Delta=(-1)^{n-1}\frac{n^2(n-3)(n+1)}{4}.$$

Since $4\le n\le p-1$, $\Delta=0$ if and only if $n<p-1$. In this case we have

$$A^{n-1}XA^{n-2}=0. \tag{4.4}$$

Return to the statement of the lemma. We must prove that $\mathrm{ad}^{n-1}_{A^{n-1}x}=0$ for any $x\in L$. Suppose on the contrary that

$$\mathrm{ad}^{n-1}_{A^{n-1}x}=(\mathbf{ad}^{n-1}_A X)^{n-1}=\sum m_{\mathbf{i}}A^{i_1}XA^{i_2}X\cdots XA^{i_n}\neq 0, \tag{4.5}$$

where the integers m_i are nonzero in $\mathbb{F}$, $0\le i_k<n$ for all $1\le k\le n$, and

$$i_1+i_2+\cdots i_n=(n-1)^2. \tag{4.6}$$

Let

$$A^{i_1}XA^{i_2}X\cdots XA^{i_n}\neq 0 \tag{4.7}$$

be the summand in (4.5) such that the sequence $(i_1,\cdots,i_n)$ is lexicographically maximal. We claim that $i_k>0$ for all $1\le k\le n$. If $i_k=0$ for some k, then it follows from (4.6) that all i_j, $j\neq k$, are equal to $n-1$. Hence, since $n\ge 4$ by assumption, the product (4.7) contains a factor $A^{n-1}XA^{n-1}$, which is equal to zero by (4.3), a contradiction.

Our next claim is that $i_k\le n-2$ for all $2\le k\le n$. Suppose on the contrary that $i_k=n-1$ for some k such that $2\le k\le n$. Since $i_{k-1}\neq 0$, we have by (4.2)

$$A^{i_{k-1}}XA^{i_k}=A^{i_{k-1}-1}(AXA^{n-1})=A^{i_{k-1}-1}(\sum_{i\ge 2}\alpha_iA^iXA^{n-i}),\ \alpha_i\in\mathbb{F},$$

which contradicts the lexicographically maximality of $(i_1,\cdots,i_n)$ in (4.4). Now it follows from (4.6) that

$$i_1=n-1,\ i_2=i_3=\cdots=i_n=n-2.$$

If $n\neq p-1$, then we have by (4.4) that $A^{i_1}XA^{i_2}=0$, a contradiction. The proof will be complete by proving that the assumption $n=p-1$ implies

$$A^{n-1}XA^{n-2}XA^{n-2}=0,$$

which is again a contradiction. Multiplying (4.2) on the right by A^{n-4} we get

$$A^{n-2}XA^{n-2}=\alpha_1A^{n-1}XA^{n-3}+\alpha_2A^{n-3}XA^{n-1}$$

with $\alpha_i\in\mathbb{F}$, $i=1,2$. Hence, by (4.3),

$$A^{n-1}XA^{n-2}XA^{n-2}=\alpha_2A^{n-1}XA^{n-3}XA^{n-1}.$$

Since the prime number p is odd, $n = p - 1$ is even, and $n - 3$ is again odd, we have

$$[X, \mathbf{ad}_A^{n-3} X] = 2XA^{n-3}X + \sum_{i+j>0} \beta_{ij} A^i X A^{n-3-i-j} X A^j.$$

Hence, by (4.3),

$$A^{n-1}XA^{n-3}XA^{n-1} = \frac{1}{2}A^{n-1}[X, \mathbf{ad}_A^{n-3}X]A^{n-1} = 0.$$

This completes the proof of the lemma. □

COROLLARY 4.23. *Let L be a Lie algebra over a field $\mathbb{F}$ of characteristic* 0 *or* $p \geq 5$, *and let $a \in L$ be an ad-nilpotent element of index n,* $4 \leq n \leq p-1$. *Then L contains a nonzero Jordan element.*

PROOF. By Lemma 4.22, $\mathrm{ad}^{n-1}_{A^{n-1}(L)} = 0$. Let b be a nonzero element in $A^{n-1}(L)$. If $n = 4$, then b is a Jordan element and we have finished. Otherwise we can repeat the process to get finally the desired Jordan element, since b is a nonzero ad-nilpotent element of index less or equal to $n - 1$. □

For $n = 4$, Corollary 4.23 remains true for any Lie algebra over a ring of scalars just assuming 6-torsion free.

LEMMA 4.24. *Let L be a* 6*-torsion free Lie Φ-algebra, and let* $0 \neq a \in L$ *be such that* $\mathrm{A}^4 = 0$. *Then L contains a nonzero Jordan element.*

PROOF. If $A^3 = 0$, then a itself is a Jordan element, so we may suppose $A^3x \neq 0$ for some $x \in L$. We claim that A^3x is a Jordan element, i.e.

$$\mathrm{ad}^3_{A^3x} = \sum_i m_i A^{i_1}XA^{i_2}XA^{i_3}XA^{i_4} = 0. \quad i_1 + i_2 + i_3 + i_4 = 9,$$

If this were not the case, we would have, arguing as in the proof of Lemma 4.22,

$$A^3XA^2XA^2XA^2 \neq 0. \tag{4.8}$$

But this leads to a contradiction. From

$$0 = \mathrm{ad}_{A^4x} = -4A^3XA + 6A^2XA^2 - 4AXA^3 \tag{4.9}$$

and just using 2-torsion free, we get

$$A^3XA^3 = 0. \tag{4.10}$$

Hence $0 = A^3[[X, A], X]A^3 = 2A^3XAXA^3$. Then, by (4.9),

$$6(A^3X)A^2XA^2 = 4(A^3X)A^3XA + 4(A^3X)AXA^3 = 0,$$

which contradicts (4.8). □

A refinement of Kostrikin's descent lemma. Let L be a Lie algebra over a ring of scalars Φ and let $a \in L$ be such that $\mathrm{ad}_a^n = 0$, $n \geq 2$. Following the paper of E. García and M. Gómez [**GGL09**], we prove in this subsection a result which extends those obtained for $n = 2$ (Proposition 3.6) and for $n = 3$ (Proposition 4.6). In fact, Kostrikin's descent lemma is sharpened in the sense that $\mathrm{ad}_a^{n-1} L$ is actually an abelian inner ideal of L and hence any $x \in \mathrm{ad}_a^{n-1} L$ is a Jordan element.

To illustrate the process of the proof of this result, which is quite technical and involves long calculations, we begin by analyzing the case that $n = 6$ (assuming $7! \in \Phi^*$). Note first that, by Lemma 2.17, the Φ-submodule $\mathrm{ad}_a^5 l$ is an abelian subalgebra, so, by Lemma 2.35, it is enough to show that $\mathrm{ad}^2_{A^5x} L \subset \mathrm{ad}_a^5 L$, $x \in L$.

Set $B = \mathrm{ad}_{A^5x} = A^5X - 5A^4XA + 10A^3XA^2 - 10A^2XA^3 + 5AXA^4 - XA^5$. We will prove that $B^2 \in A^5\,\mathrm{Ad}(L)$, i.e. $B^2 \equiv 0 \pmod{A^5\,\mathrm{Ad}(L)}$.

From $A^6 = 0$, we get

$$-6A^5XA + 15A^4XA^2 - 20A^3XA^3 + 15A^2XA^4 - 6AXA^5 = 0. \tag{4.11}$$

Multiply (4.11) by A^2, first on the left and then on the right. Then multiply (4.11) by A on both sides. We get the system of linear equations

$$\begin{aligned} -6A^5XA^3 + 15A^4XA^4 - 20A^3XA^5 &= 0\\ -20A^5XA^3 + 15A^4XA^4 - 6A^3XA^5 &= 0\\ 15A^5XA^3 - 20A^4XA^4 + 15A^3XA^5 &= 0. \end{aligned}$$

As $\begin{vmatrix} -6 & 15 & -20\\ -20 & 15 & -6\\ 15 & -20 & 15 \end{vmatrix} = -980 = -2^2.5.7^2 \in \Phi^*$, the system has a unique solution, so

$$A^5XA^3 = A^3XA^5 = A^4XA^4 = 0. \tag{4.12}$$

Hence we have

$$A^iXAXA^j = A^i[[X,A],X]A^j = 0,\ i+j \geq 0. \tag{4.13}$$

Using (4.12) and (4.13) we get that B^2 is congruent with

$$50A^4XA^3XA^3 - 25A^4XA^2XA^4 - 100A^3XA^4XA^3 + 50A^3XA^3XA^4 - 10A^3XA^2XA^5$$

We deal with each summand of the above expression separately. Multiplying (4.11) on the left by A and on the right by XA^3 we get

$$-20A^4XA^3XA^3 + 15A^3XA^4XA^3 = -15A^5XA^2XA^3.$$

Again, multiplying (4.11) on the right by AXA^3 we obtain

$$15A^4XA^3XA^3 - 20A^3XA^4XA^3 = -A^5XA^2XA^3$$

The determinant of the system formed by these two equations is 7.5^2, so this system can be solved obtaining

$$A^4XA^3XA^3,\ A^3XA^4XA^3 \in A^5\,\mathrm{Ad}(L). \tag{4.14}$$

Multiplying (4.11) on the left by A^4X and using (4.12) and (4.13), we get

$$15A^4XA^2XA^4 = 20A^4XA^3XA^3.$$

Hence, by (4.14),

$$A^4XA^2XA^4 \in A^5\,\mathrm{Ad}(L). \tag{4.15}$$

Multiplying now (4.11) on the right by XA^4 we obtain

$$15A^4XA^2XA^4 = 20A^3XA^3XA^4.$$

Hence, by (4.15),

$$A^3XA^3XA^4 \in A^5\,\mathrm{Ad}(L). \tag{4.16}$$

Finally, multiplying (4.11) on the left by A^3XA, we get

$$-20A^3XA^4XA^3 + 15A^3XA^3XA^4 - 6A^3XA^2XA^5 = 0.$$

Hence, by (4.14) and (4.16), $A^3XA^2XA^5 \in A^5\,\mathrm{Ad}(L)$, as desired.

LEMMA 4.25. *Let L be a Lie Φ-algebra, and let $a \in L$ be such that $A^n = 0$, $n > 2$. For each $0 < m < n$ and any $x \in L$, put*

$$y_1 = A^m X A^{n-1},\ y_2 = A^{m+1} X A^{n-2},\ \dots,\ y_{n-m} = A^{n-1} X A^m.$$

(1) *If $2m \geq n$ and $(2n - m - 1)! \in \Phi^*$, then all $y_1, \dots, y_{n-m}$ are zero.*
(2) *If $2m < n$ and $(n + m - 1)! \in \Phi^*$, then for any choice of m indexes*

$$1 \leq i_1 < i_2 < \cdots < i_m \leq n - m,$$

we can express the elements $y_{i_1}, y_{i_2}, \dots, y_{i_m}$ as a linear combination of the other $n - 2m$ elements of $\{y_1, \dots, y_{n-m}\}$.

PROOF. Let $a \in L$ be such that $A^n = 0$. For any $x \in L$, we have

$$\sum_{k=1}^{n-1} (-1)^k \binom{n}{k} A^{n-k} X A^k = 0. \tag{4.17}$$

Denote by ♣ the left hand side of (4.17)and consider the equalities

$$A^{m-1}♣ = 0, \quad A^{m-2}♣A = 0, \quad \dots, \quad A♣A^{m-2} = 0, \quad ♣A^{m-1} = 0, \tag{4.18}$$

regarded as a linear system of m equations in the unknowns $y_1, y_2, \dots, y_{n-m}$.

(1) If $2m \geq n$, the first $n - m$ equations form an homogeneous linear system whose matrix is invertible (see [**Kos90**, Theorem 3.1, p. 40]), since we are assuming that $(2n - m - 1)! \in \Phi^*$, so the only solution of the system is the trivial one. Therefore, $A^i X A^j = 0$ for $i + j = n + m - 1$.

(2) Suppose that $2m < n$ and let $1 \leq i_1 < i_2 < \cdots < i_m \leq n - m$. Then the m columns $(i_1, \dots, i_m)^t$ of the coefficients of the linear systems (4.18) form a matrix which is invertible. Thus we can express the unknowns $y_{i_1}, y_{i_2}, \dots, y_{i_m}$ as a linear combination of the other $n - 2m$ elements in $\{y_1, \dots, y_{n-m}\}$. □

LEMMA 4.26. *Let L be a Lie Φ-algebra and let $a \in L$ be such that $A^n = 0$, with $n > 2$ and $(n + [\frac{n}{2}] - 1)! \in \Phi^*$. Then for any $x \in L$ and any triple (i, j, k) of nonnegative integers such that $i + j + k = 2n - 2$, we have that $A^i X A^j X A^k$ belongs to $A^{n-1} \operatorname{Ad}(L)$.*

PROOF. If $j = 0$, then $i + k = 2n - 2$, and hence either $A^i X^2 A^k = 0$ (when $i \geq n$ or $k \geq n$) or both i, k are equal to $n - 1$. In the two cases,

$$A^i X A^j X A^k \in A^{n-1} \operatorname{Ad}(L).$$

We will assume from now on that n is odd, $n = 2t + 1$. For n even the proof is similar.

If $k < t$ or $i < t$, then $A^i X A^j X A^k = 0$. Since both cases are similar, let us suppose that $k < t$. Taking $m = t+1$ in (4.18), we have by 4.25(1) that $A^i X A^j = 0$ whenever $i + j = n + m - 1 = (2t + 1) + (t + 1) - 1 = 3t + 1$. But $k < t$, together with $i + j + k = 2n - 2 = 4t$, implies $i + j > 3t$. Therefore $A^i X A^j X A^k = 0$ as desired.

If $k = t$ or $i = t$ then $A^i X A^j X A^k \in A^{n-1} \operatorname{Ad}(L)$. Consider first the case of a term $A^i X A^j X A^t$ ending in A^t. Then $i + j + t = 4t$ implies $i + j = 3t$. Take $m = t$ in (4.18). Then $n - 2m = (2t + 1) - 2t = 1$ and hence it follows from 4.25(2) that $A^i X A^j$ is a scalar multiple of $A^{2t} X A^t$, so $(A^i X A^j) X A^t \in A^{n-1} X A^m$. Consider now a term of the form $A^tXA^jXA^k$. Then $j+k=3t$ and hence, by 4.25(2) again, $A^jXA^k = \lambda A^{2t}XA^t$ for some $\lambda \in \Phi$. Then

$$A^tX(A^jXA^k) = \lambda A^tX(A^{2t}XA^t) \in A^{n-1}\operatorname{Ad}(L)$$

by we have just proved.

To finish the proof we will use induction. Suppose that given $0 \le r < [\frac{t}{2}]$, every term $A^iXA^jXA^k$ such that $i \le t+r$ or $k \le t+r$ belongs to $A^{n-1}\operatorname{Ad}(L)$. We will prove that the same holds for every term beginning or ending with A^{t+r+1}. Let us first consider the case $k = t+r+1$. Then $i+j+t+r+1 = 4t$ implies $i+j = 3t-r-1$. Taking $m = t-r-1$ in (4.18), it follows from Lemma 4.25(2) that A^iXA^j can be expressed as a linear combination of any set of $n-2m = (2t+1)-2(t-r-1) = 2r+3$ elements taken among the terms

$$A^{2t}XA^{t-r-1},\ A^{2t-1}XA^{t-r},\ \ldots,\ A^{t+r}XA^{t-2r-1},\ldots,\ A^{t-r-1}XA^{2t},$$

so we can express $(A^iXA^j)XA^{t+r+1}$ as a linear combination of the $2r+3$ terms

$$A^{2t}XA^{t-r-1}XA^{t+r+1},\quad A^{t-r-1}XA^{2t}XA^{t+r+1},$$
$$A^{t-r}XA^{2t+1}XA^{t+r+1},\ldots,A^{t+r}XA^{t-2r-1}XA^{t+r+1}.$$

It is clear that $A^{2t}XA^{t-r-1}XA^{t+r+1} \in A^{n-1}\operatorname{Ad}(L)$, since $n = 2t+1$. The other $2r+2$ terms also belong to $A^{n-1}\operatorname{Ad}(L)$ by the induction hypothesis. In the same way, any term beginning with A^{t+r+1} can be written as a linear combination of the $2r+3$ terms ending in A^k, $t-r-1 \le k \le t+r+1$, which belong to $A^{n-1}\operatorname{Ad}(L)$ by we have just proved. □

THEOREM 4.27. [**GGL09**, Theorem 2.3] *Let L be a Lie Φ-algebra and let $a \in L$ be such that* $\mathrm{A}^n = 0$. *If $n > 2$ and $(n+[\frac{n}{2}]-1)! \in \Phi^*$, then $\operatorname{ad}_a^{n-1} L$ is an abelian inner ideal of L. Hence any $x \in \operatorname{ad}_a^{n-1} L$ is a Jordan element.*

PROOF. By Lemma 2.17, $\operatorname{ad}_a^{n-1} L$ is an abelian subalgebra. Thus, by Lemma 2.35, it suffices to prove that for each $x \in L$, $\operatorname{ad}^2_{A^{n-1}x} \in A^{n-1}\operatorname{Ad}(L)$. Since

$$\operatorname{ad}_{A^{n-1}x} = \sum_{k=0}^{n-1} (-1)^k \binom{n-1}{k} A^{n-1-k}XA^k,$$

$\operatorname{ad}^2_{A^{n-1}x}$ is a linear combination of monomials of the form $A^iXA^jXA^k$, where $i,j,k \ge 0$ and $i+j+k = 2n-2$. Thus all we need to see is that each one of these monomials is equal to $A^{n-1}M$ for some $M \in \operatorname{Ad}(L)$. But this is just what proves Lemma 4.26. □

REMARK 4.28. Let A be a nonassociative algebra over a field of characteristic 0 and let d be a derivation of A such that $d^n = 0$ for some $n \ge 2$. Then the subspace $D = d^{n-1}(A)$ satisfies $D^2 = 0$ and $(DA)D \subset D$. As observed in [**BFL10**, Theorem 2.2], this assertion can be proved by an adaptation of the arguments used in the proof of the (apparently very special case) [**GGL09**, Theorem 2.3].

Minimal inner ideals revisited. The structure theorem on minimal inner ideals (see Theorem 2.43) can be sharpened as follows.

THEOREM 4.29. [**Ben77**, Lemma 1.12] *Let L be a 6-torsion free Lie algebra and let B be a minimal inner ideal of L. Then either:*

(1) *$B = \Phi b$, where b is an absolute zero divisor of L and Φ is a field,*

(2) *$B = \mathrm{ad}_b^2 L$ for all $b \neq 0$ in B and $[B, B] = 0$, or*
(3) *B is an ideal of L which is innerly simple as a Lie algebra.*

PROOF. We may assume that B has no nonzero absolute zero divisors (since otherwise we would be in case (1) of Theorem 2.43). If B is as in case (2) of 2.43, then it follows from Proposition 4.6 that $B = \mathrm{ad}_b^2 L$ for all $b \neq 0$ in B. Thus we may assume that B is an ideal of L which is simple as a Lie algebra and every proper inner ideal V of B is abelian. For every $v \in V$, $\mathrm{ad}_v^4 L \subset [v, [v, [v, B]]] \subset [v, V] = 0$, so v is ad-nilpotent of index less than or equal to 4. If the index is 4, then it follows from Lemma 4.24 that there exist $x \in L$ such that $u = \mathrm{ad}_v^3 x$ is a nonzero Jordan element of L. Since $u \in B$, we have by 4.6 that $\mathrm{ad}_u^2 L$ is an abelian inner ideal of L contained in B, and hence either $\mathrm{ad}_u^2 L = B$ or $\mathrm{ad}_u^2 L = 0$. In the first case, $[B, B] = 0$, which is a contradiction since B is simple as an algebra. The case $\mathrm{ad}_u^2 L = 0$ gives $u = 0$ (since otherwise $B = \Phi u$, which has been discarded), which is again a contradiction. Therefore, $\mathrm{ad}_v^3 L = 0$. By the same reasoning applied to v we obtain $V = 0$, so B does not contain proper nonzero inner ideals. □

EXAMPLE 4.30. The following Lie algebras are innerly simple and therefore they are themselves minimal inner ideals of type (3) in the theorem above.

(i) $[\Delta, \Delta]/Z(\Delta) \cap [\Delta, \Delta]$, where Δ is a division associative algebra such that $[[\Delta, \Delta], \Delta] \neq 0$, [**Ben76**, Corollary 3.15].
(ii) The finitary orthogonal Lie algebra $\mathfrak{fo}(X, \langle\, ,\rangle)$ (Section 2.2) where the bilinear form $\langle\, ,\rangle$ is anisotropic and X has dimension (possibly infinite) greater than 4.

Finite gradings and Jordan elements. Let Λ be a torsion free abelian group and let L be a Lie algebra. Recall that a Λ-grading $L = \bigoplus_{\lambda \in \Lambda} L_\lambda$ of L is said to be finite if the set $\Lambda^* = \{\lambda \in \Lambda : L_\lambda \neq 0\}$ is finite, and nontrivial if Λ^* contains a nonzero element. Note that if a Λ-grading is finite and nontrivial, then the subgroup $G = G(\Lambda^*)$ of Λ generated by Λ^* is free of finite rank, and therefore is isomorphic to $\mathbb{Z}^r$ for some positive integer r. In fact, as pointed out by E. Zelmanov in [**Zel83a**, Proof of Lemma 14], there is no loss of generality in assuming $\Lambda = \mathbb{Z}$ and that the finite Λ-grading is actually a finite $\mathbb{Z}$-grading.

PROPOSITION 4.31. *Let L be a Lie algebra with a nontrivial finite Λ-grading, where Λ is a torsion free abelian group. Then L contains a nonzero Jordan element.*[2]

PROOF. Take a basis $\{\lambda_1, \ldots, \lambda_r\}$ of G such that for some $\alpha \in \Lambda^*$, $\alpha = n_{\alpha_1}\lambda_1 + \ldots + n_{\alpha_r}\lambda_r$ with $n_{\alpha_1} \neq 0$, and let $\pi : G \to \mathbb{Z}$ be the homomorphism defined by putting $\pi(\lambda_1) = 1$ and $\pi(\lambda_i) = 0$ for $1 < i \leq r$. We may also assume that $|\pi(\beta)| \leq |\pi(\alpha)|$ ($\beta \in \Lambda^*$). Then any $x \in L_\alpha$ is a Jordan element: for any $\beta \in \Lambda^*$, $\mathrm{ad}_x^3 L_\beta \subset L_{3\alpha+\beta} = 0$ since $|\pi(3\alpha + \beta)| > |\pi(\alpha)|$. □

COROLLARY 4.32. *Let L be a Lie algebra over an algebraically closed field $\mathbb{F}$ of characteristic 0. If L has a nonzero ad-algebraic element, then L contains a nonzero Jordan element.*

[2]The existence of nonzero Jordan elements in Lie algebras with a nontrivial finite Λ-grading is mentioned in [**Zel83a**, p. 548].

PROOF. Let a be a nonzero element of L such that ad_a is algebraic. If a is not ad-nilpotent, then we have by Example 1.4 that ad_a produces a nontrivial finite $(\mathbb{F},+)$-grading on L given by

$$L_\lambda = \{x \in L : (\mathrm{ad}_a - \lambda 1_L)^m x = 0 \text{ for some } m \geq 1\},$$

with $L_\lambda = 0$ if $\lambda \in \mathbb{F}$ is not an eigenvalue of ad_a. As $\mathrm{char}(\mathbb{F}) = 0$, the additive group of $\mathbb{F}$ is torsion free, so Proposition 4.31 applies to prove that L contains a nonzero Jordan element. If a is ad-nilpotent, then Kostrikin's descent lemma yields a nonzero Jordan element. □

4.6. Jordan elements in the Lie algebra of a ring

Let R be a semiprime ring. Following Sections 1.3 and 1.5, we denote by $\mathcal{C}(R)$ the extended centroid of R and by $\tilde{R} = \mathcal{C}(R)R$ its central closure. Sometimes we will need to work in the *unital central closure* $\hat{\tilde{R}} = \tilde{R} + \mathcal{C}(R)$, a subring of the symmetric Martindale ring of quotients $Q_s(R)$ of R. In this section we describe the Jordan elements of the Lie algebra R^- and prove a necessary and sufficient condition for the existence of minimal abelian inner ideals in the Lie algebra $\overline{R}' = [R,R]/([R,R] \cap Z(R))$.

THEOREM 4.33. [**BMM96**, Theorem 2.3.3] *Let R be a semiprime ring and let $a_1, a_2, \ldots, a_n \in R$. If $a_1 \notin \sum_{i=2}^n \mathcal{C}(R)a_i$ in $\tilde{R}$, then there exist $r_j, s_j \in R$, $j = 1, 2, \ldots, m$, such that $\sum_{j=1}^m r_j a_1 s_j \neq 0$ and $\sum_{j=1}^m r_j a_k s_j = 0$, $k = 2, \ldots, n$.*

PROPOSITION 4.34. *A semiprime 6-torsion free ring R contains a nonzero nilpotent element if and only if the Lie algebra $\overline{R} = R/Z(R)$ (equivalently, $\overline{R}'$) contains a nonzero Jordan element.*

PROOF. Let $0 \neq a \in R$ be a nilpotent element, we may assume that $a^2 = 0$. Then, as seen in Example 4.2, a is a Jordan element of R^-. Moreover, since R is semiprime, a is not a central element, so $\bar{a}$ is a nonzero Jordan element of $\overline{R}$. Suppose conversely that $a \in R$ is such that $\bar{a}$ is a nonzero Jordan element of $\overline{R}$. Since R has no 3-torsion, $(\mathrm{ad}_a^2 R)^2 = 0$ by Proposition 3.35(1), with $\mathrm{ad}_a^2 R \neq 0$ since $\overline{R}$ is nondegenerate by Proposition 3.35(2). Note finally that $\overline{R}' \cong [\overline{R}, \overline{R}]$ is an essential ideal of $\overline{R}$ and hence, by Proposition 4.14, $\overline{R}$ contains nonzero Jordan elements if and only if so does $\overline{R}'$. □

We have just seen that if $\bar{a}$ is a Jordan element of the Lie algebra $\overline{R}$, then any element of the principal inner ideal $\mathrm{ad}_a^2 R$ of R^- has square 0 in R. It would be interesting to know whether a itself has square 0 modulo $Z(R)$. The answer to this question is affirmative if R coincides with its unital central closure, or if R is a simple ring.

THEOREM 4.35. *Let R be a semiprime 6-torsion free ring and let a be a Jordan element of the Lie algebra R^-. Then there exists $\lambda \in Z(\hat{\tilde{R}})$ such that $(a-\lambda)^2 = 0$. If R is a centrally closed prime ring, then $\lambda \in Z(R)$.*

PROOF. By Proposition 3.35(1), for $x, y \in R$ we have $(\mathrm{ad}_a^2 x)(\mathrm{ad}_a^2 y) = 0$, i.e.

$$\begin{aligned} 0 = (\mathrm{ad}_a^2 x)(\mathrm{ad}_a^2 y) &= a^2x(a^2y + ya^2 - 2aya) \\ &\quad + ax(-2a^3y - 2aya^2 + 4a^2ya) \\ &\quad + x(a^4y - 2a^3ya + a^2ya^2). \end{aligned} \tag{4.19}$$

We claim that $a^2 = \mu a + \tau$ for some $\mu, \tau \in \mathcal{C}(R)$. If this were not the case, by Theorem 4.33 applied to the elements $a^2, a, 1$ in $\hat{R}$, there would exist $r_j, s_j \in R$, $j = 1, \ldots, m$, such that $\sum_{j=1}^m r_j a^2 s_j \neq 0$ and $\sum_{j=1}^m r_j a^k s_j = 0$ for $k = 0, 1$. In (4.19), for each $j = 1, \ldots m$, replace x by $s_j x$ and multiply on the left by r_j. We obtain

$$0 = \sum_{j=1}^{m} r_j(a^2 s_j x(a^2y + ya^2 - 2aya) + as_j x(-2a^3y - 2aya^2 + 4a^2ya)$$
$$+ s_j x(a^4y - 2a^3ya + a^2ya^2)) = (\sum_{j=1}^{m} r_j a^2 s_j)x(a^2y + ya^2 - 2aya)$$

which tells us that for each $y \in R$

$$a^2y + ya^2 - 2aya \in \operatorname{Ann}(I), \tag{4.20}$$

where I denotes the ideal generated by $\sum_{j=1}^m r_j a^2 s_j$. For each $j = 1, \ldots m$, replace y by s_j in formula (4.20) and multiply on the left by r_j. Then, by semiprimeness of R, we obtain

$$\sum_{j=1}^{m} r_j a^2 s_j = \sum_{j=1}^{m} r_j(a^2 s_j + s_j a^2 - 2as_j a) \in I \cap \operatorname{Ann}(I) = 0,$$

which is a contradiction. Therefore there exist $\mu, \tau \in \mathcal{C}(R)$ such that

$$a^2 = \mu a + \tau \text{ and } a^3 = (\mu^2 + \tau)a + \mu\tau. \tag{4.21}$$

Substituting these expressions of a^2 and a^3 in the equation

$$0 = \operatorname{ad}_a^3 x = a^3x - 3a^2xa + 3axa^2 - xa^3,$$

we obtain

$$0 = (\mu^2 + 4\tau)ax - (\mu^2 + 4\tau)xa = [(\mu^2 + 4\tau)a, x]$$

for every $x \in R$, which proves that $(\mu^2 + 4\tau)a$ is a central element of $\tilde{R}$.

Set $\alpha := \mu^2 + 4\tau$. Since $\mathcal{C}(R)$ is von Neumann regular (Theorem 1.17), $\alpha = \alpha\beta\alpha$ for some $\beta \in \mathcal{C}(R)$, so $e_\alpha := \alpha\beta$ is an idempotent of $\mathcal{C}(R)$ and

$$e_\alpha a = \beta\alpha a = \beta(\mu^2 + 4\tau)a \in Z(\tilde{R}). \tag{4.22}$$

Denote by f_α the idempotent $1 - e_\alpha$ of $\mathcal{C}(R)$ and set $\lambda := e_\alpha a + \frac{1}{2} f_\alpha \mu$. Then $\lambda \in Z(\hat{\tilde{R}})$ and satisfies

$$(a - \lambda)^2 = (a - e_\alpha a - \frac{1}{2} f_\alpha \mu)^2 = (f_\alpha a - \frac{1}{2} f_\alpha \mu)^2 = f_\alpha(a^2 - \mu a + \frac{1}{4}\mu^2)$$
$$= f_\alpha(\tau + \frac{1}{4}\mu^2) = \frac{1}{4} f_\alpha \alpha = 0.$$

Suppose now that R is a centrally closed prime ring. Then we have by Lemma 1.22 that either $Z(R) = 0$ or R is unital and therefore $Z(R) = \mathcal{C}(R)$. But $(a - \lambda)^2 = 0$ implies that $\lambda^2 \in Z(R)$, so in both cases $\lambda \in Z(R)$. □

When referred to a simple ring, Proposition 4.34 adopts the following refined form [**Ben76**, Theorem 3.2].

LEMMA 4.36. *Let R be a simple ring and let $a \in R$ be ad-nilpotent of index n. Then n is odd and a is algebraic of degree $r = (n+1)/2$. If the characteristic of R is 0 or $p > r$. Then $a = x + z$ where x is nilpotent of index r and $z \in Z(R)$.*

We are mainly interested in the following particular case of the above lemma.

LEMMA 4.37. *Let R be as in Lemma 4.36 and let $a \in R$ be such that* $\operatorname{ad}_a^4 R = 0$. *Then $a = x + z$ where $x^2 = 0$ and $z \in Z(R)$.*

REMARK 4.38. Lemma 4.36 was extended by Martindale and Miers [**MM89**, Corollary 1] to prime rings, and by Grzeszczuk [**Grz92**] to semiprime rings. The proof of Theorem 4.35 given above is taken from [**BGGL16**, Theorem 3.2].

4.7. Jordan elements in Lie algebras of skew-symmetric elements

Let R be a prime ring with involution $*$, extended centroid $\mathcal{C}(R)$, and central closure $\tilde{R}$. We describe in this section the Jordan elements of the Lie algebra $K = \operatorname{Skew}(R, *)$.

PROPOSITION 4.39. [**BGGL14**, Proposition 4.1(a)] *Let R be a 2-torsion free ring with involution $*$. If a is a Jordan element of K, then a^2 is a Jordan element of R^-.*

PROOF. Note first that

$$\operatorname{ad}_{a^2}^3 = (l_{a^2} - r_{a^2})^3 = (l_a^2 - r_a^2)^3 = (l_a + r_a)^3(l_a - r_a)^3 = (l_a + r_a)^3 \operatorname{ad}_a^3.$$

Let $x \in R$. Then $2x = x_k + x_s$, where

$$x_k = x - x^* \in \operatorname{Skew}(R, *) \quad \text{and } x_s = x + x^* \in \operatorname{Sym}(R, *).$$

Since $\operatorname{ad}_a^3(x_s)a = \operatorname{ad}_a^3(x_s a)$, $a \operatorname{ad}_a^3(x_s) = \operatorname{ad}_a^3(a x_s)$ and $x_s \circ a := x_s a + a x_s \in K$, we have

$$\begin{aligned}
2 \operatorname{ad}_{a^2}^3(x) = \operatorname{ad}_{a^2}^3(2x) &= \sum_{i=0}^{3} \binom{3}{i} a^i \operatorname{ad}_a^3(x_k + x_s) a^{3-i} = \sum_{i=0}^{3} \binom{3}{i} a^i \operatorname{ad}_a^3(x_k) a^{3-i} + \\
&+ \sum_{i=0}^{3} \binom{3}{i} a^i \operatorname{ad}_a^3(x_s) a^{3-i} = \sum_{i=0}^{3} \binom{3}{i} a^i \operatorname{ad}_a^3(x_s) a^{3-i} \\
&= \operatorname{ad}_a^3(x_s) a^3 + 3a \operatorname{ad}_a^3(x_s) a^2 + 3a^2 \operatorname{ad}_a^3(x_s) a + a^3 \operatorname{ad}_a^3(x_s) \\
&= \operatorname{ad}_a^3(x_s a + a x_s) a^2 + 2a \operatorname{ad}_a^3(x_s a + a x_s) a + a^2 \operatorname{ad}_a^3(x_s a + a x_s) = 0,
\end{aligned}$$

so $\operatorname{ad}_{a^2}^3(R) = 0$, since R is 2-torsion free. □

LEMMA 4.40. *Let R be a semiprime 2-torsion free ring, $x \in R$ and $\lambda \in \mathcal{C}(R)$. If $x - \lambda$ is nilpotent, then λ is uniquely determined. If R has an involution $*$ and x is symmetric (resp. skew-symmetric), then $\lambda^* = \lambda$ (resp. $\lambda^* = -\lambda$).*

PROOF. Let $\mu \in \mathcal{C}(R)$ be such that $x - \mu$ is nilpotent. As $[x - \lambda, x - \mu] = 0$, we have that $\lambda - \mu = (x - \mu) - (x - \lambda)$ is a nilpotent element of $\mathcal{C}(R)$. Hence $\lambda = \mu$, since $\mathcal{C}(R)$ has no nonzero nilpotent elements because R is semiprime.

Suppose now that R has an involution $*$ and that $x = x^*$. Then $x - \lambda^* = (x - \lambda)^*$ is nilpotent. Hence $\lambda^* = \lambda$ by the uniqueness we have just proved. For $x = -x^*$ the proof is similar. □

LEMMA 4.41. [**BFLGL16**, Propostion 6.1] *Let R be a prime ring of characteristic not 2 with involution $*$ of the first kind, and let $\langle K\rangle$ be the subring of R generated by K.*

(i) *If K is not abelian, then $\langle K\rangle$ is a prime ring and $\mathcal{C}(\langle K\rangle) = \mathcal{C}(R)$.*
(ii) *If K is abelian, then $K = 0$ or $\overline{\mathcal{C}(R)} \otimes_{\mathcal{C}(R)} \tilde{R} = \mathrm{M}_2(\overline{\mathcal{C}(R)})$, where $\overline{\mathcal{C}(R)}$ denotes the algebraic closure of the field $\mathcal{C}(R)$.*

PROOF. (i) Taking $U = K$ in [**BMM96**, Theorem 9.1.13(d)], we have that $[K,K] \neq 0$ implies $[K,K]^2 \neq 0$. Let I be the ideal of R generated by $[K,K]^2$. By [**BMM96**, Lemma 9.1.4], $0 \neq I \subset \langle K\rangle$ and hence it follows from [**Lam99**, Theorem 14.14 and subsequent Remark] that $\langle K\rangle$ is prime with $\mathcal{C}(\langle K\rangle) = \mathcal{C}(R)$.

(ii) Take $U = K$ in [**BMM96**, Theorem 9.1.13(a)]. □

LEMMA 4.42. [**BFLGL16**, Propostion 6.2] *Let R be a prime ring of characteristic 0 or $p > 5$ with involution $*$ of the first kind such that $[K,K] \neq 0$. If a is a Jordan element of K, then $a^3 = 0$.*

PROOF. By 4.41(i), $\langle K\rangle$ is a prime ring with $\mathcal{C}(\langle K\rangle) = \mathcal{C}(R)$. Since $\langle K\rangle$ is spanned by the elements of K and their squares [**BMM96**, Lemma 9.1.5], we get (using the Leibniz rule) that $\mathrm{ad}_a^3 K = 0$ implies $\mathrm{ad}_a^5\langle K\rangle = 0$. Now it follows from [**MM83**, Corollary 1] that $(a-\alpha)^3 = 0$ for some $\alpha \in \mathcal{C}(R)$ (the formula making sense in the unital hull of $\langle K\rangle$). Since the involution $*$ is of the first kind, $\alpha = 0$ by Lemma 4.40. Thus $a^3 = 0$. □

THEOREM 4.43. [**BGGL14**, Corollary 4.3] *Let R be a centrally closed prime ring of characteristic not 2 or 3 with involution $*$, and let a be a nonzero Jordan element of the Lie algebra K. Then one of the following possibilities holds:*

(i) *There exists $z \in \mathrm{Skew}(Z(R), *)$ such that $(a-z)^2 = 0$.*
(ii) *$a \in Z(K)$ and $a \notin Z(R)$.*
(iii) *$a^3 = 0$ and $a^2 \neq 0$.*

PROOF. Suppose first that $*$ is of the second kind and let ξ be a nonzero skew-symmetric element in $\Gamma(R)$, which is a field since R is a centrally closed prime ring. Then $R = K \oplus \xi K$ and hence a is a Jordan element of the Lie algebra R^-. Now it follows from Theorem 4.35 that there exist $z \in Z(R)$ such that $(a-z)^2 = 0$, with $z^* = -z$ by Lemma 4.40, so we are in case (i).

Suppose now that $*$ is of the first kind. Then it follows from Lemma 4.42 that either K is abelian or $a^3 = 0$. If the first, $a \in Z(K)$, with $a \notin Z(R)$ since $*$ is of the first kind and a is nonzero by hypothesis, so we are in case (ii); if the second, we may suppose that $a^2 \neq 0$, since otherwise we would be in case (ii). □

REMARKS 4.44. (1) Theorem 4.43 appears in [**BGGL16**, Corollary 4.3] as a corollary of a result for centrally closed semiprime 6-torsion free rings with involution. The proof given here is an adaption of that of [**BFLGL16**, Theorem 6.3].

(2) Jordan elements of type (ii) occur when R is the ring $\mathrm{M}_2(\mathbb{F})$ with transpose involution. As will be seen in Chapter 8, case (iii) takes place when the inner ideal generated by a is Clifford (see Exercise 2.99(3)).

Jordan elements of K and $[K,K]$ when R is simple. Recall that by Lemma 1.13, if R is a simple ring, then either R is unital (and hence $Z(R) \cong \Gamma(R)$), or $Z(R) = 0$.

LEMMA 4.45. [**Ben76**, Lemma 4.23] *Let R be a simple ring of characteristic not 2 or 3 with involution of the first kind. Assume that either $Z(R) = 0$ or $\dim_{Z(R)} R > 16$, and let $a \in K$ be such that $\mathrm{ad}_a^3[K,K] = 0$. Then $a^3 = 0$ and $a^2Ka^2 = 0$.*

PROOF. From $\mathrm{ad}_a^3[K,K] = 0$ we get $\mathrm{ad}_a^4 K = 0$. By [**Her69**, Theorem 2.3], $R = \langle K \rangle$ and hence, as in the proof of Lemma 4.42, we get $\mathrm{ad}_a^7 R = 0$. Now it follows from [**Ben76**, Theorem 3.2] that $(a-z)^4 = 0$ for some $z \in Z(R)$. Since a is skew-symmetric and the involution is of the first kind, we have by Lemma 4.40 that $z = 0$.

Now let $k \in [K,K]$. Then $0 = \mathrm{ad}_a^3 k = a^3k - 3a^2kz + 3aka^2 - ka^3$ implies that $a^3[K,K] \subset Ra$. Therefore $a^3[K,K]a^3 \subset Ra^4 = 0$. But $a^3[K,K]a^3 = \mathrm{ad}_{a^3}^2[K,K]$ since $(a^3)^2 = 0$, which implies by [**Ben76**, Corollary 2.12] that $a^3 = 0$. Hence, for any $k \in K$, we have $0 = \mathrm{ad}_a^4 k = 6a^2Ka^2$, which implies $a^2Ka^2 = 0$. □

COROLLARY 4.46. *Let R be as in the previous lemma and let a be a Jordan element of the Lie algebra K. Then $a^3 = 0$ and $a^2Ka^2 = 0$.*

PROOF. $\mathrm{ad}_a^3[K,K] \subset \mathrm{ad}_a^3 K = 0$. Now Lemma 4.45 applies. □

THEOREM 4.47. *Let R be a simple ring of characteristic 0 or $p > 3$ with involution $*$. Assume that either $Z(R) = 0$ or $\dim_{Z(R)} R > 16$. If a is a Jordan element of K, then one of the following holds:*

(i) *$(a-z)^2 = 0$ for some $z \in \mathrm{Skew}(Z(R), *)$,*
(ii) *$a^3 = 0$, $a^2Ka^2 = 0$, and $a^2 \neq 0$.*

The same is true if a is a Jordan element of $[K,K]$.

PROOF. Suppose first that the involution is of the second kind. Then it follows as in the proof of Theorem 4.43 that a is as in (i). If the involution is of the first kind, we have by Lemma 4.45 that $a^3 = 0$ and $a^2Ka^2 = 0$, since a is a Jordan element of K. If $a^2 = 0$ we are in (i); otherwise we are in (ii).

Now let a be a Jordan element of $[K,K]$. Suppose first that the involution is of the second kind and let ξ be a nonzero skew-symmetric element of $\Gamma(R)$. Then $R = K \oplus \xi K$. From $\mathrm{ad}_a^3[K,K] = 0$ it follows $\mathrm{ad}_a^4 K = 0$ and hence $\mathrm{ad}_a^4 R = 0$. Now it follows from [**Ben76**, Theorem 3.2] that $\mathrm{ad}_a^3 R = 0$ and there exists $z \in Z(R)$ such that $(a-z)^2 = 0$, with $z^* = -z$ by Lemma 4.40. If the involution is of the first kind, then we have by Lemma 4.45 that $a^3 = 0$ and $a^2Ka^2 = 0$. This completes the proof of the theorem. □

4.8. Exercises

EXERCISE 4.48. Let $R = \mathrm{M}_n(\mathbb{F})$, where $\mathbb{F}$ is a field. Show that every nilpotent element $a \in R$ can be expressed as a sum of Jordan elements of the Lie algebra R'.

EXERCISE 4.49. Following Section 2.2 for notation, let X be a vector space over a field $\mathbb{F}$ of characteristic not 2, $\dim_{\mathbb{F}} X > 2$, which is endowed with a nondegenerate symmetric bilinear form $\langle\,,\rangle$, and let x, z be nonzero orthogonal vectors in X such that x is isotropic. Show that $a = x^*z - z^*x$ is a Jordan element of the Lie algebra $\mathfrak{o}(X, \langle\,,\rangle)$. Note that $a^2 = 0$ if and only if z is isotropic.

EXERCISE 4.50. Let L be a 6-torsion free Lie algebra. Show that L is strongly prime if and only if for any $x \in L$ and any $I \lhd L$, $\mathrm{ad}_x^2 I = 0 \Rightarrow x = 0$ or $I = 0$.

EXERCISE 4.51. Let L be a simple Lie algebra over a field of characteristic 0 which is generated by ad-nilpotent elements. Show that L is spanned by Jordan elements.

EXERCISE 4.52. (Brešar) Let A be a 3-torsion free nonassociative Φ-algebra and let d be a derivation of A such that $d^3 = 0$. Show that the Φ-submodule $D = d^2(A)$ satisfies $D^2 = 0$ and $(DA)D \subset D$.

EXERCISE 4.53. Let $x, a \in L$ and let m be a natural number. Show that if m is even, then $[X, \mathbf{ad}_A^m X] = \sum_{i+j>0} \beta_{ij} A^i X A^{m-i-j} X A^j$. So the argument used in Lemma 4.22 for $n = p - 1$ doesn't work in the general case.

EXERCISE 4.54. Carry out all computations in Lemma 4.25 for the cases: (i) $n = 5,\ m = 3$, and (ii) $n = 5,\ m = 2$.

EXERCISE 4.55. Carry out all computations in Lemma 4.26 for $n = 5$.

EXERCISE 4.56. Give and *easier* proof of Lemma 4.24 assuming that L is 30-torsion free.

EXERCISE 4.57. Let L be a Lie algebra over a ring of scalar Φ, $30 \in \Phi^*$, and let $a \in L$ be such that $\mathrm{ad}_a^4 = 0$. Give a direct proof that $\mathrm{ad}_a^3 L$ is an abelian inner ideal.

EXERCISE 4.58. Show that any nonzero finite-dimensional Lie algebra over an algebraically closed field of characteristic 0 contains nonzero Jordan elements.

EXERCISE 4.59. Suppose that R is a 2-torsion free ring with involution $*$, $\lambda \in \mathrm{Skew}(Z(R), *)$, and let a be a Jordan element of K. Show that λa is a Jordan element of the Lie algebra R^-.

CHAPTER 5

Von Neumann Regular Elements

Let L be a Lie algebra over a ring of scalars Φ. A Jordan element $a \in L$ is called von Neumann regular if $a \in \mathrm{ad}_a^2 L$. There are good reasons for calling these elements von Neumann regular. The main one is that they play a role in Lie algebras similar to that played by von Neumann regular elements in associative algebras and Jordan systems. In fact, if a is an element of a ring R such that $a^2 = 0$ (therefore a Jordan element of the Lie algebra R^-), then all the notions of von Neumann regularity (associative, Jordan and Lie) agree for the element a. The material of this chapter is organized as follows. In Section 5.1 we give some examples of von Neumann regular elements in Lie algebras and prove that any abelian inner ideal B of an ideal of a 6-torsion free nondegenerate Lie algebra such that every element in B is von Neumann regular, remains as an inner ideal in the whole algebra. Suppose that 6 is invertible in Φ. Adapting Seligman's proof of the so-called Jacobson–Morozov Lemma to this setting, we prove in Section 5.2 that every nonzero von Neumann regular element e of L can be extended to a $\mathfrak{sl}_2$-triple (e, f, h) such that ad_h is diagonalizable with eigenvalues $0, \pm 1, \pm 2$. Furthermore, if 5 is also invertible in Φ, then the decomposition $L = L_{-2} \oplus L_{-1} \oplus L_0 \oplus L_1 \oplus L_2$ is actually a $\mathbb{Z}$-grading and f is also von Neumann regular (in Chapter 11 we will regard this $\mathbb{Z}$-grading as the $+$-part of the Peirce decomposition defined by an idempotent of a suitable Jordan pair). This leads us to the notion of an idempotent in Lie algebras, which is studied in Section 5.3 for Lie algebras coming from associative ones. In Section 5.4 we define the socle of a Lie algebra as the sum of its minimal abelian inner ideals and prove that the socle of a nondegenerate Lie algebra is a direct sum of ideals each of which is a simple nondegenerate Lie algebra coinciding with its socle. Roughly speaking, we can say that the socle of a nondegenerate Lie algebra encloses its structural data and can be regarded as a natural extension of the notion of split semisimple Lie algebra. In Section 5.5 we associate a filtration to any Jordan element a of a Lie algebra L and prove that when L is nondegenerate the $\mathbb{Z}$-graded Lie algebra $\hat{L}$ defined by this filtration is isomorphic to L if and only if a is von Neumann regular.

5.1. Definition, examples, and first results

Throughout this section, L will denote a Lie algebra over a ring of scalars Φ.

DEFINITION 5.1. A Jordan element $a \in L$ is called *von Neumann regular* [1] if $a \in \mathrm{ad}_a^2 L$, equivalently, the abelian inner ideals $[a] = \mathrm{ad}_a^2 L$ and $(a) = \Phi a + [a]$ agree.

[1] Von Neumann regular elements for Lie algebras were considered in [**Ben76**] (without giving them this name) in the definition of *Lie algebra.

EXAMPLE 5.2. Let B be a minimal abelian inner ideal of a nondegenerate 3-torsion free Lie algebra. Then every nonzero element $b \in B$ is von Neumann regular (Lemma 4.19).

EXAMPLE 5.3. Let R be an associative algebra over a ring of scalars Φ in which 2 is invertible, and let $a \in R$ be such that $a^2 = 0$. Then a is von Neumann regular in R (in the associative sense) if and only if it is von Neumann regular in the Lie algebra R^-. Moreover, if R has an involution and a is skew-symmetric, then a is von Neumann regular in the Lie algebra $\mathrm{Skew}(R, *)$.

LEMMA 5.4. *Let I be an ideal of a Lie algebra L and let $a \in I$ be von Neumann regular in L. Then a is von Neumann regular in I.*

PROOF. Let $b \in L$ be such that $a = \mathrm{ad}_a^2 b$. We have by Lemma 4.4(3)

$$a = \mathrm{ad}_a^2 b = \mathrm{ad}_{\mathrm{ad}_a^2 b}^2 b = \mathrm{ad}_a^2(\mathrm{ad}_b^2 \mathrm{ad}_a^2 b) \in \mathrm{ad}_a^2 I.$$

□

LEMMA 5.5. *Let L be a 6-torsion free Lie algebra, let I be an ideal of L, and let B be an abelian inner ideal of I such that every element in B is von Neumann regular. If I is nondegenerate, then B is an inner ideal of L.*

PROOF. Given $b \in B$, let $x \in L$ be such that $b = \mathrm{ad}_b^2 x$. By Proposition 4.13, b is a Jordan element of L, and hence, by 4.4(3),

$$\mathrm{ad}_b^2 L = \mathrm{ad}_{\mathrm{ad}_b^2 x}^2 L = \mathrm{ad}_b^2 \mathrm{ad}_x^2 \mathrm{ad}_b^2 L \subset \mathrm{ad}_b^2 I \subset B,$$

which proves that B is an abelian inner ideal of L. □

5.2. Jacobson–Morozov type results

Adapting Seligman's proof [**Sel67**, V.8.2] of the so-called Jacobson–Morozov Lemma to the setting of Lie algebras over a ring of scalars Φ in which $2, 3$, and 5 are invertible, we show in this section that the existence of a nonzero von Neumann regular element gives rise to a finite $\mathbb{Z}$-grading.

LEMMA 5.6. *Let R be a ring with 1, and let $e, z, h \in R$ such that $[e, z] = h$ and $[h, e] = 2e$. Then $[e^n, z] = n(h - (n-1))e^{n-1}$ for each positive integer n.*

PROOF. We use induction on n. For $n = 1$ it is clear. Suppose the formula is true for n. Using the fact that fixed y the map $x \mapsto [x, y]$ is a derivation of the ring R and the induction hypothesis, we get

$$\begin{aligned}[e^{n+1}, z] &= [e, z]e^n + e[e^n, z] = he^n + ne(h - (n-1)1)e^{n-1} \\ &= he^n + nehe^{n-1} - n(n-1)e^n = he^n + n[e, h]e^{n-1} + nhe^n - n(n-1)e^n \\ &= (n+1)he^n - 2ne^n - n(n-1)e^n = (n+1)(h-n)e^n.\end{aligned}$$

□

DEFINITION 5.7. A triple (e, f, h) of elements of a Lie algebra L is said to be a $\mathfrak{sl}_2$-triple if they satisfy the relations of the standard generators of the Lie algebra $\mathfrak{sl}_2(\mathbb{F})$ of 2×2 matrices of trace zero over a field $\mathbb{F}$, i.e. $[e, f] = h$, $[h, e] = 2e$ and $[h, f] = -2f$. By a $\mathfrak{sl}_2$-pair we mean a pair (e, f) such that $(e, f, [e, f])$ is a $\mathfrak{sl}_2$-triple.

LEMMA 5.8. *Let L be a Lie Φ-algebra with $6 \in \Phi^*$, and let $e \in L$ be a nonzero von Neumann regular element. For each $h = [e, z] \in [e, L]$ such that $[h, e] = 2e$, there exists an element $f \in L$ such that (e, f, h) is a $\mathfrak{sl}_2$-triple.*

PROOF. Put $E = \mathrm{ad}_e$, $H = \mathrm{ad}_h$ and $Z = \mathrm{ad}_z$. We have

$$H = [E, Z],\ [H, E] = 2E, \text{ and } E^3 = 0. \tag{5.1}$$

It follows from (5.1) and Lemma 5.6

$$[E^2, Z] = 2(H-1)E \text{ and } 0 = [E^3, Z] = 3(H-2)E^2, \tag{5.2}$$

where we have abbreviated $H - j = H - j1_L$.

Let $x \in \ker(E)$. Using (5.1) and (5.2), we get

$$\begin{aligned} EHx &= [E, H]x + HEx = -2Ex = 0, \\ E^2Zx &= E(EZ)x = EHx + EZEx = 0 \end{aligned} \tag{5.3}$$

and

$$\begin{aligned} 6(H-2)(H-1)Hx = 6(H-2)(H-1)EZx &= 3(H-2)[E^2, Z]Zx \\ &= 3(H-2)E^2Z^2x - 3(H-2)ZE^2Zx = 0, \end{aligned}$$

which implies, since $6 \in \Phi^*$,

$$(H-2)(H-1)H(x) = 0 \text{ for all } x \in \ker(E). \tag{5.4}$$

As the ideals $(\lambda+2)$, (λ), $(\lambda-1)$, and $(\lambda-2)$ of $\Phi[\lambda]$ are pairwise comaximal (because 2 and 3 are invertible in Φ), there exist two polynomials $p(\lambda), q(\lambda) \in \Phi[\lambda]$ such that

$$p(\lambda)(\lambda+2) + q(\lambda)\lambda(\lambda-1)(\lambda-2) = 1.$$

Hence, by (5.3) and (5.4), the restriction of $H + 2$ to $\ker(E)$ is a well-defined invertible linear map. Since $[e, [h, z]] = [[e, h], z] = -2[e, z]$, $[h, z] + 2z \in \ker(E)$, there exists $v \in \ker(E)$ such that $[h, v] + 2v = [h, z] + 2z$. Putting $f = z - v$, we get $[e, f] = h$ and $[h, f] = -2f$, which proves that $\{e, h, f\}$ is a $\mathfrak{sl}_2$-triple. □

LEMMA 5.9. *Let L be a Lie Φ-algebra with $6 \in \Phi^*$, and let (e, f, h) be a $\mathfrak{sl}_2$-triple, where e is a Jordan element. Then ad_h is diagonalizable with eigenvalues $0, \pm 1, \pm 2$, i.e.*

$$L = L^h_{-2} \oplus L^h_{-1} \oplus L^h_0 \oplus L^h_1 \oplus L^h_2,$$

where $L^h_i = \{x \in L : [h, x] = ix\}$ for $i = -2, -1, 0, 1, 2$.

PROOF. We follow the argument of the proof of [**Jac58**, Lemma 1]. Note first that

$$(H+i)F = [H, F] + F(H+i) = -2F + FH + iF = F(H+i-2) \tag{5.5}$$

for any $i \geq 0$. Replacing Z by F in (5.2) of the previous lemma gives

$$[E^2, F] = 2(H-1)E \text{ and } [E^3, F] = 3(H-2)E^2. \tag{5.6}$$

Using (5.6), we get

$$[[E^3, F], F] = 3[(H-2)E^2, F] = -6FE^2 + 6(H-2)(H-1)E, \tag{5.7}$$

which proves, since $E^3 = 0$ and $6 \in \Phi^*$,

$$(H-2)(H-1)E = FE^2. \tag{5.8}$$

Similarly, using (5.5), (5.6) and (5.7), we obtain

$$\begin{aligned}[[[E^3,F],F],F] &= -6[FE^2,F]+6[(H-2)(H-1)E,F] = -6F[E^2,F]\\ &\quad -12F(H-1)E-12(H-2)FE+6(H-2)(H-1)H\\ &= -24F(H-1)E-12F(H-4)E+6(H-2)(H-1)H\\ &= -24FHE+24FE-12FHE+48FE+6(H-2)(H-1)H\\ &= -36F(H-2)E+6(H-2)(H-1)H.\end{aligned}$$

Since $E^3=0$ and $6\in\Phi^*$, this proves

$$(H-2)(H-1)H = 6F(H-2)E. \tag{5.9}$$

Multiplying this last equation by E on the right and using (5.6), we get

$$(H-2)(H-1)HE = 6F(H-2)E^2 = 2F[E^3,F]=0. \tag{5.10}$$

Multiply now (5.9) by $(H+2)(H+1)$ on the left. Using repeatedly (5.5), we get by (5.10)

$$\begin{aligned}(H+2)(H+1)H(H-1)(H-2) &= 6(H+2)(H+1)F(H-2)E\\ &= 6(H+2)F(H-1)(H-2)E\\ &= 6FH(H-1)(H-2)E=0.\end{aligned}$$

Since the ideals $(\lambda-j)$, $j\in\{0,\pm1,\pm2\}$, of $\Phi[\lambda]$ are pairwise comaximal, we have that $\mathrm{ad}_h=H$ is diagonalizable with eigenvalues $0,\pm1,\pm2$, so yielding the decomposition

$$L=L^h_{-2}\oplus L^h_{-1}\oplus L^h_0\oplus L^h_1\oplus L^h_2,$$

with $L^h_i=\{x\in L:[h,x]=ix\}$ for $i=-2,-1,0,1,2$. □

REMARK 5.10. In general, the decomposition induced by ad_h is not a $\mathbb{Z}$-grading on the Lie algebra L. Consider the Lie algebra $W(5)$ with the standard basis $\{x_0,x_1,x_2,x_3,x_4\}$ (Example 2.11). We have that x_2 is von Neumann regular and $(-x_2,x_0,2x_1)$ is a $\mathfrak{sl}_2$-triple. However, x_0 is not even a Jordan element. Since $\mathrm{ad}^3_{x_0}x_3=6x_0$, the $\mathbb{Z}_5$-grading induced by ad_{2x_1} is not a $\mathbb{Z}$-grading. The situation is much better when $5\in\Phi^*$.

THEOREM 5.11. *Let L be a Lie Φ-algebra with $30\in\Phi^*$, and let (e,f,h) be a $\mathfrak{sl}_2$-triple, where e is a Jordan element. We have:*

(i) *The decomposition $L=L^h_{-2}\oplus L^h_{-1}\oplus L^h_0\oplus L^h_1\oplus L^h_2$ of the previous lemma is actually a $\mathbb{Z}$-grading on L.*
(ii) *f is a Jordan element and therefore von Neumann regular.*
(iii) *$L^h_2=\mathrm{ad}^2_e L$ and $L^h_{-2}=\mathrm{ad}^2_f L$.*
(iv) *$\mathrm{ad}^2_e: L^h_{-2}\to L^h_2$ is an isomorphism of Φ-modules whose inverse is ad^2_f.*

PROOF. (i) Since ad_h is a derivation of L, $[L^h_i,L^h_j]\subset L^h_{i+j}$. Hence the decomposition $L=L^h_{-2}\oplus L^h_{-1}\oplus L^h_0\oplus L^h_1\oplus L^h_2$ will be a $\mathbb{Z}$-grading if the Φ-submodules $L^h_3,L^h_4,L^h_{-3},L^h_{-4}$ vanish. Let $x\in L^h_{-3}$. Writing $x=x_{-2}+x_{-1}+x_0+x_1+x_2$ (according to the 5-decomposition) and applying the adjoint operator ad_h gives

$$-3x=-2x_{-2}-x_{-1}+x_1+2x_2=-3x_{-2}-3x_{-1}-3x_0-3x_1-3x_2$$

which implies

$$x_{-2}=2x_{-1}=3x_0=4x_1=5x_2=0,$$

so $x = 0$, since $2, 3, 5$ are invertible in Φ. (Note that this is the unique point of the proof where the invertibility of 5 is required. Furthermore, as we have seen in Remark 5.10, a peculiarity of characteristic 5 is that the equality $\mathrm{ad}_f^3 = 0$ need not hold.) The other cases are similar.

(ii) By the grading properties, both L_2^h and L_{-2}^h are abelian inner ideals of L. Hence $f \in L_{-2}^h$ is a Jordan element. In fact, von Neumann regular since (e, f, h) is a $\mathfrak{sl}_2$-triple.

(iii) As L_2^h is an inner ideal and $e \in L_2^h$, we have the inclusion $\mathrm{ad}_e^2 L \subset L_2^h$. Conversely, set $E = \mathrm{ad}_e$, $F = \mathrm{ad}_f$. For any $x \in L_2^h$, we have

$$\begin{aligned} E^2F^2x &= E[e,[f,[f,x]]] = E[f,[e,[f,x]]] \\ &= [h,[e,[f,x]]] = 2[e,[f,x]] = 2[h,x] = 4x. \end{aligned} \tag{5.11}$$

This proves that $x \in \mathrm{ad}_e^2(L)$ and hence $L_2^h = \mathrm{ad}_e^2 L$. The equality $L_{-2}^h = \mathrm{ad}_f^2 L$ follows similarly.

(iv) By (5.11), $\frac{1}{2}E^2\frac{1}{2}F^2 = 1_{L_2^h}$ and $\frac{1}{2}F^2\frac{1}{2}E^2 = 1_{L_{-2}^h}$. Hence, by (iii), the linear maps $\frac{1}{2}E^2 : L_{-2}^h \to L_2^h$ and $\frac{1}{2}F^2 : L_2^h \to L_{-2}^h$ are inverse each other. □

REMARK 5.12. In [**Ben77**, Lemma 2.1], G. Benkart provides the polynomial equality:

$$1 = \frac{1}{24}p_2(\lambda) - \frac{1}{6}p_1(\lambda) + \frac{1}{4}p_0(\lambda) - \frac{1}{6}p_{-1}(\lambda) + \frac{1}{24}p_{-2}(\lambda)$$

where $p_i(\lambda) = p(\lambda)/(\lambda+i)$ with $p(\lambda) = (\lambda-2)(\lambda-1)\lambda(\lambda+1)(\lambda+2)$, which produces directly the $\mathbb{Z}$-grading given in Theorem 5.11.

As a direct consequence of Theorem 5.11, we obtain a further criterion of nondegeneracy for simple Lie algebras of characteristic 0 or greater than 5.

PROPOSITION 5.13. *Let L be a simple Lie algebra over a field $\mathbb{F}$ of characteristic 0 or greater than 5. If L contains a nonzero von Neumann regular element, then L is nondegenerate.*

PROOF. Given a nonzero von Neumann regular element e in L, extend it to a $\mathfrak{sl}_2$-triple (e, f, h) (Lemma 5.8) with associated $\mathbb{Z}$-grading given by Theorem 5.11. Then L is nondegenerate by Proposition 3.28. □

COROLLARY 5.14. *For $p > 5$ the Witt algebra $W(p)$ does not contain nonzero von Neumann regular elements.*

PROOF. If $W(p)$ contained a nonzero von Neumann regular element, then it would be nondegenerate by Proposition 5.13, which is a contradiction by Example 3.2. □

REMARK 5.15. The existence of nonzero von Neumann regular elements in any semisimple finite-dimensional Lie algebra over an algebraically closed field of characteristic 0, and therefore of a finite $\mathbb{Z}$-grading (Theorem 5.11), can be here derived as follows: Such an algebra contains a nonzero Jordan element (Corollary 4.32) and hence a minimal abelian inner ideal (Proposition 4.6), whose elements are von Neumann regular (Lemma 4.19).

5.3. Idempotents in Lie algebras

By an *idempotent* of a Lie algebra we mean a $\mathfrak{sl}_2$-pair (e, f) (Definition 5.7) consisting of Jordan elements. As will be seen in Chapter 11, idempotents in Lie algebras have a behavior similar to that of the idempotents of Jordan pairs. In the light of this analogy, the $\mathbb{Z}$-grading defined in Theorem 5.11 can be regarded as a sort of Peirce decomposition. Using this terminology, this theorem can be rephrased by saying that any von Neumann regular element e in a Lie algebra over a ring of scalars Φ in which 30 is invertible extends to an idempotent (e, f). In some special cases, we do not need to invoke these results to get an idempotent out of a von Neumann regular element.

LEMMA 5.16. *Let L be a nondegenerate Lie Φ-algebra with $6 \in \Phi^*$, and let B and C be abelian inner ideals of L such that C is minimal and $\operatorname{ad}_c^2 B \neq 0$ for some $c \in C$. Then $C = \operatorname{ad}_c^2 B$ and for any $x \in B$ such that $\operatorname{ad}_c^2 x = 2c$, $(c, -\frac{1}{2}\operatorname{ad}_x^2 c)$ is an idempotent of L.*

PROOF. That $C = \operatorname{ad}_c^2 B$ follows from Proposition 4.6 since C is a minimal abelian inner ideal. Put $b = -\frac{1}{2}\operatorname{ad}_x^2 c$. As B and C are abelian inner ideals, b and c are Jordan elements. Therefore we only need to show that $[[c, b], c] = 2c$ and $[[c, b], b] = -2b$, which is in fact a standard application of 4.4(3). Indeed,

$$\begin{aligned}[[c,b],c] &= -\operatorname{ad}_c^2 b = \frac{1}{2}\operatorname{ad}_c^2\operatorname{ad}_x^2 c = \frac{1}{4}\operatorname{ad}_c^2\operatorname{ad}_x^2\operatorname{ad}_c^2 x\\ &= \frac{1}{4}\operatorname{ad}^2_{\operatorname{ad}_c^2 x} x = \frac{1}{4}\operatorname{ad}^2_{2c} x = \operatorname{ad}_c^2 x = 2c.\end{aligned}$$

Similarly,

$$\begin{aligned}[[c,b],b] &= \operatorname{ad}_b^2 c = \operatorname{ad}^2_{(-\frac{1}{2}\operatorname{ad}_x^2 c)} c = \frac{1}{4}\operatorname{ad}^2_{\operatorname{ad}_x^2 c} c = \frac{1}{4}\operatorname{ad}_x^2\operatorname{ad}_c^2\operatorname{ad}_x^2 c\\ &= -\frac{1}{2}\operatorname{ad}_x^2\operatorname{ad}_c^2 b = \operatorname{ad}_x^2 c = -2b,\end{aligned}$$

which completes the proof. □

Let R be a ring. Recall that if $x \in R$ is of square 0, then x is a Jordan element of the Lie algebra R^-, and that in this case, x is von Neumann regular in R if and only if so is in R^-. In the following two results we obtain something more: x can be regularly paired with an element $y \in R$ of square 0. Moreover, if R has an involution and x is symmetric or skew-symmetric, then y can be taken symmetric or skew-symmetric accordingly.

LEMMA 5.17. (Brox) *Let R be an associative Φ-algebra with $2 \in \Phi^*$, and let $c \in R$ be a von Neumann regular element such that $c^2 = 0$. Then there exists $d \in R$ such that $c = cdc$, $d = dcd$ and $d^2 = 0$. Moreover, if R has an involution and $c \in K$, then d can be chosen to be skew-symmetric. In both cases, c and d are Jordan elements and (c, d) is an idempotent of the corresponding Lie algebra.*

PROOF. Let $x \in R$ be such that $c = cxc$. Replacing x by $b = xcx$, we get $c = cbc$ and $b = bcb$. We claim that $d := b - b^2c$ is an element of square 0 satisfying the same conditions as b. Indeed,

$$\begin{aligned}&d^2 = (b - b^2c)(b - b^2c) = b^2 - b^3c - b(bcb) + b(bcb)bc = b^2 - b^3c - b^2 + b^3c = 0,\\ &cdc = c(b - b^2c)c = cbc = c, \text{ and}\\ &dcd = (b - b^2c)c(b - b^2c) = bc(b - b^2c) = bcb - (bcb)bc = b - b^2c = d.\end{aligned}$$

Suppose now that c is skew-symmetric. Since $\frac{1}{2} \in \Gamma$, we can take $b \in K$ such that $cbc = c$ and $b = bcb$. We claim that $d := b - \frac{1}{2}(cb^2 + b^2c) + \frac{1}{4}cb^3c$ satisfies the required properties. It is clear that $d^* = -d$. Furthermore, we have:

$$\begin{aligned} d^2 &= \left(b - \frac{1}{2}(cb^2 + b^2c) + \frac{1}{4}cb^3c\right)\left(b - \frac{1}{2}(cb^2 + b^2c) + \frac{1}{4}cb^3c\right) = b^2 - \frac{1}{2}(bcb)b \\ &- \frac{1}{2}b^3c + \frac{1}{4}(bcb)b^2c - \frac{1}{2}cb^3 + \frac{1}{4}cb(bcb)b + \frac{1}{4}cb^4c - \frac{1}{8}cb(bcb)b^2c - \frac{1}{2}b(bcb) \\ &+ \frac{1}{4}b(bcb)bc + \frac{1}{4}cb^2(bcb) - \frac{1}{8}cb^2(bcb)bc = b^2 - \frac{1}{2}b^2 - \frac{1}{2}b^3c + \frac{1}{4}b^3c - \frac{1}{2}cb^3 \\ &+ \frac{1}{4}cb^3 + \frac{1}{4}cb^4c - \frac{1}{8}cb^4c - \frac{1}{2}b^2 + \frac{1}{4}b^3c + \frac{1}{4}cb^3 - \frac{1}{8}cb^4c = 0, \end{aligned}$$

$cdc = c(b - \frac{1}{2}(cb^2 + b^2c))c = cbc = c$, and

$$\begin{aligned} dcd &= \left(b - \frac{1}{2}(cb^2 + b^2c)\right) c \left(b - \frac{1}{2}(cb^2 + b^2c)\right) = (b - \frac{1}{2}cb^2)c(b - \frac{1}{2}b^2c) \\ &= bcb - \frac{1}{2}(bcb)bc - \frac{1}{2}cb(bcb) + \frac{1}{4}cb(bcb)bc = b - \frac{1}{2}b^2c - \frac{1}{2}cb^2 + \frac{1}{4}cb^3c = d. \end{aligned}$$

Since $c^2 = 0$, c is a Jordan element of R (of K if c is skew-symmetric) satisfying $[[c, d], c] = [cd - dc, c] = 2cdc = 2c$. Similarly, d is Jordan and $[[c, d], d] = -2d$, which completes the proof. □

LEMMA 5.18. *Let R be a 6-torsion free ring and let (x, y) be a $\mathfrak{sl}_2$-pair of the Lie algebra R^-. If $x^2 = 0$ then $y^2 = 0$ and therefore (x, y) is actually an idempotent of R^-.*

PROOF. The elements x, y satisfy the relations:

$$x^2y - 2xyx + yx^2 + 2x = 0, \ -xy^2 + 2yxy - y^2x^2 - 2y = 0, \ \text{and} \ x^2 = 0.$$

But as proved in [**CIR08**, Lemma 2.2], it follows from these relations $12y^2 = 0$. Since R is 6-torsion free, $y^2 = 0$. □

5.4. The socle of a nondegenerate Lie algebra

In this section we deal with Lie algebras over a ring of scalars Φ in which 30 is invertible. We define the socle of a Lie algebra as the sum of its minimal abelian inner ideals, and prove that the socle of a nondegenerate Lie algebra is a direct sum of ideals each of which is a simple nondegenerate Lie algebra coinciding with its socle. We also characterize nondegenerate Lie algebras with essential socle as essential subdirect product of strongly prime Lie algebras with nonzero socle. We warn the reader that this definition of socle does not coincide with the usual one for finite-dimensional Lie algebras (defined as the sum of its minimal ideals), although for finite-dimensional Lie algebras over an algebraically closed field both notions agree.

DEFINITION 5.19. Let L be a Lie algebra over a ring of scalars Φ. The *socle* of L, denoted by $\mathrm{Soc}(L)$, is defined as the sum of its minimal abelian inner ideals, so $\mathrm{Soc}(L) = 0$ if and only if L does not contain minimal abelian inner ideals.

Recall (Definition 2.20) that given a distinguished generating set X of a Lie algebra, we denote by $\mathrm{Elem}(X)$ the elementary group defined by X.

PROPOSITION 5.20. *Let L be a simple nondegenerate Lie algebra of characteristic 0 or greater than 5. If L contains a minimal abelian inner ideal, then L coincides with its socle. In fact, for any minimal abelian inner ideal B of L, $L = \sum_{\xi \in \mathrm{Elem}(X)} \xi(B)$ for some distinguished generating set X of L.*

PROOF. By Lemma 4.19, L contains a nonzero von Neumann regular element and hence it has a nontrivial $\mathbb{Z}$-grading $L = L_{-2} \oplus L_{-1} \oplus L_0 \oplus L_1 \oplus L_2$. By Example 2.19(3), $X := \bigcup_{j=\pm 1 \pm 2} L_j$ is a distinguished generating set of L. Since the subspace $\sum_{\xi \in \mathrm{Elem}(X)} \xi(B)$ is clearly invariant under $\mathrm{Elem}(X)$, it is a (nonzero) ideal of L by Lemma 2.21, so $L = \sum_{\xi \in \mathrm{Elem}(X)} \xi(B)$. □

PROPOSITION 5.21. *Let L be a 6-torsion free Lie algebra. If L is nondegenerate, then every minimal abelian inner ideal B of L generates an ideal which is a simple Lie algebra.*

PROOF. Set $I := \mathrm{id}_L(B)$ and $J := \mathrm{id}_I(B)$. By Proposition 2.36, J is actually an inner ideal of L, which implies by Lemma 4.5 and Proposition 4.6 that $\mathrm{ad}_b^2 J$ is an abelian inner ideal of L for any nonzero element $b \in B$. Then, by the inner minimality of B, we have that either $\mathrm{ad}_b^2 J = 0$, and hence $b \in \mathrm{Ann}_I(J)$ by Propositions 4.7 and 4.10, or $B = \mathrm{ad}_b^2 J$. But the former cannot occur since in this case we would have $b \in \mathrm{Ann}_I(J) \cap J = 0$. Thus $B = \mathrm{ad}_b^2 J$ for any $0 \neq b \in B$. Hence $B \subset [J, J]$, which implies $J = [J, J]$. So J is actually an ideal of L and therefore $J = I$. Now let K be any nonzero ideal of I and take a nonzero element $b \in B$. Then $\mathrm{ad}_b^2 K \neq 0$. Otherwise $b \in \mathrm{Ann}_I(K)$ and hence $I = J$ would be contained in $\mathrm{Ann}_I(K)$, which is a contradiction. Consequently, $B = \mathrm{ad}_b^2 K$ and therefore $B \subset K$, so $I = J = K$. This proves that I is a simple algebra. □

THEOREM 5.22. *Let L be a nondegenerate Lie algebra over a ring of scalars Φ where 30 is invertible. We have:*

(i) *$\mathrm{Soc}(L)$ is a direct sum of ideals each of which is a simple nondegenerate Lie algebra coinciding with its socle.*
(ii) *For any ideal I of L, I is nondegenerate and $\mathrm{Soc}(I) = I \cap \mathrm{Soc}(L)$.*
(iii) *$\mathrm{Soc}(L) = [\mathrm{Soc}(L), \mathrm{Soc}(L)] = \mathrm{Soc}([L, L])$.*
(iv) *If B is an abelian inner ideal of L, then either B contains a minimal inner ideal of L or $B \subset \mathrm{Ann}_L(\mathrm{Soc}(L))$.*

PROOF. (i) Without loss of generality we may assume that L contains minimal abelian inner ideals. Define two minimal abelian inner ideals of L to be *equivalent* if they generate the same ideal and fix a minimal abelian inner ideal B_λ for each class of equivalence. It follows from Proposition 5.21, that $\mathrm{Soc}(L) = \bigoplus \mathrm{id}_L(B_\lambda)$ is a direct sum of ideals each of which is a simple Lie algebra. Since every minimal abelian inner ideal of L contained in an ideal I is a minimal inner ideal of I (Corollary 4.21), each of the simple components $\mathrm{id}_L(B_\lambda)$ coincides with its socle.

(ii) Let I be an ideal of L. We know by Proposition 4.7 that I is nondegenerate. Moreover, since by 4.21 the minimal abelian inner ideals of I are precisely the minimal abelian inner ideals of L contained in I, we have $\mathrm{Soc}(I) \subset I \cap \mathrm{Soc}(L)$. For the reverse inclusion notice that by (i), $\mathrm{Soc}(L) \cap I = \bigoplus M_\alpha$, where the M_α are the simple ideals of $\mathrm{Soc}(L)$ contained in I. Thus $I \cap \mathrm{Soc}(L) \subset \mathrm{Soc}(I)$ by 4.21 again.

(iii) The equality $\mathrm{Soc}(L) = [\mathrm{Soc}(L), \mathrm{Soc}(L)]$ is a direct consequence of (i). Now we have by (ii)

$$\mathrm{Soc}([L, L]) = [L, L] \cap \mathrm{Soc}(L) = [L, L] \cap [\mathrm{Soc}(L), \mathrm{Soc}(L)] = [\mathrm{Soc}(L), \mathrm{Soc}(L)].$$

(iv) Let B be an abelian inner ideal of L. If $\mathrm{ad}_b^2 \mathrm{Soc}(L) = 0$ for every $b \in B$, then $B \subset \mathrm{Ann}_L(\mathrm{Soc}(L))$ by Proposition 4.10. Suppose then that $\mathrm{ad}_b^2 M \neq 0$ for some simple component M of $\mathrm{Soc}(L)$ and some $b \in B$. Then $\mathrm{ad}_b^2 C \neq 0$ for some minimal abelian inner ideal C of L and hence, by Corollary 4.20, $\mathrm{ad}_b^2 C$ is a minimal abelian inner contained in B. □

LEMMA 5.23. *Let L be a nondegenerate Lie Φ-algebra with $30 \in \Phi^*$, let E be an essential ideal of L, and let K be a subalgebra of L containing E. Then K is nondegenerate and* $\mathrm{Soc}(E) = \mathrm{Soc}(K) = \mathrm{Soc}(L)$.

PROOF. Let $z \in K$ be such that $\mathrm{ad}_z^2 K = 0$. Then $\mathrm{ad}_z^2 E = 0$ and hence $z \in \mathrm{Ann}(E) = 0$ since E is essential. Thus K is nondegenerate. Then, by Theorem 5.22, $\mathrm{Soc}(E) \subset \mathrm{Soc}(K)$. Let B be a minimal abelian inner ideal of L and $0 \neq b \in B$. By Proposition 4.6, $B = \mathrm{ad}_b^2 E = \mathrm{ad}_b^2 K = \mathrm{ad}_b^2 L$. Hence it follows easily that $\mathrm{Soc}(E) = \mathrm{Soc}(K) = \mathrm{Soc}(L)$. □

Recall that a subdirect product of a family $\{L_\lambda\}_{\lambda \in \Lambda}$ of Lie algebras is any subalgebra L of the full direct product $\prod L_\lambda$ such that the canonical projections $\pi_\lambda : L \to L_\lambda$ are onto, and that a subdirect product is said to be essential if it contains an essential ideal of the full direct product. Any subdirect product of nondegenerate Lie algebras is a nondegenerate Lie algebra.

PROPOSITION 5.24. *Let L be a Lie Φ-algebra with $30 \in \Phi^*$. Then the following conditions are equivalent:*

(i) *L is nondegenerate and every nonzero ideal of L contains a minimal abelian inner ideal.*
(ii) *L is nondegenerate with essential socle.*
(iii) *L is an essential subdirect product of a family of strongly prime Lie algebras with nonzero socle.*
(iv) $\bigoplus M_\lambda \lhd L \leq \prod \mathrm{Der}(M_\lambda)$, *where $\{M_\lambda\}_{\lambda \in \Lambda}$ is a family of simple nondegenerate Lie algebras with minimal abelian inner ideals.*

PROOF. (i) $\Rightarrow$ (ii). If $\mathrm{Ann}(\mathrm{Soc}(L))$ were nonzero, then it would contain a minimal abelian inner ideal, a contradiction since $\mathrm{Soc}(L) \cap \mathrm{Ann}(\mathrm{Soc}(L)) = 0$.

(ii) $\Rightarrow$ (iii). Let $\mathrm{Soc}(L) = \bigoplus M_\lambda$, where each M_λ is a minimal ideal containing a minimal abelian inner ideal of L, and set $L_\lambda := L/\mathrm{Ann}(M_\lambda)$. By Corollary 4.11, L_λ is nondegenerate, and by Proposition 1.8, it is prime, with $\mathrm{Soc}(L_\lambda) \cong M_\lambda$. Since $\bigcap \mathrm{Ann}(M_\lambda) = \mathrm{Ann}(\mathrm{Soc}(L)) = 0$, L is a subdirect product of the Lie algebras L_λ, and this subdirect product is essential by Lemma 1.9.

(iii) $\Rightarrow$ (iv). Suppose that $E \lhd_e L \leq \prod L_\lambda$ is an essential subdirect product of a family $\{L_\lambda\}$ of strongly prime Lie algebras with nonzero socle, $\mathrm{Soc}(L_\lambda) = M_\lambda$. Via the adjoint representation, we have by primeness of L_λ that L_λ can be imbedded into $\mathrm{Der}(M_\lambda)$. Since E is an essential ideal of the full direct product, $M_\lambda \subset E$ for each index λ, and hence we may replace, without loss of generality, E by the direct sum of the M_λ. Then we can imbeds L into $\prod \mathrm{Der}(M_\lambda)$ via the map $x \mapsto \{\mathrm{ad}(x)_\lambda\}$, with $\mathrm{ad}(x)_\lambda$ denoting the restriction of $\mathrm{ad}(x)$ to M_λ.

(v) $\Rightarrow$ (i). We may assume that L is an essential subdirect product of a family of strongly prime Lie algebras (so L is nondegenerate), each of which contains a minimal abelian inner ideal. By Lemma 5.23, L has essential socle. Thus, if I is a nonzero ideal of L, we have by Theorem 5.22 that $\operatorname{Soc}(I) = \operatorname{Soc}(L) \cap I \neq 0$, so I contains a minimal abelian inner ideal. □

PROPOSITION 5.25. *Let L be a nondegenerate finite-dimensional Lie algebra over an algebraically closed field $\mathbb{F}$ of characteristic 0. Then $L = \operatorname{Soc}(L)$.*

PROOF. By Corollary 3.23, L is actually semisimple, so $L = \operatorname{Soc}(L)$. In fact (see Corollary 8.65), L is spanned by its extremal elements. □

However, nondegenerate Artinian Lie algebras do not need to coincide with their socle, even if they are finite-dimensional. In fact, by [**Pre86b**, Theorem 3], a finite-dimensional nondegenerate Lie algebra L over an algebraically closed field of characteristic $p > 5$ is a classical semisimple Lie algebra (and therefore it coincides with its socle) if and only if it is perfect, i.e. $L = [L, L]$.

REMARK 5.26. In [**DFFLGGL08**], the socle of a nondegenerate Lie algebra L (here denoted by lower case letters, $\operatorname{soc}(L)$) was defined as the sum of all minimal (not necessary abelian) inner ideals of L. With that definition, the above structure theorem remains true. Moreover, if L is finite-dimensional, then $\operatorname{soc}(L)$ is an essential ideal, this result not being true for $\operatorname{Soc}(L)$. In fact, $\operatorname{soc}(L) = \operatorname{Soc}(L) \oplus I$, where I is a direct sum of ideals each of which is an innerly simple Lie algebra (Theorem 4.29(3)). Since these simple components of I have proved to be irrelevant in structure theory (they are some sort of weird stuff in the Lie universe), it seems to be more convenient to define the socle in terms of minimal abelian inner ideals.

Conjugation. Let L be a nondegenerate Lie algebra over a ring of scalars Φ in which 30 is invertible, and let B and C be minimal abelian inner ideals of L. It follows from Proposition 5.21 that if B and C are conjugate by an inner automorphism, then they generate the same ideal. The converse also holds as will be proved now.

LEMMA 5.27. *Let L be a Lie Φ-algebra with $6 \in \Phi^*$, let (e, f) be an idempotent, and set $h = [e, f]$. We have:*

(i) *$(e + h - f, f)$ is an idempotent of L.*

(ii) *The principal inner ideals $\operatorname{ad}_e^2 L$, $\operatorname{ad}_f^2 L$, $\operatorname{ad}_{e+h-f}^2 L$ are conjugate under* $\operatorname{Int}(L)$.

PROOF. By Corollary 1.3, $\exp(\operatorname{ad}_{-f})$ and $\exp(\operatorname{ad}_e)$ are automorphisms of L, with $\exp(\operatorname{ad}_{-f})(f) = f$ and $\exp(\operatorname{ad}_{-f})(e) = e + h - f$. This proves (i) and that the principal inner ideals $\operatorname{ad}_e^2 L$ and $\operatorname{ad}_{e+h-f}^2 L$ are conjugate. Furthermore, $\exp(\operatorname{ad}_e)\exp(\operatorname{ad}_{-f})(e) = \exp(\operatorname{ad}_e)(e + h - f) = -f$, so $\operatorname{ad}_e^2 L$ and $\operatorname{ad}_f^2 L$ are also conjugate under $\operatorname{Int}(L)$. □

EXAMPLE 5.28. To illustrate the above construction, take $L = \mathfrak{sl}_2(\mathbb{F})$ where $\mathbb{F}$ is a field of characteristic not 2 or 3. Set $e = [12]$ and $f = [21]$. Then (e, f) is an idempotent of L, $h = [e, f] = [11] - [22]$, and $e' = e + h - f = [11] + [12] - [21] - [22]$.

PROPOSITION 5.29. *Let L be a nondegenerate Lie Φ-algebra with $30 \in \Phi^*$, and let B and C be minimal abelian inner ideals of L. Then $\operatorname{id}_L(B) = \operatorname{id}_L(C)$ if and only if B and C are conjugate under* $\operatorname{Int}(L)$.

PROOF. Denote by I (resp. by J) the ideal of L generated by B (resp. by C). We know by Proposition 5.21 that I and J are simple Lie algebras. Hence it is clear that if B and C are conjugate by an inner automorphism, then $I = J$. Suppose conversely that $I = J$. By Corollary 4.21 B and C are minimal inner ideals of I, so we may suppose that L is simple. Then $L = \sum_{\xi \in \mathrm{Elem}(X)} \xi(B)$ for some distinguished generating set X of L. By nondegeneracy of L, for any $0 \neq c \in C$, there exists a $\xi \in \mathrm{Elem}\, L$ such that $\mathrm{ad}_c^2 \xi(B) \neq 0$. Hence $C = \mathrm{ad}_c^2 \xi(B)$ since $\mathrm{ad}_c^2 \xi(B)$ is an inner ideal (Proposition 4.6) and C is minimal. Let $x \in \xi(B)$ such that $\mathrm{ad}_c^2 x = 2c$. Then it follows from Lemma 5.16 that $(c, -\frac{1}{2}\mathrm{ad}_x^2 c)$ is an idempotent of L. Hence $C = \mathrm{ad}_c^2 L$ and $\xi(B) = \mathrm{ad}^2(-\frac{1}{2}\mathrm{ad}_x^2 c)L$ are conjugate under an inner automorphism by Lemma 5.27(ii). Therefore B and C are also conjugate. □

REMARK 5.30. Proposition 5.29 answers a question posed by O. Loos on the relationship among the minimal abelian inner ideals of a simple nondegenerate Lie algebra. The reader is referred to [**Loo91**] for a similar question in Jordan systems.

5.5. Principal filtrations

Following [**GGL12**], we associate to any Jordan element a of a Lie algebra L a bounded $\mathbb{Z}$-filtration and proved that, when L is nondegenerate, the $\mathbb{Z}$-graded algebra $\hat{L}$ defined by the filtration is isomorphic to L if and only if a is von Neumann regular, equivalently, $\hat{L}$ is nondegenerate.

DEFINITION 5.31. Let L be a Lie Φ-algebra. A *$\mathbb{Z}$-filtration* $\{\mathcal{F}_i\}_{i\in\mathbb{Z}}$ is a collection of Φ-submodules of L

$$\cdots \subset \mathcal{F}_{-2} \subset \mathcal{F}_{-1} \subset \mathcal{F}_0 \subset \mathcal{F}_1 \subset \mathcal{F}_2 \subset \cdots$$

indexed in $\mathbb{Z}$ such that $[\mathcal{F}_i, \mathcal{F}_j] \subset \mathcal{F}_{i+j}$ for all integers i, j. A $\mathbb{Z}$-filtration $\{\mathcal{F}_i\}_{i\in\mathbb{Z}}$ is *bounded* if there exist integers $m < n$ such that $\mathcal{F}_m = 0$ and $\mathcal{F}_n = L$. it is clear that every element in $\mathcal{F}_j$, $j < 0$, is ad-nilpotent.

EXAMPLE 5.32. Let $L = L_{-n} \oplus \cdots \oplus L_0 \oplus \cdots \oplus L_n$ be a finite $\mathbb{Z}$-grading of a Lie algebra L. A bounded filtration of L is defined by setting $\mathcal{F}_i = 0$ for $i < -n$, $\mathcal{F}_i = \sum_{j=-n}^{i} L_j$ for $-n \leq i \leq n-1$, and $\mathcal{F}_i = L$ for $i \geq n$.

Conversely, let $\{\mathcal{F}_i\}_{i\in\mathbb{Z}}$ be a $\mathbb{Z}$-filtration of a Lie algebra L and let $\hat{L}$ denote the Φ-module $\bigoplus_{i\in\mathbb{Z}} \mathcal{F}_i/\mathcal{F}_{i-1}$. For $x \in \mathcal{F}_i$, $y \in \mathcal{F}_j$, set $\bar{x} = x + \mathcal{F}_{i-1}$, $\bar{y} = y + \mathcal{F}_{j-1}$ and define $[\bar{x}, \bar{y}] = \overline{[x, y]} = [x, y] + \mathcal{F}_{i+j-1}$. It is easy to verify that this product is well defined and that it endows $\hat{L}$ with a structure of $\mathbb{Z}$-graded Lie algebra.

PROPOSITION 5.33. *Let L be a Lie Φ-algebra with $6 \in \Phi^*$, and let $a \in L$ be a Jordan element. Set*

$$\mathcal{F}_i = 0,\ i \leq -3, \quad \mathcal{F}_{-2} = [a, [a, L]], \quad \mathcal{F}_{-1} = [a, \mathrm{Ker}_L\{a\}],$$

$$\mathcal{F}_0 = \{x \in L : [a, x] \in \mathcal{F}_{-2}\}, \quad \mathcal{F}_1 = \mathrm{Ker}_L\{a\}, \quad \mathcal{F}_j = L,\ j \geq 2,$$

where $\mathrm{Ker}_L\{a\} = \{z \in L : [a, [a, z]] = 0\}$. *Then $\{\mathcal{F}_i\}_{i\in\mathbb{Z}}$ is a bounded filtration called the* principal filtration *of L defined by a.*

PROOF. Since $[a, L] \subset \mathrm{Ker}_L\{a\}$, we have $\mathcal{F}_{-2} \subset \mathcal{F}_{-1} \subset \mathcal{F}_0 \subset \mathcal{F}_1$. So we mut show that $[\mathcal{F}_i, \mathcal{F}_j] \subset \mathcal{F}_{i+j}$ for all integers i, j.

(1) $[\mathcal{F}_{-2}, \mathcal{F}_{-1}] = 0$. Let $\mathrm{ad}_a^2 x \in \mathcal{F}_{-2}$, $z \in \mathrm{Ker}_L\{a\}$. By 4.4(6),

$$[\mathrm{ad}_a^2 x, \mathrm{ad}_a z] = [\mathrm{ad}_a^2 z, \mathrm{ad}_a x] = 0.$$

(2) $[\mathcal{F}_{-2}, \mathcal{F}_0] \subset \mathcal{F}_{-2}$. Let $\mathrm{ad}_a^2 y \in \mathcal{F}_{-2}$ and $z \in \mathrm{Ker}_L\{a\}$ with $[a, z] = \mathrm{ad}_a^2 x$ for some $x \in L$. Then

$$\begin{aligned}[[a,[a,y]],z] &= [[a,z],[a,y]] + [a,[[a,z],y]] + [a,[a,[y,z]]] \\ &= [\mathrm{ad}_a^2 x, \mathrm{ad}_a y] + \mathrm{ad}_a[\mathrm{ad}_a^2 x, y] + \mathrm{ad}_a^2[y,z] \\ &= 2[\mathrm{ad}_a^2 x, \mathrm{ad}_a y] + \mathrm{ad}_a^2[y,z] \\ &= (\text{ by } 4.4(6)) - 2\,\mathrm{ad}_a^2\,\mathrm{ad}_x\,\mathrm{ad}_a y + \mathrm{ad}_a^2[y,z] \subset \mathcal{F}_{-2}.\end{aligned}$$

(3) $[\mathcal{F}_{-2}, \mathcal{F}_1] \subset \mathcal{F}_{-1}$. Let $\mathrm{ad}_a^2 x \in \mathcal{F}_{-2}$, $z \in \mathrm{Ker}_L\{a\}$. Then

$$\begin{aligned}[[a,[a,x]],z] &= [[a,z],[a,x]] + [a,[[a,x],z]] \\ &= [a,[[a,z],x]] + [a,[[a,x],z]] \\ &= 2[a,[[a,z],x]] + [a,[a,[x,z]]] \in \mathcal{F}_{-1},\end{aligned}$$

since $[a,[x,z]]$, $[[a,z],x]$ belong to $\mathrm{Ker}_L\{a\}$:

$$\mathrm{ad}_a^2[[a,z],x] = [\mathrm{ad}_a z, \mathrm{ad}_a^2 x] = [\mathrm{ad}_a x, \mathrm{ad}_a^2 z] = 0$$

by 4.4(6).

(4) $[\mathcal{F}_{-1}, \mathcal{F}_{-1}] \subset \mathcal{F}_{-2}$. Let $z, w \in \mathrm{Ker}_L\{a\}$. Then $\mathrm{ad}_a^2[z,w] = 2[\mathrm{ad}_a z, \mathrm{ad}_a w]$, so $[\mathrm{ad}_a z, \mathrm{ad}_a w] = \frac{1}{2}\mathrm{ad}_a^2[z,w] \in \mathcal{F}_{-2}$. And so on. □

LEMMA 5.34. *Let L be a Lie algebra over a ring of scalars Φ in which 30 is invertible and let $f \in L$ be von Neumann regular. For any $e \in L$ such that (e, f) is an idempotent, the filtration defined by f coincides with the filtration of the $\mathbb{Z}$-grading defined by (e, f).*

PROOF. We must prove the following equalities:

(1) $\mathcal{F}_{-2} = \mathrm{ad}_f^2 L = L_{-2}^h$.

By Theorem 5.11(iii), $\mathrm{ad}_f^2 L = L_{-2}^h$.

(2) $\mathrm{Ker}_L\{f\} = L_{-2}^h \oplus L_{-1}^h \oplus L_0^h \oplus L_1^h$.

By the grading properties, $L_{-2}^h \oplus L_{-1}^h \oplus L_0^h \oplus L_1^h \subset \mathrm{Ker}_L\{f\}$. Conversely, by Theorem 5.11(iv), for any nonzero element $x \in L_2^h$, we have $\mathrm{ad}_f^2 x \neq 0$. Thus $\mathcal{F}_1 = \mathrm{Ker}_L\{f\} = L_{-2}^h \oplus L_{-1}^h \oplus L_0^h \oplus L_1^h$.

(3) $[f, \mathrm{Ker}_L\{f\}] = L_{-2}^h \oplus L_{-1}^h$.

By (1), $L_{-2}^h = \mathrm{ad}_f^2 L = \mathcal{F}_{-2} \subset \mathcal{F}_{-1} = [f, \mathrm{Ker}_L\{f\}]$. Let $x \in L_{-1}^h$. Then, by the grading properties, $-x = [h, x] = [[e, f], x] = [[e, x], f]$ with $[e, x] \in L_1^h \subset \mathrm{Ker}_L\{f\}$, which proves that $L_{-2}^h \oplus L_{-1}^h \subset [f, \mathrm{Ker}_L\{f\}]$. Conversely, it follows from (2) and the grading properties

$$[f, \mathrm{Ker}_L\{e\}] = [f, L_{-2}^h \oplus L_{-1}^h \oplus L_0^h \oplus L_1^h] = [f, L_0^h] \oplus [f, L_1^h] \subset L_{-2}^h \oplus L_{-1}^h.$$

(4) $\{z \in \mathrm{Ker}_L\{f\} : [f, z] \in \mathrm{ad}_f^2 L\} = L_{-2}^h \oplus L_{-1}^h \oplus L_0^h$.

By (2), $L_{-2}^h \oplus L_{-1}^h \oplus L_0 \subset \mathrm{Ker}_L\{f\}$, with $[f, L_{-2}^h \oplus L_{-1}^h \oplus L_0] \subset L_{-2}^h = \mathrm{ad}_f^2 L$ by (1), which proves the inclusion $\supset$. Conversely, let $z \in \mathrm{Ker}_L\{f\}$ such that $[f, z] \in \mathrm{ad}_f^2 L$. Write $z = z_{-2} + z_{-1} + z_0 + z_1$ according with (2). Then $[f, z] = [f, z_0] + [f, z_1] \in L_{-2}^h$ if and only if $[f, z_1] = 0$, equivalently, $z_1 = [h, z_1] = [[e, f], z_1] = [e, [f, z_1]] = 0$, which proves the inclusion $\subset$, so completing the proof. □

REMARK 5.35. Principal filtrations in the context of Jordan systems were studied by O. Loos. The above lemma is a version for Lie algebras of [**Loo90**, Proposition 4].

THEOREM 5.36. *Let L be a nondegenerate Lie algebra over a ring of scalars Φ in which* 30 *is invertible, and let $f \in L$ be a nonzero Jordan element. Let*

$$\hat{L} = \mathcal{F}_{-2} \oplus \mathcal{F}_{-1}/\mathcal{F}_{-2} \oplus \mathcal{F}_0/\mathcal{F}_{-1} \oplus \mathcal{F}_1/\mathcal{F}_0 \oplus \mathcal{F}_2/\mathcal{F}_1$$

be the $\mathbb{Z}$-graded Lie algebra associated to the filtration defined by f. The following conditions are equivalent:

(i) *f is von Neumann regular,*
(ii) *$\hat{L}$ is nondegenerate.*

PROOF. (i) $\Rightarrow$ (ii). By Lemma 5.8 and Theorem 5.11, the von Neumann regular element f can be extended to an idempotent (e, f), and this idempotent yields a finite $\mathbb{Z}$-grading $L = L^h_{-2} \oplus L^h_{-1} \oplus L^h_0 \oplus L^h_1 \oplus L^h_2$, where $h = [e, f]$ and $L^h_i = \{x \in L : [h, x] = ix\}$. By Lemma 5.34, the filtration defined by this grading coincides with the filtration of L defined by f. Hence $\hat{L}$ is isomorphic to L and therefore nondegenerate.

(ii) $\Rightarrow$ (i). Suppose that f is not von Neumann regular, i.e. f does not belong to $\operatorname{ad}_f^2 L = \mathcal{F}_{-2}$, where $\{\mathcal{F}_i\}_{i\in\mathbb{Z}}$ is the filtration defined by f. However, f does belong to $\mathcal{F}_{-1} = [f, \operatorname{Ker}_L\{f\}]$, since $f = \frac{1}{2}[f, h]$ and $\operatorname{ad}_f^2 h = 0$. Thus $0 \neq f + \mathcal{F}_{-2} \in \mathcal{F}_{-1}/\mathcal{F}_{-2} = \hat{L}_{-1}$. The proof will be complete by proving that $\bar{f} = f + \mathcal{F}_{-2}$ is an absolute zero divisor of $\hat{L}$. This is verified below:

(1) $[\bar{f}, [\bar{f}, \hat{L}_{-2}]] = \overline{\operatorname{ad}_f^4 L} = \bar{0}$.
(2) $[\bar{f}, [\bar{f}, \hat{L}_{-1}]] = \overline{[f, [f, \operatorname{Ker}_L\{f\}]]} = \bar{0}$.
(3) $[\bar{f}, [\bar{f}, \hat{L}_0]] = \overline{[f, [f, \mathcal{F}_0]]} = \bar{0}$, since $[f, [f, \mathcal{F}_0]] \subset [f, \operatorname{ad}_f^2 L] = \operatorname{ad}_f^3 = 0$.
(4) $[\bar{f}, [\bar{f}, \hat{L}_1]] = \overline{[f, [f, \operatorname{Ker}_L\{f\}]]} = \bar{0}$.
(5) $[\bar{f}, [\bar{f}, \hat{L}_2]] = \overline{[f, [f, L]]} = \bar{0}$, since $\operatorname{ad}_f^2 L \subset \operatorname{Ker}_L\{f\} = \mathcal{F}_1$. □

5.6. Exercises

EXERCISE 5.37. Let R be a ring such that $Rx = 0$ implies $x = 0$ (for instance, R is unital or semiprime). Show that $a \in R$ is von Neumann regular if and only if the principal left ideal Ra is generated by an idempotent.

EXERCISE 5.38. Let R be a semiprime ring. Show that $\operatorname{Soc}(R)$ is von Neumann regular.

EXERCISE 5.39. Let X be a left vector space over a division ring Δ. Show that the ring $R = \operatorname{End}_\Delta(X)$ is von Neumann regular.

EXERCISE 5.40. Let X and R be as above, and let $e \in R$ be an idempotent. Show that $X = \operatorname{im}(e) \oplus \ker(e)$.

EXERCISE 5.41. Keeping the same notation, show that for $a \in R$ the following conditions are equivalent:

(i) $X = \operatorname{im}(a) \oplus \ker(a)$,
(ii) there exists $b \in R$ such that $aba = a$, $bab = b$, $ab = ba$.

EXERCISE 5.42. Let S be a multiplicative semigroup and let $a \in S$. Show that the following conditions are equivalent:

(i) $a \in a^2Sa^2$,

(ii) there exists $b \in S$ such that $aba = a$, $bab = b$, $ab = ba$,

(iii) there exists $e = e^2 \in S$ such that a is invertible in the semigroup eSe.

The element b is uniquely determined (it is the inverse of a in eSe) and called the *group inverse* of a.

EXERCISE 5.43. In the Lie algebra $\mathfrak{sl}_2(\mathbb{Z})$, show that $2[12]$ is a Jordan element which is not von Neumann regular.

EXERCISE 5.44. Verify the assertions of Example 5.3.

EXERCISE 5.45. Let R be a simple ring of characteristic different from 2 and 3. Determine the von Neumann regular elements of the Lie algebra R^-.

EXERCISE 5.46. Following Example 2.99, show that for any $z \in H^\perp$, the linear map $a = [x, z]$ is a von Neumann regular element of $\mathfrak{fo}(X, \langle\,,\rangle)$. Consider separately the cases that z is isotropic and anisotropic.

EXERCISE 5.47. Describe the von Neumann regular elements of the finitary symplectic algebra $\mathfrak{fsp}(X, \langle\,,\rangle)$.

EXERCISE 5.48. In the ring $\mathrm{M}_3(\mathbb{Q})$, show that the pair of entry matrices $([12], [21])$ is an idempotent of the Lie algebra $L = \mathrm{M}_3(\mathbb{Q})^-$, and determine the eigenspaces of ad_h, $h = [11] - [22]$.

EXERCISE 5.49. Let R be a simple ring of characteristic 0 or $p > 5$ and let $e_1, e_2 \in R$ be two nonzero orthogonal idempotents. Show that e_1Re_2 and e_2Re_1 are abelian inner ideals of the Lie algebra $R' = [R, R]$, and find two elements $x \in e_1Re_2$. $y \in e_2Re_1$ such that the pair (x, y) is an idempotent of R'.

EXERCISE 5.50. Relate the 5-grading of the previous exercise to that induced in the ring $\mathrm{M}_3(\mathbb{Q})$ by the sequence of idempotents $e_0 = [22]$, $e_1 = [33]$, and $e_2 = [11]$ (see Exercise 1.38).

EXERCISE 5.51. In the proof of Theorem 5.11, verify that the missing summands L_3^h, L_4^h, and L_{-4}^h actually vanish.

CHAPTER 6

Extremal Elements

Let L be a Lie algebra over a field $\mathbb{F}$. An element $a \in L$ is said to be *extremal* if $\mathrm{ad}_a^2 L \subset Fa$. Every extremal element is then a Jordan element, which is von Neumann regular whenever it is not an absolute zero divisor. One of the main reasons for studying extremal elements stems from the role they play in classical Lie algebras (modular versions of the complex finite-dimensional simple Lie algebras). Over an algebraically closed field of characteristic $p > 5$, classical Lie algebras are characterized by the property of being generated by extremal elements. Following our general principle, in this chapter we will only deal with those aspects of the theory of extremal elements which have a Jordan flavor. In Section 6.1 we study the basic properties of the extremal elements in a Lie algebra L over a field $\mathbb{F}$ of arbitrary characteristic. Assuming $\mathrm{char}(\mathbb{F}) \neq 2$, it is shown in Section 6.2 that the set $\mathfrak{E}$ of the extremal elements of L spans a subalgebra, and that any Lie algebra L generated by extremal elements is locally finite and has an invariant symmetric bilinear form. This form is nondegenerate whenever L is simple and hence L is also nondegenerate. In Section 6.3 Jacobson–Morozov type results are revisited for extremal elements. Finally, in Section 6.4, a classification free proof is given of the following theorem: Let L be a simple finite-dimensional Lie algebra over a field $\mathbb{F}$ of characteristic $p > 3$. If L contains an extremal element which is not an absolute zero divisor, then either L is generated by extremal elements, or $\mathbb{F}$ has characteristic 5 and L is isomorphic to $W(5)$. When $\mathbb{F}$ is algebraically closed, the existence of such an extremal element is guaranteed by a Premet's theorem.

6.1. Definition and properties

Throughout this section, L will denote a Lie algebra over an arbitrary field $\mathbb{F}$. Following our notational convention, we will sometimes use upper case letters to denote adjoint operators.

Definition 6.1. An element $a \in L$ is said to be *extremal* if $\mathrm{ad}_a^2 L \subset \mathbb{F}a$.

Note that $a \in L$ is extremal if there is a linear functional $x \mapsto \epsilon_a(x)$ on L such that $A^2x = \mathrm{ad}_a^2 x = \epsilon_a(x)a,$ for all $x \in L$. We write $\mathfrak{E}(L)$, or simply $\mathfrak{E}$ when there is no risk of confusion, to denote the set of extremal elements of L.

Examples 6.2. The reader is referred to Section 2.2 for notation.

(1) Let $(X, Y, \langle\, ,\rangle)$ be a pair of dual vector spaces over a field $\mathbb{F}$, let $x \in X$, $y \in Y$ be nonzero vectors such that $\langle x, y\rangle = 0$. Then $y^*x \in \mathfrak{f}\mathfrak{sl}_Y(X)$ is an extremal element of $\mathfrak{gl}_Y(X)$.

(2) Let X be a vector space over a field $\mathbb{F}$ of characteristic not 2, $\dim_{\mathbb{F}} X > 2$, which is endowed with a nondegenerate symmetric bilinear form $\langle\, ,\rangle$ and let $x, z \in X$ be two nonzero isotropic and mutually orthogonal vectors. Then $[x, z] \in \mathfrak{fo}(X, \langle\, ,\rangle)$ is an extremal element of $\mathfrak{o}(X, \langle\, ,\rangle)$.

(3) Let X be a vector space over a field $\mathbb{F}$ of characteristic not 2, $\dim_{\mathbb{F}} X > 2$, which is endowed with a nondegenerate alternate bilinear form $\langle\, ,\rangle$ and let x be a nonzero vector in X. Then $x^*x \in \mathfrak{f}sp(X, \langle\, ,\rangle)$ is an extremal element of $\mathfrak{sp}(X, \langle\, ,\rangle)$.

PROPOSITION 6.3. *Let $a \in \mathfrak{E}(L)$. We have:*

(i) *a is a Jordan element of L, i.e. $A^3 = 0$.*
(ii) *If $a \neq 0$, then $\mathbb{F}a$ is an (one-dimensional) inner ideal of L.*
(iii) *If a is not an absolute zero divisor, then a is von Neumann regular.*

PROOF. (i) $\mathrm{ad}_a^3 L \subset [a, \mathbb{F}a] = 0$, so a is a Jordan element. (ii) It follows directly from the definition of extremal element. (iii) If a is not an absolute zero divisor, then $0 \neq \mathrm{ad}_a^2 L \subset \mathbb{F}a$ implies $\mathbb{F}a = \mathrm{ad}_a^2 L$, so $a \in \mathrm{ad}_a^2 L$ is von Neumann regular. □

Extremal elements are Jordan elements so good that they satisfy the identities of Lemma 4.4 without any restriction on the characteristic of the ground field.

LEMMA 6.4. *Let $a \in \mathfrak{E}(L)$. For any $x, y \in L$, we have:*

(1) $A^2XA = AXA^2$.
(2) $A^2XA^2 = 0$.
(3) $\mathrm{ad}_{A^2x}^2 = A^2X^2A^2$.
(4) $[A^2x, Ay] = -[Ax, A^2y]$.
(5) $[A^2x, A^2y] = 0$.

PROOF. Let $y \in L$. Using the Leibniz rule and $A^3 = 0$, we get

$$\begin{aligned} A^2XAy = A^2[x, Ay] &= [A^2x, Ay] + 2[Ax, A^2y] = \epsilon_a(x)[a, Ay] \\ &+ 2\epsilon_a(y)[Ax, a] = \epsilon_a(x)\epsilon_a(y)a - 2\epsilon_a(x)\epsilon_a(y)a \\ &= -\epsilon_a(x)\epsilon_a(y)a = \epsilon_a(y)A[x, a] = AX\epsilon_a(y)a = AXA^2y. \end{aligned}$$

which proves (1). The remaining identities follow from (i) as in the proof of Lemma 4.4. □

LEMMA 6.5. *Let L be a simple Lie algebra over a field $\mathbb{F}$ and let $e \in L$ be an extremal element which is not an absolute zero divisor. Then $\Gamma(L) = \mathbb{F}1_L$, so L is central over $\mathbb{F}$.*

PROOF. Since L is simple, any nonzero element in the centroid is one-to-one. Let $\gamma \in \Gamma(L)$. Then $\mathrm{ad}_e^2 L = \mathbb{F}e$ implies $\gamma(e) = \alpha e$ for some $\alpha \in \mathbb{F}$. Hence $\gamma(x) = \alpha x$ for all $x \in L$, because $0 \neq e \in \ker(\gamma - l_\alpha)$, with $\gamma - l_\alpha \in \Gamma(L)$. □

COROLLARY 6.6. *Let L be a strongly prime Lie algebra over a field $\mathbb{F}$ of characteristic 0 or greater than 5 containing a nonzero extremal element e. Then L is centrally closed.*

PROOF. Denote by M the socle of L, which is a simple ideal and contains the extremal element e (Theorem 5.22). By Lemma 6.5, $\Gamma(M) = \mathbb{F}1_M$. Let (g, I) be an essential defined centralizer of L (Section 1.3). Then $M = [M, M]$ implies $g(M) \subset M$, so $[g, I] = [g_{|M}, M] = [\alpha 1_M, M] = [\alpha 1_L, L]$, for some $\alpha \in \mathbb{F}$, which proves that L is centrally closed as an $\mathbb{F}$-algebra. □

6.2. Lie algebras generated by extremal elements

Throughout this section, L will denote a Lie algebra over a field $\mathbb{F}$. We follow the exposition of [**CSUW01**].

LEMMA 6.7. *Let $a, b \in \mathfrak{E}(L)$. Then $\epsilon_a(b) = \epsilon_b(a)$.*

PROOF. Let us first see that $\epsilon_a(b) = 0 \Leftrightarrow \epsilon_b(a) = 0$. If $\epsilon_b(a) \neq 0$, then $B^2a = \epsilon_b(a)b$ implies that $b = \beta B^2 a$, where $\beta = \epsilon_b(a)^{-1}$. Hence, using 6.4(3), we get

$$\begin{aligned} b = \beta B^2 a = B^2(\beta a) &= \operatorname{ad}^2_{\beta B^2 a}(\beta a) = \beta^2 B^2 A^2 B^2(\beta a) \\ &= \beta^2 B^2 A^2 b = \beta^2 B^2 \epsilon_a(b) a \end{aligned}$$

which implies $\epsilon_a(b) \neq 0$. Suppose then that both $\epsilon_a(b)$ and $\epsilon_b(a)$ are nonzero. By 6.4(3), we get

$$\begin{aligned} \epsilon_b(a)\epsilon_a(b)^2 a &= \epsilon_b(a)\epsilon_a(b) A^2 b = \epsilon_a(b) A^2 \epsilon_b(a) b = \epsilon_a(b) A^2 B^2 a \\ &= A^2 B^2 \epsilon_a(b) a = A^2 B^2 A^2 b = \operatorname{ad}^2_{A^2 b} b \\ &= \operatorname{ad}^2_{\epsilon_a(b) a} b = \epsilon_a(b)^2 A^2 b = \epsilon_a(b)^3 a, \end{aligned}$$

which implies $\epsilon_a(b) = \epsilon_b(a)$ since both are nonzero. □

LEMMA 6.8. *The following identities hold for any $x, y, z \in \mathfrak{E}(L)$:*

(i) $2[\operatorname{ad}_x y, \operatorname{ad}_x z] = \epsilon_x([y,z])x + \epsilon_x(z)\operatorname{ad}_x y - \epsilon_x(y)\operatorname{ad}_x z$,
(ii) $2\operatorname{ad}_x \operatorname{ad}_y \operatorname{ad}_x z = \epsilon_x([y,z])x - \epsilon_x(z)\operatorname{ad}_x y - \epsilon_x(y)\operatorname{ad}_x z$,
(iii) $\epsilon_x([z,y]) = \epsilon_y([x,z])$.

PROOF. Identity (i) follows from the Leibniz rule applied to $\operatorname{ad}_x^2[y,z]$. Now (ii) follows from (i) and the fact that ad_x is a derivation. The proof of (iii) is divided into two cases. If $[x,y] = 0$ with both x, y being nonzero, then

$$\epsilon_x([z,y])x = \operatorname{ad}_x^2[z,y] = [\epsilon_x(z)x, y] = 0$$

implies $\epsilon_x([z,y]) = 0$. The symmetry of this argument also gives $\epsilon_y([z,x]) = 0$. Thus we may suppose that $[x,y] \neq 0$. Replacing z by $[y,z]$ in (ii) we obtain

$$\operatorname{ad}_x \operatorname{ad}_y \operatorname{ad}_x \operatorname{ad}_y z = \epsilon_x(y)\epsilon_y(z)x - \epsilon_x([y,z])[x,y] - \epsilon_x(y)[x,[y,z]]. \tag{6.1}$$

Interchanging x and y in (ii) and applying ad_x to the result we obtain

$$2\operatorname{ad}_x \operatorname{ad}_y \operatorname{ad}_x \operatorname{ad}_y z = \epsilon_y([x,z])[x,y] + \epsilon_y(z)\epsilon_x(y)x - \epsilon_y(x)[x,[y,z]].$$

Combining (6.1) with this last result and applying Lemma 6.7 we obtain (iii). □

THEOREM 6.9. *Suppose that L is spanned by its extremal elements. Then L possesses a unique bilinear form ϵ with $\epsilon(x,y) = \epsilon_x(y)$ for every $x \in \mathfrak{E}(L)$ and every $y \in L$. This form ϵ is symmetric and invariant.*

PROOF. Uniqueness of ϵ is clear and symmetry follows from Lemma 6.7. The invariance is a consequence of Lemma 6.8(iii) and the symmetry of ϵ. □

REMARKS 6.10. (1) Extremal elements were used by Chernousov [**Che89**] in his proof for the Hasse principle for E_8.

(2) The connections between the ϵ-form and the Killing form is studied in Section 9 of [**CSUW01**].

(3) Extremal elements are the Lie counterpart of reduced elements in Jordan triple systems and the ϵ-form is the analogue of the Faulkner form for Jordan triple systems spanned by reduced elements [**McC85**].

(4) An application of the Faulkner form to Banach Jordan-*-triples can be found in [**FLGRSCSM94**].

LEMMA 6.11. *Let L be a Lie algebra over a field $\mathbb{F}$ of characteristic not 2 and let $a \in \mathfrak{E}(L)$. Then* $\exp(\mathrm{ad}_a) \in \mathrm{Aut}(L)$.

PROOF. Since 2 is invertible in $\mathbb{F}$, $\exp(\mathrm{ad}_a) = 1_L + \mathrm{ad}_a + \frac{1}{2}\mathrm{ad}_a^2$ is a well-defined linear map, so we only need to show that $\exp(\mathrm{ad}_a)[x,y] = [\exp(\mathrm{ad}_a)x, \exp(\mathrm{ad}_a)y]$ for all $x, y \in L$. Using the Leibniz rule, we get

$$[\exp(\mathrm{ad}_a)x,\, \exp(\mathrm{ad}_a)y] = [x + Ax + \frac{1}{2}A^2x,\; y + Ay + \frac{1}{2}A^2y] = \exp(\mathrm{ad}_a)[x,y],$$

since $[A^2x, Ay] + [Ax, A^2y] = 0$ by 6.4(4), and $[A^2x, A^2y] = 0$ by 6.4(5). □

COROLLARY 6.12. *Let L be a Lie algebra over a field of characteristic not 2. Then $\mathfrak{E}$ spans a subalgebra of L.*

PROOF. Let $a, b \in \mathfrak{E}$. By Lemma 6.11, $\exp(\mathrm{ad}_a)b \in \mathfrak{E}$. Hence

$$[a,b] = \exp(\mathrm{ad}_a)b - b - \frac{1}{2}\epsilon_a(b)a$$

is a linear combination of extremal elements. □

PROPOSITION 6.13. *Let L be a simple Lie algebra over a field $\mathbb{F}$ of characteristic not 2 which is generated by extremal elements. Then:*

(i) *L is spanned by its extremal elements.*
(ii) *The form ϵ is nondegenerate.*
(iii) *L is nondegenerate.*

PROOF. (i) It follows from Corollary 6.12.

(ii) As $\mathrm{Rad}(\epsilon)$ is an ideal of L and L is simple, if $\mathrm{Rad}(\epsilon)$ were not zero, then it would be the whole L, and hence every $a \in \mathfrak{E}$ would be an absolute zero divisor ($\mathrm{ad}_a^2 L = \epsilon(a, L)a = 0$). Then L would be generated by absolute zero divisors and therefore locally nilpotent by Corollary 3.18, which is a contradiction by Proposition 1.25.

(iii) $\mathrm{ad}_a^2 L = 0$ implies $\epsilon(a, L) = 0$ and hence $a = 0$ by (ii). □

REMARK 6.14. Note that the above proposition cannot be obtained as a consequence of Proposition 3.25 since we are not assuming $\mathrm{char}(\mathbb{F}) > 5$.

THEOREM 6.15. [**Zel83a**, Lemma 15] *Let L be a Lie algebra over a field $\mathbb{F}$ of characteristic not 2. If L is generated by a finite number of extremal elements, then L is finite-dimensional.*

PROOF. Suppose that L is generated by r extremal elements $a_1, \ldots, a_r$. Let $\mathfrak{L} = \mathfrak{L}\langle X \rangle$ be the free Lie $\mathbb{F}$-algebra on the set $X = \{x_1, \ldots, x_r\}$. Take the quotient of $\mathfrak{L}$ with respect to the ideal $\mathfrak{I}$ generated by the set $\bigcup_{i=1}^r [x_i, [x_i, \mathfrak{L}]]$. Then $\mathfrak{L}/\mathfrak{I}$ is generated by the absolute zero divisors $x_i + \mathfrak{I}$, $i = 1, \ldots, r$. So, by Theorem 3.17, it has finite dimension, say d. Then there exist d commutators $w_1, \ldots, w_d \in X$ such that $\{w_1 + \mathfrak{I}, \ldots, w_d + \mathfrak{I}\}$ is a basis of $\mathfrak{L}/\mathfrak{I}$. Let $\mathfrak{U}$ be the linear span of the w_i, $i = 1, \ldots, d$. Then $\mathfrak{L} = \mathfrak{I} \oplus \mathfrak{U}$ as a direct sum of vector subspaces. Let $y \mapsto \tilde{y}$ be the homomorphism of $\mathfrak{L}$ onto L mapping each x_i into a_i. We claim that L is linearly spanned by the set $\{\tilde{w}_1, \ldots, \tilde{w}_d\}$, i.e. $L = \widetilde{\mathfrak{U}}$. Suppose that $\widetilde{\mathfrak{U}}$ is properly contained in L. Then there exists a commutator g in X of minimal length which does not belong to $\mathfrak{U}$. Write g as a linear combination of the elements w_j, $j = 1, \ldots, d$ and the left commutators t_k of the form $[u_s, \ldots, u_1, x_i, x_i, u]$, where $u, u_1, \ldots, u_s$ are commutators in X and all the w_j and t_k are of the same length as g. Among these t_k there is one, say t, such that $\tilde{t} \notin \widetilde{\mathfrak{U}}$. Set $t_0 = [u_s, \ldots, u_1, x_i]$. Then

$$\tilde{t} = [\tilde{u}_s, \ldots, \tilde{u}_1, \operatorname{ad}^2_{a_i} \tilde{u}] = \epsilon_{a_i}(\tilde{u})[\tilde{u}_s, \ldots, \tilde{u}_1, a_i] = \epsilon_{a_i}(\tilde{u})\tilde{t}_0.$$

As $\tilde{t} \notin \widetilde{\mathfrak{U}}$, $\epsilon_{a_i}(\tilde{u}) \neq 0$ and $\tilde{t}_0 \notin \widetilde{\mathfrak{U}}$, which is a contradiction since the length of t_0 is shorter than the length of g. □

COROLLARY 6.16. *Every Lie algebra over a field $\mathbb{F}$ of characteristic not 2 generated by extremal elements is locally finite.*

6.3. Jacobson–Morozov revisited

Following [**CIR08**], we refine in this section Lemmas 5.8 and 5.9 when the von Neumann regular element e is actually an extremal element, i.e. $\operatorname{ad}^2_e L = \mathbb{F}e$.

PROPOSITION 6.17. *Let L be a Lie algebra over a field $\mathbb{F}$ of characteristic 0 or greater than 3 and let e be an extremal element which is not an absolute zero divisor. Then e can be extended to a $\mathfrak{sl}_2$-triple (e, f, h). Moreover, for each such triple, ad_h is diagonalizable with eigenvalues $0, \pm 1, \pm 2$ and satisfies $L^h_2 = \mathbb{F}e$ and $L^h_{-2} = \mathbb{F}f$.*

PROOF. By Lemmas 5.8 and 5.9, only the two equalities $L^h_2 = \mathbb{F}e$ and $L^h_{-2} = \mathbb{F}f$ need to be verified. Let $S = \mathbb{F}e + \mathbb{F}f + \mathbb{F}h$ be the subalgebra generated by these elements. Then S is isomorphic to $\mathfrak{sl}_2(\mathbb{F})$ and ad_h has eigenvalues $-2, 0, 2$ on S, each of which with multiplicity 1. Then L is an S-module and so is the $\mathbb{F}$-vector space L/S. Denote by E, F and H the actions of $\operatorname{ad}_e, \operatorname{ad}_f$ and ad_h, resp. on the S-module L/S. Regarded as elements of the associative algebra $\operatorname{End}_{\mathbb{F}}(L/S)$, (E, F) is a $\mathfrak{sl}_2$-pair of the Lie algebra $\mathfrak{gl}(L/S)$ with $E^2 = 0$. Hence, by Lemma 5.18, we also have $F^2 = 0$. It readily follows that the subalgebra of $\operatorname{End}(L/S)$ generated by E, F and H is linearly spanned by $1, E, F, H, EF, EH$, and FH, and that the relation $H^3 = H$ is satisfied, i.e. $(H^2 - 1)H = 0$. Consequently, there are subspaces U, V of L such that $L = S + U + V$ is a direct sum such that $(S + U)/S = \ker(H^2 - 1)$ and $(S + V)/S = \ker(H)$. Let $x \in L^h_2$. We can write $x = \alpha e + \beta f + \gamma h + u + v$, where $\alpha, \beta, \gamma \in \mathbb{F}$, $u \in U$ and $v \in V$. As $x \in L^h_2$, we have

$$\operatorname{ad}^2_h x = 4x = 4\alpha e + 4\beta f + 4\gamma h + 4u + 4v.$$

On the other hand,

$$\operatorname{ad}^2_h x = \operatorname{ad}^2_h(\alpha e + \beta f + \gamma h) + \operatorname{ad}^2_h(u) + \operatorname{ad}^2_h(v) = 4\alpha e + 4\beta f + u + s + t$$

where $s, t \in S$. Since the sum $S + U + V$ is direct, we get from these equalities that $u = 0$ and $v = 0$, and therefore that $L^h_2 \subset S$. Since the eigenvalue 2 has multiplicity 1 on S, $L^h_2 \subset \mathbb{F}e$, the other inclusion are trivial. Similarly it is proved that $L^h_{-2} = \mathbb{F}f$. □

REMARK 6.18. Recall that if $\mathrm{char}(\mathbb{F}) = 0$ or $p > 5$, then the decomposition $L = L^h_{-2} \oplus L^h_{-1} \oplus L^h_0 \oplus L^h_1 \oplus L^h_2$ is actually a $\mathbb{Z}$-grading and the equalities $L^h_2 = \mathbb{F}e$ and $L^h_{-2} = \mathbb{F}f$ follows from (iv) of Theorem 5.11:

$$L^h_2 = \mathrm{ad}_e^2 L = \mathbb{F}e \quad \text{and} \quad L^h_{-2} = \mathrm{ad}_f^2 L = \mathbb{F}f,$$

so f is also an extremal element and L is generated by extremal elements. However, this last result does not hold in general if $p = 5$.

PROPOSITION 6.19. *Let L be a Lie algebra over a field $\mathbb{F}$ of characteristic $p \neq 2, 3$, and let (e, f, h) be a $\mathfrak{sl}_2$-triple, where $e \in \mathfrak{E}(L)$. Let $L_i = L^h_i$ be the i-component of the $\mathbb{Z}_p$-grading defined by ad_h. Then either $p = 5$ and $\mathrm{ad}_f^2 L_1 = \mathbb{F}e$, or f is also extremal and the components L_i actually define a $\mathbb{Z}$-grading, with $L_2 = \mathbb{F}e$, $L_{-2} = \mathbb{F}f$, $[e, L_{-1}] = L_1$ and $[f, L_1] = L_{-1}$.*

PROOF. The identifications of L_2 and L_{-2} with $\mathbb{F}e$ and $\mathbb{F}f$ were proved in Proposition 6.17. Suppose that f is not extremal. As

$$\mathrm{ad}_f^2 L_2 \subset L_{-2} = \mathbb{F}f, \quad \mathrm{ad}_f^2 L_0 \subset [f, L_{-2}] = \mathbb{F}[f, f] = 0, \quad \text{and}$$
$$\mathrm{ad}_f^2 L_1 \subset [f, L_{-1}] \subset L_{-3},$$

this can only happen if $[f, L_{-1}] \neq 0$ (since $L_{-3} = L_{-2}$ would force $p = 1$). Then, again by the $\mathbb{Z}_p$-grading properties, $[f, L_{-1}] \subset L_{-3}$ and so -3 is equal to a member i of $\{-2, -1, 0, 1, 2\}$ modulo p. Since $p \geq 5$, this implies $p = 5$ and $i = 2$. Thus $[f, L_{-1}] = \mathbb{F}e$. Then, for every $u \in L_{-1}$, $\mathrm{ad}_e \mathrm{ad}_f u = 0$, so $\mathrm{ad}_f \mathrm{ad}_e u = -\mathrm{ad}_h u = u$. Therefore, $\mathrm{ad}_f^2 L_1 \supset \mathrm{ad}_f^2[e, L_{-1}] = [f, L_{-1}] = \mathbb{F}e$, and since $\mathrm{ad}_f^2 L_1 \subset [f, L_{-1}] = \mathbb{F}e$, we actually have $\mathrm{ad}_f^2 L_1 = \mathbb{F}e$, so the first case holds. To complete the proof, assume that f is also extremal. By the argument above, $[f, L_{-1}] = 0$ and $[e, L_1] = 0$, since e is extremal. Hence it is easy to see that the decomposition

$$L = L_{-2} \oplus L_{-1} \oplus L_0 \oplus L_1 \oplus L_2$$

is indeed a $\mathbb{Z}$-grading. To show the last two equalities, note that for every $v \in L_1$, $\mathrm{ad}_e \mathrm{ad}_f v = \mathrm{ad}_h v = v$, since $[e, L_1] \subset L_3 = 0$ and hence

$$L_1 = \mathrm{ad}_e \mathrm{ad}_f L_1 \subset \mathrm{ad}_e L_{-1} \subset L_1.$$

Similarly, $\mathrm{ad}_f \mathrm{ad}_e u = -\mathrm{ad}_h u = u$ for every $u \in L_{-1}$. □

REMARK 6.20. The first case in the above proposition actually occurs. Suppose that $\mathrm{char}(\mathbb{F}) = 5$ and let $W(5)$ be the five-dimensional Witt algebra over $\mathbb{F}$, i.e. $W(5)$ has a basis $\{x_0, x_1, x_2, x_3, x_4\}$ such that $[x_i, x_j] = (j - i)x_{i+j-1}$, with $x_k = 0$ if $k \neq 0, 1, 2, 3, 4$. As noted in Remark 5.10, $W(5)$ is a simple Lie algebra containing the $\mathfrak{sl}_2$-triple $(-x_2, x_0, 2x_1)$, where x_2 is extremal but x_0 is not even Jordan ($\mathrm{ad}_{x_0}^3 x_2 = 6x_0 \neq 0$). In fact we have that $L_{-1} = \mathbb{F}x_3$ and $[x_o, L_{-1}] = \mathbb{F}[x_0, x_3] = \mathbb{F}x_2$. Since x_4 is an absolute zero divisor, $W(5)$ is not generated by extremal elements by Proposition 6.13.

6.4. Simple Lie algebras with extremal elements

Let L be a simple finite-dimensional Lie algebra over a field $\mathbb{F}$ of characteristic distinct from 2 and 3 containing an extremal element which is not an absolute zero divisor. Following **[CIR08]**, we give in this section a classification-free proof of the fact that either L is generated by extremal elements, or $\mathbb{F}$ has characteristic 5 and L is isomorphic to $W(5)$. We recall (Remark 6.18) that if $\mathrm{char}(\mathbb{F})$ is 0 or greater than 5, then L is generated by extremal elements, but, as remarked in 6.20, this does not hold for $W(5)$. We begin with an extension of $W(5)$ which plays an essential role in the proof.

DEFINITION 6.21. The Lie algebra $\widetilde{W(5)}$ is defined by giving a basis and its multiplication table as follows: To the standard basis $\{x_0, x_1, x_2, x_3, x_4\}$ of $W(5)$ add the new element x_6. The only entry of the multiplication table that differs between $\widetilde{W(5)}$ and $W(5)$ is $[x_3, x_4]$: This is 0 in $W(5)$ and x_6 in $\widetilde{W(5)}$. Furthermore, x_6 commutes with all other elements. So $\widetilde{W(5)}$ is indeed an extension of $W(5)$ by a one-dimensional center, $W(5) \cong \widetilde{W(5)}/\mathbb{F}x_6$.

PROPOSITION 6.22. *Suppose that L is a finite-dimensional simple Lie algebra over a field $\mathbb{F}$ of characteristic $p \neq 2, 3$ containing a $\mathfrak{sl}_2$-triple (e, f, h) such that $e \in \mathfrak{E}(L)$ but f is not extremal. Then $p = 5$ and L is isomorphic to the Witt algebra $W(5)$.*

PROOF. Since f is not extremal, we have by Proposition 6.19 that $p = 5$ and $\mathrm{ad}_f^2 L_1 = \mathbb{F}e$. Let $v \in L_1$ be such that $[f, [v, f]] = e$. Consider the linear span W in L of $e, f, h, v, [v, f]$ and $a = [v, [v, f]]$. The multiplication on these elements is fully determined:

(1) $[e, f] = h$,
(2) $[h, e] = 2e$,
(3) $[h, f] = -2f$,
(4) $[e, v] = 0$,
(5) $[e, [v, f]] = -v$,
(6) $[e, a] = 0$,
(7) $[f, [v, f]] = e$,
(8) $[f, a] = 0$,
(9) $[h, v] = v$,
(10) $[h, [v, f]] = -[v, f]$,
(11) $[h, a] = 0$,
(12) $[v, a] = 0$,
(13) $[[v, f], a] = 0$.

Identities (1), (2) and (3) follow from the definition of $\mathfrak{sl}_2$-triple, (7) from the definition of v, and (9), (10) and (11) by the grading. Since $[e, v] \in L_3 = L_{-2} = \mathbb{F}f$, $[e, v] = \alpha f$ for some $\alpha \in \mathbb{F}$ and hence $\alpha h = [e, [e, v]] \in \mathbb{F}e \cap L_5 = L_2 \cap L_0 = 0$, so $[e, v] = 0$, which proves (4). Identity (5) follows from (4) applying Jacobi: $[e, [v, f]] = [[e, v], f] + [v, [e.f]] = [v, h] = -v$. Identities (6) and (8) are proved in a similar way:

$$[e, a] = [v, [e, [v, f]]] = -[v, v] = 0, \text{ and}$$
$$[f, a] = [v, [f, [v, f]]] = [v, e] = 0.$$

The proof of (12) requieres further calculation. We must show that $[v, a] = 0$:

$$\begin{aligned}
[v, a] &=^{(5)} [[e, [f, v]], a] =^{(\text{Jacobi})} [[e, a], [f, v]] + [e, [[f, v], a]] \\
&=^{(6)} [e, [[f, v], a]] =^{(\text{defn of } v)}= [[f, [f, v]], [[f, v], a]] \\
&=^{(\text{Jacobi})} [[f, [[f, v], a]], [f, v]] + [f, [[f, v], [[f, v], a]]].
\end{aligned}$$

Note that $[[f,v],[[f,v],a]] \in L_{-2+1-2+1+0} = L_{-2} = \mathbb{F}f$ and thus the second term is 0. For the first one, apply Jacobi once again:

$$[f,[[f,v],a]] = [[f,[f,v]],a] + [[f,v],[f,a]]$$
$$=^{(8)} [[f,[f,v]],a] =^{(\text{defn of } v)} [e,a] =^{(6)} 0.$$

Identity (13) is a consequence of (8) and (12) applying Jacobi.

It follows from the equations listed above that $a = [v,[v,f]]$ is a central element of W and the quotient of W by the ideal generated by a is a simple Lie algebra of dimension 5. We claim that if $a \neq 0$, then $W \cong W(5)$; otherwise W is isomorphic to $\widetilde{W(5)}$ as defined in 6.21. The map $\varphi : \widetilde{W(5)} \to W$, defined by

$$\varphi(x_0) = f,\ \varphi(2x_1) = h,\ \varphi(-x_2) = e,\ \varphi(x_3) = v,\ \varphi(x_4) = [v,f],\ \varphi(x_6) = a,$$

is an epimorphism of Lie algebras mapping the unique nontrivial ideal $\mathbb{F}x_6$ onto $\mathbb{F}a$. Thus φ is an isomorphism of $\widetilde{W(5)}$ onto W if $a \neq 0$, and induces an isomorphism of $W(5) \cong \widetilde{W(5)}/\mathbb{F}x_6$ onto W otherwise.

It remains to prove that L coincides with W, because then $L \cong W(5)$, since $\widetilde{W(5)}$ is not simple. Suppose that L strictly contains W and consider L as a W-module. We compute in $\mathrm{End}_{\mathbb{F}}(L/W)$ the subalgebra generated by ad_W. Write E, F, V for the actions of $\mathrm{ad}_e, \mathrm{ad}_f, \mathrm{ad}_v$, resp. on L/W. Then (E,F) is a $\mathfrak{sl}_2$-pair of the Lie algebra $\mathrm{End}_{\mathbb{F}}(L/W)^-$ with $E^2 = 0$. Hence, by Lemma 5.18,

(14) $F^2 = 0$, and
(15) $F - FEF = 0$.

It follows from the definition of v and (14) that

(16) $E - 2FVF = 0$, and
(17) $EVF - EFV - VFE + FVE + V = 0$.

Using (15), (16) and the fact that $E^2 = 0$, we get

$$0 = (E - 2FVF)(1 - EF) = E - 2FVF - E^2F + 2FV(FEF) = E,$$

which immediately implies by (15) that $F = 0$, and by (17) that $V = 0$. So the images of ad_w, for $w \in W$, in $\mathrm{End}_{\mathbb{F}}(L/W)$ are trivial. This means that W is an ideal of L. Since L is simple and W is nontrivial, $L = W$. □

Now we proceed with the remaining case, namely, we will assume that both components e and f of the $\mathfrak{sl}_2$-pair are extremal.

PROPOSITION 6.23. *Assume that L is a simple Lie algebra over a field $\mathbb{F}$ of characteristic $p \neq 2,3$ having a $\mathfrak{sl}_2$-pair (e,f) of extremal elements. If L is not isomorphic to $W(5)$, then L is generated by extremal elements.*

PROOF. As remarked in 6.18, we may assume $p = 5$. Since f is extremal, $\mathrm{ad}_f^2 L_1 = 0$ and hence, by Proposition 6.19, $L_2 = \mathbb{F}e$, $L_{-2} = \mathbb{F}f$, $[e, L_{-1}] = L_1$, and $[f, L_1] = L_{-1}$, where $L = L_{-2} \oplus L_{-1} \oplus L_0 \oplus L_1 \oplus L_2$ is the $\mathbb{Z}_5$-grading induced by ad_h, $h = [e,f]$.

Consider the subalgebra I of L generated by e, f and L_1. As $L_{-1} = [f, L_1]$, the subalgebra I contains the linear subspace $J = L_{-2} + L_{-1} + L_1 + L_2$. As $[J, L_0] \subset J$ and J generates I, we have $[I, J_0] \subset I$. This implies that I is an ideal of L, and so, by simplicity of L, it coincides with L. Therefore, it suffices to show that for each $z \in L_1$, there exists an extremal element $u \in L$ such that z lies in the subalgebra generated by e, f and u.

The following relations hold in L for some $\alpha \in \mathbb{F}$.

(1) $[h, e] = 2e$,
(2) $[h, f] = -2f$,
(3) $[h, z] = z$,
(4) $[e, z] = 0$ (by the grading),
(5) $\mathrm{ad}_f^2 z = 0$ (by the grading),
(6) $[e, [f, z]] = z$ (by (3) and (4)),
(7) $[e, \mathrm{ad}_z^2 f] = 0$ (by (4) and (6)),
(8) $[f, \mathrm{ad}_z^2 f] = 0$ (by (5)),
(9) $[e, \mathrm{ad}_z^3 f] = 0$ (by (4) and (7)),
(10) $]\ \mathrm{ad}_f^2 \mathrm{ad}_z^3 f = 0$ (by the grading),
(11) $[e, [f, \mathrm{ad}_z^3 f]] = 0$ (by the grading and (9)),
(12) $\mathrm{ad}_z^4 f = \alpha e$ (by the grading).

We claim that the subalgebra S of L generated by e, f and z is linearly spanned by the set B consisting of the following eight elements:

$$f \in L_{-2}; \quad [f, z], [f, \mathrm{ad}_z^3 f] \in L_{-1}; \quad h, \mathrm{ad}_z^2 f \in L_0; \quad z, \mathrm{ad}_z^3 f \in L_1; \quad e \in L_2.$$

To see that this is true, we verify that the image of each elements of B under any of the adjoint operators ad_e, ad_f and ad_z is a scalar multiple of some element of B. For ad_e and ad_f this is straightforward. As for ad_z, the statement is trivially verified for all the elements of B but $[f, \mathrm{ad}_z^3 f]$. Put $h_1 = \mathrm{ad}_z^2 f$. Since $[f, h_1] = 0$ by (8), we have

$$\begin{aligned}\mathrm{ad}_z[f, \mathrm{ad}_z^3 f] &= [[z, f], \mathrm{ad}_z^3 f] + [f, \mathrm{ad}_z^4 f] = [[z, f], [z, h_1]] + [f, \alpha e]\\ &= [[[z, f], z], h_1] + \mathrm{ad}_z[[z, f], h_1] - \alpha h\\ &= -[h_1, h_1] + \mathrm{ad}_z[\mathrm{ad}_z^3 f, f] - \alpha h,\end{aligned}$$

so $\mathrm{ad}_z[f, \mathrm{ad}_z^3 f] = -\frac{\alpha}{2} h$. This establishes the claim that S is linearly spanned by B.

We exhibit an element $u \in S$ as specified. Since $z \in L_1$, ad_z is nilpotent of index at most 5 and $\exp(\mathrm{ad}_z)$ is a well-defined linear map since $p \neq 2, 3$. Put

$$u = \exp(\mathrm{ad}_z) f = f + \mathrm{ad}_z f + \frac{1}{2} \mathrm{ad}_z^2 f + \frac{1}{6} \mathrm{ad}_z^3 f + \frac{1}{24} \mathrm{ad}_z^4 f.$$

A straightforward computation shows that (e, u) is a $\mathfrak{sl}_2$-pair. By (6), (7), (9) and (12) we find

$$[e, u] = [e, f] + [e, [z, f]] = h - z,$$

so, by (1) and (4),

(13) $[[e, u], e] = [h - z, e] = [h, e] - [z, e] = 2e.$

Now we compute $[[e, u], u] = [h - z, u]$. By using the grading and equations (4) and (12):

$[h - z, f] = -2f - \mathrm{ad}_z f$,

$[h - z, \mathrm{ad}_z f] = [h, \mathrm{ad}_z f] - \mathrm{ad}_z^2 f = -\mathrm{ad}_z f - \mathrm{ad}_z^2 f$,

$[h - z, \frac{1}{2} \mathrm{ad}_z^2 f] = -\frac{1}{2} \mathrm{ad}_z^3 f$,

$[h - z, \frac{1}{6} \mathrm{ad}_z^3 f] = \frac{1}{6} \mathrm{ad}_z^3 f - \frac{1}{6} \mathrm{ad}_z^4 f$,

$[h-z, \frac{1}{24}\operatorname{ad}_z^4 f] = \frac{1}{12}\operatorname{ad}_z^4 f$. So,

$$\begin{aligned}[[e,u],u] &= -2f - \operatorname{ad}_z f - \operatorname{ad}_z f - \operatorname{ad}_z^2 f - \frac{1}{2}\operatorname{ad}_z^3 f + \frac{1}{6}\operatorname{ad}_z^3 f - \frac{1}{6}\operatorname{ad}_z^4 f + \frac{1}{12}\operatorname{ad}_z^4 f \\ &= -2f - 2\operatorname{ad}_z f - \operatorname{ad}_z^2 f - \frac{2}{6}\operatorname{ad}_z^3 f - \frac{2}{24}\operatorname{ad}_z^4 f = -2u,\end{aligned}$$

which proves, together with (13), that (e,u) is a $\mathfrak{sl}_2$-pair. Furthermore, since, by assumption, L is not isomorphic to $W(5)$, we have by Proposition 6.22 that u is extremal. We verify that z lies in the subalgebra T of L generated by the three extremal elements e, f and u. Observe that

$$\operatorname{ad}_z f + \frac{1}{2}\operatorname{ad}_z^2 f + \frac{1}{6}\operatorname{ad}_z^3 f = u - f - \frac{\alpha}{24}e.$$

Acting by ad_e and using (6), (7) and (9), we get

$$-z = [e,u] - [e,f] \in T.$$

This proves that z belongs to T so completing the proof. □

REMARK 6.24. The idea behind the choice of the element u in the proof of Proposition 6.23 is the fact that $\exp(\operatorname{ad}_z)$ is an automorphism of the subalgebra S of L, and therefore $(e,u) = (\exp(\operatorname{ad}_z)e, \exp(\operatorname{ad}_z)f)$ is a $\mathfrak{sl}_2$-pair. However, as the index of ad_z can be greater than 3, we cannot use Corollary 1.3. So to prove that $\exp(\operatorname{ad}_z) \in \operatorname{Aut}(S)$ it would be necessary to verify it for each pair of elements of the basis B of S. Needless to say that it is easier to check directly that (e,u) is a $\mathfrak{sl}_2$-pair.

THEOREM 6.25. *Let L be a finite-dimensional simple Lie algebra over a field $\mathbb{F}$ of characteristic $p \neq 2, 3$. If L contains an extremal element e which is not an absolute zero divisor, then either* $\operatorname{char}(\mathbb{F}) = 5$ *and L is isomorphic to $W(5)$, or L is generated by extremal elements.*

PROOF. By Proposition 6.17, we can extend e to a $\mathfrak{sl}_2$-pair. Then Propositions 6.22 and 6.23 apply to complete the proof. □

REMARK 6.26. Let L be a finite-dimensional Lie algebra over an algebraically closed field of characteristic $p > 5$. A. A. Premet proved in [**Pre86b**, Theorem 1] ([**Pre86a**] for $p > 3$) that L contains an one-dimensional inner ideal. As a consequence, if L is nondegenerate, then it has a nontrivial finite $\mathbb{Z}$-grading (Remark 6.18). If L is in fact simple, then it is nondegenerate if and only if it is generated by extremal elements, which completes the proof of Kostrikin's conjecture that a finite-dimensional simple Lie algebra over an algebraically closed field of characteristic $p > 5$ is classical if and only if it is nondegenerate.

6.5. Exercises

EXERCISE 6.27. Verify the assertions (1), (2), and (3) in Example 6.2.

EXERCISE 6.28. Let L be a classical linear Lie algebra of type $\mathrm{A}_l, \mathrm{B}_l, \mathrm{C}_l$, or D_l (see [**Hum72**]). Show that L is spanned by its extremal elements.

EXERCISE 6.29. Let L be a Lie algebra over a field $\mathbb{F}$ and let $e \in L$ be an extremal element. Show that $[[x,e,t,e],[x,e,z,e] = 0$ for all $x, y, z \in L$.

EXERCISE 6.30. Let L be a simple finite-dimensional Lie algebra over an algebraically closed field of characteristic 0. Show that $a \in L$ is extremal if and only if $a \in L_\alpha$, where α is a long root relative to some Cartan subalgebra. (See [**Hum72**] for definitions.)

EXERCISE 6.31. [**CSUW01**, Lemma 3.1] Let L be a Lie algebra generated by two nonzero extremal element over a field $\mathbb{F}$ of characteristic not 2. Show that one the following possibilities holds:

(1) L is abelian,
(2) L is the Heisenberg algebra (see [**Hum72**]),
(3) $L \cong \mathfrak{sl}_2(\mathbb{F})$.

EXERCISE 6.32. In the Lie algebra $\mathfrak{fo}(X, \langle\, , \rangle)$ over a field $\mathbb{F}$ of characteristic not 2, show that the subspace $B = [x, S]$, where S is a totally isotropic subspace of X and x is a nonzero vector of S, is an abelian inner ideal with every element being extremal. These abelian inner ideals are called point spaces and will be studied in Section 13.3.

EXERCISE 6.33. [**CSUW01**, Lemma 9.5] Let L be a finite-dimensional Lie algebra over a field of characteristic 0, and suppose that L is spanned by its extremal elements. Show that the Killing form κ and the form ϵ (see Theorem 6.9) have the same radical.

EXERCISE 6.34. In the notation of Proposition 6.19, show that $\operatorname{ad}_f^2 L_{-1} = \mathbb{F}f$ (resp. $\operatorname{ad}_f^2 L_{-2} = \mathbb{F}f$) forces $p = 3$ (resp. $p = 2$).

EXERCISE 6.35. In the notation of Proposition 6.23, show that $\exp(\operatorname{ad}_x)$ is an automorphism of S.

CHAPTER 7

A Characterization of Strong Primeness

An associative algebra R is prime if and only if $aRb \neq 0$ for arbitrary nonzero elements $a, b \in R$. A similar characterization cannot be expected for Lie algebras, since, as we have seen, there exist simple Lie algebras in positive characteristic which are degenerate. Nevertheless, it is possible to get a characterization of strong primeness in terms of elements. Following a paper of E. García and M. Gómez, we prove in this chapter that a 6-torsion free Lie algebra L is strongly prime, i.e. L is prime and nondegenerate, if and only if $[x, [y, L]] \neq 0$ for any nonzero elements $x, y \in L$. Similar characterizations are known for Jordan systems.

7.1. Orthogonality relations of adjoint operators

The results of this chapter are taken from [**GGL07**]. Let L be a Lie algebra over a ring of scalar Φ and let $x, y \in L$. We say that x is *strongly orthogonal* to y if $x \in \mathrm{Ann}(\mathrm{id}(y))$. Since the annihilator of an ideal is an ideal, this relation of strong orthogonality between elements is symmetric. We will prove in this section that if L is nondegenerate, then x, y are strongly orthogonal if and only if $\mathrm{ad}_x \mathrm{ad}_a \mathrm{ad}_y = 0$ for every $a \in L$. Once again we will use capital letters to denote adjoint operators. Note that, according to this notation, nondegeneracy of a Lie algebra L means $X^2 = 0 \Rightarrow x = 0$, $x \in L$.

LEMMA 7.1. (`Switching lemma`) *Let L be a Lie algebra, let n be a non-negative integer and let $x, y \in L$ be such that $XA_1 \cdots A_k Y = YA_1 \cdots A_k X = 0$ for every $0 \le k \le n$ and $a_1, \ldots, a_k \in L$. Then $YA_1 \cdots A_{n+1} X = (-1)^{n+1} XA_1 \cdots A_{n+1} Y$ for $a_1, \ldots, a_{n+1} \in L$.*

PROOF.

$$\begin{aligned} YA_1 \cdots A_{n+1} X &= YA_1 \cdots [A_{n+1}, X] = Y[A_1, [\cdots, [A_{n+1}, X]]] \\ &= [Y, [A_1, [\cdots, [A_{n+1}, X]]]] + [A_1, [\cdots, [A_{n+1}, X]]]Y \\ &= [A_1, [\cdots, [A_{n+1}, X]]]Y = (-1)^{n+1} XA_{n+1}A_n \cdots A_1 Y, \end{aligned}$$

since $[Y, [A_1, [\cdots, [A_{n+1}, X]]]] = -\mathrm{ad}_{[y,[a_1,[\ldots,[x,a_{n+1}]]]]}$ and $YA_1 \cdots A_k X = 0$ for every $0 \le k \le n$ by hypothesis. $\square$

PROPOSITION 7.2. *Let L be nondegenerate and let $x, y \in L$ be such that $XY = 0$. Then any product $Z_1 Z_2 \cdots Z_n$ is 0 whenever the number of the $z_i = x$, say s, and the number of the $z_j = y$, say r, are both nonzero and satisfy $n + 1 < 2(r + s)$.*

PROOF. By induction on n. If $n = 2$ then $XY = 0$ by definition, and

$$\mathrm{ad}^2_{[x,y]} = [X, Y]^2 = (-YX)^2 = 0 \Rightarrow [x, y] = 0.$$

Since L is nondegenerate, $[x, y] = 0$ and hence $YX = \mathrm{ad}_{[y,x]} + XY = 0$.

Suppose that the proposition is true for every $2 \leq j \leq n-1$ and prove it for $j = n$. Let $Z_1 Z_2 \cdots Z_n$ be such that the number of $z_i = x$, say s, and the number of $z_j = y$, say r, are both nonzero and satisfy $n + 1 < 2(r + s)$.

(a) Consider first the case that there exist $1 \leq k_1 < k_2 < k_3 \leq n$ such that $z_{k_1} = z_{k_3} = x$ and $z_{k_2} = y$. Let s' (resp. r') be the number of $z_i = x$ (resp. $z_i = y$) such that $1 \leq i \leq k_2$, and let s'' (resp. r'') be the number of $z_j = x$ (resp. $z_j = y$) such that $k_2 \leq j \leq n$. Then $s' = s - s''$ and $r' = r - r'' + 1$. So s', s'', r', r'' are nonzero and either $k_2 + 1 < 2(s' + r')$ (in this first case $Z_1 Z_2 \cdots Z_{k_2} = 0$ by the induction hypothesis), or $k_2 + 1 \geq 2(s' + r')$. In this second case we have

$$\begin{aligned} (n - k_2 + 1) + 1 &= n + 3 - (k_2 + 1) \leq n + 3 - 2s' - 2r' \\ &= n + 3 - 2s + 2s''2r + 2r'' - 2 = n + 1 \\ &- 2(s + r) + 2s'' + 2r'' < 2s'' + 2r'', \end{aligned}$$

And since $n - k_2 + 1$ is the number of z_j such that $k_2 \leq j \leq n$, we can again apply the induction hypothesis to get $Z_{k_2} \cdots Z_{n-1} Z_n = 0$.

If $z_{k_1} = z_{k_3} = y$ and $z_{k_2} = x$, the meanings of s, s' and s'' can be exchanged with r, r' and r'' respectively to also obtain $Z_1 Z_2 \cdots Z_n = 0$.

(b) Suppose now that there exist $1 \leq k < k' \leq n$ such that $z_k = x$ and $z_{k'} = y$, if $z_i = x$ then $i \leq k$, and if $z_i = y$ then $i \geq k'$. Since we are assuming $n > 2$, $s + r$ is greater than 2 and we may suppose $s > 1$ (if $s = 1$ we exchange the roles of x and y). Note that if $k = n - 1$ then $k' = n$ necessarily, so $Z_{n-1} Z_n = XY = 0$. Thus we may suppose that the result is true when the furthest position of a X is greater than k and then prove that it holds when X appears at the position k. We have

$$\begin{aligned} Z_1 \cdots Z_{k-1} X Z_{k+1} \cdots Z_n &= Z_1 \cdots Z_{k-1} Z_{k+1} X Z_{k+2} Z_{k+3} \cdots Z_n \\ &+ Z_1 \cdots Z_{k-1} \ \mathrm{ad}_{[x, z_{k+1}]} \ Z_{k+2} Z_{k+3} \cdots Z_n \\ &= Z_1 \cdots Z_{k-1} Z_{k+1} X Z_{k+2} Z_{k+3} \cdots Z_n \\ &+ Z_1 \cdots Z_{k-1} Z_{k+2} \ \mathrm{ad}_{[x, z_{k+1}]} \ Z_{k+3} \cdots Z_n \\ &+ Z_1 \cdots Z_{k-1} \ \mathrm{ad}_{[[x, z_{k+1}], z_{k+2}]} \ Z_{k+3} \cdots Z_n. \end{aligned}$$

(i) If $k' = k + 1$ then $Z_{k+1} = Y$ and we have $X Z_{k+1} = XY = 0$.
(ii) If $k' = k + 2$ then $Z_{k+2} = Y$ and we have that the first summand in the formula above is zero (since $X Z_{k+2} = XY = 0$), the second summand is also zero by (a) (since $s > 1$), and the third one is also zero since

$$\mathrm{ad}_{[[x, z_{k+1}], z_{k+2}]} = \mathrm{ad}_{[[x, z_{k+1}], y]} = 0.$$

(iii) If $k' > k + 2$, i.e. neither z_{k+1} nor z_{k+2} is equal to y, then the first and the second summands are zero by the induction hypothesis on k, since X has moved at least one place to the right. Note that

$$Z_{k+2} \ \mathrm{ad}_{[x, z_{k+1}]} \ Z_{k+3} = Z_{k+2} X Z_{k+1} Z_{k+3} - Z_{k+2} Z_{k+1} X Z_{k+3}.$$

The nullity of the third summand follows by the induction hypothesis on the length of the expression. Indeed, in this summand the number of factors equal to X is $s' = s - 1$, while the number of factors equal to Y is $r' = r$, and its length is $n' = n - 2$. Then, since $s > 1$, $s' > 0$, we have

$$2(r' + s') = 2r + 2s - 2 > n + 1 - 2 = n - 1 = n' + 1,$$

so the induction hypothesis applies. □

PROPOSITION 7.3. (Going-down proposition) *Let L be nondegenerate, let n be a positive integer, and let $x, y \in L$ be such that $XA_1A_2\cdots A_nY = 0$ for $a_1, a_2, \ldots, a_n \in L$. Then $XA_1A_2\cdots A_kY = 0$ and $YA_1A_2\cdots A_kX = 0$ for every $0 \le k \le n$ and $a_1, a_2, \ldots, a_k \in L$. Furthermore, $[x, y] = 0$.*

PROOF. We begin by proving that for any $0 \le k \le n$ and $a_1, a_2, \ldots, a_n \in L$, we have

$$XA_1\cdots A_kYA_{k+1}\cdots A_n = 0. \tag{7.1}$$

The following relation holds for arbitrary elements $y, a_k, \ldots, b$ of any Lie algebra

$$\begin{aligned}[y, a_k, \ldots, a_n, b] &= [y, [a_k, [a_{k+1}, \ldots, a_n, b]]] \\ &= [[y, a_k], [a_{k+1}, \ldots, a_n, b]] + [a_k, [y, a_{k+1}, \ldots, a_n, b]] \\ &= -[[a_{k+1}, \ldots, a_n, b], y, a_k] + [a_k, y, a_{k+1}, \ldots, a_n, b].\end{aligned} \tag{7.2}$$

Starting from $XA_1A_2\cdots A_nY = 0$ (case $k = n$) we will reach (7.1) in a downward induction process by moving Y one place to the left in each step.

Suppose then that (7.1) holds for $0 < k \le n$. By (7.2),

$$\begin{aligned}[x, a_1, \ldots, a_{k-1}, y, a_k, \ldots a_n, b] &= -[x, a_1, \ldots, a_{k-1}, [a_{k+1}, \ldots, a_n, b], y, a_k] \\ &+ [x, a_1 \ldots, a_{k-1}, a_k, y, a_{k+1}, \ldots, a_n, b] = 0,\end{aligned}$$

since the first summand is zero because $XA_1\cdots A_{k-1}\operatorname{ad}_{[a_{k+1},\ldots,a_n,b]}Y$ begins with X and ends with Y and if we spans the factor $\operatorname{ad}_{[a_{k+1},\ldots,a_n,b]}$ the terms in the middle contains n factors of the form A_i or B (so the initial hypothesis applies), while the second summand is zero by the induction hypothesis on k. (Note that the equality $XYA_1\cdots A_n = 0$ is the case $k + 1 = 1$).

Now fix b and $a_1, \ldots, a_n \in L$, and set $t := A_1\cdots A_{n-1}Y(b)$. We have

$$XT = X\operatorname{ad}_{[a_1,\ldots,a_{n-1},y,b]} = X[A_1, \cdots, A_{n-1}, Y, B] = 0$$

(span the bracket and apply (7.1)). Hence $\operatorname{ad}^2_{[x,t]} = (XT - TX)^2 = 0$ implies (by nondegeneracy of L) $[x, t] = 0$, i.e. $XA_1\cdots A_{n-1}Y = 0$. By downward induction we get the remaining equalities $XA_1\cdots A_kY = 0$. Now, by Lemma 7.1, we also have $YA_1\cdots A_kX = 0$. □

In the following two propositions we list some orthogonal relations between adjoint operators which will be used in what follows.

PROPOSITION 7.4. (Orthogonality relations 1) *Let L be a nondegenerate Lie algebra and let $x, y \in L$ be such that $XY = 0$. Then:*

(OR1) $XY = 0$ and $[x, y] = 0$,
(OR2) $XAY = -YAX$, $a \in L$,
(OR3) $X^2AY = XAY^2 = X^2ABY^2 = 0$, $a, b \in L$,
(OR4) $X^2ABY = X^2BAY$ and $XABY^2 = XBAY^2$ $a, b \in L$,
(OR5) $X^2ABY + YABX^2 = 2(YAXBX + YBXAX)$, $a, b \in L$,
(OR6) $X^2A_1A_2A_3Y^2 = \sum_{\sigma\in S_3} 2XA_{\sigma_1}XA_{\sigma_2}YA_{\sigma_3}Y$, $a_1, a_2, a_3 \in L$.

PROOF. Assertions (OR1), (OR2) and (OR3) follow from Proposition 7.3, Lemma 7.1, and Proposition 7.2 resp. while (OR4) is a direct consequence of

(OR3): $X^2ABY - X^2BAY = X^2\,\mathrm{ad}_{[a,b]}\,Y = 0$. Finally,

$$\begin{aligned} X^2ABY &= X[X,A]BY + XAXBY = X[[X,A],B]Y + XB[X,A]Y \\ &+ XAXBY = -Y[[X,A],B]X + XBXAY + XAXBY \\ &= YAXBX + YBXAX - YABX^2 + XBXAY + XAXBY \\ &= 2YAXBX + 2YBXAX - YABX^2, \end{aligned}$$

and

$$\begin{aligned} X^2A_1A_2A_3Y^2 &= X[X,A_1]A_2A_3Y^2 + XA_1XA_2A_3Y^2 = X[[X,A_1],A_2]A_3Y^2 \\ &+ XA_2[X,A_1]A_3Y^2 + XA_1XA_2A_3Y^2 = XA_3[[X,A_1],A_2]Y^2 \\ &+ XA_2XA_1A_3Y^2 + XA_1XA_2A_3Y^2 = XA_3XA_1A_2Y^2 \\ &+ XA_2XA_1A_3Y^2 + XA_1XA_2A_3Y^2 = 2XA_3XA_1YA_2Y \\ &+ 2XA_3XA_2YA_1Y + 2XA_2XA_1YA_3Y + 2XA_2XA_3YA_1Y \\ &+ 2XA_1XA_2YA_3Y + 2XA_1XA_3YA_2Y. \end{aligned}$$

□

PROPOSITION 7.5. (Orthogonality relations 2) *Let L be a nondegenerate Lie algebra and let $x, y \in L$ be such that $XAY = 0$ for every $a \in L$. Then $XY = 0$. Furthermore,*

(OR7) $XA_1A_2Y = XA_2A_1Y, \quad a_1, a_2 \in L,$
(OR8) $XA_1A_2Y = YA_1A_2X, \quad a_1, a_2 \in L,$
(OR9) $X^2A_1A_2A_3Y = 0, \quad a_1, a_2, a_3 \in L,$
(OR10) $XA_1A_2A_3Y^2 = 0, \quad a_1, a_2, a_3 \in L,$
(OR11) $X^2A_1A_2A_3A_4Y^2 = 0, \quad a_1, a_2, a_3, a_4 \in L.$

PROOF. (OR7) follows from the hypothesis:

$$XA_1A_2Y - XA_2A_1Y = X\,\mathrm{ad}_{[a_1,a_2]}\,Y = 0$$

and (OR8) from Lemma 7.1, while (OR9) is a consequence of the previous ones:

$$\begin{aligned} X^2A_1A_2A_3Y &= X[X,A_1]A_2A_3Y + XA_1XA_2A_3Y \\ &= X[[X,A_1],A_2]A_3Y + XA_2[X,A_1]A_3Y \\ &+ (XA_1Y)A_2A_3X = XA_3[[X,A_1],A_2]Y \\ &= XA_3(XA_1A_2Y) = (XA_3Y)A_1A_2X = 0, \end{aligned}$$

and (OR10) follows from symmetry. Finally, using (OR9) and (OR10) we obtain:

$$X^2A_1A_2A_3A_4Y^2 = X[X,A_1]A_2A_3A_4Y^2 = X[[X,A_1]A_2]A_3A_4Y^2 = 0.$$

□

LEMMA 7.6. *Let L be nondegenerate and let $x, y \in L$ be such that $XABY = 0$ for any $a, b \in L$. Then $\mathrm{id}_L(x) \cap \mathrm{id}_L(y) = 0$.*

PROOF. We claim that the set $I = \{s \in L : [s,[L,[L,[y,L]]]] = 0\}$ is an ideal of L, i.e. $[[s,a],[b,[c,[y,d]]]] = 0$ for every $a, b, c, d \in L$ and every $s \in I$.

$$\begin{aligned} \mathrm{ad}_{[[s,a],[b,[c,[y,d]]]]} &= [[S,A],[B,[C,[Y,D]]]] = SABCYD - SABCDY \\ &+ SABDYC + SACDYB + ASBCDY - YDCBSA \\ &+ YDCBAS - BYDCAS - CYDBAS - DYCBAS. \end{aligned}$$

Using the equality $SABCY = -YABCS$ (7.1), we see that each summand of $\mathrm{ad}^2_{[[s,a],[b,[c,[y,d]]]]}$ can be arranged as a product of the form

$$\cdots SE_1E_2Y\cdots,$$

and hence it vanishes. So $[[s,a],[b,[c,[y,c]]]] = 0$ by nondegeneracy of L. Now $SABY = 0$ for every $s \in I$ and $a, b \in L$ implies, by Proposition 7.3, $[s,y] = 0$, i.e. $y \in \mathrm{Ann}_L(I) \subset \mathrm{Ann}_L(\mathrm{id}_L(x))$ (since $x \in I$). Then $[\mathrm{id}_L(x), \mathrm{id}_L(y)] = 0$ and hence $\mathrm{id}_L(x) \cap \mathrm{id}_L(y) = 0$ by nondegeneracy of L. □

THEOREM 7.7. *Let L be a nondegenerate 6-torsion free Lie algebra and let $x, y \in L$ be such that $XAY = 0$ for every $a \in L$. Then $\mathrm{id}_L(x) \cap \mathrm{id}_L(y) = 0$.*

PROOF. For every $c \in L$, write $x_c = \mathrm{ad}^2_x c$ and $y_d := \mathrm{ad}^2_y d$. Using Equations (OR8)-(OR11), we get, for every $a, b \in L$,

$$X_cABY_d = (X^2C - 2XCX + CX^2)AB(Y^2D - 2YDY + DY^2) = 0.$$

By Lemma 7.6, for each $d \in L$ we have $\mathrm{ad}^2_x c \in \mathrm{Ann}_L(\mathrm{id}_L(y_d))$ for $c \in L$. Hence $x \in \mathrm{Ann}_L(\mathrm{id}_L(y_d))$ (since the Lie algebra $L/\mathrm{Ann}_L(\mathrm{id}_L(y_d))$ is nondegenerate by Corollary 4.11), equivalently, $\mathrm{ad}^2_y d \in \mathrm{Ann}_L(\mathrm{id}_L(x))$ for $d \in L$, which implies that $y \in \mathrm{Ann}_L(\mathrm{id}_L(x))$, by nondegeneracy of $L/\mathrm{Ann}_L(\mathrm{id}_L(x))$. □

REMARK 7.8. J. Brox, E. García and M. Gómez have proved in [**BGGL14**, Theorem 2] that, assuming the invertibility of 2 and 3 in Φ, x, y are strongly orthogonal if and only if $XY = 0$.

7.2. A characterization of strong primeness

As in the preceding section, we use capital letters to denote adjoint operators.

PROPOSITION 7.9. (`Going-up proposition`) *Let L be a nondegenerate 3-torsion free Lie algebra, let n be a nonnegative integer, and let x, y be nonzero elements in L such that $XA_1A_2\cdots A_nY = 0$, for $a_1, a_2, \ldots, a_n \in L$ (if $n = 0$ we understand $XY = 0$). Then there exist an $x' \neq 0$ and $y' \neq 0$ in L such that $X'A_1A_2\cdots A_{n+1}Y' = 0$ for $a_1, a_2, \ldots, a_{n+1} \in L$.*

PROOF. Note that by Theorem 7.7 and Proposition 7.3, only the case $n = 0$ need to be considered. If there exists an $a \in L$ such that $\mathrm{ad}^3_x a \neq 0$, then taking $x' = \mathrm{ad}^3_x a$ and $y' = y$ we obtain by Proposition 7.2 that $X'Y' = 0$ and $X'A_1Y' = 0$, for every $a_1 \in L$. Similarly, if there exists $b \in L$ such that $\mathrm{ad}^3_y b \neq 0$, we take $x' = x$ and $y' = \mathrm{ad}^3_y b$. Thus, without loss of generality, we may assume that both x, y are Jordan elements. Using (OR3), we get

$$\begin{aligned}
\mathrm{ad}_{\mathrm{ad}^2_x a}\,\mathrm{ad}_b\,\mathrm{ad}_{\mathrm{ad}^2_y c} &= (X^2A + AX^2 - 2XAX)B(Y^2C + CY^2 - 2YCY)\\
&= X^2ABCY^2 - 2XAXBCY^2 - 2X^2ABYCY\\
&+ 4XAXBYCY + 2XAXBYCY + 2XAXCYBY\\
&+ 2XBXAYCY + 2XBXCYAY + 2XCXAYBY\\
&+ 2XCXBYAY - 4XAXBYCY - 4XAXCYBY\\
&- 4XAXBYCY - 4XBXAYCY + 4XAXBYCY\\
&= -2XAXBYCY + 2XCXBYAY - 2XAXCYBY\\
&+ 2XCXAYBY - 2XBXAYCY + 2XBXCYAY.
\end{aligned}$$

Since in the last expression the roles of a and c are skew-symmetrical, we can interchange these elements to get

$$\mathrm{ad}_{\mathrm{ad}_x^2 a}\,\mathrm{ad}_b\,\mathrm{ad}_{\mathrm{ad}_y^2 c} = -\,\mathrm{ad}_{\mathrm{ad}_x^2 c}\,\mathrm{ad}_b\,\mathrm{ad}_{\mathrm{ad}_y^2 b}\,.$$

Therefore, taking $a = \mathrm{ad}_u^2\,\mathrm{ad}_x^2\, v$ and $c = \mathrm{ad}_{u'}^2\,\mathrm{ad}_y^2\, v'$, $u, u', v, v' \in L$, we get

$$\begin{aligned} \mathrm{ad}_{\mathrm{ad}_x^2 a}\,\mathrm{ad}_b\,\mathrm{ad}_{\mathrm{ad}_y^2 c} &= \mathrm{ad}_{\mathrm{ad}_x^2\,\mathrm{ad}_u^2\,\mathrm{ad}_x^2\, v}\,\mathrm{ad}_b\,\mathrm{ad}_{\mathrm{ad}_y^2\,\mathrm{ad}_{u'}^2\,\mathrm{ad}_y^2\, v'} \\ &= -\,\mathrm{ad}_{\mathrm{ad}_x^2\,\mathrm{ad}_{u'}^2\,\mathrm{ad}_y^2\, v'}\,\mathrm{ad}_b\,\mathrm{ad}_{\mathrm{ad}_y^2\,\mathrm{ad}_u^2\,\mathrm{ad}_x^2\, v} = 0 \end{aligned} \tag{7.3}$$

because $\mathrm{ad}_x^2\,\mathrm{ad}_{u'}^2\,\mathrm{ad}_y^2\, v' = 0$ by (OR3). Finally, since x and y are nonzero and L is nondegenerate there exist $u, u', v, v' \in L$ such that both $\mathrm{ad}^2_{\mathrm{ad}_x^2 u}\, v$ and $\mathrm{ad}^2_{\mathrm{ad}_y^2 u'}\, v'$ are nonzero. And since at the beginning of the proof we showed that x and y can be taken to be Jordan elements, we have by Lemma 4.4(3) that

$$x' = \mathrm{ad}^2_{\mathrm{ad}_x^2 u}\, v = \mathrm{ad}_x^2\,\mathrm{ad}_u^2\,\mathrm{ad}_x^2\, v \neq 0 \quad \text{and} \quad y' = \mathrm{ad}^2_{\mathrm{ad}_y^2 u'}\, v' = \mathrm{ad}_y^2\,\mathrm{ad}_{u'}^2\,\mathrm{ad}_y^2\, v' \neq 0$$

satisfy the case $n = 0$. □

THEOREM 7.10. *A 6-torsion free Lie algebra L is strongly prime if and only if for $x, y \in L$, $[x, [y, L]] = 0$ implies $x = 0$ or $y = 0$.*

PROOF. Clearly, $[x, [y, L]] = 0$ implies $x = 0$ or $y = 0$ is a sufficient condition for L to be strongly prime. Conversely, suppose L is strongly prime and there exist two nonzero elements $x, y \in L$ such that $[x, [y, L]] = 0$. Then it follows from Proposition 7.9 that there exist two nonzero elements $x', y' \in L$ such that $X'AY' = 0$ for every $a \in L$, which implies, by Theorem 7.7, $\mathrm{id}_L(x) \cap \mathrm{id}_L(y) = 0$, a contradiction since L is prime. □

REMARK 7.11. The reader is referred to [**BMS87**] for an elemental characterization of strong primeness for alternative and linear Jordan algebras. A modified version of this characterization was proved in [**ACLM96**, Theorem 1.10] for quadratic Jordan systems by using structure theory [**D'A92**, **DM00**].

CHAPTER 8

From Lie Algebras to Jordan Algebras

Introduced by Meyberg, local algebras of Jordan systems have played an important role in the structure of prime nondegenerate Jordan pairs. A similar construction works for Lie algebras, but unlike the Jordan case, only to Jordan elements of a Lie algebra we can attach a Jordan algebra. This Jordan algebra has a behavior similar to that of the local algebra of a Jordan system at an element. Thus many properties can be transferred from the Lie algebra to its Jordan algebras, and in addition the nature of the Jordan element in question is reflected in the structure of the attached Jordan algebra. These facts turn out to be crucial for applications of Jordan theory to Lie algebras. The material of this chapter is organized as follows. Section 8.1 is a brief survey on linear Jordan algebras, i.e. Jordan algebras over a ring of scalars Φ containing $1/2$. We include definitions and basic results on Jordan theory with the purpose of helping the reader to go through them when required. Section 8.2 is the core of the chapter, and maybe the core of the book, since it is in this section where a Lie–Jordan connection is introduced by associating with a Jordan element a of a Lie algebra L (over a ring of scalars in which 6 is invertible) a Jordan algebra L_a. Most properties of the Lie algebra are inherited by its Jordan algebras (we do not know if the simplicity is inherited), and some of the notable elements studied in the previous chapters, as von Neumann regular or extremal elements, are characterized in terms of their associated Jordan algebra. As applications of this Lie–Jordan connection, we give in Section 8.3 a classification-free proof of the fact that every simple finitary Lie algebra over an algebraically closed field of characteristic 0 is spanned by its extremal elements; in Section 8.4 we compute the Jordan algebra at a Clifford element; in Section 8.5 we give the proof of Zelmanov's solution of the Kurosh–Lie problem for the particular case of a Lie algebra over a field of characteristic 0 (in its general version for Lie algebras over an arbitrary field, this Zelmanov's result has important implications in group theory, significatively extending the positive solution of the Restricted Burnside Problem and his work on compact torsion groups); finally, in Section 8.6, we outline a proof of a theorem due to C. Martínez and E. Zelmanov on nil Lie algebras of finite width.

8.1. Linear Jordan algebras

Throughout this section, Φ is a ring of scalars in which 2 is invertible. For basic results and terminology of linear Jordan algebras the reader is referred to [**Jac68**, **McC04**, **ZSSS82**].

Definition and examples.

DEFINITION 8.1. A (linear) *Jordan algebra* is a Φ-algebra J whose product, denoted by $\bullet$, satisfies the following conditions:

(J1) $x \bullet y = y \bullet x$,

(J2) $x^2 \bullet (y \bullet x) = (x^2 \bullet y) \bullet x$,
for all $x, y \in J$, where $x^2 = x \bullet x$.

In operator form, the defining relations (J1) and (J2) give $[l_x, l_{x^2}] = 0$, which is in turn yields the identity (see [**Jac68**, I.7(54)]):

$$[[l_a, l_b], l_c] = l_{[l_a, l_b]c}, \quad a, b, c \in J, \tag{8.1}$$

proving that the linear map $[l_a, l_b]$ is a derivation of the Jordan algebra J.

– For each $x \in J$, the linear map $U_x : J \to J$, defined by $U_x = 2l_x^2 - l_{x^2}$, satisfies the *fundamental Jordan identity*:

(J3) $U_{U_x y} = U_x U_y U_x$, for all $x, y \in J$.

– Triple product is defined by

$$V(x, y)z = V_{x,y}z = \{x, y, z\} := U_{x,z}y := U_{x+z}y - U_x y - U_z y.$$

EXAMPLE 8.2. Let R be an associative Φ-algebra. Then R with the new product defined by $x \bullet y = \frac{1}{2}(xy + yx)$ becomes a Jordan algebra R^+, with $U_x z = xzx$, $\{x, y, z\} = xyz + zyx$. If R has an involution $*$, then $\mathrm{Sym}(R, *) = \{x \in R : x = x^*\}$ is a subalgebra of the Jordan algebra R^+.

– By [**Her69**, Theorem 17], if R is simple then the Jordan algebra R^+ is simple, and the same is for the Jordan algebra $\mathrm{Sym}(R, *)$ if R is simple with involution by [**Her69**, Theorem 21].

DEFINITION 8.3. A Jordan algebra is said to be *special* if it is isomorphic to a subalgebra of R^+ for an associative algebra R. Non-special Jordan algebra are called *exceptional*.

Jordan algebras of the form $\mathrm{Sym}(R, *)$ are special. An example of exceptional Jordan algebras is the *split Albert algebra*, introduced in Section 1.6, $\mathrm{Her}_3(\mathcal{C})$ of Hermitian 3×3 matrices on the split Cayley algebra $\mathcal{C}$ (over a field $\mathbb{F}$) with standard involution (see [**Jac68**, I.5.Theorem 4]). E. Zelmanov proved in [**Zel79b**] that under a scalar extension every simple exceptional Jordan algebra is a split Albert algebra.

It could be said that exceptionality is a triune phenomenon in the heavenly kingdom of the nonassociative algebras. Starting with the split Cayley Lie algebra $\mathcal{C}$, [1] we get, as its algebra of derivations, the split simple Lie algebra of type G_2, and by means of Hermitian 3×3-matrices, the split exceptional Jordan algebra $\mathrm{Her}_3(\mathcal{C})$. Then $\mathrm{Der}(\mathrm{Her}_3(\mathcal{C}))$ is the split simple Lie algebra of type F_4, and we can still enlarge $\mathrm{Der}(\mathrm{Her}_3(\mathcal{C}))$ by addition of the multiplication operators l_X, $X \in \mathrm{Her}_3(\mathcal{C})$, to get, as a subalgebra of $\mathfrak{gl}(\mathrm{Her}_3(\mathcal{C}))$, the split simple Lie algebra of type E_6. For details of these miraculous constructions the reader is referred to [**Jac79**].

EXAMPLE 8.4. Let X be a vector space over a field $\mathbb{F}$ of characteristic not 2 which is equipped with a symmetric bilinear form $\langle\, , \rangle$. Then the vector space $\mathbb{F} \oplus X$ becomes a Jordan algebra, denoted by $J(X, \langle\, , \rangle)$, with the product defined by

$$(\alpha, x) \bullet (\beta, y) = (\alpha\beta + \langle x, y\rangle, \alpha y + \beta x)$$

for $\alpha, \beta \in \mathbb{F}$ and $x, y \in X$. This Jordan algebra is special; in fact, it is isomorphic

[1]Under a scalar extension, any simple alternative algebra that is not associative is a split Cayley algebra (Kleinfeld's theorem) [**ZSSS82**, 7.3 Corollary 1]

to a Jordan subalgebra of the Clifford (associative) algebra defined by the bilinear form $\langle\,,\rangle$ [**Jac68**, II.3.Example 4]. For this reason, $J(X,\langle\,,\rangle)$ is sometimes called a *Clifford Jordan algebra.*

Note that if R is an associative Φ-algebra, then R is unital if and only if R^+ is unital (with the same unit element). Note also that the Jordan algebra $J(X,\langle\,,\rangle)$ in Example 8.4 is unital with $(1,0)$ as unit element, and that it is simple if the symmetric bilinear form $\langle\,,\rangle$ is nondegenerate and $\dim_{\mathbb{F}} X > 2$.

As a corollary of his astonishing structure theorem for prime nondegenerate Jordan algebras [**Zel83b**] (*The Russian Revolution of Jordan Algebras* in McCrimmon's words), E. Zelmanov proved that the four types of Jordan algebras given in the above examples describe all the simple Jordan algebras.

Homotopes and Local algebras of Jordan algebras. Let J be a Jordan algebra and let $a \in J$. In the Φ-module J a new product $\bullet_a$ is defined by setting $x \bullet_a y = \frac{1}{2}\{x,a,y\}$ for all $x, y \in J$. The resulting algebra is also a Jordan algebra denoted by $J^{(a)}$ and called the *a-homotope* of J at a. Then

$$\operatorname{Ker}_J\{a\} = \{z \in J : U_a z = 0\}$$

is an ideal of $J^{(a)}$, and the algebra $J_a = J^{(a)}/\operatorname{Ker}_J\{a\}$ is called the *local algebra* of J at a. (See [**Mey72**] or [**DM95**]).

As will be seen in the next section, a similar process works for any Lie algebra, but with the restriction that the element is required to be Jordan. In Exercise 8.106, we illustrate this process of localization in the case of an associative algebra.

Macdonald Principle. [**McC04**, II.5.1.2] *Any Jordan polynomial in three variables which has degree ≤ 1 in one variable and vanishes in all special Jordan algebras necessarily vanishes in all Jordan algebras, i.e. is an identity for Jordan algebras.*

Shirshov–Cohn Principle. [**McC04**, II.5.1.2] *Let J be a unital Jordan algebra (we may always suppose that J is unital by taking its unital hull $\hat{J}$). To verify that certain relations between elements x, y in J always imply certain other relations among the elements $f_1(x,y),\dots,f_n(x,y)$ in the unital subalgebra generated by x, y, it is sufficient to establish the implication in a Jordan algebra* $\operatorname{Sym}(R,*)$, *where R is a unital associative algebra with involution.*

Power associativity, algebraic and nilpotent elements. An algebra A over an arbitrary ring of scalars Φ is called *power-associative* if the subalgebra $\langle x\rangle$ generated by an arbitrary element $x \in A$ is associative. Using Macdonald's Principle it is easily verified that Jordan algebras are power associative. Thus the notion of nilpotency makes sense for elements of a Jordan algebra J: an element $a \in J$ is nilpotent if $a^n = 0$ for some $n \geq 1$. If any $x \in J$ is nilpotent, then J is called a *nil Jordan algebra.* The same is true for the notion of algebraic element: an element a in a Jordan algebra J over a field $\mathbb{F}$ is said to be *algebraic* if it is a root of a nonzero polynomial in $\mathbb{F}[\xi]$, equivalently, the subalgebra of J generated by a is finite-dimensional. In this case, $\deg(a) = \dim_{\mathbb{F}} \mathbb{F}[a]$ is the *degree* of a, where $\mathbb{F}[a]$ denotes the *unital* subalgebra of $\hat{J}$ generated by a. Then J is said to be *algebraic* if every $x \in J$ is algebraic, and *algebraic of bounded degree* if it is algebraic and there exists a positive integer n such that $\deg(x) \leq n$ for all $x \in J$.

Idempotents and Peirce decompositions in Jordan algebras. An element e of a Jordan algebra J is called an *idempotent* if $e^2 = e$.

PROPOSITION 8.5. [**McC04**, II.5.2.4] *If e is an idempotent of J with $U_e = 1_J$, then J is unital with e as unit element.*

PROOF. Taking $x = e$ in Exercise 8.107(i), we obtain

$$l_e = l_e 1_J = l_e U_e = \frac{1}{2} U_{e,e} = U_e = 1_J,$$

which proves that e is the unit element of J. □

PROPOSITION 8.6. [**McC04**, II.8.1.2-4] *Any idempotent e of a Jordan algebra J yields the* Peirce decomposition

$$J = J_1(e) \oplus J_{\frac{1}{2}}(e) \oplus J_0(e),$$

where $J_i(e) = \{x \in J : e \bullet x = ix\}$, for $i = 1, \frac{1}{2}, 0$. The supplementary projections relative to this decomposition are given by

$$E_1 = U_e, \quad E_{\frac{1}{2}} = U_{e,1-e} = 2l_e - 2U_e, \quad E_0 = U_{1-e} = 1_J - 2l_e + U_e,$$

with $1 - e$ in the unital hull $\hat{J}$ of J.

PROOF. Use Exercise 8.108 to prove that U_e and U_{1-e} are orthogonal projections on J. Hence $U_{e,1-e} = 1_J - U_e - U_{1-e}$ is also a projection. Then $l_e U_e = U_e l_e = U_e$ (identity obtained taking $x = e$ in Exercise 8.107(i)) proves that $U_e J = J_1(e)$, and similarly, $U_{1-e} J = J_0(e)$. That $U_{e,1-e} J = J_{\frac{1}{2}}(e)$ is now easily verified. □

An alternative proof. Replacing U_e by $2l_e^2 - l_e$ in the identity $l_e U_e = U_e$ we get that l_e vanishes the polynomial $(\xi - 1)(\xi - \frac{1}{2})\xi \in \Phi[\xi]$. Since the ideals $(\xi - 1)$, $(\xi - \frac{1}{2})$, (ξ) of $\Phi[\xi]$ are comaximal (because $\frac{1}{2} \in \Phi$), a standard application of the Chinese remainder theorem yields the decomposition $J = J_1(e) \oplus J_{\frac{1}{2}}(e) \oplus J_0(e)$.

EXAMPLE 8.7. Let e be an idempotent of an associative algebra R. Then e is an idempotent of the Jordan algebra $J = R^+$ and

$$J_1(e) = eRe, \; J_{\frac{1}{2}}(e) = eR(1-e) + (1-e)Re, \text{ and } J_0(e) = (1-e)R(1-e).$$

DEFINITION 8.8. Let J be a Jordan algebra. An element $z \in J$ is said to be an *absolute zero divisor* if $U_z J = 0$, and J is said to be *nondegenerate* if it has no nonzero absolute zero divisors. Note that R^+ is nondegenerate if and only if R is semiprime.

LEMMA 8.9. [**McC04**, II.8.10.2]. *Let e be an idempotent of a nondegenerate Jordan algebra J. If $J_0(e) = 0$, then J is unital with e as unit element.*

LEMMA 8.10. [**Jac68**, III.7.Lemma 1]. *Let J be a Jordan algebra and let x be a non-nilpotent algebraic element of J. Then the subalgebra $\langle x \rangle$ of J generated by x contains a nonzero idempotent.*

Recall that an algebra A is said to be nilpotent if there exists a positive integer such that any product (in any association) of n elements in A vanishes.

THEOREM 8.11. (Albert) *Any finite-dimensional Jordan nil algebra is nilpotent.*

PROOF. It follows from [**Jac68**, V.2. Corollary 1 and V.3.Theorem 3]. □

COROLLARY 8.12. *Any finite-dimensional Jordan algebra J which is not nilpotent contains a nonzero idempotent.*

PROOF. By Albert's theorem J contains an element which is algebraic but not nilpotent. So J contains a nonzero idempotent by Lemma 8.10. □

Von Neumann regularity in Jordan algebras. Recall that an element a of an associative algebra R is von Neumann regular if there exists $b \in R$ such that $a = aba$. This associative notion of von Neumann regularity has a natural extension to Jordan algebras.

DEFINITION 8.13. Let J be a Jordan algebra. An element $a \in J$ is said to be *von Neumann regular* if there exists $b \in J$ such that $a = U_a b$.

If $J = R^+$, then both notions of von Neumann regularity, the associative and the Jordan, agree. It is well known that if $a \in R$ is von Neumann regular, a=axa, then by replacing x by $b = xax$, we get $a = aba$ and $b = bab$. The same is true for elements of a Jordan algebra.

LEMMA 8.14. *Let $a \in J$ be von Neumann regular. Then there exists $b \in J$ such that $U_a b = a$ and $U_b a = b$.*

PROOF. Let $a = U_a x$. Taking $b = U_x a$ we have:

$$U_a b = U_a U_x a = U_a U_x U_a x = U_{U_a x} x = U_a x = a \quad \text{and}$$
$$U_b a = U_{U_x a} a = U_x U_a U_x U_a x = U_x U_{U_a x} x = U_x U_a x = U_x a = b.$$

□

It is also well known that every von Neumann regular element $a = aba$ in an associative algebra R gives rise to two idempotents: $e = ab$ and $f = ba$. A similar fact does not hold in general for von Neumann regular elements of a Jordan algebra, although, as will be seen in Chapter 11, there is a natural extension of this result for Jordan pairs. Yet, in spite of what we have just said, it is still possible in some occasions to produce idempotents in Jordan algebras by means of von Neumann regular elements.

LEMMA 8.15. (E. García Rus) *Let $x \in J$. If x^2 is von Neumann regular and nonzero, then there exists an element $y \in J$ such that $U_x y$ is a nonzero idempotent of J.*

PROOF. Since x^2 is von Neumann regular, we have by Lemma 8.14 that there exists $y \in J$ such that $x^2 = U_{x^2} y$ and $y = U_y x^2$. Taking $e := U_x y$ and using Macdonald Principle, we get

$$(U_x y)^2 = U_x U_y x^2 = U_x y,$$

with $U_x e = U_x(U_x y) = U_{x^2} y = x^2 \neq 0$, which proves that e is a nonzero idempotent.
. □

Division Jordan algebras and inner ideals. Let J be a unital Jordan algebra with 1 denoting its unit element. An element $x \in J$ is called *invertible* if there exists $y \in J$ satisfying:

(Jinv) $x \bullet y = 1$ and $x^2 \bullet y = x$.

In this case, U_x is invertible and the *inverse* of x, denoted by x^{-1}, is uniquely determined: $x^{-1} = U_x^{-1} x$. (See [**McC04**, II.6.1].)

DEFINITION 8.16. A unital Jordan algebra in which every nonzero element is invertible is called a *division Jordan algebra.*

E. Zelmanov proved in [**Zel79a**] that the division Jordan algebras are precisely those of classical type:

- R^+ where R is a division associative algebra.
- $\mathrm{Sym}(R,*)$, where R is a division algebra with involution.
- $J(X,\langle\,,\rangle)$ where $\langle x,x\rangle$ is not a square in $\mathbb{F}$ for any $0\neq x\in X$.
- A form of a split Albert algebra defined by an anisotropic cubic form (see [**McC04**, I.2.3.8] for definition).

Let J be a unital Jordan algebra and let $x\in J$ be an algebraic element. Then (see [**Jac68**, page 54]), x is invertible if and only if its minimum polynomial is irreducible. Hence we easily obtain.

PROPOSITION 8.17. *Any algebraic division Jordan algebra over an algebraicaly closed field is one-dimensional.*

DEFINITION 8.18. A Φ-submodule B of a Jordan algebra J is called an *inner ideal* of J if $U_xJ\subset B$ for all $x\in B$. If actually $U_x\hat{J}\subset B$, then B is called a *strict inner ideal.*

PROPOSITION 8.19. *For any element a of a Jordan algebra J the Φ-submodule U_aJ a strict inner ideal of J, called the* principal inner ideal generated by a.

PROOF. It follows from (J3) and the Macdonald Principle $(U_xy)^2=U_xU_yx^2$. □

By [**McC04**, II.18.14], if a Jordan algebra J contains an element b with surjective U-operator, $U_bJ=J$, then J is unital and b is invertible. Hence it follows that J is a division Jordan algebra if it is nondegenerate and has no proper inner ideals. We give here another proof of this result.

PROPOSITION 8.20. *A Jordan algebra. is a division Jordan algebra if and only if it is nondegenerate and has no proper inner ideals.*

PROOF. Clearly, a division Jordan algebra satisfies the above conditions. Suppose then that J is a nondegenerate algebra and has no proper inner ideals. Then for any $0\neq x\in J$, $U_xJ=J$ and $U_{x^2}J=U_x^2J=J$, so $x^2\neq 0$. Since x^2 is von Neumann regular, we have by Lemma 8.15 that there exists $y\in J$ such that $e:=U_xy$ is a nonzero idempotent in J. Now $U_eJ=J$ implies $U_e=1_J$, so e is the unit element of J by Propostion 8.5. Let us now see that every nonzero element $x\in J$ is invertible. Let $y\in J$ be such that $U_xy=e$. By (J3), $1_J=U_eJ=U_{U_xy}=U_xU_yU_x$ which implies that U_x is invertible in $\mathrm{End}_\Phi(J)$. Now let $b\in J$ be such that $U_xb=x$. Again by (J3), $U_xU_bU_x=U_x$ and hence $U_xU_b=U_bU_x$, which proves that $U_x^{-1}=U_b$. We will show that b is the inverse of x by verifying equations (Jinv). Use Macdonald Principle to check the identity $U_x(x\bullet y)=x\bullet U_xy$ and evaluate it in $y=b$. We get $U_x(x\bullet b)=x\bullet U_xb=x^2=U_xe$, which implies $x\bullet b=e$ since U_x is invertible. Similarly we get $U_b(x^2\bullet b)=b\bullet U_bx^2=b\bullet U_x^{-1}U_xe=b\bullet e=b$, which implies $x^2\bullet b=U_xb=x$, so showing that b is the inverse of x. □

Isotopy. Suppose that L is unital and $u\in J$ invertible. Then the u-homotope $J^{(u)}$ is called the *u-isotope*. Note that $J^{(u)}=J_u$, the local of J at u, and that $J^{(u)}$ is unital with $1_{J^{(u)}}=u^{-1}$. While in the associative case, isotopy yields isomorphism, in the Jordan case this is not true in general. Nevertheless, as will be seen in Chapter 11, isotopy in Jordan algebras is connected with isomorphism in Jordan pairs.

Annihilators. The notion of annihilator $\mathrm{ann}_J(X)$ of a subset X of a Jordan algebra J, introduced by E. Zelmanov in [**Zm78**], has played a fundamental role in the structure theory of Jordan algebras. As will be shown now, $\mathrm{ann}_J(X)$ is a strict inner ideal of J and coincides with $\mathrm{Ann}(I)$ when applied to an ideal I of J.

DEFINITION 8.21. Given $X \subset J$, the annihilator of X in J is defined as the set $\mathrm{ann}_J(X) = \{a \in J : \{a, X, \hat{J}\} = 0\}$.

THEOREM 8.22. (Zelmanov) $\mathrm{ann}_J(X)$ *is a strict inner ideal of* J. *If* X *is an ideal* I, *then* $\mathrm{ann}_J(I) = \mathrm{Ann}_J(I)$ *and therefore an ideal of* J.

PROOF. Since the intersection of strict inner ideals is a strict inner ideal, we may assume that X is a single set $\{x\}$. Given $a \in \mathrm{ann}_J(x)$ and $b \in \hat{J}$, we must prove that $\{U_a b, x, \hat{J}\} = 0$. Using Macdonald Principle we get the identities

$$\{y, x, z\} + \{x, y, b\} = 4(y \bullet x) \bullet z, \tag{8.2}$$

$$V(U_y a, a) = V(a, U_a y). \tag{8.3}$$

Replacing y by $x + \lambda b$ in (8.3), we get by linearization

$$V(\{x, a, b\}, x) = V(x, U_a b) + V(b, U_a x). \tag{8.4}$$

Since $a \in \mathrm{ann}_J(x)$, $U_a x = 0$ and also $\{x, a, b\} = 0$ (applying 8.2), so $\{x, U_a b, \hat{J}\} = 0$, which again by (8.2) implies $\{U_a b, x, \hat{J}\} = 0$, as required.

Let now I be an ideal of J. Interchanging the rolles of a, b in (8.4) and using (8.29), we get $V(U_b a, x_1)\hat{J} = 0$ for any $b \in \hat{J}$, which proves that $\mathrm{ann}_J(I)$ is an ideal annihilating I, so equal to $\mathrm{Ann}_J(I)$. □

s-identities and Jordan PI-algebras. A *Jordan polynomial* $p(x_1, \ldots, x_n)$ of the free Jordan Φ-algebra $FJ(X)$ [**Jac68**, I.9] is said to be an *s-identity* if (i) it is *admissible*, i.e. the coefficient of one of the terms of higher degree is 1, and (ii) it is satisfied by all special Jordan algebras but not by all Jordan algebras. A Jordan algebra J satisfying a polynomial identity which is not an s-identity is called a *Jordan PI-algebra*. Given $x_1, x_2, \ldots, x_n \in J$, we put

$$x_1 \bullet x_2 \cdots \bullet x_n := x_1 \bullet (x_2 \bullet \cdots \bullet x_n).$$

For $n > 1$, let $\mathfrak{S}_n$ denote the group of permutations of $1, \ldots, n$.

PROPOSITION 8.23. *A nonzero Jordan polynomial of the form*

$$p(x_1, \ldots, x_n, x_{n+1}) = \sum_{\sigma \in \mathfrak{S}_n} \alpha_\sigma x_{\sigma(1)} \bullet \cdots \bullet x_{\sigma(n)} \bullet x_{n+1} \quad (\alpha_\sigma \in \Phi)$$

is never an s-identity.

PROOF. By relabeling the variables we may assume that $\alpha_\sigma = 1$ for $\sigma = 1$. Let $Y = \{y_1, y_2, \ldots\}$ be a countable set. Denote by S the free semigroup generated by $Y \cup \{0\}$ satisfying the relations $y_i y_j = 0\,(j \neq i+1)$ and $y_i 0 = 0 y_i = 0\,(i \geq 1)$. Let R be the associative algebra defined by taking $S - \{0\}$ as a basis. It is easy to verify that $2^n p(y_1, \ldots, y_n, y_{n+1}) = y_1 \bullet \cdots \bullet y_n \bullet y_{n+1} \neq 0$. Thus the special Jordan algebra R^+ does not satisfies the identity $p(x_1, \ldots, x_n, x_{n+1}) = 0$, and therefore $p(x_1, \ldots, x_n, x_{n+1})$ is not an s-identity. □

LEMMA 8.24. [**ACGGL05**, 1.9] *Let J be algebraic of bounded degree. Then J is a Jordan PI-algebra.*

PROOF. Suppose that every element of J is algebraic of degree less than or equal to a fixed number n. Then J satisfies the admissible Jordan polynomial

$$p(x,y,z) := \mathcal{A}_{n+1}(V_{x^n,y},\cdots,V_{x,y},V_{1,y})z$$

for the alternating standard identity

$$\mathcal{A}_{n+1}(x_1,\dots,x_n,x_{n+1}) := \sum_{\pi\in\mathfrak{S}_{n+1}} sg(\pi)x_{\pi(1)}\cdots x_{\pi(n)}x_{\pi(n+1)},$$

which proves that J is PI. □

The following result, due to E. Zelmanov, is the Jordan analog of Posner's theorem for prime associative PI-algebras [**Coh03**, Theorem 8.6.6].

THEOREM 8.25. [**Zel83b**, Theorem 1] *Any strongly prime Jordan PI-algebra has nonzero (associative) center $Z = Z(J)$ and the Jordan algebra of fractions $Z^{-1}J$ is either a simple finite-dimensional Jordan algebra over the field $Z^{-1}Z$, or the Jordan algebra of a nondegenerate symmetric bilinear form over $Z^{-1}Z$.*

REMARKS 8.26. (1) Using Zelmanov's classification of division Jordan algebras and computing their isotopes in each one of the four types, one checks that the isotope of division Jordan PI-algebra is again a division Jordan PI-algebra.

(2) In their recent paper [**SZ19**, Lemma 4.1], I. Shestakov and E. Zelmanov have proved that finite presentation is also an isotopy invariant.

The McCrimmon radical of a Jordan algebra. A Jordan algebra J is *strongly prime* if it is prime and nondegenerate. By a *nondegenerate* (resp. *strongly prime*) *ideal of* J, we mean an ideal I of J such that J/I is a nondegenerate (resp. strongly prime) Jordan algebra. The intersection of nondegenerate ideals of J is a nondegenerate ideal of J, so there exists a smallest nondegenerate ideal $\mathrm{Mc}(J)$ called the *McCrimmon radical* of J. The McCrimmon radical is a radical in the sense of Amitsur-Kurosh (see [**The85**, Theorem 4]).

Set $\mathrm{Mc}_0(J) = 0$ and let $\mathrm{Mc}_1(L)$ be the ideal of J generated by its absolute zero divisors ($\mathrm{Mc}_1(L)$ is actually spanned by the absolute zero divisors of J [**McC69**, Theorem 9]). Using transfinite induction we define a chain of ideals of J by

$$\mathrm{Mc}_\alpha(J) = \bigcup_{\beta<\alpha}\mathrm{Mc}_\beta(J) \quad \text{for a limit ordinal } \alpha, \text{ and}$$

$$\mathrm{Mc}_\alpha(J)/\mathrm{Mc}_{\alpha-1}(J) = \mathrm{Mc}_1(J/\mathrm{Mc}_{\alpha-1}(J)) \quad \text{otherwise,}$$

and put $\mathrm{Mc}(J) = \bigcup_\alpha \mathrm{Mc}_\alpha(J)$. It is clear from the construction that $\mathrm{Mc}(J)$ is the McCrimmon radical of J. Moreover, it is proved in [**Zel84a**] that $\mathrm{Mc}(J)$ is locally nilpotent.

DEFINITION 8.27. An *m-sequence* of a Jordan algebra J is a sequence $\{a_n\}$ of elements of J such that for $n \geq 1$, $a_{n+1} = U_{a_n}b_n$ for some $b_n \in J$. We will say that $\{a_n\}$ has (finite) length k if $a_k \neq 0$ and $a_{k+1} = 0$.

There is a beautiful characterization of the McCrimmon radical in terms of m-sequences which is similar to that of the Baer radical of a ring [**Coh03**, Proposition 8.5.6].

PROPOSITION 8.28. [**The85**, Theorem 2] *An element* $x \in J$ *belongs to* $\mathrm{Mc}(J)$ *if and only if every m-sequence of* J *beginning with* x *has finite length.*

COROLLARY 8.29. [**The85**, Theorem 3] $\mathrm{Mc}(J)$ *is equal to the intersection of the strongly prime ideals of* J.

Capacity. By a *division idempotent* of a Jordan algebra J we mean a nonzero idempotent e of J such that U_eJ is a division Jordan algebra. It follows from Proposition 8.17 that every division idempotent e of an algebraic Jordan algebra J over an algebraically closed field $\mathbb{F}$ generates an one-dimensional inner ideal, i.e. $U_eJ = \mathbb{F}e$, such an idempotent e is called *reduced.*

• A Jordan algebra J has *capacity* n if J is unital and 1 can be written as a sum $1 = e_1 + \cdots + e_n$ of n orthogonal division idempotents. It has *finite capacity* if it has capacity n for some n. By Jacobson's capacity theorem [**McC04**, I.5.2], any nondegenerate Jordan algebra having finite capacity is a direct sum of ideals each of which is a simple Jordan algebra of finite capacity.

• If $J = R^+$, where R is a semiprime associative Φ-algebra, then J has finite capacity if and only if R is Artinian.

• A Jordan algebra over a field $\mathbb{F}$ having finite capacity and such that every division idempotent is reduced is said to be *reduced.* Any finite-dimensional nondegenerate Jordan algebra over an algebraically closed field is reduced.

• A Jordan algebra is said to be *semiprimitive* if it has no quasi-invertible ideals (see [**McC04**, III.1.3.1] for definition), i.e. its Jacobson radical is zero. Every semiprimitive Jordan algebra is nondegenerate [**McC04**, III.1.6.1]. Under finiteness conditions, the converse is also true.

• A Jordan algebra is said to be an *I-algebra* if every non-nil inner ideal of J contains a nonzero idempotent. Algebraic Jordan algebras (Lemma 8.10) and von Neumann regular Jordan algebras (Lemma 8.15) are I-algebras.

THEOREM 8.30. [**McC04**, I.8.1 (I-Finite Capacity Theorem)] *Any semiprimitive I-algebra having no infinite family of nonzero orthogonal idempotents is unital and has finite capacity.*

LEMMA 8.31. (E. Zelmanov) *Let* J *be a Jordan PI-algebra over a field* $\mathbb{F}$ *of characteristic* 0 *such that every element of* J *is a sum of nilpotent elements. Then* $J = \mathrm{Mc}(J)$.

PROOF. By replacing J by $J/\mathrm{Mc}(J)$, we may suppose that J is nondegenerate. Furthermore, since a nondegenerate Jordan algebra is a subdirect product of strongly prime Jordan algebras (Corollary 8.29), we may assume that J is strongly prime. Then, by Theorem 8.25, J has nonzero (associative) center $Z = Z(J)$ and the Jordan algebra of fractions $Z^{-1}J$ is either a simple finite-dimensional Jordan algebra over the field $Z^{-1}Z$, or the Jordan algebra of a nondegenerate symmetric bilinear form over $Z^{-1}Z$. Let $\overline{Z}$ be the algebraic closure of the field $Z^{-1}Z$. Then $\tilde{J} := \overline{Z} \otimes_{Z^{-1}Z} Z^{-1}J$ is a reduced simple Jordan algebra over $\overline{Z}$. Hence, by [**Jac68**, V.4.Theorem 6], $\tilde{J}$ has a nonzero linear trace $\mathrm{tr} : \tilde{J} \to \tilde{J}$. Since every element of $\tilde{J}$ is a sum of nilpotent elements, it follows that $\mathrm{tr}(\tilde{J}) = 0$, a contradiction. This proves the lemma. □

COROLLARY 8.32. *Let J be a Jordan algebra over a field $\mathbb{F}$ of characteristic* 0 *such that every element of J is a sum of nilpotent elements. If L is algebraic of bounded degree, then $J = \mathrm{Mc}(J)$.*

PROOF. By Lemma 8.24, J is PI, so Lemma 8.31 applies. □

The socle of a Jordan algebra. The *socle* $\mathrm{Soc}(J)$ of a Jordan algebra J is defined as the sum of its minimal inner ideals. If J is nondegenerate, then $\mathrm{Soc}(J)$ is a direct sum of ideals each of which is a simple algebra containing a minimal inner ideal [**OR79**, Theorem 17]. As mentioned in the introduction of the chapter, the nature of an element x of a Jordan algebra J is reflected in the structure of the local algebra J_x. An example of this fact is the local characterization of the socle of a nondegenerate Jordan algebra: $x \in \mathrm{Soc}(J)$ if and only if the (nondegenerate) Jordan algebra J_x has finite capacity (see [**Loo89**, Theorem 1] or [**Mon99**, Lemma 0.7]).

As most of the symmetric notions in associative algebras, the socle can be characterized in Jordan terms: Let R be a semiprime associative algebra. Then $\mathrm{Soc}(R) = \mathrm{Soc}(R^+)$ [**FL85**, Proposition 2.6]. Hence it follows easily the local characterization of the socle of a semiprime associative algebra that will be used in Proposition 13.85.

PROPOSITION 8.33. *Let R be a semiprime associative algebra. Then x is in the socle if and only if R_x is Artinian.*

PROOF. Following Exercise 8.106, given $x \in R$, the associative algebra R_x inherits semiprimeness from R and keeps up the Jordan structure: $(R_x)^+ = (R^+)_x$. By the local characterization of the socle of a nondegenerate Jordan algebra quoted above, $x \in \mathrm{Soc}(R) = \mathrm{Soc}(R^+)$ if and only if the Jordan algebra $(R^+)_x \cong R_x^+$ has finite capacity, equivalently, the associative algebra R_x is Artinian. □

Primitive Jordan algebras. Let J be a Jordan Φ-algebra. Following [**HM81**], an inner ideal K of J is called *e-modular* for some $e \in J$ if $\hat{K} = K + \Phi(1-e)$ is an inner ideal of the unital hull of J. It is clear that K is proper if and only if $e \notin K$, and that if J is unital, then any inner ideal of J is 1-modular. Moreover, using (iii) of Exercise 8.107, it is easy to see that given $e \in J$, the inner ideal $U_{1-e}J$ of J is e-modular if and only if $e^2 - e \in U_{1-e}J$. In particular, if e is an idempotent, then $U_{1-e}J$ is an e-modular inner ideal of J.

DEFINITION 8.34. A Jordan algebra J is called *primitive* if it contains a proper e-modular inner ideal K such that $K + I = J$ for any nonzero ideal I of J.

EXAMPLES 8.35. (1) Let R be an associative algebra. Then R is (left or right) primitive if and only if the Jordan algebra R^+ is primitive [**AMC94**, Theorem 2.2].

(2) Let R be a prime associative algebra with involution $*$. Then R is primitive if and only if the Jordan algebra $\mathrm{Sym}(R, *)$ is primitive [**AMC94**, Theorem 2.3].

(3) Let J be a strongly prime Jordan algebra with nonzero socle. Then J is primitive [**FLRP86b**, Theorem 12].

PROPOSITION 8.36. *Let J be a Jordan algebra and let I be a nonzero ideal of J. If J is primitive, then J is strongly prime and I is primitive. Conversely, if J is strongly prime and I is primitive, then J is primitive.*

PROOF. If J is primitive, then so is I by [**AMC94**, Theorem 1.1], and J is prime by [**HM81**, Proposition 5.5]. Moreover, J is nondegenerate. Otherwise $\mathrm{Mc}(J)$ would be primitive and locally nilpotent (see Remark 3.19(2)), which leads to a contradiction, since if e is a modulus for an inner ideal, then so is any power e^n [**HM81**, Proposition 2,10]. Conversely, if J is strongly prime and I is primitive, then J is primitive by [**AMC94**, Theorem 1.5]. □

The notion of primitivity in Jordan algebras can be characterized locally.

PROPOSITION 8.37. *Let J be a Jordan algebra.*

(1) *If J is primitive, then so is J_a for any $0 \neq a \in J$.*
(2) *If J is strongly prime and J_a is primitive for some $0 \neq a \in J$, then J is primitive.*

PROOF. (1) is proved [**ACM95**, Theorem 4.1], and (2) in [**AC98**, Theorem 4.1]. □

PROPOSITION 8.38. *Any primitive Jordan PI-algebra J is simple and has finite capacity.*

PROOF. By [**Zel83b**, Theorems 7 and 8], J is either finite-dimensional over its center, or isomorphic to the Jordan algebra of a nondegenerate symmetric bilinear form. In both cases, J is simple and has finite capacity. □

Algebraic Jordan algebras of bounded degree.

LEMMA 8.39. *Let J be algebraic of bounded degree. If J is nondegenerate, then it is semiprimitive.*

PROOF. By Lemma 8.24, J is PI and hence, by Zelmanov's PI-Radical Theorem [**Zel82**, Theorem 4], J does not contain any nonzero nil ideal. This, together with the fact that J is an I-algebra, implies that J is semiprimitive, since otherwise the Jacobson radical of J would contain a nonzero idempotent, a contradiction. □

LEMMA 8.40. *Let J be a Jordan algebra over a field $\mathbb{F}$ of characteristic 0. If J is algebraic of bounded degree n, then any family of nonzero orthogonal idempotents of J has a cardinal less than or equal to n.*

PROOF. Given $m \geq 1$, let $(e_1, \ldots, e_m)$ be a sequence of nonzero orthogonal idempotents of J, let $\lambda_1, \ldots, \lambda_m$ be nonzero elements of $\mathbb{F}$ such that $\lambda_i \neq \lambda_j$ whenever $i \neq j$, and set $a := \lambda_1 e_1 + \cdots + \lambda_m e_m$. Vandermonde determinant says to us that the vectors $a, a^2, \ldots, a^m$ are linearly independent. This proves that any sequence of nonzero orthogonal idempotents of J has a cardinal less than or equal to n. □

THEOREM 8.41. *Let J be a nondegenerate algebraic Jordan algebra of bounded degree over a field $\mathbb{F}$ of characteristic 0. Then J has finite capacity.*

PROOF. L is an I-algebra (Lemma 8.10), semiprimitive (Lemma 8.39), with no infinite family of nonzero orthogonal idempotents (Lemma 8.40). This proves (Theorem 8.30) that J is unital and has finite capacity. □

REMARK 8.42. E. Zelmanov shows in [**Zel82**] that any strongly prime algebraic Jordan PI-algebra over a field of characteristic different from 2 is simple and has finite capacity. We encourage the reader to glance through the proof and observe the amazing reduction to the case of a special Jordan algebra [**Zel82**, Lemma 20] to prove the idempotent-finite condition [**Zel82**, Lemma 24].

8.2. The Jordan algebra attached to a Jordan element

Throughout this section we will deal with Lie algebras over a ring of scalars Φ in which 6 is invertible. As in preceding chapters, and in order to simplify the notation, we will often make use of capital letters to denote the adjoint operators.

Definition, examples, and basic results. Let a be a Jordan element of L. In the Φ-module L a new product, denoted by $\bullet$, is defined by setting $x \bullet y := [[x,a],y]$ for all $x,y \in L$. The resulting (nonassociative) algebra is denoted by $L^{(a)}$.

THEOREM 8.43. *Let L be a Lie algebra and let $a \in L$ be a Jordan element. We have:*

(i) $\mathrm{Ker}_L\{a\} = \{z \in L : [a,[a,z]] = 0\}$ *is an ideal of* $L^{(a)}$.
(ii) $L_a := L^{(a)}/\,\mathrm{Ker}_L\{a\}$ *is a Jordan algebra, with*
(iii) $l_{\overline{x}}\overline{y} = \overline{XAy}, \quad U_{\overline{x}}\overline{y} = \overline{X^2A^2y}, \quad \{\overline{x},\overline{y},\overline{z}\} = 2\overline{XZA^2y}$

for all $x,y \in L$, where $\overline{x}$ denotes the coset $x + \mathrm{Ker}_L\{a\}$.

PROOF. (i) It suffices to show that $[[z,a],y] \in \mathrm{Ker}_L\{a\}$ for all $z \in \mathrm{Ker}_L\{a\}$ and $y \in L$. But this follows from the Leibniz rule and 4.4(6):

$$A^2[[z,a],y] = -[Az, A^2y] = [A^2z, Ay] = 0.$$

(ii) Let $x,y \in L$. Using (L1) and (L2) we get

$$x \bullet y - y \bullet x = [[x,a],y] - [[y,a],x] = [[x,y],a] \in \mathrm{Ker}_L\{a\}, \tag{8.5}$$

which proves that L_a is commutative. Thus we only need to verify that for all $x \in L$, the operators $l_{\overline{x}}$ and $l_{\overline{x}^2}$ commute.

In the multiplication ideal $\mathrm{Ad}(L)$, define an equivalence relation by $T_1 \equiv T_2$ if $T_1 - T_2$ vanishes on $\mathrm{Ker}_L\{a\}$, i.e. if $A^2T_1 = A^2T_2$, $T_1, T_2 \in \mathrm{Ad}(L)$, and denote by $\overline{T}$ the equivalence class of T. Clearly, $T_1 \equiv T_2$ implies $T_1S \equiv T_2S$ for all $S \in \mathrm{Ad}(L)$, and since $A^3 = 0$, we have

$$\overline{AT} = \overline{0}. \tag{8.6}$$

for all $T \in \mathrm{Ad}(L)$. Moreover, by 4.4(2),

$$\overline{XA^2T} = \overline{0}. \tag{8.7}$$

Let $x \in L$. Then $l_x = [X,A]$, $\;x^2 = [[x,a],x]$ and $l_{x^2} = [[[X,A],X],A]$. Using (8.6) and (8.7), we get:

$$l_x l_{x^2} \equiv XA[[X,A],X]A \equiv 2(XA)^3 - XAX^2A^2$$

and

$$l_{x^2} l_x \equiv [[X,A],X]A(XA - AX) \equiv 2(XA)^3 - X^2A^2XA,$$

since by 4.4(1) and (8.7), $X(AXA^2)X = (XA^2)XAX \equiv 0$. Now it follows from (1) and (5) of Lemma 4.4 that

$$A^2XAX^2A^2 = A^2X^2AXA^2 = A^2X^2A^2XA,$$

so completing the proof that L_a is a Jordan algebra.

Similarly we get $l_x^2 \equiv (XA)^2$ and $l_{x^2} \equiv 2(XA)^2 - X^2A^2$, so $2l_x^2 - l_{x^2} \equiv X^2A^2$, which implies

$$U_{\overline{x}}\overline{y} = \overline{X^2A^2y}, \quad \text{and}$$
$$\{\overline{x}, \overline{y}, \overline{z}\} = U_{\overline{x}+\overline{z}}\overline{y} - U_{\overline{x}}\overline{y} - U_{\overline{z}}\overline{y} = \overline{(X+Z)^2A^2y} - \overline{X^2A^2y} - \overline{Z^2A^2y} = 2\overline{XZA^2y}$$

since $A^2[X, Z]A^2 = 0$ by 4.4(1), which completes the proof. □

DEFINITION 8.44. For any Jordan element a of a Lie algebra L, the Jordan algebra L_a defined above is called the *Jordan algebra of L at a*.

LEMMA 8.45. *Let R be a semiprime associative algebra over a ring of scalars Φ in which 6 is invertible and let $b \in R$ be such that $b^2 = 0$. Denote by π the canonical homomorphism of R^- onto $\overline{R} = R^-/Z(R)$. Then we have the following chain of isomorphisms of Jordan algebras: $R_b^+ \cong R_b^- \cong \overline{R}_{\pi(b)}$.*

PROOF. From $b^2 = 0$ it easily follows that the linear isomorphism $x \mapsto \frac{1}{2}x$ of R induces a Jordan isomorphism of R_b^+ onto R_b^-. And we also have that the Lie epimorphism $\pi : R^- \to \overline{R}$ induces the Jordan epimorphism $\overline{\pi} : R_b^- \to \overline{R}_{\pi(b)}$. We claim that $\overline{\pi}$ is actually an isomorphism. Let $x \in R$ be such that $[\pi(b), [\pi(b), \pi(x)]] = \pi(0)$, Then $[b, [b, x]] \in Z(R)$ and hence $[b, [b, x]] = -bxb = 0$ since $Z(R)$ has no nonzero nilpotente elements. Now we have by above that $R_b^+ \cong R_b^- \cong \overline{R}_{\pi(b)}$. □

COROLLARY 8.46. *Let R be a centrally closed prime ring of characteristic different from 2 and 3, and let a be a Jordan element of the Lie algebra R^-. Then $R_a^- \cong R_b^+$ for some $b \in R$, $b^2 = 0$.*

PROOF. By Theorem 4.35, there exists $z \in Z(R)$ such that $(a - z)^2 = 0$. Set $b = a - z$. It is clear that the identity on R induces an isomorphism of R_a^- onto R_b^-. Now Lemma 8.45 applies. □

REMARK 8.47. There is a natural functor from the *pointed Lie algebra category* $(\mathcal{L}, \bullet)$ (whose objects are the pairs (L, a), where L is a Lie algebra and a is a Jordan element, and whose morphisms $f : (L, a) \to (M, b)$ are the homomorphism of Lie algebras such that $f(a) = b$) to the category of the Jordan Φ-algebras, assigning to each pair (L, a) the Jordan algebra L_a, and to each morphism $f : (L, a) \to (M, b)$ the homomorphism $\overline{f} : L_a \to M_b$ given by $\overline{f}(\overline{x}) = \overline{f(x)}$, $x \in L$, which is well defined since $f(\mathrm{Ker}_L\{a\}) \subset \mathrm{Ker}_L\{b\}$.

PROPOSITION 8.48. *Let $a \in L$ be a Jordan element. We have:*

(1) *For any inner ideal M of L, $M_a = M/(\mathrm{Ker}_L\{a\} \cap M)$ is a subalgebra of L_a.*
(2) *Let $L = B \oplus C$ be a direct sum of ideals and let $a = b + c$ with respect to this decomposition, then b and c are Jordan elements of the Lie algebras B and C respectively, and $L_a \cong B_b \times C_c$.*

PROOF. (1) Regarded M as a subalgebra $M^{(a)}$ of $L^{(a)}$, denote by $x \mapsto \overline{x}$ the injection of $M^{(a)}$ into $L^{(a)}$. It is clear that $M_a = M^{(a)}/\mathrm{Ker}_M\{a\}$ is a subalgebra of L_a.

(2) $0 = \mathrm{ad}_a^3 L = \mathrm{ad}_b^3 B \oplus \mathrm{ad}_c^3 C$ implies that b is a Jordan element in B and c is a Jordan element in C. Let $x = y + z$, where $y \in B$ and $z \in C$, and denote by $\overline{y}_B$ (resp. $\overline{z}_C$) the coset of y in B_b (resp. the coset of z in C_c). Then the map $\overline{x} \mapsto (\overline{y}_B, \overline{z}_C)$, is an isomorphism of L_a onto $B_b \times C_c$. □

The process of forming the Jordan algebra of a Lie algebra at a Jordan element and then a local algebra of the resulting Jordan algebra (Section 8.1) yields a Jordan algebra of the Lie algebra in a new Jordan element.

PROPOSITION 8.49. *Let a be a Jordan element of a Lie algebra L. For any $y \in L$, $b = [[a,y],a]$ is a Jordan element and $(L_a)_{\overline{y}} \cong L_b$.*

PROOF. We already know by 4.4(4) that $b = [[a,y],a]$ is a Jordan element. For any $x \in L$, denote by $\overline{x}$ and $\widetilde{x}$ the cosets of x in L_a and L_b respectively. We have

$$\overline{x} = 0 \Leftrightarrow \operatorname{ad}_a^2 x = 0 \Rightarrow \operatorname{ad}_b^2 x = \operatorname{ad}_a^2 \operatorname{ad}_y^2 \operatorname{ad}_a^2 x = 0,$$

which proves that $\overline{x} \mapsto \widetilde{x}$ defines a linear map φ of L_a onto L_b. We claim that φ is actually a homomorphism of the $\overline{y}$-homotope $L_a^{(\overline{y})}$ onto L_b, with $\ker(\varphi) = \ker(U_{\overline{y}})$. By 4.4(2), $\varphi(\overline{x}^2) = \varphi(U_{\overline{x}}\overline{y}) = \widetilde{\operatorname{ad}_x^2 \operatorname{ad}_a^2 y} = \widetilde{-\operatorname{ad}_x^2 b} = \widetilde{[[x,b],x]} = \widetilde{x} \bullet \widetilde{x}$, with $\widetilde{x} = 0$ if and only if $\operatorname{ad}_b^2 x = \operatorname{ad}_a^2 \operatorname{ad}_y^2 \operatorname{ad}_a^2 x = 0$, equivalently, $U_{\overline{y}}\overline{x} = 0$, i.e. $\overline{x} \in \ker(U_{\overline{y}})$, which proves the claim. Hence φ induces an isomorphism of $(L_a)_{\overline{y}}$ onto L_b. □

Inner inheritance. Recall (Proposition 4.6) that any Jordan element a of a Lie algebra L yields two abelian inner ideals: the principal inner ideal $[a]_L = \operatorname{ad}_a^2 L$, and the inner ideal $(a)_L = \Phi a + [a]_L$ generated by a. Similar notations are used for the principal inner ideal $[a]_J = U_a J$, and the inner ideal $(a)_J = \Phi a + U_a J$ generated by an element of a Jordan algebra J.

LEMMA 8.50. *Let a be a Jordan element of a Lie algebra L. We have:*

(i) *The map $\overline{x} \mapsto \operatorname{ad}_a^2 x$ is a Φ-module isomorphism of L_a onto the inner ideal $[a]_L$ of L.*
(ii) *This isomorphism induces an isomorphism from the lattice of the inner ideals of the Jordan algebra L_a onto the lattice of the inner ideals of the Lie algebra L contained in $[a]_L$.*

PROOF. Only statement (ii) requires some attention. Let $\overline{M}$ be an inner ideal of L_a. For $\overline{x} \in \overline{M}$ and $b \in L$, $U_{\overline{x}}\overline{b} = \overline{\operatorname{ad}_x^2 \operatorname{ad}_a^2 b} \subset \overline{M}$, equivalently,

$$\operatorname{ad}_{\operatorname{ad}_a^2 x}^2 b = \operatorname{ad}_a^2 \operatorname{ad}_x^2 \operatorname{ad}_a^2 b \subset \operatorname{ad}_a^2 M.$$

□

Global to Local Inheritance. We have seen (Theorem 7.10) that a Lie algebra L is strongly prime if and only if $[[x,L],y] = 0$ implies $x = 0$ or $y = 0$ for any $x, y \in L$. Strongly prime Jordan algebras are characterized in a similar way [**BMS87**]: A (linear) Jordan algebra J is strongly prime if and only if $\{x, J, y\} = 0$ implies $x = 0$ or $y = 0$, $x, y \in J$.

PROPOSITION 8.51. *Let a be a Jordan element of a Lie algebra L. We have:*

(1) *If $\overline{x}$ is a nonzero absolute zero divisor of the Jordan algebra L_a, then $\operatorname{ad}_a^2 x$ is a nonzero absolute zero divisor of L. Thus, if L is nondegenerate, then so is L_a.*
(2) *If L is strongly prime, then so is L_a.*
(3) *If L is strongly prime with nonzero socle, then so is L_a.*
(4) *Let $x \in L$. Then $\overline{x}^n = \overline{\operatorname{ad}_{[x,a]}^{n-1} x}$. Hence if $[x,a]$ is ad-nilpotent, then $\overline{x} \in L_a$ is nilpotent. In particular, if L is nil, then the Jordan algebra L_a is nil.*

Suppose now that the ring of scalars of L is a field of characteristic not 2 or 3. We have:

(5) *If L is algebraic, then L_a is an algebraic Jordan algebra.*
(6) *If L is locally finite, then L_a is a locally finite Jordan algebra.*

PROOF. (1) Let $\overline{x}$ be an absolute zero divisor of L_a. By 4.4(3),

$$\mathrm{ad}^2_{\mathrm{ad}^2_a x} L = \mathrm{ad}^2_a \mathrm{ad}^2_x \mathrm{ad}^2_a L = 0,$$

which implies $\mathrm{ad}^2_a x = 0$ by nondegeneracy of L, i.e. $\overline{x} = \overline{0}$.

(2) It suffices to show that $\{\overline{x}, L_a, \overline{z}\} = 0$ implies $\overline{x} = 0$ or $\overline{z} = 0$. Suppose then that $\{\overline{x}, \overline{y}, \overline{z}\} = 0$ for all $y \in L$, i.e. $A^2XZA^2 = 0$. By 4.4(8),

$$\mathrm{ad}_{A^2x}\,\mathrm{ad}_{A^2z} = A^2XZA^2 = 0,$$

and hence, by Theorem 7.10, $\mathrm{ad}^2_a x = 0$ or $\mathrm{ad}^2_a z = 0$, i.e. $\overline{x} = 0$ or $\overline{z} = 0$.

(3) We only need to prove that L_a contains a minimal inner ideal. Since L is strongly prime with nonzero socle, there exists a minimal abelian inner ideal B of L such that $\mathrm{ad}^2_a B \neq 0$ and therefore a minimal abelian inner ideal (Corollary 4.20). Then, by Lemma 8.50(ii), $\overline{B}$ is a minimal inner ideal of L_a.

It follows by induction that for each $x \in L$ and any integer $n > 1$,

$$\overline{x}^n = \overline{\mathrm{ad}^{n-1}_{[x,a]} x}.$$

Items (4) and (5) now follow easily.

(6) It suffices to prove that $L^{(a)}$ is locally finite. Suppose that $\{x_1, \dots, x_n\}$ is a finite subset of L. Clearly, the subalgebra of $L^{(a)}$ generated by $\{x_1, \dots, x_n\}$ is contained in the subalgebra of L generated by $\{x_1, \dots, x_n, a\}$, which is finite-dimensional. □

OPEN QUESTION 8.52. Let L be a simple nondegenerate Lie algebra and let $a \in L$ be a nonzero Jordan element. Is the Jordan algebra L_a simple?

The answer is affirmative for the following types of Lie algebras: $L = \overline{R'}$ where R is a simple ring [**FLGGL07**, Theorem 3.3], $L = \overline{K'}$ where $K = \mathrm{Skew}(R, *)$ and R is a simple ring with involution [**FLGGL07**, Theorems 3.6 and 3.10], and L is a simple nondegenerate Lie algebra of characteristic not 2 or 3 which is finite-dimensional over its centroid (L is strongly prime and hence so is L_a by Proposition 8.51(2), since L_a is also finite-dimensional over the centroid of L, L_a coincides with its socle and hence it is simple). Using Zelmanov's theorem for simple Lie algebras with a finite $\mathbb{Z}$-grading (Section 14.1) and the previous three cases, it can be proved that the answer is also affirmative for simple Lie algebras of characteristic 0 or greater than 7 containing a nonzero von Neumann regular element.

REMARK 8.53. Inheritance of nondegeneracy was used in [**GGLN11**] to show that a Lie triple system is nondegenerate if and only if its standard envelope is nondegenerate and hence that a Kantor pair is nondegenerate if and only if its standard envelope is nondegenerate.

PI-inheritance. We show in this subsection that if L is a Lie PI-algebra, then L_a is a Jordan PI-algebra for any Jordan element a of L.

DEFINITION 8.54. Let $p = p(x_1, \ldots, x_n)$ be an element of a free Lie Φ-algebra $\mathfrak{L}\langle X\rangle$. We say that a Lie algebra L satisfies the identity p if $p(a_1, \ldots, a_n) = 0$ for any $a_1, \ldots, a_n$ in L. A Lie algebra satisfying a nontrivial polynomial identity is called a *Lie PI-algebra.*

Recall that left commutators $[x_1, x_2, \ldots, x_n]$ are defined recursively as follows: $[x_1] = x_1$ and $[x_1, x_2, \ldots, x_n] = [x_1, [x_2, \ldots, x_n]] = \operatorname{ad}_{x_1} \ldots \operatorname{ad}_{x_{n-1}} x_n$ for $n > 1$, $x_1, x_2, \ldots, x_n \in L$. We also recall that by the Jacobi identity any monomial of the Φ-algebra $\mathfrak{L}\langle X\rangle$ over a countable set of indeterminates X can be written as a linear combination of left commutators.

LEMMA 8.55. *For $n \geq 1$ there exists a function $\epsilon_n : \mathfrak{S}_n \to \{-1, 0, 1\}$ such that, for any $x_1, \ldots, x_n, x_{n+1}$ in X,*

$$\operatorname{ad}_{[x_1, \ldots, x_n]} x_{n+1} = \sum_{\sigma \in \mathfrak{S}_n} \epsilon_n(\sigma)[x_{\sigma(1)}, \ldots, x_{\sigma(n)}, x_{n+1}].$$

PROOF. By induction on n. The case $n = 1$ is trivial. Now

$$\begin{aligned} \operatorname{ad}_{[x_1, x_2, \ldots, x_{n+1}]} x_{n+2} &= \operatorname{ad}_{[x_1, [x_2, \ldots, x_{n+1}]]} x_{n+2} \\ &= \operatorname{ad}_{x_1} \operatorname{ad}_{[x_2, \ldots, x_{n+1}]} x_{n+2} - \operatorname{ad}_{[x_2, \ldots, x_{n+1}]} \operatorname{ad}_{x_1} x_{n+2}. \end{aligned}$$

Hence, by the induction hypothesis,

$$\begin{aligned} \operatorname{ad}_{[x_1, \ldots, x_{n+1}]} x_{n+2} &= \sum_{\sigma \in \mathfrak{S}_n} \epsilon_n(\sigma)[x_1, x_{\sigma(1)+1}, \ldots, x_{\sigma(n)+1}, x_{n+2}] \\ &- \sum_{\sigma \in \mathfrak{S}_n} \epsilon_n(\sigma)[x_{\sigma(1)+1}, \ldots, x_{\sigma(n)+1}, x_1, x_{n+2}] \\ &= \sum_{\tau \in \mathfrak{S}_{n+1}} \epsilon_{n+1}(\tau)[x_{\tau(1)}, \ldots, x_{\tau(n+1)}, x_{n+2}]. \end{aligned}$$

where ϵ_n is defined inductively by

$$\epsilon_{n+1}(\tau) = \begin{cases} 0 & \text{if } \tau(1) \neq 1 \text{ and } \tau(n+1) \neq 1 \\ \epsilon_n(\sigma) & \text{if } \tau(1) = 1 \text{ and } \sigma \in \mathfrak{S}_n \text{ is defined by } \sigma(i) = \tau(i+1) - 1 \\ -\epsilon_n(\sigma) & \text{if } \tau(n+1) = 1 \text{ and } \sigma \in \mathfrak{S}_n \text{ is defined by } \sigma(i) = \tau(i) - 1. \end{cases}$$

□

PROPOSITION 8.56. *Any Lie PI-algebra L satisfies the multilinear identity of the form*

$$p(x_1, \ldots, x_n, x_{n+1}) = \sum_{\sigma \in \mathfrak{S}_n} \alpha_\sigma[x_{\sigma(1)}, \ldots, x_{\sigma(n)}, x_{n+1}] \quad (\alpha_\sigma \in \Phi).$$

PROOF. As pointed out above, we may assume that L satisfies a polynomial identity $p = 0$, where p is a linear combination of left commutators $[x_{i_1}, \ldots, x_{i_m}]$. Moreover, by [**ZSSS82**, Corollary of Theorem 1.5.7], we can assume that p is multilinear. Finally, by Lemma 8.55, we can replace p by a polynomial having the required form. □

PROPOSITION 8.57. *Let a be a Jordan element of a Lie algebra L. If L is PI, then L_a is a Jordan PI-algebra.*

PROOF. By Proposition 8.56, L satisfies a multilinear identity of the form

$$p(x_1,\dots,x_n,x_{n+1}) = \sum_{\sigma\in\mathfrak{S}_n} \alpha_\sigma[x_{\sigma(1)},\dots,x_{\sigma(n)},x_{n+1}].$$

For each integer $1\le i\le n$, replace x_i by $[x_i,a]$. We obtain the Jordan polynomial

$$\begin{aligned} q(x_1,\dots,x_n,x_{n+1}) &= \sum_{\sigma\in\mathfrak{S}_n} \alpha_\sigma x_{\sigma(1)}\bullet\cdots\bullet x_{\sigma(n)}\bullet x_{n+1} \\ &= \sum_{\sigma\in\mathfrak{S}_n} \alpha_\sigma[[x_{\sigma(1)},a],\dots,[x_{\sigma(n)},a],x_{n+1}], \end{aligned}$$

which vanishes on the Jordan algebra L_a. Since by Proposition 8.23, this Jordan polynomial is never an s-identity, L_a is PI. □

REMARK 8.58. Let W_m denote the Lie algebra of derivations of $\mathbb{F}[x_1,\dots,x_m]$, where $\mathbb{F}$ is a field of characteristic 0 (see Exercise 2.88). It is proved in [**Bah76**, 2.1.3(iv and v)]:

(1) $s_n(x_1,\dots,x_n,x_{n+1}) = \sum_{\sigma\in\mathfrak{S}_n}\epsilon_\sigma[x_{\sigma(1)},\dots,x_{\sigma(n)},x_{n+1}]$, where $\epsilon_\sigma=1$ if σ is an even permutation and $\epsilon_\sigma=-1$ otherwise, is a nontrivial polynomial identity,

(2) W_m satisfies the identity $s_n(x_1,\dots,x_n,x_{n+1})$ for $n=m^2+2m+2$.

Von Neumann regular elements in Lie algebras revisited. Recall that a Jordan element a in a Lie algebra L is von Neumann regular if $a\in \operatorname{ad}_a^2 L$. And that an element x of a Jordan algebra J is von Neumann regular if $x\in U_xJ$.

PROPOSITION 8.59. *Let a be a Jordan element of a Lie algebra L. For any $x\in L$ the following conditions are equivalent:*

(i) *$\operatorname{ad}_a^2 x$ is a von Neumann regular element of L.*

(ii) *$\overline{x}$ is a von Neumann regular element of the Jordan algebra L_a.*

PROOF. By Theorem 8.43, for any $x,y\in L$,

$$U_{\overline{x}}\overline{y}=\overline{x} \text{ if and only if } A^2X^2A^2y=A^2x,$$

so $\overline{x}$ is von Neumann regular in the Jordan algebra L_a if and only if $\operatorname{ad}_a^2 x = A^2x$ is von Neumann regular in L. □

PROPOSITION 8.60. *Let L be a nondegenerate Lie algebra, let I be an ideal of L, and let $a\in I$ be von Neumann regular. Then the Jordan algebras I_a and L_a agree.*

PROOF. By Proposition 8.48, I_a is contained in L_a, so it suffices to show that any coset $\overline{x}$ in L_a is equal to a coset $\overline{y}$ in I_a. Since a is von Neumann regular, $a=\operatorname{ad}_a^2 b$ for some $b\in L$. Then we have by 4.4(3) $\operatorname{ad}_a^2=\operatorname{ad}^2_{\operatorname{ad}_a^2 b}=\operatorname{ad}_a^2\operatorname{ad}_b^2\operatorname{ad}_a^2$. Hence, for any $x\in L$, $\operatorname{ad}_a^2 x=\operatorname{ad}_a^2 y$, where $y=\operatorname{ad}_b^2\operatorname{ad}_a^2 x\in I$. □

We have already seen the important role played by von Neumann regular elements in the structure of Lie algebras, so it is very useful to have criterions guaranteeing the existence of nontrivial von Neumann regular elements. The next proposition provides us with such a criterion in terms of Jordan algebras. As an example of the usefulness of this Jordan criterion, we will prove later that every locally finite simple Lie algebra over an algebraically closed field of characteristic 0 containing a nonzero Jordan element is linearly spanned by von Neumann regular elements.

PROPOSITION 8.61. *Let $a \in L$ be a Jordan element. We have:*

(i) *If $a = [[a, e], a]$ for some $e \in L$, i.e. a is von Neumann regular, then L_a is unital with $\overline{e}$ as unit element.*

(ii) *If L is nondegenerate and L_a is unital, then a is von Neumann regular.*

PROOF. (i) Let $a = [[a, e], a]$. Then $A = [[A, E], A] = 2AEA - EA^2 - A^2E$ and hence $A^2 = 2A^2EA - AEA^2 = A^2EA$ since $A^3 = 0$ and $AEA^2 = A^2EA$ by 4.4(1), which proves $l_{\overline{e}} = 1_{L_a}$.

(ii) Suppose now that L is nondegenerate and that L_a is unital, with $\overline{e}$ as unit element, i.e. $l_{\overline{e}} = U_{\overline{e}} = 1_{L_a}$. We claim that $a = [[a, e], a]$. Set $z = [[a, e], a] - a$. Since L is nondegenerate, it suffices to show that $Z^2 = 0$. But $Z = 2AEA - EA^2 - A^2E - A$ implies $Z^2 = 0$, since $A^3 = 0$, $A^2 = A^2EA$ and $A^2 = A^2E^2A^2$ by hypothesis, and $AEA^2 = A^2EA$, $A^2EA^2 = 0$ and $A^2E^2A^2 = \mathrm{ad}^2_{A^2(e)} = \mathrm{ad}^2_{-a} = A^2$ by 4.4(1-3). $\square$

Next we compute the Jordan algebra L_a of a Lie algebra L when a is actually a von Neumann regular element. This particular case was already known in [**Ben77**, Lemma 2.2].

PROPOSITION 8.62. *Let $a \in L$ be von Neumann regular, $a = [[a, b], a]$ for some $b \in L$. Then L_a is isomorphic to the Jordan algebra, $J(a, b)$, defined in the Φ-module $\mathrm{ad}^2_a L$ by the product $x \bullet y = [[x, b], y]]$ for all $x, y \in \mathrm{ad}^2_a L$.*

PROOF. The map $\varphi : L_a \to J(a, b)$, defined by $\varphi(\overline{x}) := -A^2x$, where $\overline{x}$ denotes the coset of x in L_a, is an algebra-isomorphism. Clearly, φ is a linear isomorphism, and since both algebras are commutative and $\frac{1}{2} \in \Phi$, it suffices to check that $\varphi(\overline{x})^2 = \varphi(\overline{x}^2)$ for every $x \in L$, which follows from the definitions of the involved products:

$$\varphi(\overline{x})^2 = [[A^2x, b], A^2x] = -\mathrm{ad}^2_{A^2x} b = -A^2X^2A^2b = A^2X^2a = -A^2XAx = \varphi(\overline{x}^2),$$

where we have used the identity $XAx = [x, [a, x]] = -X^2a$. $\square$

Division elements of Lie algebras. By a *division element* of a Lie algebra L we mean a nonzero Jordan element $a \in L$ such that L_a is a division Jordan algebra.

PROPOSITION 8.63. *Let L be a nondegenerate Lie algebra and let $a \in L$ be a Jordan element. Then a is a division element of L if and only if $\mathrm{ad}^2_a L$ is a minimal inner ideal.*

PROOF. Since L_a is nondegenerate (Proposition 8.51(i)), it suffices to prove, by the inner characterization of division Jordan algebras (Proposition 8.20), that L_a contains no nontrivial inner ideal if and only if the principal inner ideal $\mathrm{ad}^2_a L$ is minimal, but this follows from Lemma 8.50(ii). $\square$

Extremal elements revisited. In this subsection, L denotes a Lie algebra over a field $\mathbb{F}$ of characteristic not 2 or 3. *By an extremal element we will mean here one which is not an absolute zero divisor, i.e. a nonzero element $e \in L$ such that $\mathrm{ad}^2_e L = \mathbb{F}e$.*

PROPOSITION 8.64. *A nonzero Jordan element $e \in L$ is extremal if and only if the Jordan algebra L_e is one-dimensional.*

PROOF. It follows from the vector space isomorphism $\overline{x} \mapsto \mathrm{ad}^2_e x$ from L_e onto $\mathrm{ad}^2_e L$ (Lemma 8.50(i)). $\square$

COROLLARY 8.65. *Let L be algebraic over an algebraically closed field. Then every division element of L is extremal.*

PROOF. Let $b \in L$ be a division element. By Proposition 8.51(5), L_b is algebraic and hence, by Proposition 8.17, one-dimensional, so b is an extremal element by Proposition 8.64. □

Primitive Lie algebras. As we have seen in Proposition 8.37, a Jordan algebra J is primitive if and only if J is strongly prime and J_a is primitive for some $0 \neq a \in J$. In this subsection we take this local approach to define a notion of primitivity for Lie algebras. All the algebras considered here are over a ring of scalars Φ in which 6 is invertible.

DEFINITION 8.66. Let L be a Lie algebra and let $0 \neq a \in L$ be a Jordan element. Then L is called *primitive at a* if L is strongly prime and the Jordan algebra L_a is primitive. By a *primitive Lie algebra* we will mean a primitive Lie algebra L at some nonzero Jordan element.

PROPOSITION 8.67. *Let I be a nonzero ideal of a Lie algebra L. If L is primitive, then so is I. Conversely, if L is strongly prime and I is primitive, then L is primitive.*

PROOF. Let L be primitive. By Corollary 4.9, I is strongly prime. Let a be a nonzero Jordan element of L such that L_a is primitive. By Proposition 4.10, there exists $y \in I$ such that $b = [[a, y], a]$ is a nonzero *Jordan* (4.4(4)) element of I. Since I_b can be regarded as an ideal of the Jordan algebra $L_b \cong (L_a)_{\overline{y}}$ (Proposition 8.49), it follows from Proposition 8.36 that I_b is a primitive Jordan algebra, so I is primitive.

Conversely, suppose that L is strongly prime and I primitive at some nonzero Jordan element $b \in I$. Then b is as a Jordan element of L (Proposition 4.13), I_b is a nonzero ideal of L_b, and L_b is strongly prime (Proposition 8.51(2)). Hence it follows from Proposition 8.36 that L_b is primitive, so L is primitive. □

As mentioned in Proposition 8.37, in any strongly prime Jordan algebra, primitivity of one its local algebras implies that of the global algebra, and therefore of all its local algebras. Since at present we don't know whether a similar result holds for Lie algebras, we will call those Lie algebras enjoying this property *strongly primitive.*

PROPOSITION 8.68. *Let L be a strongly prime Lie algebra with nonzero socle. Then L is strongly primitive.*

PROOF. Let a be a nonzero Jordan element of L (whose existence is guaranteed by Lemma 4.5). By Proposition 8.51(3), L_a is a strongly prime Jordan algebra with nonzero socle. Then L_a is primitive by Example 8.35(3), which proves that L is strongly primitive. □

A partial converse of this result holds for primitive Lie PI-algebras.

PROPOSITION 8.69. *Let L be a primitive Lie PI-algebra. Then it has nonzero socle.*

PROOF. Let a be a nonzero element of L such that the Jordan algebra L_a is primitive. By Proposition 8.57, L_a is PI. So, by Proposition 8.38, L_a is simple and has finite capacity. Hence, by Lemma 8.50, L has nonzero socle. □

In the remainder of this section, R will be an associative algebra over ring of scalars Φ in which 6 is invertible, and $\overline{R}$ will stand for the Lie algebra $R^-/Z(R)$. Primitive associative algebra are understood to be left or right primitive.

PROPOSITION 8.70. *If R is primitive, then $\overline{R}$ is strongly prime and all its Jordan algebras are primitive. Conversely, if R is prime and $\overline{R}$ is primitive, then R is primitive.*

PROOF. Let $\pi : R^- \to \overline{R}$ be the canonical Lie epimorphism and suppose that R is primitive. By [**Coh03**, Proposition 8.5.1], R is prime, and hence $\overline{R}$ is strongly prime (Proposition 3.35(3)). Let $\pi(a)$ be a nonzero Jordan element of $\overline{R}$, i.e. $a \in R \setminus Z(R)$ and $\mathrm{ad}_a^3 R \subset Z(R)$. It follows from Proposition 3.35(1) that there exists $x \in R$ such that $b = [[a,x],a]$ is nilpotent of index 2. Since R_b is primitive [**ACM95**, Proposition 0.1(iii)], and therefore so is the Jordan algebra R_b^+ (Example 8.35(1)), we have by Lemma 8.45 that $\overline{R}_{\pi(b)} \cong R_b^+$ is primitive.

Conversely, suppose that R is prime and $\overline{R}$ is primitive. Let $\pi(a)$ be a nonzero Jordan element of $\overline{R}$ such that the Jordan algebra $\overline{R}_{\pi(a)}$ is primitive. As before, there exists $x \in R$ such that $b = [[a,x],a]$ is nilpotent of index 2. Now it follows from Lemma 8.45, Proposition 8.49, and Proposition 8.37(1) that the Jordan algebra $R_b^+ \cong \overline{R}_{\pi(b)} \cong (\overline{R}_{\pi(a)})_{\overline{\pi(x)}}$ is also primitive, which implies by Example 8.35(1) that R_b is primitive. Then R is primitive by [**AC98**, Theorem 1.1]. □

COROLLARY 8.71. *If R is primitive and contains nonzero nilpotent elements, then $\overline{R}$ is strongly primitive.*

PROOF. It follows from Proposition 8.70 since R contains nilpotent elements of index 2, and therefore $\overline{R}$ has nonzero Jordan elements (Example 4.2). □

Suppose in addition that R has an involution $*$ and contains nonzero nilpotent skew-symmetric elements. Set $K = \mathrm{Skew}(R,*)$ and $\overline{K} = K/K \cap Z(R)$.

PROPOSITION 8.72. *If R is primitive, then $\overline{K}$ is primitive whenever one of the following two conditions holds:*

(i) *The involution is of the second kind.*
(ii) *R is not an order in a simple ring of dimension less than 16 over its center.*

PROOF. $\overline{K}$ is strongly prime by Propositions: 3.36 if (i), and 3.39 if (ii). Choose a nonzero element $a \in K$ such that $a^2 = 0$. Denote by π the canonical epimorphism of R^- onto $\overline{R}$ and its corresponding restriction of K onto $\overline{K}$. By [**ACM95**, Proposition 0.1] and Example 8.35(2), the Jordan algebra $\mathrm{Sym}(R_a,*)$ is primitive. Since $\overline{R}_{\pi(a)} \cong R_a^+$ by Lemma 8.45, and hence $\overline{K}_{\pi(a)} \cong \mathrm{Sym}(R_a,*)$, we have that $\overline{K}_{\pi(a)}$ is primitive, which proves that $\overline{K}$ is primitive. □

8.3. Extremal elements and finitary Lie algebras

The results of this section are taken from [**FL16**]. Unless specified otherwise, $\mathbb{F}$ will stand for an arbitrary field. Given a vector space X over $\mathbb{F}$, we denote by $\mathcal{L}(X)$ the associative $\mathbb{F}$-algebra of all linear maps of X, and by $\mathcal{F}(X)$ the ideal of $\mathcal{L}(X)$ consisting of all finite rank linear maps. As in the above subsection, by an extremal element we will mean one which is not an absolute zero divisor.

DEFINITION 8.73. A Lie $\mathbb{F}$-algebra L is called *finitary* if it is isomorphic to a subalgebra of the Lie algebra $\mathfrak{fgl}(X) = \mathcal{F}(X)^-$ for some vector space X over $\mathbb{F}$.

PROPOSITION 8.74. *Every finitary Lie algebra $L \leq \mathcal{F}(X)^-$ is locally finite.*

PROOF. It is enough to show that the associative algebra $\mathcal{F}(X)$ is locally finite. But this is a direct consequence of Lifoff' theorem [**BMM96**, Theorem 4.3.11]. $\square$

Let L be a simple infinite-dimensional Lie algebra over an algebraically closed field $\mathbb{F}$ of characteristic 0. According to Baranov's structure theorem [**Bar99**, Corollary 1.2], L is finitary if and only if it is isomorphic to one of the following (see Section 2.2 for notation): $\mathfrak{fsl}_Y(X)$, $\mathfrak{fo}(X, \langle\,,\rangle)$, or $\mathfrak{fsp}(X, \langle\,,\rangle)$. As mentioned in Example 6.2 (or in [**FLGGL04**, Proposition 6.4]), each one of these Lie algebras contains an extremal element. In fact, they are spanned by extremal elements. In this section we provide a classification-free proof of this result by using Jordan theory instead of representation theory.

- For any $a \in \mathcal{F}(X)$, $\operatorname{rank}(a) := \dim(Xa)$ denotes the rank of a. The following properties of the rank are immediate. Let $a, b \in \mathcal{F}(X)$ and $c \in \mathcal{L}(X)$. Then:
  (i) $\operatorname{rank}(a+b) \leq \operatorname{rank}(a) + \operatorname{rank}(b)$,
  (ii) $\max\{\operatorname{rank}(ac), \operatorname{rank}(ca)\} \leq \operatorname{rank}(a)$,
  (iii) $\operatorname{rank}([a,c]) \leq 2\operatorname{rank}(a)$.

LEMMA 8.75. *Let R be an associative $\mathbb{F}$-algebra and let a, b be algebraic elements of R such that $ab = ba$. Then for any $c \in \langle a, b\rangle$, we have that c is algebraic with $\deg(c) \leq \deg(a)\deg(b)$.*

PROOF. Let $r = \deg(a)$ and $s = \deg(b)$. Then $\dim_{\mathbb{F}}\langle a, b\rangle \leq rs$. $\square$

LEMMA 8.76. *Let $a \in \mathcal{F}(X)$ with $\operatorname{rank}(a) = r > 0$. We have:*
(i) *a is algebraic with $\deg(a) \leq r^2 + 1$.*
(ii) *l_a and r_a are algebraic with $\max\{\deg(l_a), \deg(r_a)\} \leq r^2 + 1$.*
(iii) *ad_a is algebraic with $\deg(\operatorname{ad}_a) \leq (r^2+1)^2$.*

PROOF. (i) The restriction $\hat{a}$ of a to the r-dimensional subspace Xa is algebraic with $\deg(\hat{a}) \leq r^2$, so a is algebraic with $\deg(a) \leq r^2 + 1$.

(ii) For any associative $\mathbb{F}$-algebra R, the map $x \mapsto l_x$ (resp. $x \mapsto r_x$) is a homomorphism (resp. anti-homomorphism) of R into $\mathcal{L}(A)$. Hence, by (i), both l_a and r_a are algebraic of degree less than or equal to $r^2 + 1$.

(iii) Since $[l_a, r_a] = 0$, $\deg(\operatorname{ad}_a) = \deg(l_a - r_a) \leq (r^2+1)^2$ by Lemma 8.75. $\square$

PROPOSITION 8.77. *Let $L \leq \mathfrak{fgl}(X)$ be a finitary Lie algebra over a field $\mathbb{F}$ of characteristic different from 2 and 3, and let $a \in L$ be a Jordan element. Then L_a is algebraic of bounded degree.*

PROOF. Let $\operatorname{rank}(a) = r$. By Proposition 8.51(4), for each $c \in L$ and any integer $n \geq 1$, $\overline{c}^n = \overline{\operatorname{ad}^{n-1}_{[c,a]} c}$. Then, by Lemma 8.76(iii), we have that $\overline{c}$ is algebraic with $\deg(\overline{c}) \leq (4r^2+1)^2 + 1$. This proves that the Jordan algebra L_a is algebraic of bounded degree. $\square$

THEOREM 8.78. *Let L be a nondegenerate finitary Lie algebra over an algebraically closed field $\mathbb{F}$ of characteristic 0. Then L contains extremal elements.*

PROOF. By Lemma 8.76(iii), the adjoint of every $a \in L$ is algebraic, so L contains a nonzero Jordan element by Corollary 4.32. Let $a \in L$ be a nonzero Jordan element. By Propositions 8.51(1) and 8.77, L_a is nondegenerate and algebraic of bounded degree. Thus L_a has finite capacity (Theorem 8.41). Let $\bar{e}$ be a division idempotent of L_a. Since $\mathbb{F}$ is algebraically closed, L_e is one-dimensional, so e is an extremal element of L by Proposition 8.64. □

COROLLARY 8.79. *Let L be a simple finitary Lie algebra over an algebraically closed field $\mathbb{F}$ of characteristic* 0. *Then L is spanned by its extremal elements.*

PROOF. Any simple Lie algebra over a field of characteristic 0 is nondegenerate (Corollary 3.24), so L contains extremal elements by Theorem 8.78. Let $e \in L$ be an extremal element. By Proposition 6.17, L has a nontrivial finite grading and therefore it is generated by ad-nilpotent elements. Thus any subspace of L which is invariant under automorphisms is an ideal. In particular, the linear span of the extremal elements of L is a nonzero ideal and therefore equal to L. □

8.4. Clifford elements

Let L be a Lie algebra over a field $\mathbb{F}$ of characteristic not 2 or 3. A Jordan element $a \in L$ is called a *Clifford element of* L if L_a is a Clifford Jordan algebra (see Example 8.4 for definition).

DEFINITION 8.80. Let R be a centrally closed prime ring of characteristic different from $2, 3$ with involution $*$, and let K be the Lie algebra of its skew-symmetric elements. By a *Clifford element of* R we mean an element $c \in K$ such that

$$c^3 = 0,\ c^2 \neq 0 \text{ and } c^2kc = ckc^2 \quad \text{for all} \quad k \in K.$$

Any Clifford element of R is a Jordan element of K: for any $k \in K$ we have $\operatorname{ad}_c^3 k = c^3k - 3c^2kc - 3ckc^2 - kc^3 = 0$. Conversely, if c is a Jordan element of K such that $c^3 = 0$, then $c^2kc = ckc^2$ since $\operatorname{char}(R) \neq 3$. Note also that by Lemma 4.42, if K is not abelian and $*$ is of the first kind, then any Jordan element of K is of cube 0.

Following [**BFLL17**], we prove in this section that every Clifford element c of the ring R is a Clifford element of the Lie algebra K.

EXAMPLE 8.81. [**FLGGL06**, Lemma 3.7] Following the notation of Exercise 2.99, let X be a vector space of dimension greater than 2 with a nondegenerate symmetric bilinear form $\langle\, , \rangle$ over a field $\mathbb{F}$ of characteristic not 2, let $H = \mathbb{F}x \oplus \mathbb{F}y$ be a hyperbolic plane of X: $\langle x, x\rangle = \langle y, y\rangle = 0$, $\langle x, y\rangle = 1$, let $z \in H^\perp$, and set $c := [x, z] = x^*z - z^*x$. We have:

(i) $c^3 = 0$, with $c^2 = 0$ if and only if the vector z is isotropic.
(ii) If z is anisotropic, then c is a Clifford element of the ring $\mathcal{L}_X(X)$ with the adjoint involution, and $\operatorname{ad}_c^2 \mathfrak{fo}(X, \langle\, , \rangle) = [x, H^\perp]$.

PROOF. (i) $c^2 = (x^*z - z^*x)^2 = -x^*\langle z, z\rangle x = 0$ if and only if z is isotropic, and $c^3 = -\langle z, z\rangle x^*x(x^*z - z^*x) = 0$, since x, y are isotropic and $z \in H^\perp$.

(ii) Suppose that $\alpha := -\langle z, z\rangle \neq 0$ (the existence of such a vector z is guaranteed since $\dim_{\mathbb{F}} X > 2$ and $\langle\, , \rangle$ is nondegenerate), and let $a \in \mathfrak{o}(X, \langle\, , \rangle)$. We have:

$$c^2ac - cac^2 = \alpha((x^*x)ac - ca(x^*x)) = 0, \tag{8.8}$$

since

$$\begin{aligned}(x^*x)ac &= (x^*x)a(x^*z - z^*x) = -x^*\langle xa, z\rangle x = x^*\langle za, x\rangle x \\ &= (x^*z - z^*x)a(x^*x) = ca(x^*x),\end{aligned}$$

because $\langle xa, x\rangle = \langle x, xa^*\rangle = -\langle x, xa\rangle = -\langle xa, x\rangle$ implies $\langle xa, x\rangle = 0$. This, together with (i), proves that c is a Clifford element of $\mathcal{L}_X(X)$.

Now we show that for every $v \in H^\perp$ there exists $a \in \mathfrak{fo}(X, \langle\,,\rangle)$ such that $\mathrm{ad}_c^2\, a = [x, v]$. For any $a \in \mathfrak{o}(X, \langle\,,\rangle)$, we have

$$\begin{aligned}\mathrm{ad}_c^2\, a &= c^2a - 2cac + ac^2 = \alpha x^*xa + 2\langle xa, z\rangle[x, z] - \alpha(xa)^*x \\ &= \alpha[x, xa] + 2\langle xa, z\rangle[x, z].\end{aligned} \tag{8.9}$$

Taking $a = [y, \lambda z + \mu v] \in \mathfrak{fo}(X, \langle\,,\rangle)$, $\lambda, \mu \in \mathbb{F}$, in (8.9), we get

$$\mathrm{ad}_c^2\, a = (2\mu\langle v, z\rangle - \alpha\lambda)[x, z] + \alpha\mu[x, v] = [x, v]$$

for $\mu = \alpha^{-1}$ and $\lambda = 2\alpha^{-2}\langle v, z\rangle$, which completes the proof. □

REMARK 8.82. As will be shown in Proposition 13.53, every Clifford element of the ring $\mathcal{L}_X(X)$ is actually one of those described in the example above. Note also that in this example it is proved that $[x, H^\perp]$ is an abelian inner ideal of $\mathfrak{o}(X, \langle\,,\rangle)$ contained in $\mathfrak{fo}(X, \langle\,,\rangle)$ (question proposed in Exercise 2.99(iii)). For another proof of this fact, using (8.9) one can prove that $[x, xa] \in [x, H^\perp]$ as follows: Write $xa = \beta x + \gamma y + w$ according to the decomposition $X = H \oplus H^\perp$. From $\langle xa, x\rangle = 0$, we get $\beta = 0$ and hence $[x, xa] = [x, \beta x + w] = [x, w] \in H^\perp$.

The square of a Clifford element.

PROPOSITION 8.83. *Let R be a centrally closed prime ring of characteristic different from 2, 3 with involution $*$, denote by $\mathcal{C}$ the extended centroid of R (equals its centroid), and let $c \in K$ be a Clifford element of R. We have:*

(1) $c^2Kc^2 = 0$.
(2) $(c^2xc^2)^* = c^2x^*c^2 = c^2xc^2$ *for all* $x \in R$.
(3) $c^2Rc^2 = \mathcal{C}c^2$, *so* Rc^2 *is a minimal left ideal of* R.
(4) *The involution $*$ is of the first kind.*
(5) *R is a subring of $\mathcal{L}_(X)$ containing $\mathcal{F}_X(X)$ where X is a vector space with a nondegenerate symmetric bilinear form over the field $\mathcal{C}$.*
(6) *Regarded c^2 as an element of $\mathcal{L}_X(X)$,* $\mathrm{rank}(c^2) = 1$

PROOF. (1) $c^2kc^2 = c(ckc^2) = c(c^2kc) = c^3kc = 0$.

(2) By (1), $c^2(x - x^*)c^2 = 0$ and hence $c^2xc^2 = c^2x^*c^2 = (c^2xc^2)^*$.

(3) Let $x, y \in R$. Since c^2 is symmetric it follows from (2) that

$$(c^2xc^2)yc^2 = c^2x^*c^2y^*c^2 = c^2(yc^2x)^*c^2 = c^2y(c^2xc^2),$$

with $c^2 \neq 0$. Then, by Martindale's Lemma [**BMM96**, Theorem 2.3.4], for each $x \in R$ there is a $\lambda_x \in \mathcal{C}$ such that $c^2xc^2 = \lambda_x c^2$. Since $c^2 \neq 0$ and R is prime, $c^2Rc^2 \neq 0$ and hence $c^2Rc^2 = \mathcal{C}c^2$, since $\mathcal{C}$ is a field.

(4) By (3), given $\alpha \in \mathcal{C}$ there exists $a \in R$ such that $\alpha c^2 = c^2ac^2$. Then, by (2), $\alpha^*c^2 = c^2a^*c^2 = c^2ac^2 = \alpha c^2$, so $\alpha^* = \alpha$, proving that $*$ is of the first kind.

(5) By (3), $c^2 = c^2ac^2$ for some $a \in R$ and hence $c^2R = eR$, where $e = c^2a$ is an idempotent of R. Then $eRe = c^2Rc^2a = \mathcal{C}c^2a = \mathcal{C}e$, which proves by Exercise 2.96 that eR is a minimal right ideal of R, so R has nonzero socle with associated

division ring isomorphic to the field $\mathcal{C}$ ([**BMM96**, Theorem 4.3.7]). Now it follows from Kaplansky's Theorem (2.31) that the involution $*$ of R is either of transpose type or of symplectic type; but the latter cannot occur because c^2 is a symmetric rank-one linear map, so $*$ is of transpose type.

(6) By Lemma 13.40, c^2R is a minimal right ideal of R, so $\operatorname{rank}(c^2) = 1$. □

Let c be a Clifford element of R. Since c^2 is symmetric, of square zero and von Neumann regular (Proposition 8.83(3)), we have by Lemma 5.17 that there exists $d \in R$ such that $d^* = d$, $d^2 = 0$, $c^2dc^2 = c^2$ and $d = dc^2d$. Such an element d will be called a *regular partner* of c^2. Note that then $e := dc^2$ is a $*$*-orthogonal idempotent*, i.e. $e^2 = e$ and $ee^* = e^*e = 0$.

PROPOSITION 8.84. *Let c be a Clifford element of R, let d be a regular partner of c^2, and put $e = dc^2$. We have:*

(1) $dKd = 0$.
(2) $dRd = \mathcal{C}d$.
(3) $eRe = \mathcal{C}e$, $e^*Re = \mathcal{C}c^2$, $eRe^* = \mathcal{C}d$ *and* $eKe^* = e^*Ke = 0$.
(4) $ec = ce^* = 0$, $e^*c^2 = c^2e = c^2$ *and* $de^* = ed = d$.
(5) $[K, K] \neq 0$.
(6) $e + e^* \neq 1$ *in the unital hull* $\hat{R} = \mathcal{C}1 + R$ *of* R.

PROOF. Note that by Proposition 8.83(3), $c^2Mc^2 = \mathcal{C}c^2$ for any subspace M of R such that $c^2Mc^2 \neq 0$. A fact which will be used in what follows without further mention.

(1) $dKd = dc^2(dKd)c^2d = 0$, where we have used Proposition 8.83(1) and the fact that dkd is skew-symmetric for every $k \in K$.

(2) Similarly, we have

$$dRd = (dc^2d)R(dc^2d) = dc^2(dRd)c^2d = d\mathcal{C}c^2d = \mathcal{C}dc^2d = \mathcal{C}d,$$

since $c^2dc^2 = c^2$ and $dc^2d = d$ imply $c^2(dRd)c^2 \neq 0$.

(3) $eRe = dc^2(Rd)c^2 = d\mathcal{C}c^2 = \mathcal{C}e$, since $c^2 = c^2(dc^2d)c^2 \in c^2(Rd)c^2$ and therefore the latter is not zero. In a similar way it is proved that $e^*Re = \mathcal{C}c^2$ and $eRe^* = \mathcal{C}d$. Now $eKe^* = dc^2Kc^2d = 0$ by 8.83(1), and $e^*Ke = 0$ is obtained in a similar way.

(4) The identities of this item follow straightforwardly from the very definition of e.

(5) By (4), $[c, e - e^*] = ce + e^*c = cdc^2 + c^2dc \neq 0$. Otherwise $cdc^2 = -c^2dc$ would lead to the contradiction $c^2 = c^2dc^2 = -c^3dc = 0$. Since $[c, e - e^*] \in [K, K]$, $[K, K] \neq 0$.

(6) It follows from (3) and (4) that $(e + e^*)c(e + e^*) = 0$, so $e + e^* \neq 1$. □

As we have seen in the above proposition, any Clifford element c of the ring R gives rise to two nonzero orthogonal elements e and e^* associated to any regular partner d of c^2. Furthermore, the idempotent $e + e^*$ is no complete (8.84(6)), i.e. the symmetric idempotent $g := 1 - e - e^*$ in the unital hull $\hat{R} = \mathcal{C}1 + R$ of R is not zero. We next prove that the complete system $\{e, e^*, g\}$ induces a 3-grading in the Lie algebra K.

- For any nonempty subset S of R, set $\kappa(S) = \{x - x^* : x \in S\}$.

PROPOSITION 8.85. *Let c be a Clifford element of R, $e = dc^2$ and $g = 1-e-e^*$, where d is a regular partner of c^2. Then $K = K_{-1} \oplus K_0 \oplus K_1$ is a 3-grading of K, with $K_{-1} = \kappa((1-e)Ke) = \kappa((1-e)Re) = \kappa(gRe)$, $K_0 = \kappa(eRe) \oplus gKg$ and $K_1 = \kappa(eK(1-e)) = \kappa(eR(1-e)) = \kappa(eRg)$.*

PROOF. Consider the complete system $\{e_0 = e^*,\ e_1 = g,\ e_2 = e\}$ of orthogonal idempotents of $\hat{R}$ and set $R_i = \oplus_{m-n=i} e_m R e_n$, $-2 \le i \le 2$. Then (see [**Smi97**, p.174] or Exercise 1.38), $R = \oplus_{-2\le i\le 2} R_i$ is an (associative) 5-grading of R. Explicitly,

$$R = e^*Re \oplus (e^*Rg \oplus gRe) \oplus (e^*Re^* \oplus gRg \oplus eRe) \oplus (gRe^* \oplus eRg) \oplus eRe^*.$$

Since all the components R_i are $*$-invariant subspaces, $K = \oplus_{-2\le i\le 2} K_i$, where $K_i = R_i \cap K = \mathrm{Skew}(R_i, *)$ for each index i and $[K_i, K_j] \subset \mathrm{Skew}[R_i, R_j] \cap [K, K] \subset R_{i+j} \cap K = K_{i+j}$. Thus $K = \oplus_{-2\le i\le 2} K_i$ is (a priori) a 5-grading of the Lie algebra K. But $K_{-2} = \kappa(e^*Re) = e^*\kappa(R)e = e^*Ke = 0$ and similarly $K_2 = e^*Ke = 0$. Moreover, the i-th homogenous component k_i of any $k \in K$ coincides with $\oplus_{m-n=i}\kappa(e_m k e_n)$, so $k \in K_{-1}$ if and only if

$$\begin{aligned} gke + e^*kg &= (1-e-e^*)ke + e^*k(1-e-e^*) = (1-e)ke + e^*k(1-e^*) \\ &= (1-e)ke - ((1-e)ke)^* = \kappa((1-e)ke), \end{aligned}$$

since $e^*Ke = 0$ by Proposition 8.84(3), which proves $K_{-1} = \kappa(gRe) = \kappa((1-e)Ke)$. Similarly, $K_1 = \kappa(eRg) = \kappa(eK(1-e))$. Therefore

$$K = \kappa((1-e)Ke) \oplus (\kappa(eRe) \oplus gKg) \oplus \kappa(eK(1-e))$$

is a 3-grading of K. Now, for any $x \in R$,

$$\kappa(gxe) = \kappa((1-e)xe) - \kappa(e^*xe) = \kappa((1-e)xe) - e^*\kappa(x)e = \kappa((1-e)xe)$$

since $e^*\kappa(x)e \in e^*Ke = 0$, which proves that $K_{-1} = \kappa((1-e)Re)$. Similarly we obtain that $K_1 = \kappa(eR(1-e))$. □

Although the 3-grading of K has been defined by choosing a regular partner d of c^2, it will be seen now that the component K_{-1} only depends on the Clifford element c.

- Set $x \circ y = xy + yx$ for $x, y \in R$.

PROPOSITION 8.86. *Let c be a Clifford element of R, let $e = dc^2$, where d is a regular partner of c^2, and set $B = \kappa((1-e)Ke)$. We have:*

(1) *If $b \in B$ then $eb = 0$ and $b = e^*b + be = \kappa((1-e)be)$.*
(2) *$B = c^2 \circ K$.*
(3) *$c = e^*c + ce = c^2dc + cdc^2$.*
(4) *$c \in B$.*
(5) *$cKc = \mathcal{C}c$.*

PROOF. (1) Let $b = (1-e)ke + e^*k(1-e^*) \in B$. Then

$$eb = e((1-e)ke + e^*k(1-e^*)) = 0 \quad\text{and}\quad e^*b = e^*k(1-e^*),$$

since $e^*e = 0$ and $e^*Ke = 0$. By symmetry, we also have that $be^* = 0$ and $be = (1-e)ke$. Hence $b = e^*b + be = e^*b(1-e^*) + (1-e)be = \kappa((1-e)be)$.

(2) It follows directly from the definitions:

$$\begin{aligned} c^2 \circ k &= \kappa(kc^2) = \kappa(ke^*c^2 - (eke^*)c^2) = \kappa((1-e)k(e^*c^2)) \\ &= \kappa((1-e)k(c^2e)) \in \kappa((1-e)Ke) = B. \end{aligned}$$

Conversely, let $b \in B$. Then

$$b = e^*b + be = (c^2d)b + b(dc^2) = c^2(d \circ b) + (d \circ b)c^2 = c^2 \circ (d \circ b) \in c^2 \circ K,$$

since $e^* = c^2d$, $c^2 = c^2e$ and $c^2b = (c^2e)b = c^2(eb) = 0$.

(3) Set $z := c - c^2dc - cdc^2$. We must prove that $z = 0$. For any $k \in K$ we have

$$c^2kz = c^2kc - (c^2kc^2)dc - (c^2kc)dc^2 = ckc^2 - ck(c^2dc^2) = ckc^2 - ckc^2 = 0$$

since $c^2kc = ckc^2$ and $c^2kc^2 = 0$ by Proposition 8.83(1), and d is a regular partner of c^2. We also have $zkc^2 = (c^2kz)^* = 0$, and hence $c^2xz = c^2x^*z$ and $zxc^2 = zx^*c^2$ for every $x \in R$. Let $x, y \in R$. Then $c^2\kappa(xzy)c^2 = 0$ since $c^2Kc^2 = 0$. Thus

$$\begin{aligned} 0 &= c^2(xzy + y^*zx^*)c^2 = c^2xzyc^2 + c^2y^*zx^*c^2 \\ &= c^2xzyc^2 + c^2yzxc^2 = (c^2xz)y(c^2) + (c^2)y(zxc^2), \end{aligned}$$

with $c^2 \neq 0$. By Martindale's Lemma, for every $x \in R$ there is $\lambda_x \in \mathcal{C}$ such that $c^2xz = \lambda_x c^2$. But

$$z(1-e) = (c - e^*c - ce)(1-e) = c - ce - e^*c + e^*ce - ce + ce = c - ce - e^*c = z$$

since $e^*ce \in e^*Ke = 0$. Hence $c^2xz = c^2xz(1-e) = \lambda_x c^2(1-e) = 0$, so $c^2Rz = 0$. This implies $z = 0$ because R is prime and $c^2 \neq 0$. Thus

$$c = e^*c + ce = c^2dc + cdc^2$$

as desired.

(4) By (3), $c = c^2dc + cdc^2 = c^2(dc + cd) + (dc + cd)c^2 \in c^2 \circ K = B$ by (2).

(5) Note that $cd + dc = \kappa(cd) \in K$ and $c(cd + dc)c = c^2dc + cdc^2 = c$ by (3). Hence $\mathcal{C}c \subset cKc$. Conversely, for any $k \in K$ we have

$$ckc = (e^*c + ce)k(e^*c + ce) = e^*cke^*c + cekce,$$

since $eKe^* = 0$ by Proposition 8.84(3) and $ckc \in K$. Now, again by 8.84(3), $ekce = \lambda e$ for some $\lambda \in \mathcal{C}$, and hence $e^*cke^* = (ekce)^* = (\lambda e)^* = \lambda e^*$ because the involution $*$ is of the first kind by Proposition 8.83(4). Then $ckc = \lambda e^*c + \lambda ce = \lambda c$, which completes the proof. $\square$

The square root of d. Given a Clifford element c of R and a regular partner d of c^2, put $\sqrt{d} := cd + dc$. As will be seen now, the square-root notation is absolutely justified.

PROPOSITION 8.87. *Let c be a Clifford element of R and let d be a regular partner for c^2. For any $b \in B = \kappa((1-e)Ke)$, we have:*

(1) *$\sqrt{d} \in K_1$ in the 3-grading of Proposition 8.85. In particular, $\sqrt{d}$ is a Jordan element.*
(2) $(\sqrt{d})^2 = d$.
(3) $(\sqrt{d})^3 = 0$.
(4) $\sqrt{d}K\sqrt{d} = \mathcal{C}\sqrt{d}$.
(5) $\sqrt{d}c\sqrt{d} = \sqrt{d}$.
(6) $c\sqrt{d}c = c$.
(7) $c^2 \circ \sqrt{d} = c$.
(8) $d \circ c = \sqrt{d}$.
(9) $[[c, \sqrt{d}], c] = c$.
(10) $[[\sqrt{d}, c], \sqrt{d}] = \sqrt{d}$.
(11) $[[c, \sqrt{d}], b] = b$.

PROOF. (1) Note first that $\sqrt{d} = cd + dc \in K$, since $c \in K$ and $d \in H$. Then

$$\begin{aligned}\kappa(e\sqrt{d}(1-e)) &= e(cd+dc)(1-e) + (1-e^*)(dc+cd)e^* \\ &= edc(1-e) + (1-e^*)cde^* = edc - edce + cde^* - e^*cde^* \\ &= (dc^2d)c - e(dcd)c^2 + c(dc^2d) - c^2(dcd)e^* = dc + cd = \sqrt{d},\end{aligned}$$

because $ec = 0$, $e = dc^2$, $dc^2d = d$ and $dcd \in dKd = 0$. We have thus proved (Proposition 8.85) that $\sqrt{d} \in \kappa(eK(1-e)) = K_1$. As K_1 is an abelian inner ideal (because it is an extreme of a finite grading), $\sqrt{d}$ is a Jordan element of K.

(2) $(\sqrt{d})^2 = (cd+dc)(cd+dc) = c(dcd) + cd^2c + dc^2d + (dcd)c = dc^2d = d$.

(3) $(\sqrt{d})^3 = (\sqrt{d})^2\sqrt{d} = d(cd+dc) = dcd + d^2c = 0$.

(4) If follows from (1), (2) and (3) that $\sqrt{d}$ is a Clifford element of R. Hence, by Proposition 8.86(5), $\sqrt{d}K\sqrt{d} = \mathcal{C}\sqrt{d}$.

(5) It follows directly from the definition of regular partner:

$$\begin{aligned}\sqrt{d}c\sqrt{d} &= (cd+dc)c(cd+dc) \\ &= c(dc^2d) + c(dcd)c + dc^3d + (dc^2d)c = cd + dc = \sqrt{d}.\end{aligned}$$

(6) $c\sqrt{d}c = c(cd+dc)c = c^2dc + cdc^2 = c$, by 8.86(3).

(7) $c^2 \circ \sqrt{d} = c^2(cd+dc) + (cd+dc)c^2 = c^2dc + cdc^2 = c$.

(8) $d \circ c = dc + cd = \sqrt{d}$.

(9) By (6) and (7),

$$[[c, \sqrt{d}], c] = 2c\sqrt{d}c - c^2 \circ \sqrt{d} = 2c - c = c.$$

(10) It follows from (2), (5) and (8):

$$[[\sqrt{d}, c], \sqrt{d}] = 2\sqrt{d}c\sqrt{d} - (\sqrt{d})^2 \circ c = 2\sqrt{d} - \sqrt{d} = \sqrt{d}.$$

(11) By 8.86(1),

$$[[c, \sqrt{d}], b] = [[c, cd+dc], b] = [c^2d - dc^2, b] = [e^* - e, b] = e^*b + be = b.$$

□

Trace and bilinear form. Statement 8.86(5) can be rephrased by saying that there exists a linear form tr on the $\mathcal{C}$-vector space K such that $\mathrm{tr}(k)c = ckc$ for all $k \in K$. Since $c\sqrt{d}c = c$ by Proposition 8.87(6), we have $\mathrm{tr}(\sqrt{d}) = 1$ and hence

$$K = \mathcal{C}\sqrt{d} \oplus \ker(\mathrm{tr}).$$

Moreover, it follows from Proposition 8.83 that there exists a symmetric bilinear form $\langle\, ,\rangle$ on the $\mathcal{C}$-vector space K such that $\langle k_1, k_2\rangle c^2 = c^2k_1k_2c^2$ for all $k_1, k_2 \in K$.

The theorem. Everything is now ready to prove the main result of the section: If c is a Clifford element of R, then the abelian inner ideal $c^2 \circ K = \kappa((1-e)Ke)$ (see Proposition 8.86) can be endowed with a structure of Jordan algebra of Clifford type (Example 8.4) and that this Jordan algebra is isomorphic to K_c.

PROPOSITION 8.88. *Let c be a Clifford element of R and set $B = c^2 \circ K$. We have:*

(1) *$B = \mathcal{C}c \oplus X$, where $X := \{c^2 \circ k :\ k \in \ker(\mathrm{tr})\}$.*
(2) *$B = \mathrm{ad}_c^2 K$.*

PROOF. (1) We have seen that $K = \ker(\mathrm{tr}) \oplus \mathcal{C}\sqrt{d}$. Hence

$$B = c^2 \circ K = c^2 \circ \ker(\mathrm{tr}) + \mathcal{C}c^2 \circ \sqrt{d} = c^2 \circ \ker(\mathrm{tr}) + \mathcal{C}c$$

because $c^2 \circ \sqrt{d} = c$ by Proposition 8.87(7), where the sum $c^2 \circ \ker(\mathrm{tr}) + \mathcal{C}c$ is direct since $c^2 \circ k_0 = \alpha c$, with $\mathrm{tr}(k_0) = 0$ and $\alpha \in \mathcal{C}$, implies

$$\alpha c^2 = c(c^2 k_0 + k_0 c^2) = (ck_0 c)c = \mathrm{tr}(k_0)c = 0.$$

and hence $\alpha = 0$ because $c^2 \neq 0$ by the very definition of Clifford element.

(2) For any $k \in K$ we have $\mathrm{ad}_c^2\, k = c^2 k - 2ckc + kc^2 = c^2 \circ k - 2\,\mathrm{tr}(k)c \in B$. Conversely, let $c^2 \circ k_0 + \alpha c \in B$, with $k_0 \in \ker(\mathrm{tr})$ and $\alpha \in \mathcal{C}$. Then

$$c^2 \circ k_0 + \alpha c = \mathrm{ad}_c^2\, k_0 - \alpha\, \mathrm{ad}_c^2 \sqrt{d} = \mathrm{ad}_c^2 (k_0 - \alpha\sqrt{d}),$$

since $ck_0c = 0$ and $\mathrm{ad}_c^2 \sqrt{d} = -c$ by Proposition 8.87(9). □

LEMMA 8.89. *The formula* $\langle c^2 \circ k, c^2 \circ k' \rangle_0 := -\langle k, k' \rangle$ *defines a symmetric* $\mathcal{C}$*-bilinear form on the vector space* X.

PROOF. Suppose that $c^2 \circ k_1 = c^2 \circ k_1'$. By multiplying the two members of this equality on the right by $k_2 c^2$ we obtain $c^2 k_1 k_2 c^2 = c^2 k_1' k_2 c^2$ since $c^2 K c^2 = 0$. This proves that $\langle\, ,\rangle_0$ is well defined. □

REMARK 8.90. Consider the 3-grading $K = K_{-1} \oplus K_0 \oplus K_1$ yielded by the idempotent $e = dc^2$ (Proposition 8.85(3)), with

$$K_{-1} = B,\ K_0 = \kappa(eKe) \oplus gKg, \quad \text{and} \quad K_1 = \kappa(eKg).$$

(1) Since the pair $(d, \sqrt{d})$ plays a role symmetric to that played by (c^2, c), we also have that $K_1 = d \circ K = \{d \circ k : \ k \in K, \sqrt{d}k\sqrt{d} = 0\} \oplus \mathcal{C}\sqrt{d} = ad^2_{\sqrt{d}} K$. (2) X could be zero in Proposition 8.88; in this case, we would have $B = \mathcal{C}c$. But this case can only happen if R is 9-dimensional over $\mathcal{C}$. Let $X := H \oplus \mathbb{F}z$ be the orthogonal sum of a hyperbolic plane $H = \mathbb{F}x \oplus \mathbb{F}y$ and the line $\mathbb{F}z = H^{\perp}$, with z being an anisotropic vector, and let R be the simple ring $\mathcal{L}(X)$ with the adjoint as involution. Following the notation of Example 8.81, the linear map $c = x^*z - z^*x$ is a Clifford element of R such that $\mathrm{ad}_c^2\, K = \mathbb{F}c$.

THEOREM 8.91. *Let* R *be a centrally closed ring with involution of characteristic different from* $2, 3$ *and let* c *be a Clifford element of* R. *Then* K_c *is a Clifford Jordan algebra.*

PROOF. Since $c = [[c, \sqrt{d}], c]$ (8.87(9)), by Proposition 8.62, $K_c \cong J(c, \sqrt{d})$ is the Jordan algebra defined on the $\mathcal{C}$-vector space $\mathrm{ad}_c^2\, K = c^2 \circ K = \mathcal{C}c \oplus X$ (Proposition 8.88) by the product

$$(\alpha_1 c + c^2 \circ k_1) \bullet (\alpha_2 c + c^2 \circ k_2) = [[\alpha_1 c + c^2 \circ k_1, \sqrt{d}], \alpha_2 c + c^2 \circ k_2],$$

for all $\alpha_1, \alpha_2 \in \mathcal{C}$ and $k_1, k_2 \in K$ such that $ck_1c = ck_2c = 0$. Endow the $\mathcal{C}$-vector space X with the symmetric bilinear form $\langle\, ,\rangle_0$ defined in Lemma 8.89, and consider the Clifford Jordan algebra $C \oplus X$ defined by $\langle\, ,\rangle_0$ (Example 8.4). We claim that the linear isomorphism $(\alpha c + c^2 \circ k) \mapsto (\alpha, c^2 \circ k)$ of $J(c, \sqrt{d})$ onto $\mathcal{C} \oplus X$ is actually an isomorphism of Jordan algebras. Since $\frac{1}{2} \in \Phi$, it suffices to check the identity

$$(\alpha c + c^2 \circ k)^2 = [[\alpha c + c^2 \circ k, \sqrt{d}], \alpha c + c^2 \circ k] = \alpha^2 c + \langle c^2 \circ k, c^2 \circ k \rangle_0 + 2\alpha(c^2 \circ k).$$

Using the bilinearity of the Lie product reduces the checking to three products: (i) scalar by scalar, (ii) scalar by vector, and (iii) vector by vector.

(i) $[[\alpha c, \sqrt{d}], \alpha c] = \alpha^2[[c, \sqrt{d}], c] = \alpha^2 c$, by Proposition 8.87(9).

(ii) $[[\alpha c, \sqrt{d}], c^2 \circ k] = \alpha[[c, cd+dc], c^2k+kc^2] = \alpha[c^2d-dc^2, c^2k+kc^2] = \alpha(c^2 \circ k)$, where we have used $c^2dc^2 = c^2$, $c^4 = 0$ and $c^2kc^2 = c^2(dk+kd)c^2 = 0$, the latter because $c^2Kc^2 = 0$ and $(dk+kd)^* = -(kd+dk)$, since $d^* = d$ and $k^* = -k$.

(iii) $[[c^2 \circ k, \sqrt{d}], c^2 \circ k] = 2(c^2 \circ k)\sqrt{d}(c^2 \circ k) - (c^2 \circ k)^2 \circ \sqrt{d}$, with

$$(c^2 \circ k)\sqrt{d}(c^2 \circ k) = (c^2k+kc^2)(cd+dc)(c^2k+kc^2) = (c^2kdc+kc^2dc)(c^2k+kc^2) = 0$$

since $c^3 = 0$ and $ckc = 0$ ($\mathrm{tr}(k) = 0$), and

$$\begin{aligned}(c^2 \circ k)^2 \circ \sqrt{d} &= c^2k^2c^2(cd+dc) + (cd+dc)c^2k^2c^2 = c^2k^2c^2dc \\ &\quad + cdc^2k^2c^2 = \langle k, k\rangle(c^2dc + cdc^2) = \langle k, k\rangle c\end{aligned}$$

since $c = c^2dc + cdc^2$ by Proposition 8.86(1). Therefore,

$$(c^2 \circ k) \bullet (c^2 \circ k) = -\langle k, k\rangle c = \langle c^2 \circ k, c^2 \circ k\rangle_0 c,$$

which completes the proof. □

REMARK 8.92. As $\sqrt{d}$ is a Clifford element of R (Proposition 8.87(1)), the above theorem also proves that $K_{\sqrt{d}}$ is a Clifford Jordan algebra.

8.5. The Kurosh problem for Lie algebras

Kurosh's problem for associative algebras. In 1941 A. G. Kurosh formulated the following Burnside-type problem for algebras: *Is any finitely generated associative nil algebra nilpotent?*

E. S. Golod showed that this is not always the case. However, the Kurosh problem has a positive solution in the class of PI-algebras. In fact, one of the high points of the theory of PI-algebras was the solution of the Kurosh problem given by I. Kaplansky [**Kap48b**], J. Levitzki [**Lev53**], and A. I. Shirshov [**Shi57**] in the following form:

THEOREM 8.93. *Let R be an associative PI-algebra generated by the elements $a_1, \ldots, a_m$. Let S be the multiplicative semigroup generated by $a_1, \ldots, a_m$. Suppose that every element of S is nilpotent. Then R is nilpotent.*

Kurosh's problem for Jordan algebras. In 1971, Shirshov posed the Jordan version of the Kurosh problem: *Must a Jordan nil algebra of bounded index be locally nilpotent?* This problem was solved affirmatively in the case of linear Jordan algebras by E. Zelmanov in [**Zel82**]. An extension of this solution to Jordan systems over an arbitrary ring of scalars has been recently obtained in [**ACZ15**].

Kurosh's problem for Lie algebras. Let L be a Lie algebra over a field $\mathbb{F}$. Recall that a subset $S \subset L$ is called a Lie set if $[a, b] \in S$ for any $a, b \in S$. For a subset $X \subset L$, the Lie set of L generated by X is denoted by $S\langle X\rangle$. The following Kurosh-type theorem for Lie algebras can be regarded as an infinite-dimensional extension of the Engel–Jacobson Theorem (2.54).

THEOREM 8.94. [**Zel17**, Theorem 1.1] *Let L be a Lie PI-algebra over an arbitrary field $\mathbb{F}$ generated by elements $a_1, \ldots, a_m$. If every element of the Lie set $S\langle a_1, \cdots, a_m\rangle$ is ad-nilpotent, then L is nilpotent.*

This Zelmanov's theorem has important implications in group theory (see [**Zel17**, Theorems 1.2 and 1.3]) extending significatively his positive solution of the Restricted Burnside Problem [**Zel90, Zel91**], and his work [**Zel92a**] on compact torsion groups.

The Jordan approach. As another example of the use of Jordan techniques in Lie theory, we give the proof (due also to E. Zelmanov) of Theorem 8.94 in the particular case that $\mathrm{char}(\mathbb{F}) = 0$.

Let L be a Lie algebra as in 8.94, $\mathrm{char}(\mathbb{F}) = 0$. Our first aim is to show that L contains a nonzero absolute zero divisor. Choose a nonzero element s in $\langle a_1, \cdots, a_m\rangle$ and suppose that the index of ad_s is $n > 2$. By Theorem 4.27, any $x \in \mathrm{ad}_s^{n-1} L$ is a Jordan element. Since $S\langle a_1, \cdots, a_m\rangle$ spans L, we can assume that s is a Jordan element of L.

LEMMA 8.95. *The Jordan algebra L_s is McCrimmon radical, i.e. $L_s = \mathrm{Mc}(L_s)$.*

PROOF. Since L is PI, the Jordan algebra L_s is also PI (Proposition 8.57). Moreover, for any $t \in S$, $[s, t]$ is ad-nilpotent. Hence, by Proposition 8.51(4), L_s is spanned by nilpotent elements. Then $L_s = \mathrm{Mc}(L_s)$ by Lemma 8.31. □

LEMMA 8.96. *The Lie algebra L contains a nonzero absolute zero divisor.*

PROOF. Let s be a nonzero Jordan element of S. If s is not an absolute zero divisor, then L_s is nonzero and hence, by Lemma 8.95, it contains a nonzero absolute zero divisor. By Proposition 8.51(1), L itself contains a nonzero absolute zero divisor. □

Just infinite Lie algebras. In this subsection we will assume that L is a Lie algebra over an arbitrary field $\mathbb{F}$.

LEMMA 8.97. *Let L be finitely generated by elements $a_1, \ldots, a_m$ such that every $s \in S\langle a_1, \cdots, a_m\rangle$ is ad-nilpotent, and let I be an ideal of L of finite codimension, then I is finitely generated as an algebra.*

PROOF. The finite-dimensional Lie algebra L/I is spanned by a Lie set for which every element is ad-nilpotent. By the Engel-Jacobson theorem (2.54), L/I is nilpotent. In other words, there exists $k \geq 1$ such that $L^k \subset I$.

Suppose that every commutator ρ in $a_1, \ldots, a_m$ of length less than k is ad-nilpotent of index at most t. Let $N = ktm^k$. It follows from Lemma 2.63 that every product $\mathrm{ad}(a_{i_1}) \cdots \mathrm{ad}(a_{i_N})$, $1 \leq i_1, \cdots, i_N \leq m$, can be represented as

$$\mathrm{ad}(a_{i_1}) \cdots \mathrm{ad}(a_{i_N}) = \sum_j v_j \, \mathrm{ad}(\rho_j),$$

where the v_j are (possibly empty) products of the $\mathrm{ad}(a_i)$'s and the ρ_j's are commutators in $a_1, \ldots, a_m$ of length greater than or equal to k. Furthermore, each summand on the right hand side has the same degree in each a_i as the left hand side. It follows that the algebra L^k is generated by commutators ρ in $a_1, \ldots, a_m$ such that $k < \mathrm{length}(\rho) < 2N$.

Since L is finitely generated, $\dim_{\mathbb{F}}(L/L^k) < \infty$. Let $b_1, \ldots, b_r \in I$ be a basis of the vector space I/L^k. Then the Lie algebra I is generated by $b_1, \ldots, b_r$ and all commutators ρ such that $k < \mathrm{length}(\rho) < 2N$, which proves the lemma. □

DEFINITION 8.98. A Lie algebra L over a field $\mathbb{F}$ is called *just infinite* if it is infinite-dimensional and every nonzero ideal of L has finite codimension.

LEMMA 8.99. *Let L be an infinite-dimensional Lie algebra over a field $\mathbb{F}$ generated by elements $a_1, \ldots, a_m$ such that every element of $S\langle a_1, \cdots, a_m\rangle$ is ad-nilpotent. Then L has a just infinite homomorphic image.*

PROOF. Let $I_1 \subset I_2 \subset \cdots$ be an ascending chain of ideals of infinite codimension. We claim that the union $I = \bigcup_i I_i$ has also infinite codimension. Indeed, if $\dim_{\mathbb{F}}(L/I) < \infty$, then we have by Lemma 8.97 that I is finitely generated, hence I is equal to one the terms of the ascending chain, a contradiction. By Zorn's Lemma, the algebra L has a maximal ideal M of infinite codimension. Then L/M is just infinite, which proves the lemma. $\square$

Proof of Theorem 8.94 (char($\mathbb{F}$) $= 0$)**.** Suppose that L is not nilpotent. Then it is necessarily infinite-dimensional and by Lemma 8.99 we may also suppose that it is just infinite. Then (8.96) L contains a nonzero absolute zero divisor and hence, by Theorem 3.20, a nonzero locally nilpotent ideal. Let I be a nonzero locally nilpotent ideal of L. Since L is just infinite, I has finite codimension. Hence I is finitely generated as an algebra by Lemma 8.97, and therefore nilpotent. This implies that I is finite-dimensional, which contradicts the assumption that L is infinite-dimensional and proves that L is nilpotent. $\square$

E. Zelmanov proved in [**Zel83a**, Proposition 1] the following early version of Theorem 8.94.

THEOREM 8.100. *Any nil Lie PI-algebra L over a field of characteristic 0 is locally nilpotent.*

PROOF. Suppose that L is not locally nilpotent. By factorizing L by its locally nilpotent radical (Proposition 2.66), we may assume that L is nonzero and has no nonzero locally nilpotent ideals. Then L is nondegenerate (Theorem 3.20). Let $a \in L$ be a nonzero Jordan element (whose existence is guaranteed by Kostrikin's descent lemma). Then the Jordan algebra L_a is nondegenerate (Proposition 8.51(1)), PI (Proposition 8.57), and nil (Proposition 8.51(4)). Thus, by Lemma 8.31, $L_a = \mathrm{Mc}(L_a) = 0$, and hence $\mathrm{ad}_a^2 L = 0$, which is a contradiction, so L is locally nilpotent. $\square$

8.6. Nil Lie algebras of finite width

As a further application of the Jordan techniques in Lie theory, we give in this section an outline of the proof of the following theorem due to C. Martínez and E. Zelmanov. All the algebras considered here are over a field $\mathbb{F}$ of characteristic 0.

THEOREM 8.101. [**MZ99**, Theorem 1] *Let $L = \bigoplus_{\alpha\in\Gamma} L_\alpha$ be a Lie algebra graded by an abelian group Λ over a field $\mathbb{F}$ of characteristic 0. Suppose that*

(i) *there exists $d > 0$ such that $\dim_{\mathbb{F}} L_\alpha \leq d$ for all $\alpha \in \Gamma$,*
(ii) *every homogeneous element $a \in L_\alpha$ is ad-nilpotent.*

Then the Lie algebra L is locally nilpotent.

Let A be an arbitrary $\mathbb{F}$-algebra generated by a subset X, let $x_1, \dots, x_k$ be elements of X, and let $l_1 \geq 1, \dots, l_k \geq 1$ be natural numbers. Denote by

$$\begin{pmatrix} x_1, \dots, x_k \\ l_1, \dots, l_k \end{pmatrix}$$

the $\mathbb{F}$-linear span of all products in $x_1, \dots, x_k$ (with all possible arrangements of brackets) involving l_1 elements x_1, ..., l_k elements x_k.

Following [**MZ99**], we say that the pair (A, X), satisfies the condition C_d if for arbitrary elements $x_1, \dots, x_k \in X$, $k \geq 1$, and arbitrary natural numbers $l_1, \dots, l_k$, we have:

(i) $\dim_{\mathbb{F}} \binom{x_1, \dots, x_k}{l_1, \dots, l_k} \leq d$,

(ii) every element of $\binom{x_1, \dots, x_k}{l_1, \dots, l_k}$ is nilpotent.

If A is associative or Jordan, the nilpotency of an element is understood in the usual sense. If A is Lie, we are meaning ad-nilpotency.

THEOREM 8.102. *If L is a Lie algebra generated by a subset X and the pair (L, X) satisfies C_d, then L is locally nilpotent.*

PROOF. As in the original proof [**MZ99**, Lemma 1.9], we can assume that L is nondegenerate, X is closed under commutators, and then to prove that $L = 0$. If L were nonzero, then by Kostrikin's descent lemma X would contain a nonzero Jordan element. Let $a \in X$ be a nonzero Jordan element of L. It is clear that the Jordan algebra L_a has $\overline{X}$ as a set of generators and that the pair $(L_a, \overline{X})$ satisfies the condition C_d. Hence, by [**MZ99**, Lemma 1.4], L_a coincides with its McCrimmon radical, but L_a is nondegenerate, a contradiction. □

Now we return to the proof of Theorem 8.101. Clearly, the pair (L, X), where $X = \bigcup_{\alpha \in \Gamma} L_\alpha$, satisfies the condition C_d, so Theorem 8.102 applies. □

8.7. Exercises

EXERCISE 8.103. Let A be an alternative algebra over a ring of scalars Φ containing $\frac{1}{2}$. Show that A with the bullet product $x \bullet y = \frac{1}{2}(xy + yx)$ becomes a Jordan algebra.

EXERCISE 8.104. Let J be a Jordan algebra. Verify that the operator $[l_a, l_b]$, $a, b \in J$, is a derivation, assuming that J is special (as quoted in (8.1) this is true in general).

EXERCISE 8.105. Let J be a Jordan algebra and let I be a Φ-submodule of J such that $U_x I \subset I$ for all $x \in \hat{J}$. Show that I is an ideal of J.

EXERCISE 8.106. Let R be an associative Φ-algebra and let $a \in R$. Define in R a new product by $x \cdot_a y = xay$, and set $\operatorname{Ker}\{a\} = \{z \in R : aza = 0\}$. Show:

(1) R with the product $\cdot_a$ is an associative algebra, denoted by $R^{(a)}$, and $\operatorname{Ker}\{a\} = \{z \in R : aza = 0\}$ is an ideal of $R^{(a)}$.
(2) Set $R_a = R^{(a)} / \operatorname{Ker}\{a\}$. If R is semiprime, prime, left (right) primitive, or simple, then so is R_a.
(3) $R_a^+ = (R^+)_a$.
(4) If R has an involution $*$ and $a^* = \pm a$, then the map $x \mapsto \pm x^*$ induces an involution in R_a. If $a^* = -a$, then $\operatorname{Skew}(R, *)_a \cong \operatorname{Sym}(R_a, -*)$.

(5) Suppose that a is von Neumann regular, $a = aba$ for some $b \in R$. In the Φ-module aRa, define a product by $axa \cdot_b aya := axabaya = axaya$. Then the map $\overline{x} \mapsto axa$ is an isomorphism of R_a onto $(aRa, \cdot_b)$. Furthermore, R_a is unital with $1_{R_a} = \overline{b}$.
(6) If e is an idempotent of R, then $R_e \cong eRe$.
(7) If R is unital and $u \in R$ invertible, then $R_u \cong R$.

EXERCISE 8.107. Prove (J3) and the following three identities using Macdonald principle:

(i) $l_x U_x = U_x l_x = \frac{1}{2} U_{x,x^2}$,
(ii) $(U_x y)^2 = U_x U_y x^2$,
(iii) $\{x, y, U_x z\} = U_z\{x, y, z\}$,
(iv) $V(U_x y, y) = V(x, U_y x)$.

EXERCISE 8.108. Let J be unital, $x \in J$, and y in the subalgebra of J generated by $\{1, x\}$. Use Shirshov-Cohn principle to prove that $U_{x \bullet y} = U_x U_y$.

EXERCISE 8.109. Let $e \in J$ be an idempotent of a Jordan algebra J. Show:

(1) $U_e J$ is a unital Jordan algebra with e as unit element.
(2) $J_e \cong U_e J$.

EXERCISE 8.110. Let I be an ideal and e an idempotent of a nondegenerate Jordan algebra J. Using (J3) show that the Jordan algebras I and $U_e J$ are nondegenerate.

EXERCISE 8.111. Let e be an idempotent of a Jordan algebra J, and let M be an inner ideal of $U_e J$. Show that M is an inner ideal of J.

EXERCISE 8.112. Let R be a unital associative Φ-algebra, $2 \in \Phi^*$. Show:

(1) u is invertible in R if and only if it is invertible in the Jordan algebra R^+.
(2) R is a division (associative) algebra if and only if R^+ is a division Jordan algebra.
(3) R^+ is a division Jordan algebra if and only if for any nonzero elements $a, b \in R$ there exists $x \in R$ such that $axa = b$.
(4) If R is a division algebra with involution, then $\mathrm{Sym}(R, *)$ is a division Jordan algebra.

EXERCISE 8.113. Show that the Clifford Jordan algebra $J(X, \langle\, , \rangle)$, defined by a symmetric bilinear form $\langle\, , \rangle$ on a vector space X over a field $\mathbb{F}$ of characteristic not 2, is a division Jordan algebra if and only if $\langle x, x\rangle$ is not a square in $\mathbb{F}$ for any nonzero vector $x \in X$.

EXERCISE 8.114. Let J be a division Jordan algebra. Show that for any nonzero element $u \in J$, the isotope $J^{(u)}$ is a division Jordan algebra.

EXERCISE 8.115. Using just the definition, show that the U-operator of L_a satisfies the fundamental Jordan identity (J3).

EXERCISE 8.116. [**CGGL12**] Let L be a G-graded Lie algebra and let $a \in L$ be a homogeneous Jordan element. Show that the Jordan algebra L_a is G-graded, and give an example where the grading induced in L_a by a nontrivial grading of L is trivial.

EXERCISE 8.117. Let $L = L_{-n} \oplus \cdots \oplus L_n$ be a $2n+1$-graded Lie Φ-algebra, $\frac{1}{6} \in \Phi$, and let $a \in L_{-n}$. Show that the Φ-submodule L_n, regarded as a subalgebra of $L^{(a)}$, is a Jordan algebra, and that this Jordan algebra $L_n^{(a)}$ factored out the ideal $\mathrm{Ker}_L\{a\} \cap L_n$ is isomorphic to L_a.

CHAPTER 9

The Kostrikin Radical

The nondegenerate radical of an associative algebra (Baer radical), Jordan algebra (McCrimmon radical), and Lie algebra (Kostrikin radical), say A in all the cases, is always contained in the intersection of all strongly prime ideals of A; that is, those ideals P of A such that algebra A/P is prime and nondegenerate (the second condition being automatic in the associative case, but not in the Jordan or Lie cases in general). But while in the associative and Jordan cases A is nondegenerate if and only if the intersection of strongly prime (nondegenerate plus prime) ideals of A is zero, it is not known whether this remains true for arbitrary Lie algebras. A Lie algebra such that the intersection of its strongly prime ideals is zero will be called strongly nondegenerate. In this chapter we find conditions under which a nondegenerate Lie algebra is strongly nondegenerate. The material is organized as follows: In Section 9.1 the Kostrikin radical, denoted by $\mathrm{K}(L)$, is defined for any Lie algebra L over an arbitrary ring of scalars Φ. In Section 9.2 it is shown that if L contains a distinguished generating set, then its Kostrikin radical can be characterized locally in terms of the McCrimmon radicals of its Jordan algebras and hence in terms of m-sequences; moreover, if every nonzero ideal of L contains a nonzero Jordan element, then $\mathrm{K}(L)$ coincides with the intersection of all strongly prime ideals of L. This last result is also obtained in Section 9.3 just assuming that L is a Lie algebra over a field of characteristic 0. In Section 9.4 we study the relationship between the Baer radical of a 2-torsion free ring R and the Kostrikin radical of the Lie algebra R^-. Finally, in Section 9.5, we consider a local extension of the Kostrikin radical.

9.1. Definition y basic results

Let L be a Lie algebra over an arbitrary ring of scalars Φ. By a *nondegenerate ideal* we mean an ideal I of L such that L/I is a nondegenerate Lie algebra. Since the intersection of nondegenerate ideals of L is a nondegenerate ideal, there exists a smallest nondegenerate ideal of L, called the *Kostrikin radical of* L.

Set $\mathrm{K}_0(L) = 0$ and let $\mathrm{K}_1(L)$ be the ideal generated by all absolute zero divisors of L. Using transfinite induction we define a nondecreasing chain of ideals $\mathrm{K}_\alpha(L)$ by putting $\mathrm{K}_\alpha(L) = \bigcup_{\beta<\alpha} \mathrm{K}_\beta(L)$ for a limit ordinal α, and $\mathrm{K}_\alpha(L)/\mathrm{K}_{\alpha-1}(L) = \mathrm{K}_1(L/\mathrm{K}_{\alpha-1})$ otherwise. It is clear from this construction that $\mathrm{K}(L) := \bigcup_\alpha \mathrm{K}_\alpha(L)$ is a nondegenerate ideal of L.

LEMMA 9.1. *Let I be an ideal of L such that L/I is nondegenerate. Then $\mathrm{K}(L) \subset I$, so $K(L)$ is the Kostrikin radical of L.*

PROOF. We use transfinite induction. Let x be an absolute zero divisor of L. Then $\mathrm{ad}_x^2 L = 0$ implies $x \in I$ (since L/I is nondegenerate), so $\mathrm{K}_1(L) \subset I$. Fix an index α and suppose that $\mathrm{K}_\beta(L) \subset I$ for every $\beta < \alpha$. If α is a limit ordinal, then

$\mathrm{K}_\alpha(L) = \bigcup_{\beta<\alpha} \mathrm{K}_\beta(L) \subset I$. Otherwise $\mathrm{ad}_x^2 L \subset \mathrm{K}_{\alpha-1}(L) \subset I$ implies $x \in I$ since L/I is nondegenerate. Therefore, $\mathrm{K}(L) \subset I$. □

LEMMA 9.2. *Suppose that L is 6-torsion free and that I is an ideal of L. Then $\mathrm{K}(I) \subset \mathrm{K}(L) \cap I$.*

PROOF. $I/\mathrm{K}(L){\cap}I \cong (I{+}\mathrm{K}(L))/\mathrm{K}(L)$ can be regarded as an ideal of $L/\mathrm{K}(L)$, so it is a nondegenerate Lie algebra by Proposition 4.7. Then we have by Lemma 9.1 that $\mathrm{K}(I) \subset \mathrm{K}(L) \cap I$. □

In Corollary 9.19 we will see that the equality $\mathrm{K}(I) = \mathrm{K}(L) \cap I$ holds for any Lie algebra over a field of characteristic 0.

DEFINITION 9.3. Let L be a Lie algebra. An *m-sequence* is a sequence $\{a_n\}$ of L such that for each n, $a_{n+1} = [a_n, [a_n, b_n]]$ for some $b_n \in L$. An m-sequence $\{a_n\}$ has length k if $a_k \neq 0$ and $a_{k+1} = 0$. We will say that an m-sequences *terminates* if it has finite length.

PROPOSITION 9.4. *Let $x \in L$ be such that every m-sequence beginning with x terminates. Then $x \in \mathrm{K}(L)$.*

PROOF. Suppose that x does not lie in $\mathrm{K}(L)$. Then we can construct an infinite m-sequence as follows: Take $a_1 = x$. Since $L/\mathrm{K}(L)$ is nondegenerate, we have that for some $y \in L$, $[x, [x, y]]$ does not belong to $\mathrm{K}(L)$. Take $a_2 = [x, [x, y]]$ and so on. □

DEFINITION 9.5. A Lie algebra L will be called *strongly nondegenerate* if the intersection of its strongly prime ideals is zero, equivalently, L is a subdiret product of strongly prime Lie algebras. We warn the reader that in classical Lie theory (see for instance [**Pre86b**]) strong degeneration just means degeneration in our sense.

It follows from Proposition 5.24 that any nondegenerate Lie Φ-algebra, $30 \in \Phi^*$, with essential socle is strongly nondegenerate.

9.2. Lie algebras with enough Jordan elements

The results of this section are taken from [**GGL11**].

Jordan m-sequences versus Lie m-sequences. Throughout this subsection, L will denote a Lie algebra over a ring of scalars Φ in which 6 is invertible.

LEMMA 9.6. *Let $a \in L$ be a Jordan element and let $\{a_n\}$ be an m-sequence beginning with a. Then for any $n > 1$, a_n lies in the principal inner ideal $\mathrm{ad}_a^2 L$ generated by a.*

PROOF. It follows by induction: $a_2 = [a, [a, b_1]] \in \mathrm{ad}_a^2 L$ and if $a_n \in \mathrm{ad}_a^2 L$, then $a_{n+1} \in \mathrm{ad}_{a_n}^2 L \subset \mathrm{ad}_a^2 L$ since $\mathrm{ad}_a^2 L$ is an inner ideal. □

PROPOSITION 9.7. *Let $a, x \in L$ with a being a Jordan element. Then every m-sequence of L beginning with $[a, [a, x]]$ of length k gives rise to an m-sequence of the Jordan algebra L_a beginning with $\overline{x}$ and of length k, and vice versa.*

PROOF. Let $\{c_n\}$ be an m-sequence in L with $c_1 = [a, [a, x]]$ for some $x \in L$. By Lemma 9.6, for every n, $c_n = \mathrm{ad}_a^2 b_n$ for some $b_n \in L$. We claim that $\{\bar{b}_n\}$ is an m-sequence of the Jordan algebra L_a. Given $n \geq 1$, there exists $y_n \in L$ such that

$$\operatorname{ad}_a^2 b_{n+1} = c_{n+1} = \operatorname{ad}_{c_n}^2 y_n = \operatorname{ad}_{\operatorname{ad}_a^2 b_n}^2 y_n = \operatorname{ad}_a^2 \operatorname{ad}_{b_n}^2 \operatorname{ad}_a^2 y_n,$$

i.e. $\bar{b}_{n+1} = \overline{\operatorname{ad}_{b_n}^2 \operatorname{ad}_a^2 y_n} = U_{\bar{b}_n} \bar{y}_n$. Thus $\{\bar{b}_n\}$ is an m-sequence of L_a (Definition 8.27).

Conversely, let $\{\bar{b}_n\}$ be an m-sequence of the Jordan algebra L_a beginning with $\bar{x}$. Then, for every $n \geq 1$,

$$\bar{b}_{n+1} = U_{\bar{b}_n} \bar{y}_n = \overline{\operatorname{ad}_{b_n}^2 \operatorname{ad}_a^2 y_n},$$

for some $y_n \in L$. Take $c_n := \operatorname{ad}_a^2 b_n$. Then $c_1 = \operatorname{ad}_a^2 b_1 = [a, [a, x]]$ (since $\bar{b}_1 = \bar{x}$), and

$$c_{n+1} = \operatorname{ad}_a^2 b_{n+1} = \operatorname{ad}_a^2 \operatorname{ad}_{b_n}^2 \operatorname{ad}_a^2 y_n = \operatorname{ad}_{\operatorname{ad}_a^2 b_n}^2 y_n = \operatorname{ad}_{c_n}^2 y_n.$$

This proves that $\{c_n\}$ is an m-sequence of L. Moreover, $c_n = \operatorname{ad}_a^2 b_n = 0$ if and only if $\bar{b}_n = 0$, so both sequences have the same length. □

Kostrikin radical versus McCrimmon radical. In this subsection we will assume that L is a Lie algebra over a ring of scalars Φ such that: (i) 6 is invertible in Φ, and (ii) L contains a distinguished generating set (Definition 2.18), so, by Lemma 2.21, every Φ-submodule of L which is invariant under automorphisms is an ideal of L. We begin with a lemma (suggested by A. Golubkov to the authors of [**GGL11**]) which is the first step to prove a partial converse of Proposition 9.4, i.e. that every m-sequence beginning with a Jordan element which lies in the Kostrikin radical terminates.

LEMMA 9.8. *Let C_1 be the linear span of the absolute zero divisors of L, and let $a = z_1 + \cdots + z_n \in C_1$ be a Jordan element. Then every m-sequence of L beginning with a terminates.*

PROOF. Since C_1 is invariant under automorphisms, it is an ideal of L. Note also that for any Jordan element a of L and any absolute zero divisor z of L, $\bar{z}$ is an absolute zero divisor of L_a: $\operatorname{ad}_z^2 = 0 \Rightarrow U_{\bar{z}} = \overline{\operatorname{ad}_z^2 \operatorname{ad}_a^2} = 0$. Therefore, $\overline{C_1} \subset \operatorname{Mc}(L_a)$. Now, for any $x, y \in L$ we have $U_{\bar{x}}\bar{y} = \overline{\operatorname{ad}_x^2 \operatorname{ad}_a^2 y} \in \overline{C_1} \subset \operatorname{Mc}(L_a)$. Since x, y are arbitrary elements of L this proves that $L_a = \operatorname{Mc}(L_a)$. Then, by Proposition 8.28, any m-sequence of L_a beginning with $\bar{x}$ terminates. Hence, by Proposition 9.7, every m-sequence of L beginning with a terminates. □

LEMMA 9.9. *Let $a \in K(L)$ be a Jordan element. Then every m-sequence of L beginning with a terminates.*

PROOF. We will use transfinite induction. In Lemma 9.8, we have shown that the statement is true when $a \in \mathrm{K}_1(L)$. Assume that it holds when $a \in \mathrm{K}_\beta(L)$ for every $\beta < \alpha$. If α is a limit ordinal, then $a \in \mathrm{K}_\alpha(L) = \bigcup_{\beta<\alpha} \mathrm{K}_\beta(L)$, so there exists some $\gamma < \alpha$ such that $a \in \mathrm{K}_\gamma(L)$ and by the induction hypothesis, every m-sequence beginning with a terminates. Suppose then that α is not a limit ordinal. The Φ-submodule generated by the elements $x \in L$ such that $[x, [x, L]] \subset \mathrm{K}_{\alpha-1}(L)$ is clearly invariant by automorphisms of L and therefore an ideal, so equal to $\mathrm{K}_\alpha(L)$. Let $\overline{L} = L/\mathrm{K}_{\alpha-1}(L)$. Then $\mathrm{K}_1(\overline{L}) = \mathrm{K}_\alpha(L)/\mathrm{K}_{\alpha-1}(L)$ and hence, again by Lemma 9.8, for any m-sequence $\{a_n\}$ of L beginning with a, there exists an $m \geq 1$ such that $a_m \in \mathrm{K}_{\alpha-1}(L)$. Since a_m is a Jordan element, this implies by the induction hypothesis that $a_n = 0$ for some $n \geq m$. □

THEOREM 9.10. *Let a be a Jordan element of L. Then*

$$\mathrm{Mc}(L_a) = \{\bar{x} \in L_a : [a,[a,x]] \in \mathrm{K}(L)\}$$

and the following conditions are equivalent:

(i) $a \in \mathrm{K}(L)$.
(ii) $\mathrm{Mc}(L_a) = L_a$.
(iii) *Every m-sequence of L beginning with a terminates.*

PROOF. By Proposition 8.28, if $\bar{x} \in \mathrm{Mc}(L_a)$ then every m-sequence beginning with $\bar{x}$ is finite. Hence, by Proposition 9.7, every m-sequence beginning with $[a,[a,x]]$ is finite, so $[a,[a,x]] \in \mathrm{K}(L)$ by Proposition 9.4.

Conversely, since a is a Jordan element, $[a,[a,x]]$ is a Jordan element. Thus, by Lemma 9.9, if $[a,[a,x]] \in \mathrm{K}(L)$, then every m-sequence beginning with $[a,[a,x]]$ has finite length, which implies by Proposition 9.7 that every m-sequence beginning with $\bar{x}$ is finite, so $\bar{x} \in \mathrm{Mc}(L_a)$.

The equivalence (i) $\Leftrightarrow$ (ii) $\Leftrightarrow$ (iii) follows now from above and the fact that $a \in \mathrm{K}(L)$ if and only if $[a,[a,L]] \subset \mathrm{K}(L)$. □

LEMMA 9.11. *Let $\{a_n\}$ be an infinite m-sequence of elements of L beginning with a Jordan element. Then every ideal P of L which is maximal with respect to the property $P \cap \{a_n\} = \emptyset$ is strongly prime.*

PROOF. Set $M := \{a_n : n \in \mathbb{N}\}$. We first show that the Lie algebra L/P is nondegenerate. Suppose otherwise that $\mathrm{K}(L/P) \neq 0$ and let K be the ideal of L such that $\mathrm{K}_1(L/P) = K/P$. Since P is maximal with respect to $P \cap M = \emptyset$, there exists $x \in K \cap M$, $x = a_n$ for some $n \geq 1$. We have then an infinite m-sequence which begins with the Jordan element $x \in K$ (every element of M is a Jordan element of L by 4.4(4)), which is a contradiction by Lemma 9.9. Thus L/P is nondegenerate.

Now we see that L/P is prime. Let I, J be ideals of L containing strictly the ideal P. Since L/P is nondegenerate, it suffices to check that P is strictly contained in $I \cap J$. As P is maximal with respect to the property of not meeting M, there exist $i, j \in \mathbb{N}$ such that $a_i \in I$ and $a_j \in J$. Let $k \geq \max\{i,j\}$. Then $a_k \in I \cap J$ with $a_k \notin P$. This proves that L/P is prime. □

THEOREM 9.12. *If every nonzero ideal of L contains a nonzero Jordan element, then $\mathrm{K}(L)$ is equal to the intersection of the strongly prime ideals of L. Hence L is nondegenerate if and only if it is a subdirect product of strongly prime Jordan algebras.*

PROOF. We only need to show that for any $x \in L$ which does not lie in $\mathrm{K}(L)$ there exists a strongly prime ideal P of L such that x does not belong to P. Working in the Lie algebra $L/\mathrm{K}(L)$, we may assume that L is nondegenerate. Let I be the ideal generated by x. By hypothesis, there exists a nonzero Jordan element $a \in I$. Since L is nondegenerate, we can construct an infinite m-sequence $\{a_n\}$ beginning with a. Denote by M the set $\{a_n : n \in \mathbb{N}\}$ and let P be an ideal of L which is maximal with respect to $M \cap P = \emptyset$. By Lemma 9.11, P is strongly prime, and since $a = a_1$ does not belong to P, x does not belong to P either. □

COROLLARY 9.13. *Let $L = L_{-n} \oplus \cdots \oplus L_n$ be a nondegenerate Lie algebra with a nontrivial $(2n+1)$-grading over a field $\mathbb{F}$ of characteristic 0 or $p \geq 4n+1$. Then L is strongly nondegenerate.*

PROOF. It follows from Theorem 9.12 taking into account the following two facts: (1) the set $L^* = \bigcup_{k=1}^{n} L_{\pm k}$ is a distinguished generating set of L (Example 2.19(2)), and (2) any nonzero ideal I of L contains a nonzero Jordan element: Since L is nondegenerate, $Z(L) = 0$ and hence $[L_k, I] \neq 0$ for some k such that $0 < |k| < n$. Then, by the quadratic characterization of the annihilator of an ideal (Preposition 4.10), there exists an element $x \in L_k$ such that $\mathrm{ad}_x^2 I \neq 0$, so I contains a nonzero homogeneous element. Then the Kostrikin Descent Lemma (4.22) applies to get that I contains a nonzero Jordan element. □

9.3. Lie algebras over a field of characteristic zero

Following [**GGL11**], we prove in this section that any nondegenerate Lie algebra over a field of characteristic 0 is strongly nondegenerate.

Generalized m-sequences. The results contained here were outlined by E. Zelmanov in a private communication to E. García and M. Gómez.

LEMMA 9.14. *Given a Lie algebra L over a field of characteristic 0 and an ad-nilpotent element $a \in L$ of index $s > 3$, there exists $x_1 \in L$ such that $[a, x_1]$ is a nonzero Jordan element.*

PROOF. By Theorem 4.27, for any $x \in L$, $\mathrm{ad}_a^{s-1} x \in [a, L]$ is a Jordan element. □

PROPOSITION 9.15. *Let L be a Lie algebra over a field of characteristic 0 and let $a \in L$. Suppose that there exists a positive integer q such that*

$$\mathrm{ad}_a^q = \mathrm{ad}_{[a,x_1]}^q = \mathrm{ad}_{[[a,x_1],x_2]}^q = 0$$

for all $x_1, x_2 \in L$. Then $a \in K(L)$.

PROOF. We can work in $L/\mathrm{K}(L)$, assume that L is nondegenerate, and show that $a = 0$. Suppose that $a \neq 0$ and let s be the index of ad_a. If $s > 3$ take $x_1 \in L$ given in Lemma 9.14 and let $b = [a, x_1] \neq 0$, which is a Jordan element; if $s \leq 3$, let $b = a$. In any case b is a Jordan element. We claim that for every $x \in L$, $\bar{x} \in L_b$ is ad-nilpotent of index at most $q+1$. Indeed, since $[x, b]$ is ad-nilpotent of index at most q (if $b = [a, x_1]$, $[x, b] = [x, [a, x_1]]$ which is ad-nilpotent of index at most q by hypothesis), $\bar{x}^{q+1} = \mathrm{ad}_{[x,b]}^q x = 0$. Hence, by Corollary 8.32, L_b coincides with its McCrimmon radical. But since the Jordan algebras of a nondegenerate Lie algebra inherit nondegeneracy (Proposition 8.51), $L_b = \mathrm{Mc}(L_b) = 0$, which implies $\mathrm{ad}_b^2 L = 0$, a contradiction since L is nondegenerate and $b \neq 0$. □

Given a Lie algebra L and a natural number n, let

$$B_n(L) = \left\{ \sum_{i=1}^{n} [b_{i,k_i}, \cdots, b_{i,1}, a_i] : 0 \leq k_i \leq n,\ b_{i,j} \in L,\ \mathrm{ad}_{a_i}^2 = 0 \right\}$$

be the sum of left commutators in L whose distance to an absolute zero divisor of L is less than or equal to n. Note that $B_1 \subset B_2 \subset \cdots \subset B_n$ and $\mathrm{K}_1(L) = \bigcup_{m \geq 1} B_m$.

PROPOSITION 9.16. *For each $n, r \in \mathbb{N}$, there exists $f(n,r) \in \mathbb{N}$, $f(n,r) \geq 3$, such that for every Lie algebra L over a field $\mathbb{F}$ of characteristic 0 and for every $a \in B_n(L)$, we have*

$$\mathrm{ad}_{[b_k, \cdots, b_1, a]}^{f(n,r)} L = 0$$

for every $b_1, \ldots, b_k \in L$, $0 \leq k \leq r$.

PROOF. The proof is inspired by that of [**Zel83a**, Lemma 8]. Let X be the set $\{x_i, y_i, x_{ij} : i, j \in \mathbb{N}\} \cup \{x_0\}$. Consider the free Lie $\mathbb{F}$-algebra $\mathcal{L} = \mathcal{L}\langle X\rangle$ on the set X, and denote by $\overline{\mathcal{L}} := \mathcal{L}/\mathcal{I}$ the quotient of $\mathcal{L}$ by the ideal $\mathcal{I}$ generated by $\bigcup_{i\in\mathbb{N}}[x_i, x_i, \mathcal{L}]$. For every $n, r \in \mathbb{N}$, define

$$A_{n,r} := \left\{\sum_{i=1}^{n}[\bar{x}_i, \bar{x}_{i1}, \cdots, \bar{x}_{ik_i}, \bar{y}_1, \cdots, \bar{y}_k] : 0 \le k_i \le n,\ 0 \le k \le r\right\}.$$

We note that $A_{n,r} \cup [A_{n,r}, \bar{x}_0] \subset \mathrm{K}_1(\overline{\mathcal{L}})$ and that it is a finite set. Hence, by Theorem 3.20, the subalgebra $D_{n,r}$ of $\overline{\mathcal{L}}$ generated by $A_{n,r} \cup [A_{n,r}, \bar{x}_0]$ is nilpotent, so there exists $f(n,r) \ge 3$ such that $D_{n,r}^{f(n,r)} = 0$. Now, for $1 \le k \le r$, let $a, b_1, \ldots, b_k, c$ be elements of a Lie algebra L with $a \in B_n(L)$. We must show that $\mathrm{ad}_{[a,b_1,\cdots,b_k]}^{f(n,r)} c = 0$. Since $a \in B_n(L)$, $a = \sum_{i=1}^{n}[b_{i,k_i}, \cdots, b_{i,1}, a_i]$, where the a_i are absolute zero divisors of L and the b_{ij} are arbitrary elements of L, with $1 \le i \le n$. Let $\varphi : \mathcal{L} \to L$ be the unique homomorphism such that $\varphi(x_0) = c$; $\varphi(x_i) = a_i$ if $1 \le i \le n$, and $\varphi(x_i) = 0$ otherwise; $\varphi(x_{ij}) = b_{ij}$ if $1 \le i \le n$, $1 \le j \le k_i$, and $\varphi(x_{ij}) = 0$ otherwise; and $\varphi(y_i) = b_i$ if $1 \le i \le k$, and $\varphi(y_i) = 0$ otherwise. Furthermore, as the a_i are absolute zero divisors, φ vanishes on the ideal $\mathcal{I}$, so it can be lifted to a homomorphism $\bar{\varphi} : \overline{\mathcal{L}} \to L$. It is routine to check that $\mathrm{ad}_{[b_k,\cdots,b_1,a]}^{f(n,r)} c \subset \bar{\varphi}(D_{n,r}^{f(n,r)}) = 0$, as desired. □

DEFINITION 9.17. A sequence $\{c_i\}_{i\ge 1}$ of elements of a Lie algebra L is called a *generalized m-sequence* if for each $i \ge 1$, c_{i+1} is an element of the form

$$\mathrm{ad}_{c_i}^{q_i} x_0, \quad \mathrm{ad}_{[c_i,x_1]}^{q_i} x_0, \quad \text{or} \quad \mathrm{ad}_{[[c_i,x_1],x_2]}^{q_i} x_0$$

for some $x_0, x_1, x_2 \in L$, with $q_i := f(i, 3i+2)$ and where $f(n,r)$ is the function fixed above.

Since $q_i \le 3$, we have

$$\begin{aligned}
\mathrm{ad}_{c_i}^{q_i} x_0 &\in [c_i, [c_i, [c_i, L]]] \subset [[[c_i, L], L], L],\\
\mathrm{ad}_{[c_i,x_1]}^{q_i} x_0 &\in [[c_i, x_1], [[c_i, x_1], L]] \subset [[[c_i, L], L], L],\\
\mathrm{ad}_{[[c_i,x_1],x_2]}^{q_i} x_0 &\in [[[c_i, x_1], x_2], L] \subset [[[c_i, L], L], L],
\end{aligned}$$

so in each step $c_{i+1} \in [[[c_i, L], L], L]$. As before, we will say that the sequence $\{c_i\}$ *terminates* if $c_i = 0$ for some index i, and therefore for all indexes $j \ge i$.

THEOREM 9.18. *The Kostrikin radical of a Lie algebra L over a field of characteristic 0 consists of all elements $a \in L$ such that every generalized m-sequence beginning with a terminates.*

PROOF. Suppose first that a does not belong to $K(L)$. Using Proposition 9.15 we can construct an infinite generalized m-sequence beginning with a. Conversely, let $a \in K(L)$. We must show that every generalized m-sequence $\{c_i\}$ of L beginning with a terminates.

Suppose first that $c_i \in \mathrm{K}_1(L) = \bigcup_{m\ge 1} B_m(L)$, so c_i belongs to some $B_n(L)$. Since the sequence of the subspaces $B_m(L)$ is not decreasing, we may assume $n \ge i$. We claim that $c_{n+1} = 0$. Since c_{i+1} is an element of the form $\mathrm{ad}_{c_i}^{q_i} x_0$, $\mathrm{ad}_{[c_i,x_1]}^{q_i} x_0$

or $\mathrm{ad}^{q_i}_{[[c_i,x_1],x_2]}x_0$, we have by above that $c_{i+1}\in[[[c_i,L],L],L]$. Similarly,

$$c_{i+2}\in[[[c_{i+1},L],L],L]\subset[[[[[[c_i,L],L],L],L],L],L]:=[[[c_i,\underbrace{L],\cdots L}_{3.2}]],$$

so $c_n\in[[[c_i,\underbrace{L],\cdots L}_{3.(n-i)}]]$. Since $q_n=f(n,3n+2)$, it follows from Proposition 9.16:

$$\mathrm{ad}^{q_n}_{c_n}x_0=0,\quad \mathrm{ad}^{q_n}_{[c_n,x_1]}x_0=0,\quad \mathrm{ad}^{q_n}_{[[c_n,x_1],x_2]}x_0=0$$

for all $x_0,x_1,x_2\in L$, so $c_{n+1}=0$ as claimed.

Now we will show by transfinite induction that if $c_j\in\mathrm{K}_\alpha(L)$ for some $j\geq 1$, then the generalized m-sequence $\{c_i\}$ terminates. For $\alpha=1$ it has just seen. Asume that our assertion is true for every $\beta<\alpha$. If α is a limit ordinal, then $c_j\in\bigcup_{\beta<\alpha}\mathrm{K}_\beta(L)$, so there is some $\beta<\alpha$ such that $c_j\in\mathrm{K}_\beta(L)$ and the sequence terminates by the induction hypothesis. Otherwise $\alpha=\beta+1$ for some β and we can consider the corresponding generalized m-sequence $\{\bar{c}_n\}$ in $L/\mathrm{K}_\beta(L)$, for which $\bar{c}_j\in\mathrm{K}_1(L/\mathrm{K}_\beta(L)$. By the case $\alpha=1$, $\bar{c}_n=0$ for some $n\in\mathbb{N}$, so $c_n\in\mathrm{K}_\beta(L)$ and the result follows by the induction hypothesis. □

COROLLARY 9.19. [**Zel83a**, Corollary 1] *Let L be a Lie algebra over a field of characteristic 0 and let S be a subalgebra of L. Then $\mathrm{K}(L)\cap S\subset\mathrm{K}(S)$.*

PROOF. Note first that every generalized m-sequence of the Lie algebra S is a generalized m-sequence of L. Let $a\in\mathrm{K}(L)\cap S$ and let $\{c_i\}$ be a generalized m-sequence of the Lie algebra S beginning with a. By Theorem 9.18, $\{c_i\}$ terminates and hence (again by 9.18), $a\in\mathrm{K}(S)$, so $\mathrm{K}(L)\cap S\subset\mathrm{K}(S)$. □

PROPOSITION 9.20. *Let $\{c_i\}$ be an infinite generalized m-sequence of L. If P is an ideal of L which is maximal among all ideals I of L such that $I\cap\{c_i : i\in\mathbb{N}\}=\emptyset$, then P is a strongly prime ideal of L.*

PROOF. Let I_1/P and I_2/P be nonzero ideals of L/P. By the maximality of P, there exist $i,j\in\mathbb{N}$ such that $c_i\in I_1$ and $c_j\in I_2$, so $c_k\in I_1\cap I_2$ for $k\geq\max\{i,j\}$. By the definition of generalized m-sequence, this implies $c_{k+1}\in[I_1,I_2]$, and s proves that $[I_1/P,I_2/P]\neq 0$. Therefore P is a prime ideal of L.

Now we see that L/P is nondegenerate. Suppose on the contrary that $\mathrm{K}(L/P)$ is nonzero and let I be the ideal of L such that $I/P=\mathrm{K}(L/P)$. Since I contains P strictly, there is $n\in\mathbb{N}$ such that $c_i\in I$ for all $i\geq n$. This provides the generalized m-sequence $\{\bar{c}_l\}$ in $\mathrm{K}(L/P)$. By Theorem 9.18, this sequence terminates, so $c_i\in P$ for some $i\geq n$, which is a contradiction. Thus P is a nondegenerate ideal. □

THEOREM 9.21. *The Kostrikin radical of a Lie algebra L over a field of characteristic 0 is equal to the intersection of the strongly prime ideals of L.*

PROOF. By the very definition of the Kostrikin radical, $\mathrm{K}(L)$ is contained in the intersection of the strongly prime ideals of L. Conversely, let a be an element of L which does not belong to $\mathrm{K}(L)$. By Theorem 9.18, there exists an infinite generalized m-sequence $\{c_i\}$ of L beginning with a. Now we can apply Preposition 9.20 to get a strongly prime ideal P of L such that a does not belong to P, which completes the proof. □

COROLLARY 9.22. *Any nondegenerate Lie algebra over a field of characteristic 0 is a subdirect product of strongly prime Lie algebras.*

The Graded Kostrikin radical. A G-graded (G is an arbitrary group) Lie algebra is said to be *graded nondegenerate* if it has no nonzero *homogeneous* absolute zero divisors. Let L be a Lie G-graded algebra. As in the non-graded case, the *graded Kostrikin* radical of L is defined as the least graded ideal $\mathrm{K}^G(L)$ giving a graded nondegenerate quotient. Let G be an arbitrary group and let L be a G-graded Lie algebra over a field of characteristic 0. It is shown in [**CGGL12**, Theorem 4.7] that L is nondegenerate if and only if it is graded nondegenerate, i.e. $\mathrm{K}^G(L) = \mathrm{K}(L)$.

9.4. Kostrikin radical versus Baer radical

Following [**GGL11**] we study in this section the relationship between the Baer radical of a 2-torsion free ring and the Kostrikin radical of its associated Lie. Recall that a ring R is semiprime if it has no nonzero nilpotent ideals, equivalently, no nonzero absolute zero divisors ($xRx = 0$ implies $x = 0$), and that a semiprime ideal is an ideal I of R such that the ring R/I is semiprime. The smallest semiprime ideal of R, denoted here by $\mathrm{B}(R)$, is called the *Baer radical* (or lower nil radical) of R (see [**Coh03**, Proposition 8.5.6]) and it is constructed by transfinite induction in a way similar to that seen in the cases of a Jordan or a Lie algebra (Section 8.1 and Lemma 9.1). The local nilpotency of the ideal $\mathrm{B}_1(R)$ generated by the absolute zero divisors of R can be easily obtained: this ideal is spanned by the absolute zero divisors and if $x = x_1 + \cdots + x_k$ is a sum of absolute zero divisors, then any product of elements of R in which x appears at least $k+1$ times is 0. Hence any m-sequence of R^- beginning with x has at most length n, for $2^n \leq k$. Using transfinite induction as in the proof of Theorem 9.18, one can prove.

LEMMA 9.23. *Let R be a ring and let $x \in \mathrm{B}(R)$. Then every m-sequence of R^- beginning with x terminates.*

THEOREM 9.24. *Let R be a 2-torsion free ring and let $\mathrm{K}(R^-)$ be the Kostrikin radical of its associated Lie algebra R^-. Then:*

(i) $\mathrm{K}(R^-) = \{x \in R : [x, R] \subset \mathrm{B}(R)\}$.
(ii) $\mathrm{K}(R^-) = \{x \in R :$ *every m-sequence beginning with x terminates*$\}$.
(iii) $\mathrm{K}(R^-)$ *coincides with the intersection of all strongly prime ideals of R^-.*

PROOF. (i) Set $\overline{R} := R/\mathrm{B}(R)$ and let $\pi : R \to \overline{R}$ be the canonical projection. Since $\overline{R}$ is semiprime, we have by Proposition 3.35(2) that $(\overline{R})^-/Z(\overline{R})$ is a nondegenerate Lie algebra, whence the Lie isomorphism $R/\pi^{-1}(Z(\overline{R})) \cong (\overline{R})^-/Z(\overline{R})$ implies that the set $\{x \in R : [x, R] \subset \mathrm{B}(R)\} = \pi^{-1}(Z(\overline{R}))$ is a nondegenerate ideal of R^-, so $\mathrm{K}(R^-) \subset \{x \in R : [x, R] \subset \mathrm{B}(R)\}$. Conversely, if $[x, R] \subset \mathrm{B}(R)$, it follows from Lemma 9.23 that every m-sequence of R^- beginning with x terminates. Thus, by Proposition 9.4, $x \in \mathrm{K}(R^-)$.

(ii) Again by 9.4, it suffices to show that if $x \in \mathrm{K}(R^-)$, then every m-sequence of R^- beginning with x terminates. But this follows from (i).

(iii) It is enough to show that $\mathrm{K}(R^-)$ contains the intersection of a family of strongly prime ideals of R^-. Let $\{P_\lambda\}$ the family of the prime ideals of R. By [**Coh03**, Theorem 8.5.7], $\mathrm{B}(R) = \bigcap P_\lambda$. Set $R_\lambda := R/P_\lambda$ and let $\pi_\lambda : R \to R_\lambda$ be the canonical projection. Since R_λ is prime, we have by Proposition 3.35(3) that the Lie algebra $R_\lambda^-/Z(R_\lambda)$ is strongly prime, whence the Lie-isomorphism

$R^-/\pi_\lambda^{-1}(Z(R_\lambda)) \cong R_\lambda^-/Z(R_\lambda)$ implies that

$$Q_\lambda := \pi_\lambda^{-1}(Z(R_\lambda)) = \{x \in R : [x, R] \subset P_\lambda\}$$

is a strongly prime ideal of R^-. Therefore, if $x \in \bigcap Q_\lambda$, then $[x, R] \subset \cap P_\lambda = \mathrm{B}(R)$, so $x \in \mathrm{K}(R^-)$ by (i). This completes the proof of the theorem. □

COROLLARY 9.25. *Let R be a semiprime 2-torsion free ring. Then the Lie algebra $R^-/Z(R)$ is strongly nondegenerate, equivalently, it is a subdirect product of strongly prime Lie algebras.*

In [**GGL11**] the authors also study the relationship between the Baer radical of a 2-torsion free ring R with involution and the Kostrikin radical of the Lie algebra $\mathrm{Skew}(R, *)$ of its skew-symmetric elements proving the following result analogous to that of Theorem 9.24.

THEOREM 9.26. [**GGL11**, Theorem 4.7] *Let R be a 2-torsion free ring with involution and let $L = \mathrm{Skew}(R, *)$. Then:*

(i) $\mathrm{K}(L)$ *coincides with the intersection of all strongly prime ideals of L.*
(ii) $\mathrm{K}(L) = \pi^{-1}(Z(R/(\mathrm{B}(R) \cap L)))$, *where $\pi : L \to L/(\mathrm{B}(R) \cap L)$ denotes the canonical projection.*
(iii) $\mathrm{K}(L) = \{x \in R :$ *every m-sequence beginning with x terminates*$\}$.

9.5. Locally nondegenerate Lie algebras

The results of this section are taken from Hennig's paper [**Hen14**]. For any finite subset S of a Lie algebra L, an *S-sequence* is a sequence $\{x_n\}$ in L such that for each n, $x_{n+1} = [x_n, [x_n, s_n]]$ for some $s_n \in S$. Similarly, for a finite subset S of a Jordan algebra J, an S-sequence is a sequence $\{x_n\}$ in J such that for each n, $x_{n+1} = U_{x_n} s_n$ for some $s_n \in S$. In both cases, an S-sequence $\{x_n\}$ is said to have length k if $x_k \neq 0$ and $x_{k+1} = 0$. We will say that an S-sequence *terminates* if it has finite length.

DEFINITION 9.27. A Lie algebra L is *locally nondegenerate* if there are no nonzero elements $x \in L$ such that for any *finite* subset S of L, every S-sequence beginning with x terminates. By a *locally nondegenerate ideal* we mean an ideal I of L such that such that L/I is a locally nondegenerate Lie algebra.

It is clear that the intersection of locally nondegenerate ideals of a Lie algebra L is a nondegenerate ideal, so there exists a smallest locally nondegenerate ideal, $\mathrm{K}_{loc}(L)$, called the *local Kostrikin radical* of L.

LEMMA 9.28. *If L is locally nondegenerate, then L is nondegenerate.*

PROOF. If $z \in L$ is an absolute zero divisor, then any m-sequence (hence any S-sequence) of L beginning with z has length less than or equal to 2, so $z = 0$. □

Nondegeneracy and locally nondegeneracy are in fact equivalent conditions in finite-dimensional Lie algebras.

LEMMA 9.29. *If L is finite-dimensional and nondegenerate, then L is locally nondegenerate. Therefore, in a finite-dimensional Lie algebra, the Kostrikin radical coincides with the local Kostrikin radical.*

PROOF. Let S be a basis of L and let x be a nonzero element of L. Since L is nondegenerate, there exists $s_1 \in S$ such that $[x,[x,s_1]] \neq 0$. Proceeding in this way we construct an S-sequence beginning with x of infinite length, so L is locally nondegenerate □

PROPOSITION 9.30. *Let L be a simple locally finite Lie algebra over a field of characteristic 0. Then L is locally nondegenerate.*

PROOF. Suppose that there is a nonzero element $x \in L$ such that for any finite subset S of L, every S-sequence beginning with x terminates. We will show that L is locally solvable, which is a contradiction by Remark 2.84. Let L_1 be a finite-dimensional subalgebra of L containing x. Since $L = \mathrm{id}_L(x)$ (because L is simple), there exists a finite-dimensional subalgebra L_2 of L such that $L_1 \subset \mathrm{id}_{L_2}(x)$. Thus it suffices to show that $\mathrm{id}_{L_1}(x)$ is solvable. Let $\overline{L_1} = L_1/\operatorname{Rad}(L_1)$, where $\operatorname{Rad}(L_1)$ denotes the radical (maximal solvable ideal) of L_1. As $\overline{L_1}$ is a semisimple finite-dimensional Lie algebra over a field of characteristic 0, its Killing form is nondegenerate, so $\overline{L_1}$ is nondegenerate by Proposition 3.22, whence $\overline{L_1}$ is locally nondegenerate by Lemma 9.29. Since $\bar{x}$ is an element of $\overline{L_1}$ such that every $\overline{S}$-sequence beginning with $\bar{x}$ terminates, $x \in \operatorname{Rad}(L_1)$ by Proposition 9.4. Therefore $\mathrm{id}_{L_1}(\{x\}) \subset \operatorname{Rad}(L_1)$ is solvable, as desired. □

PROPOSITION 9.31. *Let $a \in L$ be a Jordan element. If the Jordan algebra L_a is locally nilpotent, then $a \in \mathrm{K}_{loc}(L)$.*

PROOF. We must prove that for any finite subset S of L, every S-sequence $\{x_n\}$ beginning with a terminates. Let $x_{n+1} = [x_n,[x_n,s_n]]$. We will show by induction that for every $n \geq 1$ there exists $c_n \in \langle S \cup \{a\}\rangle$ such that $x_{n+1} = \mathrm{ad}_a^2 c_n$. For $n = 1$, set $c_1 = s_1$. Suppose that for some $n > 1$, $x_n = \mathrm{ad}_a^2 c_{n-1}$. As a is a Jordan element, we have

$$x_{n+1} = \mathrm{ad}_{x_n}^2 s_n = \mathrm{ad}_{\mathrm{ad}_a^2 c_{n-1}}^2 s_n = \mathrm{ad}_a^2 \mathrm{ad}_{c_{n-1}}^2 \mathrm{ad}_a^2 s_n.$$

Setting $c_n = \mathrm{ad}_{c_{n-1}}^2 \mathrm{ad}_a^2 s_n$ finishes the argument. Now consider the sequence $\{\bar{c}_n\}$ of the Jordan algebra L_a. Since for every n, $\bar{c}_{n+1} = \overline{\mathrm{ad}_{c_n}^2 \mathrm{ad}_a^2 s_{n+1}} = U_{\bar{c}_n}\bar{s}_{n+1}$, the set $\{\bar{c}_n : n \in \mathbb{N}\}$ is contained in the subalgebra of L_a generated by the finite subset $\overline{S}$, which is finite-dimensional and nilpotent, since we have assumed that L_a is locally nilpotent. Then $\bar{c}_m = 0$ for some m, i.e. $x_{m+1} = \mathrm{ad}_a^2 c_m = 0$. Thus any S-sequence beginning with a terminates, which proves that $a \in \mathrm{K}_{loc}(L)$. □

9.6. Exercises

EXERCISE 9.32. [**GGL11**, Proposition 5.3] Let L be a nondegenerate Lie algebra over a ring of scalars Φ in which 6 is invertible. Show:

(1) If any annihilator ideal of L is contained in a maximal annihilator ideal, then L is strongly nondegenerate.
(2) If L satisfies in fact the chain condition on annihilator ideals (both conditions, descending and ascending are equivalent, why?), then L is an essential subdirect product of finitely many strongly prime Lie algebras.

EXERCISE 9.33. Let R be a ring. Show that the ideal of R generated by its absolute zero divisors ($xRx = 0 \Rightarrow x = 0$) is locally nilpotent.

EXERCISE 9.34. Prove Theorem 9.26.

EXERCISE 9.35. [**CGGL12**, Lemma 2.7] Let G be a group and L a G-graded Lie algebra over a ring of scalars Φ in which 6 is invertible. Show that L is graded nondegenerate if and only if L_x is nonzero for every homogeneous Jordan element $x \in L$, and that in this case, L_x is a graded nondegenerate Jordan algebra.

EXERCISE 9.36. [**Hen14**, Lemma 20] Let R be a simple finite-dimensional associative algebra over an algebraically field $\mathbb{F}$. Show that the Lie algebra $\overline{R}'$ is locally nondegenerate.

CHAPTER 10

Algebraic Lie Algebras and Local Finiteness

Recall that a Lie $\mathbb{F}$-algebra L is algebraic if for any $x \in L$ the adjoint operator ad_x is algebraic over $\mathbb{F}$, and that an arbitrary $\mathbb{F}$-algebra A is locally finite if every finitely generated subalgebra of A is finite-dimensional. A celebrated theorem due to E. Zelmanov proves that any algebraic Lie PI-algebra over a field $\mathbb{F}$ of characteristic 0 is locally finite. In particular, any algebraic Lie $\mathbb{F}$-algebra of bounded degree is locally finite. Using the transference of properties from a Lie algebra to its Jordan algebras we give in this chapter an independent proof of the second Zelmanov's theorem quoted above. We also prove that any strongly prime algebraic Lie $\mathbb{F}$-algebra of bounded degree is simple and finite-dimensional over its centroid, which is an algebraic extension of $\mathbb{F}$.

10.1. Strongly prime algebraic Lie PI-algebras

Throughout this section, we will deal with Lie algebras over an algebraically closed field $\mathbb{F}$ of characteristic 0. We begin with result [**Bah76**, Theorem 2] proved by Y. Bahturin in 1976, but now it is an almost obvious fact from [**BS94**].

LEMMA 10.1. *Let L be a simple locally finite Lie PI-algebra. Then L is finite-dimensional.*

PROOF. Suppose otherwise that L is infinite-dimensional. Following [**Bar99**, Lemma 4.1], for any finite-dimensional subspace Q of L, there exists a finite-dimensional subalgebra P of L and a maximal ideal M of P such that $M \cap Q = 0$, so Q is embedded into the simple finite-dimensional Lie algebra $S = P/M$. Taking Q large enough, we get that S is one of the classical simple finite-dimensional Lie algebras over $\mathbb{F}$, $\mathfrak{sl}_n$, $\mathfrak{o}_{2n}$, $\mathfrak{o}_{2n+1}$, $\mathfrak{sp}_{2n}$ for large n. All these algebras contain a copy of $\mathfrak{sl}_n$, so L cannot satisfy a nontrivial identity. □

LEMMA 10.2. *Suppose that L is strongly prime and PI. If L contains a nonzero Jordan element a such that the Jordan algebra L_a is algebraic, then L is simple and finite-dimensional.*

PROOF. By Propositions 8.51 and 8.57, L_a is a strongly prime algebraic Jordan PI-algebra. Hence it follows (see Remark 8.42) that L_a is a simple Jordan algebra of finite capacity. Let $\bar{e}$ be a division idempotent of $\overline{L_a}$. Since the field $\mathbb{F}$ is algebraically closed and L_a is algebraic, $U_{\bar{e}}L_a = \overline{\mathrm{ad}_e^2\,\mathrm{ad}_a^2\,L} = \mathbb{F}\bar{e}$, which implies $\mathrm{ad}_a^2\,\mathrm{ad}_e^2\,\mathrm{ad}_a^2\,L = \mathrm{ad}^2_{\mathrm{ad}_a^2 e}\,L = \mathbb{F}\,\mathrm{ad}_a^2\,e$, i.e. $\mathrm{ad}_a^2\,e$ is an extremal element of L. Let S be the ideal of L generated by $\mathrm{ad}_a^2\,e$. By Proposition 5.21, S is a simple Lie algebra spanned by extremal elements (Vandermonde argument), so S is locally finite (Corollary 6.16). Since S is also PI, we have by Lemma 10.1 that S is actually finite-dimensional. By primeness, L can be embedded in $\mathrm{Der}(S)$ via the adjoint representation. Hence $L = S$ because every derivation of a simple finite-dimensional Lie algebra over a field of characteristic 0 is inner [**Hum72**]. □

COROLLARY 10.3. *Suppose that L is strongly prime, algebraic, and PI. Then L is simple and finite-dimensional.*

PROOF. By Corollary 4.32, L contains a nonzero Jordan element a. Let $a \in L$ be a nonzero Jordan element. By Proposition 8.51(5), L_a is an algebraic Jordan algebra. Now Lemma 10.2 applies. □

10.2. Algebraic Lie algebras of bounded degree

Following [**FLG13**], we give here a proof of the local finiteness of any algebraic Lie algebra of bounded degree over a field of characteristic 0 [**Zel83a**, Theorem 2], which is independent of the more general PI-case proved in [**Zel83a**, Theorem 1]. We also prove here that any strongly prime algebraic Lie algebra L of bounded degree over a field of characteristic 0 is simple and finite-dimensional over its centroid, which is an algebraic extension of the base field. The more general case that L is algebraic and PI was proved in [**FLG13**, Theorem 1.1].

As in the classical associative case [**Her94**, Lemma 6.2.3], any algebraic Lie algebra of bounded degree is a Lie PI-algebra.

LEMMA 10.4. *Let L be an algebraic Lie algebra of bounded degree over a field $\mathbb{F}$. Then L is PI.*

PROOF. By hypothesis, there exists a positive integer n such that for any $x \in L$ the set of linear maps $\{1_L, \mathrm{ad}_x, \ldots, \mathrm{ad}_x^n\}$ is linearly dependent over $\mathbb{F}$. Denoting the adjoint operators by capital letters, for any $y \in L$, we have

$$[X^n, Y] \in \mathrm{span}_{\mathbb{F}}\{[X,Y], [X^2,Y], \ldots, [X^{n-1}, Y]\},$$
$$[[X^n, Y], [X,Y]] \in \mathrm{span}_{\mathbb{F}}\{[[X^2,Y],[X,Y]], \ldots, [[X^{n-1},Y],[X,Y]]\}.$$

Proceeding in this way get a polynomial identity $f(x, y, z) = 0$ for L. □

LEMMA 10.5. *Let $\overline{L}$ be a Lie algebra over a field $\overline{\mathbb{F}}$ of characteristic 0, let $\mathbb{F}$ be a subfield of $\overline{\mathbb{F}}$, and let L be a $\mathbb{F}$-subalgebra of $\overline{L}$ such that L is algebraic of bounded degree and $\overline{L}$ is $\overline{\mathbb{F}}$-spanned by L. Then the Jordan $\overline{\mathbb{F}}$-algebra $\overline{L}_a$ is algebraic for any Jordan element $a \in \overline{L}$.*

PROOF. Let a be a Jordan element of $\overline{L}$ and let y be an arbitrary element of $\overline{L}$. In virtue of the formula $\overline{y}^m = \overline{\mathrm{ad}_{[y,a]}^{m-1} y}$, $m \geq 1$, proving that $\overline{y}$ is an algebraic element of $\overline{L}_a$ is equivalent to seeing that the set $\{\mathrm{ad}_{[y,a]}^s y : s \geq 0\}$ is linearly dependent over $\overline{\mathbb{F}}$.

Write a as $a = \alpha_1 a_1 + \ldots + \alpha_k a_k$, where $\alpha_i \in \overline{\mathbb{F}}$ and $a_i \in L$, and let x be an arbitrary element of L. For every $s \geq 1$ and $\beta_1, \ldots, \beta_k \in \overline{\mathbb{F}}$, we have

$$\begin{aligned}\mathrm{ad}_{[x,\beta_1 a_1 + \ldots + \beta_k a_k]}^s x &= (\beta_1 \,\mathrm{ad}_{[x,a_1]} + \ldots + \beta_k \,\mathrm{ad}_{[x,a_k]})^s x \\ &= \sum_{\substack{\{i_1,\ldots,i_s\},\\ 1 \leq i_1 \leq i_2 \leq \ldots \leq i_s \leq k}} \beta_{i_1} \cdots \beta_{i_s} \Big(\sum_{\sigma \in \mathfrak{S}_s} \mathrm{ad}_{[x,a_{i_{\sigma(1)}}]} \cdots \mathrm{ad}_{[x,a_{i_{\sigma(s)}}]} \Big) x \,.\end{aligned}$$

Since $\mathbb{F}$ is infinite, for each $s \geq 1$ we can find a finite subset T_s of $\mathbb{F}$ such that the finite-dimensional $\mathbb{F}$-subspace of L spanned by all the elements $\operatorname{ad}^s_{[x,\gamma_1 a_1+\ldots+\gamma_k a_k]} x$, $\gamma_1, \ldots, \gamma_k \in T_s$, contains all the elements

$$\Big(\sum_{\sigma \in \mathfrak{S}_s} \operatorname{ad}_{[x,a_{i_{\sigma(1)}}]} \cdots \operatorname{ad}_{[x,a_{i_{\sigma(s)}}]}\Big) x \quad (1 \leq i_1 \leq \ldots \leq i_s \leq k).$$

Hence we get that the sets

$$\begin{aligned}
&\{\operatorname{ad}^s_{[x,\beta_1 a_1+\ldots+\beta_k a_k]} x \,:\, (\beta_1, \ldots, \beta_k) \in \overline{\mathbb{F}}^k\},\\
&\{\operatorname{ad}^s_{[x,\theta_1 a_1+\ldots+\theta_k a_k]} x \mid (\theta_1, \ldots, \theta_k) \in T_s^k\},\\
&\{\operatorname{ad}^s_{[x,\gamma_1 a_1+\ldots+\gamma_k a_k]} x \,:\, (\gamma_1, \ldots, \gamma_k) \in \mathbb{F}^k\}
\end{aligned}$$

span the same (finite-dimensional) $\overline{\mathbb{F}}$-subspace $W(x,s)$ of $\overline{L}$.

Note also that we can choose the finite subset T_s on the base of standard ideas connected with Vandermonde determinant. Hence we can assume that the number of elements of T_s is less than or equal to some integer $l(s,k)$, $l(s,k) \leq (s+1)^k$, and, therefore, $\dim_{\overline{\mathbb{F}}} W(x,s) \leq l(s,k)$. Since L is algebraic of bounded degree, say n, for all $s \geq n$,

$$W(x,s) \;\subset\; \sum_{t=0}^{n-1} W(x,t).$$

Hence the $\overline{\mathbb{F}}$-subspace $W(x)$ of $\overline{L}$ spanned by the set

$$\{\operatorname{ad}^s_{[x,\beta_1 a_1+\ldots+\beta_k a_k]} x \,:\, (\beta_1, \ldots, \beta_k) \in \overline{\mathbb{F}}^k,\ s \geq 0\}$$

has finite dimension, with $\dim_{\overline{\mathbb{F}}} W(x) \leq 1 + l(1,k) + \ldots + l(n-1,k)$. In particular, the $\overline{\mathbb{F}}$-subspace $U(x)$ of $W(x)$ spanned by all the elements

$$\operatorname{ad}^s_{[x,a]} x \;=\; \operatorname{ad}^s_{[x,\alpha_1 a_1+\ldots+\alpha_k a_k]} x \quad (s \geq 0)$$

has finite dimension, with $\dim_{\overline{\mathbb{F}}} U(x) \leq d$, where d is a positive integer depending on the degree n of the algebraic adjoint representation of L and on the number k determined by the chosen representation of the Jordan element a, but which is independent of the element x. In fact, since $\operatorname{ad}^s_{[x,a]} x = \operatorname{ad}^t_{[x,a]} \operatorname{ad}^{s-t}_{[x,a]} x$ for any positive integers s, t such that $s \geq t$, we have

$$U(x) \;=\; \sum_{t=0}^{d-1} \overline{\mathbb{F}} \operatorname{ad}^t_{[x,a]} x. \tag{10.1}$$

Now let y be an arbitrary element of $\overline{L}$. We can write $y = \psi_1 y_1 + \ldots + \psi_l y_l$, where $\psi_i \in \overline{\mathbb{F}}$ and $y_i \in L$. For every $t \geq 1$ and $(\phi_1, \ldots, \phi_l) \in \overline{\mathbb{F}}^l$, we have

$$\begin{aligned}
&\operatorname{ad}^t_{[\phi_1 y_1+\ldots+\phi_l y_l,a]}(\phi_1 y_1 + \ldots + \phi_l y_l)\\
&= (\phi_1 \operatorname{ad}_{[y_1,a]} + \ldots + \phi_l \operatorname{ad}_{[y_l,a]})^t(\phi_1 y_1 + \ldots + \phi_l y_l)\\
&= \sum_{\substack{\{i_1,\ldots,i_{t+1}\},\\ 1 \leq i_1 \leq i_2 \leq \ldots \leq i_{t+1} \leq l}} \phi_{i_1} \cdots \phi_{i_{t+1}} \Big(\sum_{\sigma \in \mathfrak{S}_{t+1}} (\operatorname{ad}_{[y_{i_{\sigma(1)}},a]} \cdots \operatorname{ad}_{[y_{i_{\sigma(t)}},a]}) y_{i_{\sigma(t+1)}}\Big).
\end{aligned}$$

Set $V_0 = \mathbb{F} y_1 + \ldots + \mathbb{F} y_l$. Using the same arguments as before, we can find for each $t \geq 1$ a finite subset H_t of $\mathbb{F}$ such that the finite-dimensional $\mathbb{F}$-subspace of $\overline{L}$

spanned by the set

$$\{\mathrm{ad}^t_{[\tau_1 y_1+\ldots+\tau_l y_l, a]}(\tau_1 y_1 + \ldots + \tau_l y_l), \quad (\tau_1, \ldots, \tau_l) \in H^l_t\}$$

contains all the elements of the form

$$\sum_{\sigma \in \mathfrak{S}_{t+1}} (\mathrm{ad}_{[y_{i_{\sigma(1)}}, a]} \cdots \mathrm{ad}_{[y_{i_{\sigma(t)}}, a]}) y_{i_{\sigma(t+1)}} \quad (1 \leq i_1 \leq \ldots \leq i_{t+1} \leq l).$$

Hence the subsets $\{\mathrm{ad}^t_{[\tau_1 y_1+\ldots+\tau_l y_l, a]}(\tau_1 y_1 + \ldots + \tau_l y_l) : (\tau_1, \ldots, \tau_l) \in H^l_t\}$,

$$\{\mathrm{ad}^t_{[v,a]} v : v \in V_0\} \text{ and } \{\mathrm{ad}^t_{[w,a]} w : w \in \overline{\mathbb{F}} V_0\}$$

span the same (finite-dimensional) $\overline{\mathbb{F}}$-subspace V_t of $\overline{L}$. In particular, for each $s \geq 0$ there exists $v_1, \ldots v_{r_s} \in V_0$ such that $\mathrm{ad}^s_{[y,a]} y \in \sum_{j=1}^{r_s} U(v_j)$, and hence, by 10.1,

$$\mathrm{ad}^s_{[y,a]} y \in \sum_{j=1}^{r_s} U(v_j) \subset \sum_{j=1}^{r_s} (\sum_{t=0}^{d-1} \overline{\mathbb{F}} \, \mathrm{ad}^t_{[v_j,a]} v_j) = \sum_{t=0}^{d-1} (\sum_{j=1}^{r_s} \overline{\mathbb{F}} \, \mathrm{ad}^t_{[v_j,a]} v_j) \subset \sum_{t=0}^{d-1} V_t .$$

Since each V_t is a finite-dimensional $\overline{\mathbb{F}}$-subspace of $\overline{L}$, the set $\{\mathrm{ad}^s_{[y,a]} y : s \geq 0\}$ is linearly dependent over $\overline{\mathbb{F}}$, as desired. □

LEMMA 10.6. *Let $\overline{L}$ be a nondegenerate Lie algebra over an algebraically closed field $\overline{\mathbb{F}}$ of characteristic* 0, *let $\mathbb{F}$ be a subfield of $\overline{\mathbb{F}}$, and let L be an algebraic $\mathbb{F}$-subalgebra of $\overline{L}$ of bounded degree such that $\overline{L}$ is $\overline{\mathbb{F}}$-spanned by L. Then $\overline{L}$ is a subdirect product of a family of finite-dimensional simple Lie algebras of bounded dimension. As a consequence, $\overline{L}$ satisfies all the identities that hold in a finite-dimensional Lie algebra.*

PROOF. By Corollary 9.22, we may reduce the proof to the case that $\overline{L}$ is strongly prime. Since L is algebraic of bounded degree, it satisfies a multilinear identity of degree, say n, and since $\overline{L} = \mathrm{span}_{\overline{\mathbb{F}}}(L)$ $\overline{L}$ also satisfies that identity. By hypothesis, any element of L is algebraic over $\mathbb{F}$ and therefore also over *overline*$\mathbb{F}$, and as *overline*$\mathbb{F}$ is algebraically closed, we have by Corollary 4.32 that $\overline{L}$ contains nonzero Jordan elements. Let $a \in \overline{L}$ be a nonzero Jordan element of $\overline{L}$. The Jordan algebra $\overline{L}_a$ is algebraic over $\mathbb{F}$ (Lemma 10.5) and strongly prime (Propositions 8.51(2). Then it follows from Lemma 10.2 that $\overline{L}$ is simple and finite-dimensional over $\mathbb{F}$. Since no matrix algebra $\mathrm{M}_r(\overline{\mathbb{F}})$ satisfies an identity of degree less than $2r$ and the Lie algebra $\mathrm{M}_r(\overline{\mathbb{F}})^-$ can be embedded in each one of the simple Lie algebras A_r, B_r, C_r and D_r, $\overline{L}$ is isomorphic to one of the algebras $G_2, F_4, E_6, E_7, E_8, A_r$, B_r, C_r, or D_r, where $r \leq [n/2]$. □

Any Lie algebra over a field of characteristic 0 satisfying all the identities that hold in a finite-dimensional Lie algebra is locally finite. This result (due to E. Zelmanov [**Zel83a**, Lemma 7]) is reproduced in [**Kos90**, Theorem 5.4.6] where the proof is divided into a number of comparatively elementary steps.

THEOREM 10.7. [**Zel83a**, Theorem 2]. *Let L be an algebraic Lie algebra of bounded degree over a field $\mathbb{F}$ of characteristic* 0. *Then L is locally finite.*

PROOF. Assume that L is not locally finite. By taking the quotient of L by its locally finite radical $\mathrm{LocF}(L)$, we may suppose that L is nonzero and that it does not contain nonzero locally finite ideals (Corollary 2.82). Since $\mathrm{K}_1(L)$ is a locally nilpotent, and therefore locally finite, ideal (Theorem 3.20), L is nondegenerate under our assumption. Let $\overline{\mathbb{F}}$ be the algebraic closure of $\mathbb{F}$, $\overline{L} = \overline{\mathbb{F}} \otimes_{\mathbb{F}} L$, and $\tilde{L} = \overline{L}/\mathrm{K}(\overline{L})$. As L is a nondegenerate $\mathbb{F}$-subalgebra of $\overline{L}$, we have by Corollary 9.19 that $L \cap K(\overline{L}) = 0$, so L can be regarded as an $\mathbb{F}$-subalgebra of $\tilde{L}$. Thus by Lemma 10.6 L satisfies all the identities which hold in some finite-dimensional Lie algebra, which implies that L locally finite. [1] □

REMARK 10.8. Let L be an algebraic Lie PI-algebra over an algebraically closed field $\mathbb{F}$ of characteristic 0. Factoring L by its Kostrikin radical, we have by Corollary 9.22 that L is a subdirect product of strongly prime algebraic Lie PI-algebras each of which contains a nonzero Jordan element (Corollary 4.32). Arguing as in the proof of Lemma 10.6, we prove that L satisfies all the identities which hold in some finite-dimensional Lie algebra and hence that L is locally finite: This gives an alternative proof of [**Zel83a**, Theorem 1] when the base field is algebraically closed.

Now we approach the second goal of this section, i.e. any strongly prime algebraic Lie algebra of bounded degree over a field $\mathbb{F}$ of characteristic 0 is simple and finite-dimensional over its centroid, which is an algebraic extension of $\mathbb{F}$.

LEMMA 10.9. (Zelmanov) *Let L be a Lie algebra over a field $\mathbb{F}$ of characteristic 0 and let $\overline{\mathbb{F}}$ be the algebraic closure of $\mathbb{F}$. If L is algebraic of bounded degree, then the $\overline{\mathbb{F}}$-algebra $\overline{L} := \overline{\mathbb{F}} \otimes_{\mathbb{F}} L$ is algebraic.*

PROOF. Let $a = \alpha_1 a_1 + \cdots + \alpha_n a_n \in \overline{L}$, where α_i and a_i belong to $\overline{\mathbb{F}}$ and L respectively. By Theorem 10.7, the subalgebra M of L generated by $a_1, \ldots, a_n$ is finite-dimensional, and therefore so is its $\overline{\mathbb{F}}$-scalar extension $\overline{M} = \overline{\mathbb{F}} \otimes_{\mathbb{F}} M$. Let $e_1, \ldots, e_s$ be a basis of $\overline{M}$. By the Poincaré–Birkhoff–Witt Theorem (see [**Hum72**, Corollary 17.3(C)]), the associative $\overline{\mathbb{F}}$-subalgebra $\mathrm{Ad}(\overline{M})$ of the algebra of linear transformations of $\overline{L}$ is $\overline{\mathbb{F}}$-spanned by $\mathrm{ad}_{e_1}^{k_1} \cdots \mathrm{ad}_{e_s}^{k_s}$. Since the operators ad_{e_i} are algebraic, it follows that $\mathrm{Ad}(\overline{M})$ is finite-dimensional, which implies that ad_a is algebraic over $\overline{\mathbb{F}}$ and proves that $\overline{L}$ is algebraic over $\overline{\mathbb{F}}$. □

LEMMA 10.10. (Zelmanov) *Let $\overline{L}$ be a simple finite-dimensional Lie algebra over an algebraically closed field $\overline{\mathbb{F}}$ of characteristic 0, let $\mathbb{F}$ be a subfield of $\overline{\mathbb{F}}$, and let L be a $\mathbb{F}$-subalgebra of $\overline{L}$ such that $\overline{L}$ is $\overline{\mathbb{F}}$-spanned by L. If the $\mathbb{F}$-subalgebra L is locally finite, then L is simple, its centroid is an algebraic field extension of $\mathbb{F}$, and L is finite-dimensional over its centroid.*

PROOF. Let $\dim_{\overline{\mathbb{F}}} \overline{L} = n$. By the Jacobson Density Theorem (without unit), the associative $\overline{\mathbb{F}}$-subalgebra $\mathrm{Ad}(L)$ of $\mathrm{End}_{\overline{\mathbb{F}}}(\overline{L})$, which can be regarded as $\mathrm{M}_n(\overline{\mathbb{F}})$, is the whole of $\mathrm{End}_{\overline{\mathbb{F}}}(\overline{L})$. Let $c_n(x_1, \ldots, x_m)$ be a central polynomial of $n \times n$ matrices, i.e. c_n is multilinear in $x_1, \ldots, x_m$, all its coefficients are ± 1, c_n is not the identity on $\mathrm{M}_n(\overline{\mathbb{F}})$, and all values of c_n are scalar matrices. Let I be a nonzero ideal of L. Since $\overline{L}$ is simple and $\overline{\mathbb{F}}$-spanned by L, $\overline{\mathbb{F}}I = \overline{L}$ and $\mathrm{End}_{\overline{\mathbb{F}}}(\overline{L})$ is generated over $\overline{\mathbb{F}}$ by the adjoint operators ad_a, $a \in L$. Then $c_n(\mathrm{End}_{\overline{\mathbb{F}}}(\overline{L}), \cdots, \mathrm{End}_{\overline{\mathbb{F}}}(\overline{L}))$ is spanned over $\overline{\mathbb{F}}$ by $0 \neq c_n(T_1, \cdots, T_m)a_o$ for $a_0 \in I$ and monomials $T_i = \mathrm{ad}_{a_{i1}} \cdots \mathrm{ad}_{a_{ik(i)}}$ in elements $a_{ij} \in L$ and therefore some

$$c_n\big(\mathrm{ad}_{a_{i1}} \cdots \mathrm{ad}_{a_{ik(i)}}, \cdots, \mathrm{ad}_{a_{m1}} \cdots \mathrm{ad}_{a_{mk(m)}}\big)a_0 \neq 0.$$

[1] I have recently found in [**Cha90**] that Theorem 10.7 does not actually require any restriction on the ground field. Nevertheless, I have decided to keep it in the text because its "Jordan proof" is of independent interest.

Thus $c := c_n\big(\operatorname{ad}_{a_{i1}}\cdots\operatorname{ad}_{a_{ik(i)}},\cdots,\operatorname{ad}_{a_{m1}}\cdots\operatorname{ad}_{a_{mk(m)}}\big)$ is a linear operator which acts as a nonzero scalar multiplication on both $\overline{L}$ and L, hence $\ker c = 0$.

Now let b be any nonzero element of L. Since the $\mathbb{F}$-subalgebra of L generated by b, a_{ij_i}, $1 \le i \le m$, $1 \le j_i \le k_i$ is finite-dimensional, there exists a nonzero polynomial $f(t) \in \mathbb{F}[t]$ such that $f(c)b = 0$. Set $f(t) = t^r g(t)$, where the constant term of $g(t)$ not zero. Note also $f(c)b = c^r g(c)b = 0$ implies by injectivity of c that $g(c)b = 0$ for $g(c)$ with nonzero constant term, in particular b lies in the image of c on L, which belongs to I. We have proved that $I = L$, so L is simple. Hence the centroid Γ of L is a field. Note also that $f(c)b =$ implies $f(c) = 0$ since c is a nonzero scalar multiple of the identity on $\overline{L}$, so c is algebraic over $\mathbb{F}$. Let $\gamma \in \Gamma$. Then $\gamma c = c_n(\operatorname{ad}_{\gamma a_{11}}\cdots\operatorname{ad}_{a_{1i_1}},\dots,\operatorname{ad}_{a_{m1}}\cdots\operatorname{ad}_{a_{mk_m}})$, which proves that γc is also algebraic over $\mathbb{F}$, hence γ is algebraic over $\mathbb{F}$ and therefore Γ is an algebraic field extension of $\mathbb{F}$. Finally, since L is clearly PI and locally finite by hypothesis, it is finite-dimensional over its centroid by [**Bah76**, Theorem 2]. □

THEOREM 10.11. *Let L be an algebraic Lie algebra of bounded degree over a field $\mathbb{F}$ of characteristic 0. If L is strongly prime, then L is simple and finite-dimensional over its centroid, which is an algebraic extension of $\mathbb{F}$.*

PROOF. Let $\overline{\mathbb{F}}$ be the algebraic closure of $\mathbb{F}$. By Lemma 10.9, the $\overline{\mathbb{F}}$-algebra $\overline{L} = \overline{\mathbb{F}} \otimes_{\mathbb{F}} L$ is algebraic, and PI (because so is L). Let I be an ideal of $\overline{L}$ which is maximal with respect to the property that I has zero intersection with L, and set $\tilde{L} = \overline{L}/I$. The $\overline{\mathbb{F}}$-algebra $\tilde{L}$ is clearly prime. We now see that $\tilde{L}$ is also nondegenerate and therefore strongly prime. Let $\mathrm{K}(\tilde{L}) = \overline{K}/I$ be the Kostrikin radical of $\tilde{L}$. By Corollary 9.19, $\mathrm{K}(\tilde{L}) \cap L \subset \mathrm{K}(L) = 0$. Hence

$$\bar{0} = \frac{L \oplus I}{I} \cap \frac{\overline{K}}{I} = \frac{\overline{K} \cap (L \oplus I)}{I} = \frac{(\overline{K} \cap L) \oplus I}{I}$$

implies $L \cap \overline{K} \subset L \cap I = 0$, and hence $\overline{K} = I$ by maximality of I with respect to the property of having zero intersection with L. Thus $\mathrm{K}(\tilde{L}) = 0$, which proves that $\tilde{L}$ is nondegenerate. Summarizing, $\tilde{L}$ is a strongly prime PI-Lie algebra over the algebraically closed field $\overline{\mathbb{F}}$. Then it follows from Proposition 10.3 that $\tilde{L}$ is simple and finite-dimensional. But L is locally finite (Theorem 10.7), so we can apply Lemma 10.10 to conclude that L is a finite-dimensional simple Lie algebra over its centroid, which is an algebraic extension of $\mathbb{F}$. □

10.3. Exercises

EXERCISE 10.12. Let L be a Lie algebra which is algebraic of bounded degree 4 over a field $\mathbb{F}$. Find a polynomial identity for L.

EXERCISE 10.13. Let L be a strongly prime algebraic Lie algebra over a field $\mathbb{F}$ of characteristic 0 or $p > 5$ containing a nonzero von Neumann regular element. Show that the centroid of L is an algebraic field extension of $\mathbb{F}$.

EXERCISE 10.14. Let L be a strongly prime algebraic Lie algebra over an algebraically closed field $\mathbb{F}$ of characteristic 0 or $p > 5$ with nonzero socle. Show that L is centrally closed, i.e. its extended centroid is one-dimensional over $\mathbb{F}$.

CHAPTER 11

From Lie Algebras to Jordan Pairs

Mimicking the Jordan pair case, we associate a Jordan pair to every abelian inner ideal B of a Lie algebra L. This Jordan pair, called the subquotient of L with respect to B, inherits regularity conditions from L and shares with it the inner ideal structure within B. For instance, the subquotient is a nondegenerate Artinian Jordan pair if L is nondegenerate and B has finite length. Since we will only deal with linear Jordan pairs, the condition $\frac{1}{6} \in \Phi$ is assumed throughout the chapter. Section 11.1 is a brief survey on linear Jordan pairs. We include definitions, examples, and basic results with the purpose of helping the reader to go through them when required. In Section 11.2 we go from Jordan pairs to Lie algebra through the Tits–Kantor–Koecher construction. Let $L = L_{-n} \oplus \cdots \oplus L_n$ be a simple Lie algebra with a nontrivial finite grading, $L_n + L_{-n} \neq 0$, $n \geq 1$, over a field $\mathbb{F}$ of characteristic 0 or $p \geq 4n + 1$. In Section 11.3 it is shown: (i) the associated Jordan pair $V = (L_n, L_n)$ is simple, (ii) any element of the centroid of V is induced by the action on an element of the centroid of L, and (iii) if the Jordan pair V is exceptional, then L is the TKK-algebra of V. Section 11.4 is the core of the chapter: here the notion of subquotient of a Lie algebra with respect to an abelian inner ideal is defined, which allows us to reduce questions on Lie algebras to questions on Jordan pairs. In Section 11.5 we revisited the notions of von Neumann regularity, idempotent and socle in Lie algebras showing that they are actually Jordan notions, i.e. they can be expressed in Jordan terms by means of subquotients, and use this Jordan approach to get new insights into the structure of the socle of a nondegenerate Lie algebra.

11.1. Linear Jordan pairs

Throughout this section, Φ will denote a ring of scalars in which 6 is invertible. The reader is referred to [**Loo75**] for general results and terminology of Jordan pairs.

Definition and examples. A (linear) *Jordan pair* is a pair of Φ-modules $V = (V^+, V^-)$ with trilinear maps $\{\cdots\} : V^\sigma \times V^{-\sigma} \times V^\sigma \to V^\sigma$ $(\sigma = \pm)$, satisfying the following two conditions for $x, z, u \in V^\sigma$, $y, v \in V^{-\sigma}$:

(JP1) $\{x, y, z\} = \{z, y, x\}$
(JP2) $\{u, v, \{x, y, z\}\} + \{x, \{v, u, y\}, z\} = \{\{u, v, x\}, y, z\} + \{x, y, \{u, v, z\}\}$

- Given $x, z \in V^\sigma$, $y \in V^{-\sigma}$, set

$$D(x, y)z = Q_{x,z}y = \{x, y, z\}, \quad Q_x y = \frac{1}{2}\{x, y, x\}.$$

The Q-operators, the analogous of the U-operators in Jordan algebras, satisfy the identity (see [**Loo75**, JP3]):

(JP3) $Q_{Q_x y} = Q_x Q_y Q_x$.

- For $x, z \in V^\sigma$, $y \in V^{-\sigma}$, the Bergmann operator

$$B(x,y) := 1_{V^\sigma} - D(x,y) + Q_x Q_y$$

satisfies the identity (see [**Loo75**, JP26]:

(JPB) $Q_{B(x,y)z} = B(x,y) Q_z B(y,x)$.

EXAMPLES 11.1. (1) Every Jordan Φ-algebra J gives rise to a Jordan pair $V = (J, J)$, with $\{x, y, z\}$ being the triple product of J. (In particular, every associative Φ-algebra R yields a Jordan pair $V = (R^+, R^+)$, with Jordan triple products defined by $\{x, y, z\} = xyz + zyx$.) Thus every notion defined for Jordan pairs makes sense for Jordan algebras.

(2) A *linear Jordan triple system* is a Φ-module T with a trilinear map $\{x, y, z\}$ satisfying (JP1) and (JP2) and quadratic operator $P_x y = \frac{1}{2}\{x, y, x\}$. Any Jordan triple system T defines a Jordan pair $V = (T, T)$, and conversely, any Jordan pair $V = (V^+, V^-)$ gives rise to a Jordan triple $T = V^+ \oplus V^-$ with triple product defined by

$$\{x^+ \oplus x^-, y^+ \oplus y^-, z^+ \oplus z^-\} := \{x^+, y^-, z^+\} \oplus \{x^-, y^+, z^-\}.$$

(3) Any associative Φ-algebra R with involution $*$ gives rise to a Jordan triple system R^t with triple product $\{x, y, z\} = xy^*z + zy^*x$. Note that $\mathrm{Skew}(R, *)$ is both a Lie algebra (for the bracket product) and a Jordan subtriple system of R^t, $P_x y = -xyx$.

(4) Let X_1, X_2 be Φ-modules. Then $V = (\mathrm{Hom}_\Phi(X_1, X_2), \mathrm{Hom}_\Phi(X_2, X_1))$ is a Jordan pair for the triple products $\{a, b, c\} = abc + cba$.

(5) Let $L = L_{-n} \oplus \cdots \oplus L_n$ be a $2n+1$-graded Lie Φ-algebra. Then the pair of Φ-modules $V = (L_n, L_{-n})$ is a Jordan pair with triple product defined by $\{x, y, z\} = [[x, y], z]$ for $x, z \in L_{\pm n}$, $y \in L_{\mp n}$.

(6) An *associative pair* is a pair $\mathcal{A} = (\mathcal{A}^+, \mathcal{A}^-)$ of Φ-modules with trilinear maps $\mathcal{A}^\sigma \times A^{-\sigma} \times A^\sigma \to A^\sigma$, $(x, y, z) \mapsto xyz$ ($\sigma = \pm$) satisfying the following identity:

$$uv(xyz) = u(vxy)z = (uvx)yz.$$

Every associative algebra R gives rise to an associative pair (R, R) under the triple product abc. The same is true for the pair $(\mathrm{Hom}_\Phi(X_1, X_2), \mathrm{Hom}_\Phi(X_2, X_1))$.

Any associative pair $\mathcal{A}$ becomes a Jordan pair $\mathcal{A}^J$: $\{x, y, z\} = xyz + zyx$.

(7) A *polarized involution* of an associative pair $\mathcal{A}$ is a pair of involutive linear maps $* : \mathcal{A}^\sigma \to \mathcal{A}^\sigma$ such that $(xyz)^* = z^* y^* x^*$, for all $x, z \in \mathcal{A}^\sigma$, $y \in \mathcal{A}^{-\sigma}$. Note that if $*$ is a polarized involution, then $-*$ is again a polarized involution.

Let $*$ is a polarized involution of an associative pair $\mathcal{A}$. Then the pairs of Φ-modules

$$\mathrm{Sym}(\mathcal{A}, *) = (\mathrm{Sym}(\mathcal{A}^+, *),\ \mathrm{Sym}(\mathcal{A}^-, *))$$

and

$$\mathrm{Skew}(\mathcal{A}, *) = (\mathrm{Skew}(\mathcal{A}^+, *),\ \mathrm{Skew}(\mathcal{A}^-, *))$$

are subpairs of the Jordan pair $\mathcal{A}^J$.

Derivations of Jordan pairs. A *derivation* of a Jordan pair $V = (V^+, V^-)$ is a pair (δ^+, δ^-) of linear maps $\delta^\sigma : V^\sigma \to V^\sigma$ such that

$$\delta^\sigma(\{x, y, z\}) = \{\delta^\sigma(x), y, z\} + \{x, \delta^{-\sigma}(y), z\} + \{x, y, \delta^\sigma(z)\}$$

for $x, z \in V^\sigma$, $y \in V^{-\sigma}$. It is routine to check:

- $\mathrm{Der}(V)$ is a subalgebra of the Lie algebra $\mathfrak{gl}(V^+) \times \mathfrak{gl}(V^-)$.

Note that, on the one hand, Axiom (JP2) can be rephrased by saying that for $u \in V^+$, $v \in V^-$, the pair of linear maps $(D(u, v), -D(v, u))$ is a derivation of V, called an *inner derivation of* V. And on the other hand, rewritten as

$$[D(u, v)D(x, y)] = D(D(u, v)x, y) - D(x, D(v, u)y)$$

expresses that the commutator of two inner derivations is an inner derivation, and therefore that inner derivations span a subalgebra $\mathfrak{D}$ of $\mathrm{Der}(V)$.

Basic notions on Jordan pairs. Let V be a Jordan pair over a ring of scalars Φ containing $\frac{1}{6}$.

- The *opposite* of V is the Jordan pair $V^{op} = (V^-, V^+)$.
- An *inner ideal* of $V = (V^+, V^-)$ is a Φ-submodule $B \subset V^\sigma$ such that $Q_b V^{-\sigma} \subset B$ for any $b \in B$, equivalently, since 2 is invertible in Φ, $\{B, V^{-\sigma}, B\} \subset B$. By (JP3), for any $x \in V^\sigma$,

$$[x]_V := Q_x V^{-\sigma} \text{ and } (x)_V := \Phi x + Q_x V^{-\sigma}$$

are inner ideals of V. As for Lie and Jordan algebras, $[x]$ is called the *principal inner ideal determined by* x

- A *subpair* of $V = (V^+, V^-)$ is a a pair $S = (S^+, S^-)$ of Φ-submodules, $S^\sigma \subset V^\sigma$, such that $\{S^\sigma, S^{-\sigma}, S^\sigma\} \subset S^\sigma$.
- An *ideal* of V is a pair $I = (I^+, I^-)$ of Φ-submodules, $I^\sigma \subset V^\sigma$, such that

$$\{I^\sigma, V^{-\sigma}, V^\sigma\} + \{V^\sigma, I^{-\sigma}, V^\sigma\} \subset I^\sigma.$$

- Let $V = (V^+, V^-)$ and $W = (W^+, W^-)$ be Jordan pairs over the same ring of scalars Φ. A *homomorphism* of V to W is a pair of Φ-linear maps $\varphi = (\varphi^+, \varphi^-)$, where $\varphi^\sigma : V^\sigma \to W^\sigma$ such that

$$\varphi^\sigma\{x, y, z\} = \{\varphi^\sigma(x), \varphi^{-\sigma}(y), \varphi^\sigma(z)\}$$

for all $x, z \in V^\sigma$, $y \in V^{-\sigma}$. As in the case of algebras, the image of a homomorphism of Jordan pairs $\varphi : V \to W$ is a subpair of W, and the kernel is an ideal of V.

- Let I be an ideal of V and let r be an odd number. The r-power of I is the pair of Φ-submodules $I^r = (I^{+^r}, I^{-^r})$ given by $I^{\sigma^r} = \{I^{\sigma^{r-2}}, I^{-\sigma}, I^\sigma\}$ for $r \geq 3$.
- V is *semiprime* if $I^3 = 0$ implies $I = 0$ for any ideal $I = (I^+, I^-)$ of V.
- V is *prime* if $\{I^\sigma, J^{-\sigma}, I^\sigma\} = 0$ implies $I = 0$ or $J = 0$ for any ideals I, J of V.
- V is *simple* if it has only the trivial ideals V and 0, and $\{V^\sigma, V^{-\sigma}, V^\sigma\} \neq 0$.

DEFINITION 11.2. A Jordan pair $V = (V^+, V^-)$ is called *special* if it is a subpair of the Jordan pair (R^+, R^+) defined by an associative algebra R. As in the case of Jordan algebras, Jordan pairs which are not special are called *exceptional.* Note that the Jordan pairs of Example 4 are special. On the other hand, Jordan pairs defined by exceptional Jordan algebras are exceptional.

Annihilators. Let $I = (I^+, I^-)$, $J = (J^+, J^-)$ be ideals of a Jordan pair V. We will say that I *is orthogonal to* J and will write $I \perp J$ if the following conditions holds:

$$\{I^\sigma, J^{-\sigma}, V^\sigma\} = \{J^\sigma, I^{-\sigma}, V^\sigma\} = \{I^\sigma, V^{-\sigma}, J^\sigma\} = 0. \tag{11.1}$$

Note that the relation of orthogonality between ideals is symmetric, and that given an ideal I there exists a largest ideal, denoted by $\mathrm{Ann}(I)$ and called the *annihilator* of I, such that $I \perp \mathrm{Ann}(I)$. It is also clear that given I, J ideals of V, $I \subset \mathrm{Ann}(J)$ if and only if $J \subset \mathrm{Ann}(I)$, and that the map $I \mapsto \mathrm{Ann}(I)$ defines a Galois connection on the lattice of ideals of V (see the proof of Proposition 1.5 for definition).

• For any ideal $I = (I^+, I^-)$ of a Jordan pair V, denote by $\mathrm{ann}(I^\sigma)$ the set of all elements $x \in V^{-\sigma}$ such that

$$\{x, I^\sigma, V^{-\sigma}\} = \{x, V^\sigma, I^{-\sigma}\} = 0, \quad \{I^\sigma, x, V^\sigma\} = 0.$$

LEMMA 11.3. *For any ideal I of a Jordan pair V,* $\mathrm{Ann}(I) = (\mathrm{ann}(I^-), \mathrm{ann}(I^+))$.

PROOF. We only need to verify that the pair of Φ-modules $(\mathrm{ann}(I^-), \mathrm{ann}(I^+))$ is an ideal of V; but this easily follows using (JP1) and (JP2). □

LEMMA 11.4. *Let I, J be ideals of a semiprime Jordan pair V. Then $I \perp J$ if and only if $I \cap J = 0$.*

PROOF. Clearly, $I \cap J = 0$ implies $I \perp J$. Suppose that $I \perp J = 0$ and set $K = I \cap J$. Then $\{K^\sigma, K^{-\sigma}, K^\sigma\} \subset \{I^\sigma, J^{-\sigma}, V^\sigma\} = 0$, so $K = 0$. □

Nondegenerate Jordan pairs and McCrimmon radical. A Jordan pair V is *nondegenerate* if $Q_x V^{-\sigma} = 0$ implies $x = 0$ for $x \in V^\sigma$, and *strongly prime* if it is prime and nondegenerate. Using (JP3), one can prove that if V is nondegenerate, then any ideal $I = (I^+, I^-)$ is a nondegenerate Jordan pair. Moreover, $I^+ = 0$ if and only if $I^- = 0$ (if $I^- = 0$, then for any $x \in I^+$, $Q_x I = 0$ and hence $x = 0$ by nondegeneracy of I). Nondegenerate Jordan pairs are semiprime, but the converse is not true: there exist prime Jordan pairs which are degenerate [**Pch86**]. However, simple Jordan pairs are nondegenerate [**ACZ16**].

As in the case of a Jordan algebra, the *McCrimmon radical* of a Jordan pair V is defined as the smallest ideal $\mathrm{Mc}(V)$ such that the quotient $V/\mathrm{Mc}(V)$ is a nondegenerate Jordan pair.

The definition of m-sequence of a Jordan pair V mimics the analogue for Jordan algebras: a sequence $\{a_n\} \subset V^\sigma$ is called a m-sequence if $a_{n+1} = Q_{a_n} b_n$ for some $b_n \in V^{-\sigma}$, $n \geq 1$. The characterization of the McCrimmon radical of a Jordan algebra in terms of m-sequences remains true for Jordan pairs.

LEMMA 11.5. *$x \in \mathrm{Mc}(V)^\sigma$ if and only if every m-sequence beginning with x terminates.*

Centroid of a (linear) Jordan pair. The *centroid* $\Gamma(V)$ of a (linear) Jordan pair V is defined as the set of pairs of linear maps $\alpha = (\alpha^+, \alpha^-)$, $\alpha^\sigma : V^\sigma \to V^\sigma$, satisfying

$$\alpha^\sigma\{x, y, z\} = \{\alpha^\sigma x, y, z\} = \{x, \alpha^{-\sigma} y, z\} = \{x, y, \alpha^\sigma z\}$$

for all $x, z \in V^\sigma$, $y \in V^{-\sigma}$. If V is semiprime, then $\Gamma(V)$ is a reduced commutative ring, a domain acting faithfully on V if V is prime, and a field if V is simple (see [**Loo75**, Proposition 1.17]). See also [**McC99**] for an exhaustive study of the centroid of a Jordan system.

Structural transformations. Following [**Loo89**], a *structural transformation* between Jordan pairs V and W (over the same ring of scalars Φ) is a pair (f, g) where $f : V^+ \to W^+$ and $g : W^- \to V^-$ (in the opposite direction!) are Φ-linear maps satisfying

$$Q_{f(x)} = fQ_x g, \quad Q_{g(y)} = gQ_y f,$$

for all $x \in V^+$ and $y \in W^-$. We indicate this by $(f, g) : V \rightleftarrows W$. The main examples are the *inner structural transformations*.

Let $u \in V^+$ and $v \in V^-$. It follows from (JP3) and (JPB) that

$$(Q_u, Q_u) : V^{op} \rightleftarrows V, \quad (Q_v, Q_v) : V \rightleftarrows V^{op}, \quad \text{and} \quad (B(u, v), B(v, u)) : V \rightleftarrows V$$

are structural, and so is the composition of structural transformations. Finally, if $(f, g) : V \rightleftarrows W$ is structural, so is $(g, f) : W^{op} \rightleftarrows V^{op}$.

Kernels, subquotients, homotopes, and local algebras. As introduced in [**LN94**], given a Jordan pair $V = (V^+, V^-)$ and an inner ideal M of V contained in V^+, we define the *kernel* of M as the set

$$\operatorname{Ker}_V M = \{x \in V^- : Q_M x = 0\}.$$

Then $(0, \operatorname{Ker}_V M)$ is an ideal of the Jordan pair (M, V^-) and the quotient Jordan pair

$$\operatorname{Sub}_V M := (M, V^-)/(0, \operatorname{Ker}_V M) = (M, V^-/\operatorname{Ker}_V M)$$

is called the *subquotient of V with respect to M*. The kernel of an inner ideal $N \subset V^-$ and the corresponding subquotient are defined similarly. The reader is referred to [**LN94**] for examples of subquotients. We can also construct Jordan algebras by means of Jordan pairs.

PROPOSITION 11.6. *Let $V = (V^+, V^-)$ be a Jordan pair and let $v \in V^{-\sigma}$. On the Φ-module V^σ we define a (binary) product by $x \bullet y = \{x, v, y\}$ for all $x, y \in V^\sigma$.*

(i) *$V^{(v)} := (V^\sigma, \bullet)$ is a Jordan algebra with $U_x^{(v)} y = Q_x Q_v y$, $x, y \in V^\sigma$.*
(ii) *$\operatorname{Ker}_V\{v\} = \{z \in V^\sigma : Q_v z = 0\}$ is an ideal of $V^{(v)}$.*

PROOF. See [**Loo75**, 1.9] and [**DM95**, Proposition 1.2]. □

The Jordan algebra $V^{(v)}$ constructed above is called the *homotope* of V at v. The resulting Jordan algebra obtained by taking the quotient of $V^{(v)}$ with respect to the ideal $\operatorname{Ker}_V\{v\}$ is denoted by V_v and called the *local algebra* of V at v.

Invertible element and division Jordan pairs. Let $V = (V^+, V^-)$ be a Jordan pair. Following [**Loo75**], an element $u \in V^\sigma$ is called *invertible* if the linear map $Q_u : V^{-\sigma} \to V^\sigma$ is invertible. In this case the *inverse* of u is defined by $u^{-1} = Q_u^{-1} u$. From (JP3) it follows easily that $Q_u^{-1} = Q_{u^{-1}}$ and that u^{-1} is invertible, $(u^{-1})^{-1} = u$. It is also clear that if J is a unital Jordan algebra with u as unit element, then u is an invertible element of the Jordan pair (J, J). The converse is also true (see [**Loo75**, Proposition 1.11]). In fact, Jordan pairs with invertible elements are the Jordan pairs defined by unital Jordan algebras, and we have the announced connection (Section 8.1) between isotopy in Jordan algebras and isomorphism in Jordan pairs.

PROPOSITION 11.7. [**Loo75**, Proposition 1.8] *Let J and K be unital Jordan algebras with associated Jordan pairs V and W. Then the map $f \mapsto (f, U^{-1}_{f(1)}(f))$ is a bijection between the set of isotopies from J to K and the set of isomorphisms from V to W. In particular, V is isomorphic to W if and only if J is isotopic to K.*

By a *division Jordan pair* we mean a nonzero Jordan pair such that every nonzero element of V is invertible. Again, J is a division Jordan algebra if and only if (J, J) is a division Jordan pair. Hence, by Proposition 8.20, V is a division Jordan pair if and only if it is nondegenerate and has no proper inner ideals different from 0.

Primitive Jordan pairs. Introduced by E. Zelmanov in [**Zel84b**], a Jordan pair $V = (V^+, V^-)$ is said to be *primitive* at $b \in V^{-\sigma}$ if there exists a proper inner ideal $K \subset V^\sigma$ and an element $c \in V^\sigma$, such that K is a c-modular inner ideal of the Jordan algebra $V^{(b)}$, and for any ideal $I = (I^+, I^-)$ of V with $I^\sigma \neq 0$, $K + I^\sigma = V^\sigma$. As in the case of Jordan algebras, primitivity in Jordan pairs can be characterized locally.

PROPOSITION 11.8. [**AC96**, Corollary 3.9] *A Jordan pair $V = (V^+, V^-)$ is primitive at some $b \in V^{-\sigma}$ if and only if V is strongly prime and V_b is a primitive Jordan algebra.*

Local inheritance of the primitivity in Jordan pairs had been previously proved by A. D'Amour and K. McCrimmon in [**DM95**, Theorem 6.1].

Von Neumann regularity and Jordan pair idempotents. Recall that an element a of a Jordan algebra J is von Neumann regular if there exists $b \in J$ such that $a = U_a b$. This notion has a natural extension to Jordan pairs.

DEFINITION 11.9. Let $V = (V^+, V^-)$ a Jordan pair. An element $a \in V^\sigma$ is *von Neumann regular* if there exists $b \in V^{-\sigma}$ such that $Q_a b = a$, equivalently, a is an idempotent in the b-homotope Jordan algebra $V^{(b)}$.

PROPOSITION 11.10. [**DM95**, 1.7] Let $a = Q_a b \in V^+$ be von Neumann regular and $b \in V^-$. Then $V_a \cong Q_a V^-$, regarded $Q_a V^-$ as a subalgebra of $V^{(b)}$.

PROOF. The map $\bar{x} \mapsto Q_a x$ from V_a to $Q_a V^-$ is well defined and one-to-one. So it suffices to show that it keeps up squares:

$$\bar{x}^{(2,a)} = \overline{Q_x a} \mapsto Q_a Q_x a = Q_a Q_x Q_a b = Q_{Q_a x} b = (Q_a x)^{(2,b)}.$$

□

Any von Neumann regular element $a = aba$ in an associative algebra yields the idempotents $e = ab$ and $f = ba$ ($e^2 = (aba)b = ab = e$ and similarly for f). As already noted (see the paragraph previous to Lemma 8.15), an analogous construction does not work in general for von Neumann regular elements in a Jordan algebra. Yet there exists a suitable generalization of the notion of idempotent which makes even sense for elements of a Jordan pair.

DEFINITION 11.11. A pair $(e^+, e^-) \in V^+ \times V^-$ is said to be a *Jordan pair idempotent* or an *idempotent of the Jordan pair* V if $Q_{e^\sigma} e^{-\sigma} = e^\sigma$.

Note that (e^+, e^-) is an idempotent of V if and only if e^+ is an idempotent of the Jordan algebra $V^{(e^-)}$ and e^- is an idempotent of $V^{(e^+)}$ (see Proposition 11.6).

LEMMA 11.12. *Every von Neumann regular element $a \in V^\sigma$ can be extended to an idempotent of V.*

PROOF. Let $a = Q_a c$ for some $c \in V^{-\sigma}$. Then taking $b = Q_c a$ and using (JP3), we get

$$Q_a b = Q_a Q_c a = Q_a Q_c Q_a c = Q_{Q_a c} c = Q_a c = a,$$
$$Q_b a = Q_{Q_c a} a = Q_c Q_a Q_c a = Q_c (Q_a Q_c Q_a) c = Q_c Q_a c = Q_c a = b.$$

□

Peirce decomposition. As already noted, if (e^+, e^-) is an idempotent of a Jordan pair $V = (V^+, V^-)$, then e^+ is an idempotent of the Jordan algebra $V^{(e^-)}$ and e^- is an idempotent of the Jordan algebra $V^{(e^+)}$. Thus we can apply the Jordan algebra case to yield Peirce decompositions in V^+ and V^-.

PROPOSITION 11.13. [**Loo75**, Theorem 5.4] *Let (e^+, e^-) be an idempotent of a Jordan pair $V = (V^+, V^-)$. We have:*

(i) $V^\sigma = V_2^\sigma(e^+, e^-) \oplus V_1^\sigma(e^+, e^-) \oplus V_0^\sigma(e^+, e^-)$,
with Peirce spaces
(ii) $V_i^\sigma(e^+, e^-) = \{x \in V^\sigma : D(e^\sigma, e^{-\sigma})x = ix\}$, $i = 2, 1, 0$,
and supplementary projections
(iii) $E_2^\sigma = Q_{e^\sigma} Q_{e^{-\sigma}}$, $\quad E_1^\sigma = D(e^\sigma, e^{-\sigma}) - 2E_2^\sigma$, $\quad E_0^\sigma = B(e^\sigma, e^{-\sigma})$.
Moreover,
(iv) $V_2^\sigma(e^+, e^-)$ *and* $V_0^\sigma(e^+, e^-)$ *are inner ideals of* V.

PROOF. Set $J := V^{(e^{-\sigma})}$. Since e^σ is an idempotent of J, we have by Proposition 8.6,

$$J = J_1(e^\sigma) \oplus J_{\frac{1}{2}}(e^\sigma) \oplus J_0(e^\sigma).$$

Setting $V_{2i}^\sigma(e) := J_i(e^\sigma)$, $i = 1, \frac{1}{2}, 0$, we get

$$V_{2i}^\sigma(e) = \{x \in V^\sigma : D(e^\sigma, e^{-\sigma})x = 2ix\}, \ E_2^\sigma = U_{e^\sigma} = Q_{e^\sigma} Q_{e^{-\sigma}},$$
$$E_1^\sigma = 2l_{e^\sigma} - 2U_{e^\sigma} = D(e^\sigma, e^{-\sigma}) - 2Q_{e^\sigma} Q_{e^{-\sigma}},$$
$$E_0^\sigma = 1_{V^\sigma} - 2l_{e^\sigma} + U_{e^\sigma} = 1_{V^\sigma} = D(e^\sigma, e^{-\sigma}) + Q_{e^\sigma} Q_{e^{-\sigma}} = B(e^\sigma, e^{-\sigma}).$$

Finally, (iv) follows from Exercise 11.71(1). □

Chain conditions. The *length* of a properly ascending chain

$$M_0 \subset M_1 \subset \cdots \subset M_n$$

of inner ideals of a Jordan pair V is defined by the number of inclusions of the chains (so the length of the above chain is n). The *length* of an inner ideal M, denoted by length(M), is the supremum of the lengths of chains of inner ideals of V contained in M. A fact by no means trivial is the following proposition (see [**LN94**, Corollary 4.8]).

PROPOSITION 11.14. *If V is nondegenerate, then $l(V^+) = l(V^-)$. In this case, the common value $l(V)$ is called the length of V.*

• A Jordan pair V is *Artinian* if has descending chain condition (dcc) on inner ideals. As proved in [**LN94**, Theorem 5.1].

PROPOSITION 11.15. *A nondegenerate Jordan pair V is Artinian if and only if it has finite length.*

• A Jordan pair V has *finite capacity* if it satisfies both chain conditions on principal inner ideals.

The socle of a nondegenerate Jordan pair. Given a Jordan pair V, the *socle* of V is the defined by the pair of submodules $\mathrm{Soc}(V) = (\mathrm{Soc}(V^+), \mathrm{Soc}(V^-))$, where $\mathrm{Soc}(V^\sigma)$ is the sum of all minimal inner ideals of V contained in V^σ. It follows easily from Exercise 11.71 that $\mathrm{Soc}(V)$ is an ideal of V. In fact, [**Loo89**, Theorem 2], if V is nondegenerate, then $\mathrm{Soc}(V)$ is a direct sum of ideals each of which is a simple nondegenerate Jordan pair coinciding with its socle. By [**Loo89**, Theorems 1], we have

PROPOSITION 11.16. *For $x \in V^+$, the following conditions are equivalent:*

(i) $x \in \mathrm{Soc}(V^+)$,
(ii) *x is von Neumann regular, and if $x = e^+$ for some idempotent e, then $V_2(e)$ has finite capacity,*
(iii) *the inner ideal (x) generated by x has dcc on principal inner ideals.*

As for Jordan algebras (Example 8.35 and Proposition 8.38), we have.

PROPOSITION 11.17. *Let $V = (V^+, V^-)$ be a strongly prime Jordan pair with nonzero socle. Then V is primitive at any nonzero element $b \in V^{-\sigma}$.*

PROOF. See [**BFL94**, Theorem 1] or [**Mon99**, Proposition 4.4(i)]. □

PROPOSITION 11.18. [**Mon99**, Theorem 4.10] *Let V be a primitive Jordan pair.*

(1) *If for some nonzero element $b \in V^{-\sigma}$ the local algebra V_b is PI, then V has nonzero socle.*
(2) *If all the local algebras of V are PI, then V is simple and coincides with its socle.*
(3) *If all the local algebras of V satisfy the same polynomial identity, then V is simple and has finite capacity.*

Simple Jordan pairs with minimal inner ideals. Following [**FLTB03**], we describe in this subsection the simple Jordan pairs with minimal inner ideals (over a ring of scalars Φ in which 6 is invertible), equivalently, those simple Jordan pairs having the dcc on principal inner ideals. We begin by recalling some notation on pair of dual vector spaces and related Jordan pairs of adjointable operators (Section 2.2).

(1) Let $\mathcal{P}_1 = (X_1, Y_1, \langle\, ,\rangle_1)$ and $\mathcal{P}_2 = (X_2, Y_2, \langle\, ,\rangle_2)$ be pairs of dual vector spaces over the same division Φ-algebra Δ. Denote by $\mathcal{L}(X_1, X_2)$ the Φ-module of *adjointable* linear maps $a : X_1 \to X_2$, i.e. there exists $a^\# : Y_2 \to Y_1$ such that

$$\langle x_1 a, y_2\rangle_2 = \langle x_1, a^\# y_2\rangle_1.$$

Then

$$\mathcal{L}(\mathcal{P}_1, \mathcal{P}_2) := (\mathcal{L}(X_1, X_2),\ \mathcal{L}(X_2, X_1))$$

is an associative pair with associated Jordan pair $\mathcal{L}(\mathcal{P}_1, \mathcal{P}_2)^J$ (Example 11.1(4)), and the ideal

$$\mathcal{F}(\mathcal{P}_1, \mathcal{P}_2)^J = (\mathcal{F}(X_1, X_2),\ \mathcal{F}(X_2, X_1))$$

of finite rank adjointable linear maps is a simple Jordan pair with minimal inner ideals.

(2) Let Δ be a division Φ-algebra with involution $\alpha \mapsto \bar{\alpha}$. Following [**FLTB03**, 3.11], by a *pair of skew dual vector spaces* we mean a pair (X, Y) of left vector spaces over Δ with a map $\langle\, , \rangle : X \times Y \to \Delta$ such that the triple $(X, \overline{Y}, \langle\, , \rangle)$ is a pair of dual vector spaces over Δ, where $\overline{Y}$ denotes the abelian group $(Y, +)$ regarded as a right vector space over Δ by defining $y \cdot \alpha = \bar{\alpha} y$, for $y \in Y$, $\alpha \in \Delta$.

Given a pair of skew dual vector spaces $\mathcal{P} = (X, Y, \langle\, , \rangle)$ over $(\Delta, -)$, the *opposite* of $\mathcal{P}$ is defined as the pair of skew dual vector spaces

$$\mathcal{P}^{op} = (Y, X, \langle\, , \rangle^{op}),$$

where $\langle y, x\rangle^{op} = \overline{\langle x, y\rangle}$.

For any $a \in \mathcal{L}(X, Y)$, its adjoint $a^{\#}$ also lies in $\mathcal{L}(X, Y)$:

$$\langle xa, x'\rangle^{op} = \langle x, x'a^{\#}\rangle.$$

Then $\mathcal{L}(\mathcal{P}, \mathcal{P}^{op})$ is an associative pair with polarized involution $\#$. And

$$\mathrm{Sym}(\mathcal{F}(\mathcal{P}, \mathcal{P}^{op}), \#) = (\mathrm{Sym}(\mathcal{F}(X, Y), \#),\ \mathrm{Sym}(\mathcal{F}(Y, X), \#))$$

is a simple Jordan pair with minimal inner ideals.

(3) Let $\mathcal{P} = (X, Y, \langle\, , \rangle)$ be a pair of skew dual vector spaces over a field $\mathbb{F}$ of characteristic 0 or $p > 3$ with the identity as involution. Then

$$\mathrm{Skew}(\mathcal{F}(\mathcal{P}, \mathcal{P}^{op}), \#) = (\mathrm{Skew}(\mathcal{F}(X, Y), \#), \mathrm{Skew}(\mathcal{F}(Y, X), \#))$$

is a simple Jordan pair with minimal inner ideals.

(4) Let X be a vector space over a field $\mathbb{F}$ (of characteristic 0 or $p > 2$) and let $q : X \to \mathbb{F}$ be a quadratic form with associated bilinear form

$$q(x, y) = q(x + y) - q(x) - q(y).$$

The pair of vector spaces (X, X) becomes a Jordan pair with

$$Q_x y = q(x, y)x - q(x)y,$$

called the *Clifford Jordan pair* defined by q and denoted by $C(X, q)$. If q is nondegenerate, then $C(X, q)$ is nondegenerate and coincides with its socle [**Loo75**, 12.8]. Furthermore, it is simple if $\dim_{\mathbb{F}} X \neq 2$. Clifford Jordan pairs are special.

Strongly prime Jordan pairs with nonzero socle, and in particular simple Jordan pais with minimal inner ideals, were classified in [**FLTB03**, 5.1]. Under our present restriction of the ring of scalars Φ, the list reads as follows.

THEOREM 11.19. *A Jordan pair V is simple and satisfies the dcc on principal inner ideals if and only if it is isomorphic to one of the following:*

(1) *A Jordan pair $\mathcal{F}(\mathcal{P}_1, \mathcal{P}_2)^J$ defined by two pairs $\mathcal{P}_1$ and $\mathcal{P}_2$ of dual vector spaces over the same division algebra Δ.*
(2) *A Jordan pair $\mathrm{Sym}(\mathcal{F}(\mathcal{P}, \mathcal{P}^{op}), \#)$ defined by a pair $\mathcal{P}$ of skew dual vector spaces over a division algebra with involution.*

(3) *A Jordan pair* $\mathrm{Skew}(\mathcal{F}(\mathcal{P},\mathcal{P}^{op}),\#)$ *defined by a pair* $\mathcal{P}$ *of skew dual vector spaces over a field* $\mathbb{F}$ *of characteristic* 0 *or* $p>3$. [1]
(4) *A Clifford Jordan pair* $C(X,q)$*, where* q *is nondegenerate and* $\dim_{\mathbb{F}} X>2$.
(5) *An Albert pair defined by a simple exceptional Jordan algebra, which is 27-dimensional over its centroid.*
(6) *A Bi-Cayley pair* $(\mathrm{M}_{1\times 2}(C),\mathrm{M}_{2\times 1}(C))$, $Q_a b=(ab)a$*, where* C *is a Cayley algebra.*

REMARKS 11.20. (a) Let $V=(\mathcal{F}(X_1,X_2),\ \mathcal{F}(X_2,X_1))$ be a Jordan pair of type (1). Any inner ideal $I\subset\mathcal{F}(X_1,X_2)$ of V is of the form $I=W^*V$, where $W\le Y_1$ and $V\le X_2$. And I is a principal inner ideal if and only if the subspaces W and V have the same finite-dimension (see [**FLGRGLSM98**]). Hence it easily follows that V has finite capacity if and only if one of the pairs of dual vector spaces is finite-dimensional, and it is Artinian if and only if both pairs of dual vector spaces are finite-dimensional (in this last case $V\cong(\mathrm{M}_{n\times m}(\Delta),\ \mathrm{M}_{m\times n}(\Delta))$.

(b) Let $V=(\mathrm{Sym}(\mathcal{F}(X,Y),\#),\ \mathrm{Sym}(\mathcal{F}(Y,X),\#))$ be a Jordan pair of type (2). Following arguments similar to those of [**FLGR99**], it is shown that any inner ideal $I\subset\mathrm{Sym}(\mathcal{F}(X,Y)$ of V is of the form $I=\mathrm{Sym}(W^*W,\#)$ for some $W\le Y$. Hence V has finite capacity, equivalently it is Artinian, if and only if the pair of skew dual vector spaces is finite-dimensional. In this case, V is the Jordan pair defined by a Jordan algebra $\mathrm{Her}_n(\Delta,-)$ of Hermitian matrices over a division algebra with involution (see [**Loo75**, Theorem 12.12]).

(c) Let $V=(\mathrm{Skew}(\mathcal{F}(X,Y),\#),\ \mathrm{Skew}(\mathcal{F}(Y,X),\#))$ be a Jordan pair of type (3). Following [**FLGR99**] again, it is proved that for any subspace $W\le Y$, $\mathrm{Skew}(W^*W,\#)$ is an inner ideal of V. Thus V is Artinian if and only if the pair of skew dual vector spaces is finite-dimensional. In this case $V=(\mathrm{A}_n(\mathbb{F}),\mathrm{A}_n(\mathbb{F}))$ is a Jordan pair of skew-symmetric matrices. As will be seen in Chapter 13, V contains another type of inner ideals called point spaces.

(d) Let $V=C(X,q)$ be a Clifford Jordan pair. It is not difficult to see I is a proper inner ideal of V if and only if I is a totally isotropic subspace of X. Thus V is Artinian if and only it has no infinite-dimensional totaly isotropic subspace.

(e) Jordan pairs of type (5) and (6) are finite-dimensional over their centroid, so they satisfy the dcc and acc on all inner ideals.

As a direct consequence of Theorem 11.19 and Remark 11.20, we obtain.

COROLLARY 11.21. *Let* $V=(V^+,V^-)$ *be a simple Jordan pair with dcc on principal inner ideals. Then* V *is Artinian if and only if* V^+ *has the dcc on all inner ideals.*

11.2. From Jordan pairs to Lie algebras

Throughout this section we will deal with Jordan pairs and Lie algebras over the ring of scalars Φ in which 6 is invertible.

PROPOSITION 11.22. *Every Jordan pair* $V=(V^+,V^-)$ *gives rise to a 3-graded Lie algebra* $L=L_{-1}\oplus L_0\oplus L_1$*, called Tits–Kantor–Koecher algebra of the Jordan*

[1] Simple Jordan pairs with dcc on principal inner ideals of type (3) in Theorem 11.19 have been recently considered in connection with Kantor pairs of skew transformations [**AFS17**].

pair V and denoted by TKK(V), *which is uniquely determined by the following three conditions:*

(TKK1) $V = (L_1, L_{-1})$, *as defined in Example* 11.1(5),
(TKK2) $L_0 = [L_1, L_{-1}]$,
(TKK3) *for any* $x_0 \in L_0$, $[x_0, L_1 + L_{-1}] = 0$ *implies* $x_0 = 0$.

PROOF. Following [**Neh96**, 1.5], for $(x, y) \in V$ let $\delta(x, y) = (D(x, y), -D(y, x))$ be the inner derivation of V defined by the pair (x, y). As previously noted, the linear span of the inner derivations is a subalgebra $\mathfrak{D}$ of $\mathrm{Der}(V)$. Then, on the Φ-module $\mathfrak{K} = V^- \oplus \mathfrak{D} \oplus V^+$, define a product by

$$[x^- \oplus \zeta \oplus x^+, y^- \oplus \eta \oplus y^+] = (\zeta_-(y^-) - \eta_-(x^-)) \oplus ([\zeta, \eta] \\ + \delta(x^+, y^-) - \delta(y^+, x^-) \oplus (\zeta_+(y^+) - \eta_+(x^+)),$$

where $x^\sigma, y^\sigma \in V^\sigma$ and $\zeta = (\zeta_+, \zeta_-)$, $\eta = (\eta_+, \eta_-) \in \mathfrak{D}$. It is not difficult to see that this product defines a structure of Lie algebra on $\mathfrak{K}$ satisfying conditions (TKK1), (TKK2), and (TKK3), so $\mathfrak{K} \cong \mathrm{TKK}(V)$. □

In general, by a TKK-algebra we mean a Lie algebra of the form TKK(V) for some Jordan pair V.

- A Jordan pair $V = (V^+, V^-)$ is said to be *perfect* if $V^\sigma = \{V^\sigma, V^{-\sigma}, V^\sigma\}$, $\sigma = \pm$.

COROLLARY 11.23. *Let L be a Lie algebra with a nontrivial* 3*-grading and let $V = (L_1, L_{-1})$ be its associated Jordan pair. If V is perfect and $L' = [L, L]$ is simple, then $L' \cong$* TKK(V).

PROOF. By the Jacobi identity, $M; = L_{-1} \oplus [L_{-1}, L_1] \oplus L_1$ is a nonzero ideal of L, and since V is perfect, M is contained in $[L, L]$, hence equal to $L' = [L, L]$ by simplicity of L'. It is clear that the graded Lie algebra $L' = L_{-1} \oplus [L_{-1}, L_1] \oplus L_1$ satisfies conditions (TKK1) and (TKK2). Now let $x_0 \in L'_0 = [L_{-1}, L_1]$ be such that $[x_0, L_1 + L_{-1}] = 0$. Then $x_0 \in Z(L') = 0$ by simplicity of L'. So $L' \cong \mathrm{TKK}(V)$. □

As quoted in [**Gar04**], information flows forward and backward from a Jordan pair to its TKK-algebra.

LEMMA 11.24. *Let V be a semiprime Jordan pair and let I be a nonzero ideal of* TKK(V). *Then $(I \cap L_1, I \cap L_{-1})$ is a nonzero ideal of V.*

PROOF. Denote by $L = L_{-1} \oplus L_0 \oplus L_1$ the TKK-algebra of the Jordan pair V. For any ideal I of L, denote by π_i the projection of L onto L_i, $i \in \{-1, 0, 1\}$. Then $I \cap V = (I \cap L_1, I \cap L_{-1})$ and $\pi(I) = (\pi_1(I), \pi_{-1}(I))$ are ideals of V, with $\pi(I)^3 \subset I \cap V \subset \pi(I)$. If $I \cap V = 0$, then $\pi(I)^3 = 0$ and hence $\pi(I) = 0$ (since V is semiprime), so $I \subset L_0$ and $[I, L_{\pm 1}] = 0$. By (TKK 3), this implies $I = 0$, which proves the lemma. □

PROPOSITION 11.25. *Let V be a Jordan pair. We have:*

(1) *If V is semiprime, then so is* TKK(V).
(2) *V is prime if and only if* TKK(V) *is prime and V is semiprime.*
(3) *V is simple if and only if* TKK(V) *is simple.*
(4) *V is nondegenerate if and only if* TKK(V) *is nondegenerate.*
(5) *V is strongly prime if and only if* TKK(V) *is strongly prime.*

PROOF. As above, let $L = L_{-1} \oplus L_0 \oplus L_1$ be the TKK-algebra of V. That V semiprime (resp. prime) implies L semiprime (resp. prime) is a direct consequence of Lemma 11.24. Suppose now that V is semiprime and L is prime, and let $I = (I_1, I_{-1})$, $J = (J_1, J_{-1})$ be ideals of V such that $\{I_\sigma, J_{-\sigma}, I_\sigma\} = 0$. Then $(I \cap J)^3 = 0$, and hence $I \cap J = 0$ by semiprimines of V. Therefore $[[I_\sigma, J_{-\sigma}], L_\sigma] = \{I_\sigma, J_{-\sigma}, L_\sigma\} \subset I_\sigma \cap J_\sigma = 0$, which implies by (TKK3) that $[I_\sigma, J_{-\sigma}] = 0$. Hence we have

$$\tilde{I} := I_{-1} + [L_1, I_{-1}] + [L_{-1}, I_1] + I_1$$

and

$$\tilde{J} := J_{-1} + [L_1, J_{-1}] + [L_{-1}, J_1] + J_1$$

are ideals of L such that $[\tilde{I}, \tilde{J}] = 0$. By primeness of L, $\tilde{I} = 0$ or $\tilde{J}$, so $I = 0$ or $J = 0$, which completes the proof of (2).

(3) Suppose that V is simple and let I be a nonzero ideal of L. By Lemma 11.24, $I \cap V = (I \cap L_1, I \cap L_{-1})$ is a nonzero ideal of V, so equal to V. Hence $L_1 + L_{-1} \subset I$, and therefore $I = L$ by (TKK 2), which proves that L is simple.

Suppose conversely that L is simple, and let $I = (I_1, I_{-1})$ be a nonzero ideal of V. Then $\tilde{I} = I_{-1} + [L_1, I_{-1}] + [L_{-1}, I_1] + I_1$ is a nonzero ideal of L, so equal to L, which implies that $I = V$, so V is simple.

(4) and (5) are [**Gar04**, Proposition 2.6 and Corollary 2.7]. □

COROLLARY 11.26. *Let V be a Jordan pair. Then V is strongly prime if and only if for any $x, y \in V^\sigma$, $\{x, V^{-\sigma}, y\} = 0$ implies $x = 0$ or $y = 0$.*

PROOF. Clearly, the condition $\{x, V^{-\sigma}, y\} = 0$ implies $x = 0$ or $y = 0$ is sufficient for V to be strongly prime. Conversely, suppose that V is strongly prime, and let $x, y \in V^\sigma$ be such that $\{x, V^{-\sigma}, y\} = 0$. Then we have by the grading relations that $[x, [y, \mathrm{TKK}(V)]] = 0$, so $x = 0$ or $y = 0$ by Theorem 7.10, since $\mathrm{TKK}(V)$ is strongly prime by Proposition 11.25(5). □

COROLLARY 11.27. *Let L be a simple Lie algebra with a nontrivial 3-grading. Then L is isomorphic to the TKK-algebra of its associated Jordan pair V.*

PROOF. By Proposition 11.25(3), the Jordan pair V is simple and therefore perfect. Now Corollary 11.23 applies to prove that $L \cong \mathrm{TKK}(V)$. □

EXAMPLES 11.28. TKK-algebras of simple finite-dimensional Jordan pairs have been described in [**Neh96**]. In particular, the TKK-algebra of a simple exceptional Jordan pair is a Lie algebra of type E_6 or E_7. Following [**FLGGL04**], we compute the TKK-algebras of the Jordan pairs of types (1), (2), (3) and (4) listed in Theorem 11.19.

(1) Let $V = (\mathcal{F}(X_1, X_2), \mathcal{F}(X_2, X_1))$ be the Jordan pair of finite rank adjointable linear maps defined by the pairs of dual vector spaces $(X_1, Y_1, \langle\,,\rangle_1)$, $(X_2, Y_2, \langle\,,\rangle_2)$ over the same division algebra Δ. Set $\mathcal{P} = (X, Y, \langle\,,\rangle)$, where

$$X = X_1 \oplus X_2, \quad Y = Y_1 \oplus Y_2, \quad \text{and} \quad \langle x_1 + x_2, y_1 + y_2\rangle = \langle x_1, y_1\rangle_1 + \langle x_2, y_2\rangle_2.$$

Then $\mathfrak{fsl}_Y(X)/Z(\mathfrak{fsl}_Y(X))$ is a simple 3-graded Lie algebra with associated Jordan pair V. In fact, $\mathrm{TKK}(V) \cong \mathfrak{fsl}_Y(X)/Z(\mathfrak{fsl}_Y(X))$ by Corollary 11.27.

Conversely, any simple Lie algebra of the form $\mathfrak{fsl}_Y(X)/Z(\mathfrak{fsl}_Y(X))$, where $\mathcal{P} = (X, Y, \langle\,,\rangle)$ is a pair of dual vector spaces over a division algebra Δ, can be regarded as the TKK-algebra of a Jordan pair $V = (\mathcal{F}(X_1, X_2), \mathcal{F}(X_2, X_1))$.

Indeed, let X_1 be a finite-dimensional subspace of X. Then there exists a subspace $Y_1 \leq Y$ such that $\mathcal{P}_1 = (X_1, Y_1)$ is a subpair of dual vector spaces of $\mathcal{P}$, $X = X_1 \oplus Y_1^{\perp}$, $Y = Y_1 \oplus X_1^{\perp}$, and $\mathcal{P}_1^{\perp} = (Y_1^{\perp}, X_1^{\perp})$ is also a subpair of dual vector spaces of $\mathcal{P}$.

(2) Let V be the Jordan pair $(\mathrm{Sym}(\mathcal{F}(X,Y),\#),\ \mathrm{Sym}(\mathcal{F}(Y,X),\#))$ defined by a pair $(X, Y, \langle\,,\rangle)$ of skew dual vector spaces over a division algebra with involution $(\Delta, -)$. On the vector space $X \oplus Y$ we define a nondegenerate skew-Hermitian form h by

$$h(x+y, x'+y') = \langle x, y'\rangle - \langle y, x'\rangle^{op} = \langle x, y'\rangle - \overline{\langle x', y\rangle}.$$

The elements of the associative algebra $\mathcal{F}(X \oplus Y)$ can be represented by 2×2 matrices $a = (a_{ij})$, where

$$a_{11} \in \mathcal{F}(X), \quad a_{12} \in \mathcal{F}(X,Y), \quad a_{21} \in \mathcal{F}(Y,X), \text{ and } a_{22} \in \mathcal{F}(Y),$$

with the adjoint given by $a^* = (b_{ij})$, where

$$b_{11} = a_{22}^*, \quad b_{12} = a_{12}^*, \quad b_{21} = a_{21}^*, \quad \text{and} \quad b_{22} = a_{11}^*.$$

For $x, x' \in X$, we have

$$\langle x, x' a_{12}^{\#}\rangle = \langle x a_{12}, x'\rangle = -h(x a_{12}, x') = -h(x, x' a_{12}^*).$$

Therefore the adjoint involution $*$ in the associative algebra $\mathcal{F}(X \oplus Y)$ correponds to $-\#$ in the associative pair $(\mathcal{F}(X,Y), \mathcal{F}(Y,X))$.

If $\dim_{Z(\Delta)}(X \oplus Y) > 4$, then $[K,K]/[K,K] \cap Z$, where $K = \mathrm{Skew}(\mathcal{F}(X \oplus Y), *)$ and Z is the center of $\mathcal{F}(X \oplus Y)$, is a simple Lie algebra, and hence isomorphic to the TKK-algebra of the Jordan pair V by Corollary 11.27.

Conversely, let $(M, \langle\,,\rangle)$ be a nondegenerate skew-Hermitian inner product space over a division algebra with involution $(\Delta, -)$, where $\dim_{Z(\Delta)}(M) > 4$ and $M = X \oplus Y$ is the direct sum of two totally isotropic subspaces. Then the simple Lie algebra $[K,K]/([K,K] \cap Z)$, where $K = \mathrm{Skew}(\mathcal{F}(X), *)$ and Z is the center of $\mathcal{F}(X)$, is 3-graded and isomorphic to the TKK-algebra of a Jordan pair $(\mathrm{Sym}(\mathcal{F}(X,Y),\#), \mathrm{Sym}(\mathcal{F}(Y,X),\#))$ as in Theorem 11.19(2).

(3) Let $V = (\mathrm{Skew}(\mathcal{F}(X,Y),\#),\ \mathrm{Skew}(\mathcal{F}(Y,X),\#))$, where $(X, Y, \langle\,,\rangle)$ is a pair of skew dual vector spaces over a field $\mathbb{F}$. Define a nondegenerate symmetric bilinear form $\langle\,,\rangle$ on the vector space $X \oplus Y$ by

$$\langle x+y, x'+y'\rangle = \langle x, y'\rangle + \langle x', y\rangle.$$

We obtain as in (2) that if $\dim_{\mathbb{F}}(X \oplus Y) > 4$, then the Lie algebra $\mathrm{TKK}(V)$ is isomorphic to the finitary orthogonal algebra $\mathfrak{fo}(X \oplus Y, \langle\,,\rangle)$.

Conversely, let $L = \mathfrak{fo}(M, \langle\,,\rangle)$ be a finitary orthogonal algebra over a field $\mathbb{F}$ such that $M = X \oplus Y$ is the direct sum of two totally isotropic subspaces. Then this decomposition induces a 3-grading in L and with respect to this 3-grading, L is isomorphic to the TKK-algebra of the Jordan pair

$$V = (\mathrm{Skew}(\mathcal{F}(X,Y),\#),\ \mathrm{Skew}(\mathcal{F}(Y,X),\#))$$

as in Theorem 11.19(3).

(4) Let $V = C(X, q)$ be a Clifford Jordan pair over a field $\mathbb{F}$, $\dim_{\mathbb{F}} X > 2$, and let $H = \mathbb{F}y_1 \oplus \mathbb{F}y_2$ be a two-dimensional vector space over $\mathbb{F}$. Set $Y = H \oplus X$ and extend q to a symmetric bilinear forma $\langle\,,\rangle$ on Y by defining

$$\langle y_1, y_2\rangle = 1, \quad \langle y_i, y_i\rangle = 0,\ (i = 1, 2), \quad \text{and} \quad \langle H, X\rangle = 0.$$

Then the finitary orthogonal Lie algebra $\mathfrak{fo}(Y, \langle\,,\rangle)$ is simple and has a 3-grading with associated Jordan pair isomorphic to V. Hence, $\mathrm{TKK}(V) \cong \mathfrak{fo}(Y, \langle\,,\rangle)$ by Corollary 11.27.

Conversely, any finitary orthogonal $\mathbb{F}$-algebra $\mathfrak{fo}(X, \langle\,,\rangle)$ such that $\dim_{\mathbb{F}} X > 4$ and X contains a nonzero isotropic vector is the TKK-algebra of a simple Clifford Jordan pair. (If X contains a nonzero isotropic vector, then $X = H \oplus H^{\perp}$, where H is a hyperbolic plane.)

REMARK 11.29. Although the purpose of this monograph is to show how Jordan techniques can be used to solve Lie questions, originally the river flowed upstream. In [**Mar01**], via a TKK-construction, C. Martínez provides a necessary and sufficient Ore-type condition for an arbitrary Jordan algebra to have a ring of fractions, so answering a question posed by N. Jacobson.

11.3. Finite $\mathbb{Z}$-gradings and Jordan pairs

Following the exposition of [**Zel83a**], it is proved in this section that the Jordan pair V associated to a finite $\mathbb{Z}$-grading of a simple Lie algebra L is simple, and that the centroid of V is the restriction of the centroid of L to V.

We will deal with a Lie algebra $L = L_{-n} \oplus \cdots \oplus L_n$ endowed with a $(2n+1)$-grading over a field $\mathbb{F}$ of characteristic 0 or $p \geq 4n+1$. We denote by V the associated Jordan pair (L_n, L_{-n}). Recall that to any finite sequence $a_1, \ldots, a_{n-1}, a_n$ of elements of L we associate the left commutator $[a_n, \cdots, a_2, a_1] = \operatorname{ad} a_n \cdots \operatorname{ad}_{a_2} a_1$. With this notation, the Jordan triple products of V are expressed as

$$\{a_{\pm n}, b_{\mp n}, c_{\pm n}\} = [a_{\pm n}, b_{\mp n}, c_{\pm n}],$$

and the r-th power (r being an odd number) $I^r = (I^r_n, I^r_{-n})$ of an ideal $I = (I_n, I_{-n})$ of the Jordan pair V is the ideal of V given by $I^r{}_{\pm n} = [I_{\pm n}, I_{\mp n}, I^{r-2}{}_{\pm n}]$.

LEMMA 11.30. *Let $I = (I_n, I_{-n})$ be an ideal of the Jordan pair $V = (L_n, L_{-n})$. We have:*

(i) $(\operatorname{ad}(L)^k X) \cap L_n \subset \sum_{i=1}^m \operatorname{ad}(L_0)^i X$, *for any $k \geq 1$ and any nonempty subset $X \subset L_n$.*
(ii) $\operatorname{ad}(L_0)^k I^r_n \subset I_n$, *for r, k natural numbers such that r is odd and $k < r$.*
(iii) $(\operatorname{ad}(L)^k I^r_n) \cap L_n \subset I_n$, *for r, k natural numbers such that r is odd and $k < r$.*

PROOF. (i) By the grading properties,

$$(\operatorname{ad}(L)^m X) \cap L_n = \sum_{(i_1, \ldots, i_m)} \prod_{j=1}^{m} \operatorname{ad}(L_{i_j}) X$$

where $-n \leq i_j \leq n$ for all $1 \leq j \leq m$, and $\sum_{j=k}^m i_j = 0$ for any $0 \leq k \leq m$. Note that, by the second condition, if $i_j = 0$ for $1 \leq j \leq k$ and $i_{k+1} < 0$, then $\prod_{j=1}^m \operatorname{ad}(L_{i_j}) X = 0$, so we may assume that there exists $0 \leq k \leq m$ such that $i_j = 0$ for $1 \leq j \leq k$ and $i_{k+1} > 0$. By transposition of adjacent factors, we can move $\operatorname{ad}(L_{i_{k+1}})$ forward enough to reach a position where the product vanishes, yielding a sum of factors of length less than m (see Exercise 11.78 below for examples). An induction process completes the proof.

(ii) Let $r = 2m + 1$, $m \geq 1$ and $1 \leq k < r$. Consider first the cases $r = 3$, $k = 1, 2$.

$$[L_0, I_n^3] = [L_0, I_n, I_{-n}, I_n] \subset [L_n, I_{-n}, I_n] + [I_n, L_{-n}, I_n] + [I_n, I_{-n}, L_n] \subset I_n,$$

and hence

$$[L_0, L_0, I_n^3] \subset [L_n, L_{-n}, I_n] + [L_n, I_{-n}, L_n] + [I_n, L_{-n}, L_n] \subset I_n.$$

In general, we have

$$\begin{aligned}&\mathrm{ad}(L_0)^k I_n^{2m+1}\\ &= [\overbrace{L_0, \cdots, L_0}^{k}, \overbrace{I_n, I_{-n}, \cdots, I_n, I_{-n}}^{m}, I_n] \subset \sum [\overbrace{J_n, J_{-n}, \cdots, J_n, J_{-n}}^{m}, J_n],\end{aligned}$$

where in every commutator each letter J is either I or L, and the number of letters I is at least one. If in any of these commutators one of the last three letters is I, then this commutator is contained in I_n. If on the contrary the last three letters are L, then we can replace them by an L_n, obtaining a commutator of length $2m - 1$, and repeat the process.

(iii) It follows from (i) and (ii). $\square$

LEMMA 11.31. *Let* $(\mathrm{Mc}(L_n), \mathrm{Mc}(L_{-n}))$ *be the McCrimmon radical of the Jordan pair* $V = (L_n, L_{-n})$. *Then* $\mathrm{Mc}(L_n) + \mathrm{Mc}(L_{-n}) \subset \mathrm{K}(L)$.

PROOF. Let $x \in \mathrm{Mc}(L_n)$ and let $\{x_k\}$ be an m-sequence of L beginning at x. Regarded as an m-sequence of V, $\{x_k\}$ terminates (Lemma 11.5) and hence $x \in \mathrm{K}(L)$ by Proposition 9.4. $\square$

The details of the proofs of the next two results were communicated to the author by E. Zelmanov.

THEOREM 11.32. [**Zel84b**, Lemma 1.5] *If* $L = L_{-n} \oplus \cdots \oplus L_n$ *is simple, then the Jordan pair* $V = (L_n, L_{-n})$ *is simple.*

PROOF. As L is simple and the grading is nontrivial, $L_0 = \sum_{i=1}^n [L_i, L_{-i}]$ and L is nondegenerate by Proposition 3.27. Then the Jordan pair V is nondegenerate and therefore any nonzero ideal $I = (I_n, I_{-n})$ of V is a nondegenerate Jordan pair, which implies that both components I_n and I_{-n} are nonzero. By simplicity of L we also have $L_n = \mathrm{id}_L(I_n) \cap L_n$, so, by Lemma 11.30, any nonzero element $b_n \in L_n$ can be represented as

$$b_n = \sum_i \mathrm{ad}([x_{i1}, y_{i1}]) \cdots \mathrm{ad}([x_{ir_i}, y_{ir_i}]) a_{ni},$$

where $a_{ni} \in I_n$, $x_{ij} \in L_{k_{ij}}$, $y_{ij} \in L_{-k_{ij}}$, $1 \leq k_{ij} \leq n$. Again by simplicity of L, $\mathrm{Ad}(L) b_n = L$, and given that $L_0 = \sum_{i=1}^n [L_i, L_{-i}]$, the associative algebra $\mathrm{Ad}(L)$ is generated by the set $\{\mathrm{ad}(L_i),\ 1 \leq |i| \leq n\}$. Hence

$$b_n = w(\mathrm{ad}(w_1(T_{11}, \cdots T_{1r_1}) b_n), \cdots, \mathrm{ad}(w_s(T_{s1}, \cdots T_{sr_s}) b_n)), \tag{11.2}$$

where $w, w_1, \ldots, w_s$ are element of the free associative $\mathbb{F}$-algebra generated by the variables ξ_i, ξ_{jk}, $1 \leq i \leq s$, $1 \leq j_i \leq r_i$ for each i, and where each T_{ij} belongs to $\mathrm{ad}(L_{k_{ij}})$ for some nonzero k_{ij} with $1 \leq |k_{ij}| \leq n$.

Let L' be the subalgebra of L generated by the elements $a_{n,i}$, x_{ij}, y_{ij} and all from $\bigcup_{i \neq o} L_i$ involved in the generators T_{ij}. Since $L' = \sum_{i=-n}^n L'_i$ is generated by a finite collection of elements from $\bigcup_{i \neq 0} L'_i$, it follows from Lemma 2.70 that

$\mathrm{Ad}(L') = \sum_{i=1}^{d} \mathrm{ad}(L'_i)^k$ for some positive integer d. Set $I'_{-n} = L'_{-n} \cap I_{-n}$ and $I'_n = L'_n \cap I_n$. Then $I' = (I'_n, I'_{-n})$ is an ideal of the Jordan pair $V' = (L'_n, L'_{-n})$. Let r be an odd number, $r > d$ and let $I'^r = (I'^r_n, I'^r_{-n})$ be the r-th power of I' in V'. By Lemma 11.30,

$$\mathrm{id}_{L'}((I'^r)_n) \cap L'_n \subset \sum_{i=1}^{d} \mathrm{ad}(L'_0)^i (I'^r)_n \subset I'_n. \tag{11.3}$$

Set $\mathrm{id}_{L'}((I'^r) = \mathrm{id}_{L'}((I'^r)_{-n} + (I'^r)_n))$ and consider the graded Lie algebra

$$L'' = L' / \mathrm{id}_{L'}((I'^r),$$

where $L''_n = L'_n / \mathrm{id}_{L'}((I'^r)_n) \cap L'_n$ and $L''_{-n} = L'_n / \mathrm{id}_{L'}((I'^r)_{-n}) \cap L'_{-n}$. By (11.3),

$$I'' = (I'_{-n} / Id_{L'}((I'^r)_{-n}) \cap L'_{-n}, 1_{L'}((I'^r)_n) \cap L'_n)$$

is a nilpotent ideal of the Jordan pair $V'' = (L''_n, L''_{-n})$. Hence I''_n lies in the McCrimmon radical of V'' and therefore in the Kostrikin radical of L'' by Lemma 11.31. Since by Propositions 3.29 and 2.75,

$$\mathrm{K}(L'') \subset \widetilde{\mathrm{LocSN}}(L''), \quad \text{with } [\widetilde{\mathrm{LocSN}}(L''), L''] \subset \mathrm{LocSN}(L''),$$

the image b''_n of b_n (as an element of the Jordan pair V'') is nilpotent, which implies (reiterating equation 11.2) that $b''_n = 0$, so $b_n \in \mathrm{id}_{L'}((I'^r)_n) \cap L'_n \subset I_n$ by (11.3), which proves that $L_n = \mathrm{id}_L(I_n) \cap L_n \subset I_n$. Similarly it is proved that $I_{-n} = L_{-n}$. □

LEMMA 11.33. [**Zel84b**, Lemma 1.6] *Let $L = L_{-n} \oplus \cdots \oplus L_n$ be simple with associated Jordan pair $V = (L_n, L_{-n})$. We have:*

(i) $\Gamma(L)L_i = L_i$, $-n \leq i \leq n$,

(ii) $\Gamma(V) \cong \Gamma(L)$.

PROOF. (i) $L = \Gamma(L)L = \sum_{i=-n}^{n} \Gamma(L)L_i$ defines a pregrading in L, which is actually a grading since L is simple (Corollary 2.69). Thus the sum is direct, which implies that $\Gamma(L)L_i = L_i$ for each $-n. \leq i \leq n$

(ii) By Theorem 11.32, the Jordan pair V is simple and hence $\Gamma(V)$ is a field [**Loo75**, Proposition 1.17]). Via the map $\gamma \mapsto (\gamma_n, \gamma_{-n})$, where $\gamma_{\pm n}$ denotes the restriction of γ to $L_{\pm n}$, $\Gamma(V)$ is a field extension of $\Gamma(L)$, since V is a Jordan pair over $\Gamma(L)$ by (i). So it suffices to show that $\Gamma(V)$ is not bigger than $\Gamma(L)$. By [**Jac79**, X.1.Theorem 3], the tensor product $L \otimes_{\Gamma(L)} \Gamma(V)$ is a simple Lie algebra with centroid $\Gamma(V)$. Then $V \otimes_{\Gamma(L)} \Gamma(V)$ is a simple Jordan pair (because it is the Jordan pair of the simple graded Lie algebra $L \otimes_{\Gamma(L)} \Gamma(V)$). This implies as in [**Jac79**, X.1.Theorem 3] for the case of an arbitrary algebra that $\Gamma(L) = \Gamma(V)$. □

We finish this section with a result which is part of a work in progress with E. Zelmanov.

THEOREM 11.34. *Let $L = L_{-n} \oplus \cdots \oplus L_n$ be simple. If its associated Jordan pair $V = (L_n, L_{-n})$ is exceptional, then $L \cong \mathrm{TKK}(V)$.*

PROOF. Let $i \in \mathbb{Z}$, $0 < |i| < n$. Consider the Jordan pairs (L_n, L_{-n}) and $(\mathrm{Hom}_{\mathbb{F}}(L_{-i}, L_{n-i}), \mathrm{Hom}_{\mathbb{F}}(L_{n-i}, L_{-i}))$ defined in Examples 11.1, and the pair of linear maps

$$(\varphi_{-i}, \varphi_{n-i}) : (L_n, L_{-n}) \to (\mathrm{Hom}_{\mathbb{F}}(L_{-i}, L_{n-i}), \mathrm{Hom}_{\mathbb{F}}(L_{n-i}, L_{-i}))$$

given by

$$a_n \mapsto \mathrm{ad}(a_n) : L_{-i} \to L_{n-i}, \quad b_{-n} \mapsto \mathrm{ad}(b_{-n}) : L_{n-i} \to L_{-i},$$

for all $a_n \in L_n$ and $b_{-n} \in L_{-n}$. It is routine to verify that $(\varphi_{-i}, \varphi_{n-i})$ is a homomorphism of Jordan pairs, i.e.

$$\varphi_{-i}(\{a_n, b_{-n}, c_n\}) = \{\varphi_{-i}(a_n), \varphi_{n-i}(b_{-n}), \varphi_{-i}(c_n)\}$$

and

$$\varphi_{n-i}(\{a_{-n}, b_n, c_{-n}\}) = \{\varphi_{n-i}(a_{-n}), \varphi_{-i}(b_n), \varphi_{n-i}(c_{-n})\},$$

for all $a_n, b_n, c_n \in L_n$ and $a_{-n}, b_{-n}, c_{-n} \in L_{-n}$. Since the Jordan pair (L_n, L_{-n}) is simple (Theorem 11.32), $\ker(\varphi_{-i}, \varphi_{n-i}) = (\ker(\varphi_{-i}), \ker(\varphi_{n-i}))$ is either the zero ideal or the whole pair V. But the former is not possible since $V = (L_n, L_{-n})$ is exceptional. Thus $[L_n, L_i] = [L_{-n}, L_i] = 0$ for $-n \le i < n$. Then

$$L_{-n} + [L_n, L_{-n}] + L_n$$

is a (nonzero) ideal of L and hence equal to L by simplicity of L. Therefore $L \cong \mathrm{TKK}(V)$ by Corollary 11.27. □

11.4. Subquotient with respect to an abelian inner ideal

In this section L will denote a Lie algebra over a ring of scalars Φ in which 6 is invertible. Most of the results gathered here are taking from [**FLGGLN07**].

- Let $x, y, z, u, v \in L$. Using (L1) and (L2) we get

(L3) $\mathrm{ad}_{[u,v]}[[x,y],z] = [[\mathrm{ad}_{[u,v]}\, x, y], z] - [[x, \mathrm{ad}_{[v,u]}\, y], z] + [[x,y], \mathrm{ad}_{[u,v]}\, z]$,

which is the Lie analogue of (JP2). Moreover, if $[x,z] = 0$, then

(L4) $[[x,y],z] = [[z,y],x]$,

which is the Lie analogue of (JP1).

PROPOSITION 11.35. *Let B, C be abelian inner ideals of a Lie algebra L. Then $V = (B, C)$ is a Jordan pair with Jordan triple products defined by*

$$\{b_1, c_1, b_2\} := [[b_1, c_1], b_2], \quad \{c_1, b_1, c_2\} := [[c_1, b_1], c_2]$$

for all $b_1, b_2 \in B$ and $c_1, c_2 \in C$.

PROOF. The Jordan triples products are well defined because B and C are inner ideals, (JP2) is (L3), and (JP1) is (L4) since B and C are abelian. □

Note that Example 11.1(5) is a particular case of this proposition, since L_n and L_{-n} are abelian inner ideals.

A single abelian inner ideal is enough to yield a Jordan pair. Mimicking the Jordan pair case, given a subset X of a Lie algebra L, we define the *kernel of X* as the Φ-submodule

$$\mathrm{Ker}_L\, X := \{z \in L : [X, [X, z]] = 0\}.$$

LEMMA 11.36. *For every abelian inner ideal B of L, its kernel $K := \mathrm{Ker}_L\, B$ satisfies:*

(i) $K = \{z \in L : [b, [b, z]] = 0$ *for every* $b \in B\}$,
(ii) $[B, L] + [[L, B], K] + [[K, B], L] \subset K$.

PROOF. (i) Let $z \in L$ be such that $[b,[b,x]] = 0$ for every $b \in B$. Since B is abelian, for any $b, b' \in B$ we have $[b,[b',z]] = [b',[b,z]]$. Hence

$$2[b,[b',z]] = [b+b',[b+b',z]] - [b,[b,z]] - [b',[b',z]] = 0,$$

and therefore $[b,[b',z]] = 0$, since L has no 2-torsion.

(ii) Since B is an abelian inner ideal, $[B,[B,[B,L]]] \subset [B,B] = 0$, which implies that $[B,L] \subset K$. The inclusion $[[L,B],K] \subset K$ follows from (i) and the Leibniz rule. For any $b \in B$,

$$\mathrm{ad}_b^2[[L,B],K] \subset [\mathrm{ad}_b^2[L,B],K] + [\mathrm{ad}_b[L,B],\mathrm{ad}_b K] + [[L,B],\mathrm{ad}_b^2(K)],$$

where the three summands on the right hand side vanish: the first and the third ones by definition of kernel, and the second since $[[B,[L,B]],[B,K]] \subset [B,[B,K]] = 0$. Finally, $[[K,B],L] = [[K,L],B] + [K,[B,L]] \subset K$ follows from above. □

As we have pointed out, kernels of inner ideals of Jordan pairs were used in [**LN94**] to construct new Jordan pairs. A similar construction works for abelian inner ideals of a Lie algebra.

PROPOSITION 11.37. *Let B be an abelian inner ideal of L. Then the pair of Φ-modules $(B, L/\operatorname{Ker}_L B)$ with the triple products given by*

$$\{b,\overline{x},c\} := [[b,x],c], \quad \{\overline{x},b,\overline{y}\} := \overline{[[x,b],y]}$$

for all $b,c \in B$ and $x,y \in L$, where bars denote the cosets relative to the submodule $\operatorname{Ker}_L B$, is a Jordan pair.

PROOF. It is clear that $\{b,\overline{x},c\} = [[b,x],c]$, for every $x \in L$ and $b,c \in B$, is well defined; and it follows from Lemma 11.36 that $\{\overline{x},b,\overline{y}\} = \overline{[[x,b],y]}$, for every $b \in B$, $x,y \in L$, is well defined as well. Let us now see that both triple products satisfy (JP1). Since B is abelian, it follows from (L4)

$$\{b,\overline{x},c\} = [[b,x],c] = [[c,x],b] = \{c,\overline{x},b\}$$

for all $b,c \in B$ $x \in L$. Similarly,

$$\{\overline{x},b,\overline{y}\} = \overline{[[x,b],y]} = \overline{[y,[b,x]]} = \overline{[[y,b],x]} = \{\overline{y},b,\overline{x}\}$$

for all $x,y \in L$, $b \in B$, since $[b,[x,y]] \in \operatorname{Ker}_L B$ by Lemma 11.36. That both triple products also satisfy (JP2) follows from (L3). □

DEFINITION 11.38. Given an abelian inner ideal B of L, the Jordan pair $(B, L/\operatorname{Ker}_L B)$ defined above is called the *subquotient* of L with respect to B.

Finite gradings and subquotients. We study in this subsection the relationship between the Jordan pair associated to a finite $\mathbb{Z}$-grading of a Lie algebra and the subquotient of one of the extremes of the grading. As will be seen, they are isomorphic Jordan pairs when the Lie algebra is nondegenerate.

PROPOSITION 11.39. *Let $L = L_{-n} \oplus \cdots \oplus L_n$ be a $(2n+1)$-graded Lie algebra L with associated Jordan pair $V = (L_n, L_{-n})$, and let $B \subset L_n$ be an inner ideal of V. We have:*

(i) *B is an abelian inner ideal of L with*

$$\operatorname{Ker}_L B = \operatorname{Ker}_V B \oplus L_{-(n-1)} \oplus \cdots \oplus L_n.$$

If L is nondegenerate, then

(ii) $\operatorname{Ker}_L L_n = L_{-(n-1)} \oplus \cdots \oplus L_n$,

(iii) *the Jordan pairs V and $\operatorname{Sub}_L L_n$ are isomorphic.*

PROOF. (i) By the grading relations, B is an abelian inner ideal of L and

$$L_{-(n-1)} \oplus \cdots \oplus L_n \subset \operatorname{Ker}_L B.$$

Then, by the Modular Law,

$$\operatorname{Ker}_L B = L_{-n} \cap \operatorname{Ker}_L B \oplus (L_{-(n-1)} \oplus \cdots \oplus L_n) = \operatorname{Ker}_V B \oplus L_{-(n-1)} \oplus \cdots \oplus L_n.$$

Suppose now that L is nondegenerate. Then so is V and hence $\operatorname{Ker}_V L_n = 0$ by [**LN94**, 1.4]. Now (ii) and (iii) follow from (i). □

COROLLARY 11.40. *Any nondegenerate Jordan pair V can be regarded as a subquotient of its Tits–Kantor–Koecher algebra.*

It is possible to adopt a different approach to the Jordan algebra attached to a Jordan element of a Lie algebra (see Definition 8.43) when the Lie algebra is nondegenerate.

PROPOSITION 11.41. *Let a be a Jordan element of a Lie algebra L and let $V = ((a), L/\operatorname{Ker}_L(a))$ the subquotient of L with respect to the abelian inner ideal (a) generated by a. We have:*

(i) $\operatorname{Ker}_L\{a\} \subset \operatorname{Ker}_L(a) \subset \operatorname{Ker}_L[a]$.

(ii) *If L is nondegenerate, then the inclusions are equalities and $L_a \cong V^{(a)}$.*

PROOF. (i) Let $z \in \operatorname{Ker}_L\{a\}$. For any $\lambda \in \Phi$ and $x \in L$, we have:

$$\begin{aligned}[\lambda a + [a,[a,x]], [\lambda a + [a,[a,x]], z]] &= \lambda^2[a,[a,z]] + 2\lambda[[a,[a,x]],[a,z]] \\ &+ [[a,[a,x]],[[a,[a,x]],z]],\end{aligned}$$

where all the summands above vanish, since

$$\operatorname{ad}_a^3 = 0, \ \operatorname{ad}_a^2 z = 0, \quad \text{and} \quad \operatorname{ad}^2_{\operatorname{ad}_a^2 x} z = \operatorname{ad}_a^2 \operatorname{ad}_x^2 \operatorname{ad}_a^2 z = 0.$$

Thus $z \in \operatorname{Ker}_L(a)$. The second inclusion is trivial.

(ii) Suppose that L is nondegenrate and let $z \in \operatorname{Ker}_L[a]$. For every $x \in L$, we have $\operatorname{ad}_a^2 x \in [a]$, so

$$\operatorname{ad}_a^2 \operatorname{ad}_x^2 \operatorname{ad}_a^2 z = \operatorname{ad}^2_{\operatorname{ad}_a^2 x} z = 0$$

which implies $U_{\overline{x}}\overline{z} = \overline{\operatorname{ad}_x^2 \operatorname{ad}_a^2 z} = 0$. Hence $\overline{z} = 0$ since L_a inherits nondegeneracy from L. Then $\operatorname{ad}_a^2 z = 0$ and therefore $z \in \operatorname{Ker}_L\{a\}$. This proves that

$$\operatorname{Ker}_L\{a\} = \operatorname{Ker}_L(a) = \operatorname{Ker}_L[a].$$

Hence it easily follows the isomorphism $L_a \cong V^{(a)}$. □

PROPOSITION 11.42. *Let $L = L_{-n} \oplus \cdots \oplus L_n$ be a Lie algebra with a $(2n+1)$-grading and let $V = (L_n, L_{-n})$ be its associated Jordan pair. Then any $b \in L_{-n}$ is a Jordan element of L and L_b is isomorphic to the local algebra V_b of V at b.*

PROOF. By the grading properties, b is a Jordan element of L. Denote by π the projection of L onto L_n. For any $x \in L$ we have

$$[b,[b,x]] = [b,[b,\pi(x)]] = -\{b, \pi(x), b\},$$

so $x \in \operatorname{Ker}_L\{b\}$ if and only if $\pi(x) \in \operatorname{Ker}_V\{b\}$. Hence $\pi : L \to L_n$ induces an isomorphism of L_b onto V_b. □

COROLLARY 11.43. *Let $L = L_{-n} \oplus \cdots \oplus L_n$ be a Lie algebra with a $(2n+1)$-grading and let $V = (L_n, L_{-n})$ be its associated Jordan pair. If L is primitive in some $b \in L_{-n}$, then the Jordan pair V is primitive at b.*

PROOF. It follows from the elemental characterization of strong primeness in Lie algebras (Theorem 7.10 and Proposition 11.42). □

COROLLARY 11.44. *Let $V = (V^+, V^-)$ be a Jordan pair. Then V is primitive at some $b \in V^\sigma$ if and only if $\mathrm{TKK}(V)$ is primitive.*

PROOF. It is a direct consequence of Corollary 11.43 and Proposition 11.25. □

COROLLARY 11.45. *Let J be a unital Jordan algebra with unit element e, let $V = (J, J)$ its associated Jordan pair, and let $L = \mathrm{TKK}(V)$. Then $L_e \cong J$.*

Von Neumann regular elements which are homogeneous with respect to a grading can be completed with a homogeneous element to form an idempotent. A fact that will be very useful in the next chapter.

LEMMA 11.46. *Let G be a group with e as identity element and let $L = \bigoplus_{g\in G} L_g$ be a G-grading of a nondegenerate Lie algebra L. Let (x, z) be an idempotent of L where x is homogeneous, $x \in L_f$. Then there exists $y \in L_{f^{-1}}$ such that (x, y) is an idempotent and $[x, y] = \pi_e([x, z])$, the e-homogeneous component of $[x, z]$.*

PROOF. Let $z = \sum_{g\in G} z_g \in L$ be such that (x, z) is an idempotent, where $x \in L_f$ for some $f \in G$. By the grading relations, $2x = [[x, z], x] = [[x, z_{f^{-1}}], x]$. Put $h := \pi_e([x, z]) = [x, z_{f^{-1}}] \in L_e$. By Theorem 5.11, there exists $y \in L$ such that (x, y) is an idempotent with $h = [x, y]$. Let $L = L^h_{-2} \oplus L^h_{-1} \oplus L^h_0 \oplus L^h_1 \oplus L^h_2$ be the 5-grading induced by (x, y) in L, with $L^h_2 = [x]$ and $L^h_{-2} = [y]$ (Theorem 5.11). With respect to this 5-grading, $z_{f^{-1}} = z_{-2} + z_{-1} + z_0 + z_1 + z_2$. Since $h \in L_e$ and each $z_i = p_i(\mathrm{ad}_h)(z_{f^{-1}})$ for certain polynomials $p_i(\lambda)$ over the ring of integers (Remark 5.12), $z_i \in L_{f^{-1}}$, $i \in \{-2, -1, 0, 1, 2\}$. And since $\mathrm{ad}_x^2(z_{f^{-1}} - y) = 0$, it follows from Lemma 11.36 and Proposition 11.39(ii) that

$$z_{f^{-1}} - y \in \mathrm{Ker}_L\{x\} = \mathrm{Ker}_L[x] = \ker_L L^h_2 = L^h_{-1} \oplus L^h_0 \oplus L^h_1 \oplus L^h_2.$$

Hence

$$y - z_{-2} = (y - z_{f^{-1}}) + (z_{-1} + z_0 + z_1 + z_2) \in (L^h_{-1} \oplus L^h_0 \oplus L^h_1 \oplus L^h_2) \cap L^h_{-2} = 0,$$

so $y = z_{-2} \in L_{f^{-1}}$. Moreover, $\pi_e([x, z]) = [x, z_{f^{-1}}] = [x, y]$. □

Innerness correspondence and Lie-to-Jordan inheritance. As for Jordan pairs, the length of an inner ideal B of a Lie algebra L, denoted by $\mathrm{length}(B)$, is the supremum of the lengths of chains of inner ideals of L contained in B.

PROPOSITION 11.47. *Let B be an abelian inner ideal of a Lie algebra L. Put $K := \mathrm{Ker}_L B$ and $V := (B, L/K)$ to denote the kernel and the subquotient of L with respect to B. We have:*

- (i) *Let C be a Φ-submodule of B. Then C is an inner ideal of L if and only if it is an inner ideal of V.*
- (ii) *If C is an inner ideal of L, then $\overline{C} = (C + K)/K$ is an inner ideal of V.*
- (iii) *If L is nondegenerate (strongly prime), then V is nondegenerate (strongly prime).*

(iv) *Let L be nondegenerate. Then B is minimal if and only if V is a division Jordan pair.*

(v) *Let L be nondegenerate. Then B has finite length if and only if V is Artinian. In this case, $V \cong \operatorname{Sub}_I B$ for any ideal I of L containing B.*

(vi) *If L is strongly prime and B has the dcc on all inner ideals, then B has finite length.*

PROOF. Statements (i) and (ii) are trivial.

(iii) Suppose that L is strongly prime. If $\{a, L/K, b\} = 0$ for some $a, b \in B$, then $[[a, L], b] = 0$ and hence $a = 0$ or $b = 0$ by the elemental characterization of strong prime Lie algebras (Theorem 7.10). Suppose now that $\{\overline{x}, B, \overline{y}\} = 0$ for some $x, y \in L$. i.e. $\operatorname{ad}_b^2 \operatorname{ad}_x \operatorname{ad}_y B = 0$ for all $b \in B$. By 4.4(8),

$$\operatorname{ad}_{\operatorname{ad}_b^2 x} \operatorname{ad}_{\operatorname{ad}_b^2 y} L = \operatorname{ad}_b^2 \operatorname{ad}_x \operatorname{ad}_y \operatorname{ad}_b^2 L \subset \operatorname{ad}_b^2 \operatorname{ad}_x \operatorname{ad}_y B = 0$$

which implies by Theorem 7.10 again that $\operatorname{ad}_b^2 x = 0$ or $\operatorname{ad}_b^2 y = 0$, i.e. $\overline{x} = 0$ or $\overline{y} = 0$. A similar argument applies when L is nondegenerate to prove that V is nondegenerate.

(iv) If V is a division Jordan pair, then V has no proper inner ideals, so B is minimal. Conversely, assume that B is minimal and let x be a nonzero element of B. Since L is nondegenerate, $B = V^+ = Q_x V^-$ in Jordan notation, and hence x is invertible [**Loo75**, Proposition 10.4]. Then, by [**Loo75**, Proposition 1.11], $V \cong (J, J)$, where J is a division Jordan algebra, i.e. V is a division Jordan pair.

(v) By Proposition 11.15, V is Artinian if and only if B has finite length. Let I be an ideal of L containing B. The injection $j : I \to L$ induces the Jordan pair monomorphism $(1_B, \bar{j}) : (B, I/\operatorname{Ker}_I B) \to (B, L/\operatorname{Ker}_L B)$. Since Artinian nondegenerate Jordan pairs are von Neumann regular, $(1_B, \bar{j})$ is actually an isomorphism: $L/K = \{L/K, B, L/K\} = \overline{[[L, B], L]} = \overline{I}$. So $V \cong \operatorname{Sub}_I B$,

(vi) By (i) and (iii), V is a strongly prime Jordan pair with nonzero socle. Then $\operatorname{Soc}(V) = (\operatorname{Soc}(V^+), \operatorname{Soc}(V^-))$ is a simple Jordan pair with $\operatorname{Soc}(V^+) = B$ satisfying the dcc on all inner ideals, which implies by Corollary 11.21 that $\operatorname{Soc}(V)$ is Artinian. By (v), B has finite length. □

OPEN QUESTION 11.48. Let L be a simple Lie algebra over a field of characteristic 0 or $p > 3$ and let $B \subset L$ be a nonzero abelian inner ideal. Is the Jordan pair $\operatorname{Sub}_L B$ simple?

Proposition 11.47 and the structure theorem of the simple nondegenerate Artinian Jordan pairs, [**Loo75**, 12.12 and 12.13] and [**DM00**, 2.1], allow us to describe the subquotients of a simple nondegenerate Lie algebra relative to an abelian inner ideal of finite length.

COROLLARY 11.49. *Let L be a simple nondegenerate Lie algebra and let B be a nonzero abelian inner ideal of finite length. Then $V = (B, L/\operatorname{Ker}_L B)$ is one of the following:*

(1) *A Jordan pair of rectangular matrices over a division associative algebra.*
(2) *A Jordan pair of skew-symmetric matrices over a field.*
(3) *A Jordan pair of Hermitian matrices over a division associative algebra with involution.*

(4) *The Clifford pair defined by a nondegenerate quadratic form on a vector space over a field which does not contain totally isotropic subspaces of infinite-dimension.*

(5) *The Albert pair defined by a simple exceptional Jordan algebra, which is 27-dimensional over its centroid.*

(6) *A Bi-Cayley pair* $(\mathrm{M}_{1\times 2}(C), \mathrm{M}_{2\times 1}(C))$ *with* C *a Cayley algebra.*

REMARK 11.50. As pointed out in Corollary 11.40, any simple nondegenerate Artinian Jordan pair can be seen as a subquotient of its Tits–Kantor–Koecher algebra. In particular, looking into the classifications of the finite-dimensional simple Lie algebras over an algebraically closed field of characteristic 0 [**Jac79**], we find: (1) Jordan pairs of rectangular matrices occur as subquotients of Lie algebras of type A_l, (2) Jordan pairs of skew-symmetric matrices are the subquotients of orthogonal algebras of type D_l, (3) Jordan pairs of symmetric matrices occur as subquotients of Lie algebras of type $\mathcal{C}_l$, (4) Clifford pairs are subquotients of orthogonal algebras of types B_l and D_l, (5) Albert pairs appear as subquotients of a Lie algebra of type E_7, and (6) any Bi-Cayley pair is the subquotient of a Lie algebra of type E_6.

11.5. Lie notions by the Jordan approach

In this section we revisited the notions of von Neumann regularity, idempotent and socle in Lie algebras showing that they can be expressed in Jordan terms by means of subquotients.

Lie Von Neumann regularity is a Jordan notion. Recall that a Jordan element a in a Lie algebra L is called von Neumann regular if $a \in \mathrm{ad}_a^2 L$.

PROPOSITION 11.51. *Let* a *be a Jordan element of a Lie algebra* L. *Then the following conditions are equivalent:*

(i) a *is von Neumann regular in* L.

(ii) a *is von Neumann regular in the Jordan pair* $\mathrm{Sub}_L B$ *for any abelian inner ideal* B *of* L *containing* a.

(iii) a *is von Neumann regular in* $\mathrm{Sub}_L(a)$, *where* $(a) = \Phi a + \mathrm{ad}_a^2 L$ *is the inner ideal generated by* a.

PROOF. Straightforward. □

It is clear that any element of a minimal abelian inner ideal of a nondegenerate Lie algebra (Theorem 4.29(2)) is von Neumann regular. But this is just a particular case of a more general result. For brevity we will say that an abelian inner ideal B of a Lie algebra L has descending chain condition (dcc) on principal inner ideals if L has dcc on principal inner ideals $\mathrm{ad}_x^2 L$ for all $x \in B$.

PROPOSITION 11.52. *Let* L *be a nondegenerate Lie algebra and let* $b \in L$ *be a Jordan element of* L. *If the inner ideal* $B = (b)$ *of* L *generated by* b *has dcc on principal inner ideals, then* b *is von Neumann regular. In particular, any Jordan element of a finite-dimensional nondegenerate Lie algebra over a field* $\mathbb{F}$ *of characteristic greater than* 3 *is von Neumann regular.*

PROOF. Let $V = (B, L/\operatorname{Ker}_L B)$ be the subquotient of L with respect to B. Since the inner ideals of V contained in B are precisely the inner ideals of L contained in B (Proposition 11.47(i)), it follows from Propositions 11.16 and 11.51 that b is von Neumann regular. □

Lie idempotents revisited. Recall that an idempotent of a Lie algebra L is a pair (e, f) of Jordan (in fact, von Neumann regular) elements of L such that $[[e, f], e] = 2e$ and $[[f, e]f] = 2f$. It follows from the definition:

(i) Let B and C be abelian inner ideals of L and let $V = (B, C)$ the Jordan pair defined by them (Proposition 11.35). Then a pair $(b, c) \in B \times C$ is an idempotent of V if and only if it is an idempotent of L.

(ii) If (e, f) is an idempotent of L, then $(e, \bar{f})$ is an idempotent of the Jordan pair $\operatorname{Sub}_L B$ for any abelian inner ideal B of L such that $e \in B$.

DEFINITION 11.53. An idempotent (e, f) of a Lie algebra L is called a *division idempotent* if $V(e, f) := ([e], [f])$ is a division Jordan pair.

It is clear that if (e, f) is a division idempotent, then the inner ideals $[e]$, $[f]$ are minimal. The converse is true if L is nondegenerate.

PROPOSITION 11.54. *Let (e, f) be an idempotent of a nondegenerate Lie algebra L. Then (e, f) is a division idempotent if and only if $[e]$, and therefore $[f]$ as well, are minimal inner ideals.*

PROOF. By Proposition 11.39(iii), $V(e, f) = (L^h_2, L^h_{-2}) \cong \operatorname{Sub}_L[e]$, which is a division Jordan pair if and only if $[e]$ is minimal (Proposition 11.47(iv)). □

NOTATION 11.55. Let L be a simple nondegenerate Lie algebra L containing minimal abelian inner ideals. It was proved in Proposition 5.29 that any two minimal abelian inner ideals of L are conjugate under an elementary automorphism of L. This fact allows us to associate an invariant, namely, the isomorphism class of division Jordan pairs defined by its division idempotents, equivalently (Proposition 11.7) the isotopy class of its associated division Jordan algebras . We will write $\mathrm{DJP}(L)$ to denote the division Jordan pair defined by a division idempotent of a simple nondegenerate Lie algebra L, and by $\mathrm{DJA}(L)$ the corresponding division Jordan algebra, both uniquely determined by isomorphism and isotopy respectively.

Lie–Peirce decomposition. When restricted to an abelian inner ideal, the Jacobson–Morozov decomposition defined by an idempotent of a Lie algebra (Theorem 5.11) agrees with the Peirce (Jordan) decomposition induced in the subquotient.

PROPOSITION 11.56. [**FLGGL09**, Lemma 2.1] *Let (e, f) be an idempotent of a Lie algebra L with associated decomposition $L = L^h_{-2} \oplus L^h_{-1} \oplus L^h_0 \oplus L^h_1 \oplus L^h_2$, and let B be an abelian inner ideal of L such that $e \in B$. Put $V = \operatorname{Sub}_L B$. We have:*

(i) *$(e, \bar{f})$ is an idempotent of V,*

(ii) *$V^+_k(e, \bar{f}) = L^h_k \cap B$ for $k = 0, 1, 2$,*

(iii) *the Jordan pairs (L^h_2, L^h_{-2}) and $V_2(e, \bar{f})$ are isomorphic,*

(iv) *$B = B_2 \oplus B_1 \oplus B_0$, with $B_i = B \cap L^h_i$ for each index i, and B_2, B_0 are inner ideals of L.*

(v) *If* $0 \neq b_0 \in B_0$ *is von Neumann regular, then so is* $e' = e + b_0$ *and both* $e, b_0 \in [e']$.

(vi) *If* L *is nondegenerate and* $B_0 = 0$, *then* B_1 *is an inner ideal of* L.

PROOF. The statements (i) and (ii) follow from the definition of the Jordan triple product: $\{e, \bar{f}, x\} = [h, x]$ for any $x \in B$.

(iii) Let $\pi : L \to L/\operatorname{Ker}_L B$ denote the canonical linear projection, so $\pi(x) = \bar{x}$ for any $x \in L$. Now, $\pi(L^h_{-2}) = \pi([f]) = [\bar{f}] = Q_{\bar{f}} V^+$ is the inner ideal of V generated by $\bar{f}$, but $[\bar{f}] = V_2^-(e, \bar{f})$ by Peirce relations [**Loo75**, 5.4(3)]. Therefore $\pi(L^h_{-2}) = V_2^-(e, \bar{f})$. Note also that if $\pi(\operatorname{ad}_f^2(a)) = \bar{0}$, then

$$\operatorname{ad}_f^2(a) \in \operatorname{Ker}_L B \subset \operatorname{Ker}_L[e],$$

so $\operatorname{ad}_e^2 \operatorname{ad}_f^2(a) = 0$. By 4.4(3), $4 \operatorname{ad}_f^2(a) = \operatorname{ad}^2_{\operatorname{ad}_f^2(e)}(a) = \operatorname{ad}_f^2 \operatorname{ad}_e^2 \operatorname{ad}_f^2(a) = 0$ and hence it is easy to verify that the pair of maps

$$(1_{[e]}, \varphi) : (L^h_2, L^h_{-2}) \to (V_2^+(e, \bar{f}), V_2^-(e, \bar{f}))$$

where $1_{[e]}$ is the identity on $L^h_2 = [e]$ and φ is the restriction of π to $L^h_{-2} = [f]$, is an isomorphism of Jordan pairs.

(iv) It follows from (ii) and the Peirce decomposition of the Jordan pair V with respect to the idempotent $(e, \bar{f})$ (Proposition 11.13).

(v) Let $0 \neq b_0 \in B_0$ be von Neumann regular in L (equivalently in V). By Theorem 5.11, we can extend b_0 to an idempotent (b_0, c_0) of L, where $c_0 \in L_0$ by the grading properties. Then $(b_0, \bar{c_0})$ is an idempotent of the Jordan pair V, with $(b_0, \bar{c_0})$, $(e, \bar{f})$ being orthogonal [**Loo75**, 5.12]. Hence, by [**Loo75**, Lemma 5.11], $(e + b_0, \overline{f + c_0})$ is also an idempotent of V, so $e' = e + b_0$ is von Neumann regular and both $e, b_0 \in [e']$.

(vi) Suppose now that L is nondegenerate and $B_0 = 0$. Given $b_1 \in B_1$, by the grading properties we have

$$\operatorname{ad}_{b_1}^2(L_2 + L_1 + L_{-2}) \subset B \cap L_0 = B_0 = 0.$$

Since B is abelian, b_1 is a Jordan element, so, by 4.4(3), we have

$$\operatorname{ad}^2_{\operatorname{ad}^2_{b_1} L_0} L = \operatorname{ad}^2_{b_1} \operatorname{ad}^2_{L_0} \operatorname{ad}^2_{b_1} L \subset \operatorname{ad}^2_{b_1} \operatorname{ad}^2_{L_0} \operatorname{ad}^2_{b_1} L_{-2} = 0$$

and hence $\operatorname{ad}^2_{b_1} L_0 = 0$ by nondegeneracy of L, which proves

$$\operatorname{ad}^2_{b_1} L = \operatorname{ad}^2_{b_1} L_{-1} \subset B \cap L_1 = B_1,$$

so completing the proof. □

Lie socle revisited. Recall that the socle $\operatorname{Soc}(L)$ of a Lie algebra L is defined as the sum of all its minimal abelian inner ideals.

PROPOSITION 11.57. *Let* L *be a nondegenerate Jordan algebra. For every Jordan element* $a \in L$, *the following conditions are equivalent:*

(i) a *is von Neumann regular and* $a \in \operatorname{Soc}(L)$.

(ii) L *satisfies the dcc on principal inner ideals within* (a).

PROOF. (i) $\Rightarrow$ (ii). Let $a = x_1 + \ldots + x_r$ where each summand x_i belongs to a minimal abelian inner ideal of L, and suppose also that a is von Neumann regular. Using Theorem 5.11, extend a to an idempotent (a, b) of L. We have:

$$a = \frac{1}{4}\operatorname{ad}_a^2 \operatorname{ad}_b^2 a = \frac{1}{4}\sum_{i=1}^{r} \operatorname{ad}_a^2 \operatorname{ad}_b^2 x_i,$$

where, by Corollary 4.20, each summand $\operatorname{ad}_a^2 \operatorname{ad}_b^2 x_i$ belongs to a minimal abelian inner ideal of L. Letting $V = \operatorname{Sub}_L(a)$, we have that $a \in \operatorname{Soc}(V^+)$. Hence, by Jordan theory (Proposition 11.16), we get that L satisfy dcc on principal inner ideals within (a).

(ii) $\Rightarrow$ (i). Again by Proposition 11.16, and with the same notation as before, we get that $a \in \operatorname{Soc}(V^+) \subset \operatorname{Soc}(L)$. □

PROPOSITION 11.58. *Let B be an abelian inner ideal of a nondegenerate Lie algebra L. Set $S = \operatorname{Soc}(L)$ and $V = \operatorname{Sub}_L B$. Then*

$$[[B \cap S, S], B \cap S] \subset [[B, S], B] \subset \operatorname{Soc}(V^+) \subset B \cap S.$$

Hence $\operatorname{Soc}(V^+) = B \cap S$ if and only if $B \cap S$ is von Neumann regular.

PROOF. The inclusion $\operatorname{Soc}(V^+) \subset B \cap S$ is clear since any minimal inner ideal of V^+ is a minimal abelian inner ideal of L, while the inclusion $[[B, S], B] \subset \operatorname{Soc}(V^+)$ follows from Corollary 4.20. Finally, if $B \cap S$ is von Neumann regular, then

$$[[B \cap S, S], B \cap S] = B \cap S = \operatorname{Soc}(V^+) :$$

the converse is clear since every element of $\operatorname{Soc}(V^+)$ is von Neumann regular by [**Loo89**, Theorem 1]. □

REMARK 11.59. Using Zelmanov's theorem for simple Lie algebras with a finite $\mathbb{Z}$-grading (see Chapter 14), it can be proved (see [**FLGGL07**, Theorem 4.2]) that the condition of von Neumann regularity is redundant in Proposition 11.57. It would be useful to have a structural-independent proof of this fact.

When the abelian inner ideal B is one of the extremes of a finite grading, say $B = L_n$, von Neumann regularity of $B \cap \operatorname{Soc}(L)$ is automatic.

PROPOSITION 11.60. *Let $L = L_{-n} \oplus \ldots \oplus L_n$ be a nondegenerate Lie algebra with a $(2n+1)$-grading and let $V = (L_n, L_{-n})$ be its associated Jordan pair.*

(i) *If B is a minimal inner ideal of L and π_i denotes the projection onto L_i, then $\pi_n(B)$ is either zero or a minimal inner ideal of L contained in L_n.*
(ii) $\operatorname{Soc}(V^+) = \operatorname{Soc}(L) \cap L_n$.

PROOF. (i) Assume that $\pi_n(B) \neq 0$ and let $x \in B$ be such that $\pi_n(x) \neq 0$. Then

$$0 \neq [\pi_n(x), [\pi_n(x), L]] = [\pi_n(x), [\pi_n(x), L_{-n}]] = \pi_n[x, [x, L_{-n}]] = \pi_n(B),$$

because $[x, [x, L_{-n}]]$ is a nonzero inner ideal of L contained in B so it is equal to B by minimality of B, whence $\pi_n(B)$ is a minimal inner ideal of L contained in L_n.

(ii) Any minimal inner ideal $B \subset L_n$ of V is an (abelian) minimal inner ideal of L, so $\operatorname{Soc}(V^+) \subset \operatorname{Soc}(L) \cap L_n$. Conversely, if $x \in \operatorname{Soc}(L) \cap L_n$, x can be expressed as a sum of elements $x_1 + \cdots + x_m$, where each x_i belongs to a minimal abelian inner ideal B_i of L. Therefore, $x = \pi_n(x) = \pi_n(x_1) + \cdots + \pi_n(x_m)$, where each $\pi_n(x_i)$ is either zero or belongs to the minimal abelian inner ideal $\pi_n(B_i)$ of L, equivalently, of V. Thus $\mathrm{Soc}(L) \cap L_n \subset \mathrm{Soc}(V^+)$, which completes the proof. □

PROPOSITION 11.61. *Let L be nondegenerate and let B be an abelian inner ideal of L.*

(i) *If B has dcc on principal inner ideals, then $B = \oplus B_\alpha$, where for each index α, $B_\alpha = B \cap M_\alpha$ with M_α being a simple component of* $\mathrm{Soc}(L)$.
(ii) *If L has actually finite length, then only finitely many B_α are nonzero and for such an inner ideal B_α, $\mathrm{Sub}_L B_\alpha \cong \mathrm{Sub}_{M_\alpha} B_\alpha$ is a simple Artinian Jordan pair.*

PROOF. (i) By Proposition 11.57, $B \subset \mathrm{Soc}(L) = \oplus M_\alpha$ and any $b \in B$ is von Neumann regular in $\mathrm{Soc}(L)$. Hence we can write

$$B = [B, [B, \mathrm{Soc}(L)]] = [B, [B, \oplus M_\alpha]] = \oplus [B, [B, M_\alpha]] \subset \oplus B_\alpha \subset B.$$

(ii) Since B has finite length, $B_\alpha = 0$ up to a finite number of indexes. Moreover, if $B_\alpha \neq 0$ then $\mathrm{Sub}_L B_\alpha \cong \mathrm{Sub}_{M_\alpha} B_\alpha$ (Proposition 11.47(v)) and hence $\mathrm{Sub}_L B_\alpha$ a simple Artinian Jordan pair (Proposition 11.47(vi)). □

PROPOSITION 11.62. *Let L be a nondegenerate Lie algebra. If* $\mathrm{Soc}(L)$ *is an essential ideal and satisfies both dcc and acc chain conditions on principal inner ideals, then L and* $\mathrm{Soc}(L)$ *have the same abelian inner ideals.*

PROOF. Let B be an abelian inner ideal of $\mathrm{Soc}(L)$. Since B has dcc on principal inner ideals, every element of B is von Neumann regular (Proposition 11.52), whence B is an inner ideal of L by Lemma 5.5.

Suppose conversely that B is a nonzero abelian inner ideal of L. Since $\mathrm{Soc}(L)$ is essential $C := B \cap \mathrm{Soc}(L)$ is a nonzero abelian inner ideal of $\mathrm{Soc}(L)$. Let V be the subquotient of $\mathrm{Soc}(L)$ with respect to C. Then V is nondegenerate Jordan pair of finite capacity and hence it contains an idempotent $(e, \bar{f})$ such that $V_0(e, \bar{f}) = 0$. Let $B = B_2 \oplus B_1 \oplus B_0$ be the Peirce decomposition of B (Proposition 11.56(iv)) relative to the idempotent (e, f) of L lifted from $(e, \overline{f})$ (Theorem 5.11). We claim that $B_0 = 0$ and hence that $B = B_2 \oplus B_1 \subset \mathrm{Soc}(L)$. Indeed, $B_0 \cap \mathrm{Soc}(L) \subset V_0(e, \bar{f}) = 0$ implies $B_0 = 0$ since $\mathrm{Soc}(L)$ is essential. □

We give next an example of a Lie algebra as those studied in the above proposition.

PROPOSITION 11.63. *Let L be a central simple finite-dimensional Lie algebra over an algebraically closed field $\mathbb{F}$ of characteristic 0, and let $\mathbb{F}(x)$ be the field of fractional functions over $\mathbb{F}$. Denote by $\tilde{L}$ (resp. $\tilde{L}_{\mathbb{F}}$) the Lie algebra $\mathbb{F}(x) \otimes_{\mathbb{F}} L$ over $\mathbb{F}(x)$ (resp. over $\mathbb{F}$). We have:*

(i) $\mathrm{Der}(\tilde{L}_{\mathbb{F}})$ *is a strongly prime Lie algebra.*
(ii) $\mathrm{Soc}(\mathrm{Der}(\tilde{L}_{\mathbb{F}})) = \mathrm{ad}(\tilde{L}_{\mathbb{F}})$ *and it is strictly contained in* $\mathrm{Der}(\tilde{L}_{\mathbb{F}})$.
(iii) $\tilde{L}$ *and* $\tilde{L}_{\mathbb{F}}$ *share the same abelian inner ideals.*
(iv) $\mathrm{Der}(\tilde{L}_{\mathbb{F}})$ *has both dcc and acc chain conditions on abelian inner ideals.*

PROOF. (i) By [**Jac79**, X.1.Theorem 3], $\tilde{L}$ is a finite-dimensional and central simple Lie algebra over the field $\mathbb{F}(x)$, so nondegenerate by Corollary 3.23. Since both properties, simplicity and nondegeneracy, remain when regarded as $\mathbb{F}$-algebra, $\mathrm{Der}(\tilde{L}_{\mathbb{F}})$ is prime by Proposition 2.14, and nondegenerate by Corollary 4.16.

(ii) As L is finite-dimensional over an algebraically closed field, it contains nonzero Jordan elements (Corollary 4.32). It is clear that any Jordan element of L remains Jordan in $\tilde{L}$, and since $\tilde{L}$ is finite-dimensional and simple, it coincides with its socle. Also it is clear that every minimal inner ideal of $\tilde{L}$ is a minimal inner ideal of $\tilde{L}_{\mathbb{F}}$, so $\tilde{L}_{\mathbb{F}}$ also coincides with its socle. Regarded $\tilde{L}_{\mathbb{F}}$ as an ideal of $\mathrm{Der}(\tilde{L}_{\mathbb{F}})$ (via the adjoint representation), it follows from Theorem 5.22 and primeness of $\mathrm{Der}(\tilde{L}_{\mathbb{F}})$ that $\mathrm{Soc}(\mathrm{Der}(\tilde{L}_{\mathbb{F}})) = \mathrm{ad}(\tilde{L}_{\mathbb{F}})$. It remains to prove that $\tilde{L}_{\mathbb{F}}$ has outer derivations. Consider the usual derivation $f(x) \mapsto f'(x)$ of $\mathbb{F}(x)$. We extend this derivation to one derivation δ of $\tilde{L}_{\mathbb{F}}$ by putting $\delta(f(x) \otimes a) := f'(x) \otimes a$, for all $f(x) \in \mathbb{F}(x)$, $a \in \tilde{L}_{\mathbb{F}}$. Since δ is not $\mathbb{F}(x)$-linear, it cannot be an inner derivation of $\tilde{L}_{\mathbb{F}}$.

(iii) $\tilde{L}$ and $\tilde{L}_{\mathbb{F}}$ share the same Jordan elements and the same von Neumann regular elements. But every Jordan element in $\tilde{L}$ is von Neumann regular (Proposition 11.52). Hence every abelian inner ideal of $\tilde{L}_{\mathbb{F}}$ is an inner ideal of $\tilde{L}$ (the converse is clear). Since $\tilde{L}$ is finite-dimensional, it satisfies both chain conditions on inner ideals, so $\tilde{L}_{\mathbb{F}}$ satisfies both dcc and acc chain conditions on abelian inner ideals. Now (iv) follows from Preposition 11.62. □

11.6. Exercises

EXERCISE 11.64. Let V^+ be a left module and V^- be a right module over an associative $\mathbb{F}$-algebra R, and let $f : V^+ \times V^- \to R$ be an R-bilinear form. Show that $V = (V^+, V^-)$ is a Jordan pair over $\mathbb{F}$ with $Q_x y = f(x,y)x$ and $Q_y x = yf(x,y)$, for $x \in V^+$, $y \in V^-$.

EXERCISE 11.65. Let X be a left vector space over a division ring Δ and let X^* denote the dual of X. Show that $V = (X, X^*)$ is a Jordan pair with triple products $\{x_1, \varphi, x_2\} = (x_1\varphi)x_2 + (x_2\varphi)x_1$, $\{\varphi_1, x, \varphi_2\} = \varphi_1(x\varphi_2) + \varphi_2(x\varphi_1)$, $x, x_1, x_2 \in X$, $\varphi, \varphi_1, \varphi_2 \in X^*$.

EXERCISE 11.66. Let $V = (V^+, V^-)$ be a Jordan pair. Show that $\mathrm{Der}(V)$ is a subalgebra of the Lie algebra $\mathfrak{gl}(V^+) \times \mathfrak{gl}(V^-)$.

EXERCISE 11.67. Let $C(X,q)$ be a Clifford Jordan pair over a field $\mathbb{F}$. Show:

(i) The centroid of $C(X,q)$ is just $\mathbb{F}$ itself.
(ii) A subspace I of X is a proper inner ideal of $C(X,q)$ if and only if I is totally isotropic.
(iii) $C(X,q)$ is a division Jordan if and only if (X,q) is anisotropic.

EXERCISE 11.68. Given a Clifford Jordan pair $C(X,q)$, let S,T be totally isotropic subspaces of X. Show that the Jordan pair $V = (S,T)$ defined by the bilinear form $f(s,t) := q(s,t)$, $s \in S$, $t \in T$, is a subpair of $C(X,q)$.

Conversely, let $V = (V^+, V^-)$ be the Jordan pair defined by a bilinear form f over a field $\mathbb{F}$. Set $X = V^+ \oplus V^-$ and $q(x) = q(u+v) := f(u,v)$, $u \in V^+, v \in V^-$. Show that q is a quadratic form of X and V is a subpair of $C(X,q)$.

EXERCISE 11.69. Show that the centroid of a Bi-Cayley pair

$$(\mathrm{M}_{1\times 2}(C), \mathrm{M}_{2\times 1}(C)),$$

where C is a Cayley algebra over a field $\mathbb{F}$ is just $\mathbb{F}$ itself.

EXERCISE 11.70. Let V be the Jordan pair $(\mathcal{F}(X_1, X_2), \mathcal{F}(X_2, X_1))$ defined by the dual pairs of vector spaces $(X_1, Y_1, \langle\, ,\rangle_1)$ and $(X_2, Y_2, \langle\, ,\rangle_2)$ over the division ring Δ. Show:

(1) V has dcc on principal inner ideals.
(2) V has finite capacity whenever one of the pairs of dual vector spaces is finite-dimensional.
(3) V is Artinian if and only if both dual pairs are finite-dimensional.

EXERCISE 11.71. Let $(f, g) : V \rightleftarrows W$ be a structural transformation and let $M \subset V^+$ an inner ideal of V. Show:

(1) $f(M) \subset W^+$ is an inner ideal of W.
(2) If V is nondegenerate and M is a minimal inner, then either $f(M) = 0$ or a minimal inner ideal of W.
(3) A pair of Φ-submodules (I^+, I^-) of a Jordan pair V is an ideal of V if and only if I^+ and I^- are invariant under all inner structural transformations.

EXERCISE 11.72. Let V be a Jordan pair, let $v \in V^-$ be invertible, and $J = V^{(v)}$. Show that $(1_J.Q_v) : (J, J) \to (V^+, V^-)$ is an isomorphism of Jordan pairs.

EXERCISE 11.73. Let $L = L_{-n} \oplus \cdots \oplus L_{-n}$ be a $(2n + 1)$-graded Lie algebra with associated Jordan pair $V = (L_n, L_{-n})$. Show that if L is nondegenerate, so is V and $\mathrm{Soc}(V^+) + \mathrm{Soc}(V^-) \subset \mathrm{Soc}(L)$.

EXERCISE 11.74. Let R be a ring. Show that the Lie algebra $L = [\mathrm{M}_2(R), M_2(R)]$ has a short grading $L = L_{-1} \oplus L_0 \oplus L_1$, where

$$L_{-1} = R[12],\ L_0 = \{a[11] + b[22] :\ a + b \in [R, R]\},\ L_1 = R[21].$$

EXERCISE 11.75. Let R be an associative algebra over a ring of scalars Φ in which 6 is invertible and let $V = (R^+, R^+)$ be the Jordan pair defined by R. Show:

(1) $\mathrm{TKK}(V) \cong \overline{\mathrm{M}_2(R)'}$.

Suppose that R has an involution $*$. Determine the TKK-algebra of the following Jordan pairs:

(2) $V = (\mathrm{Sym}(R, *), \mathrm{Sym}(R, *))$,
(3) $V = (\mathrm{Skew}(R, *), \mathrm{Skew}(R, *))$.

EXERCISE 11.76. Suppose that $L = L_{-1} \oplus L_0 \oplus L_1$, $L_0 = [L_1, L_{-1}]$, is a nondegenerate finite-dimensional 3-graded Lie algebra over a field of characteristic $p > 3$. Show that L coincides with its socle.

EXERCISE 11.77. In Lemma 11.30, let $n = 2$ and $\emptyset \neq X \subset L_2$. Show:

(1) $\mathrm{ad}(L_2)\,\mathrm{ad}(L_0)\,\mathrm{ad}(L_{-1})^2 X \subset \mathrm{ad}(L_0)^2 X$.
(2) $\mathrm{ad}(L_1)\,\mathrm{ad}(L_0)^2\,\mathrm{ad}(L_{-1}) X \subset \mathrm{ad}(L_0)^2 X + \mathrm{ad}(L_0)^3 X$.

EXERCISE 11.78. In relation with Proposition 11.39, give a direct proof of the fact that if L is nondegenerate, then $L_{-n} \cap \mathrm{Ker}_L L_n = 0$.

EXERCISE 11.79. Let $(X, Y, \langle\, ,\rangle)$ be an infinite-dimensional pair of dual vector spaces over a field $\mathbb{F}$ of characteristic 0 and let L denote the Lie algebra $\mathcal{F}_Y(X)^-$. Let $V \leq X$ and $W \leq Y$ be finite-dimensional vector space such that $\langle V, W\rangle = 0$, and denote by B the abelian inner ideal W^*V of L. Take $V' \leq Y$ and $W' \leq X$

dual complements of V and W respectively, i.e. (V, V') and (W', W) are subpairs of dual vector spaces of $(X, Y, \langle\,,\rangle)$. Show:

(1) $X = W' \oplus W^{\perp}$ and $Y = V' \oplus V^{\perp}$.
(2) $\operatorname{Ker}_L B = (V^{\perp})^* W' + Y^* W^{\perp}$.
(3) $\operatorname{Sub}_L B \cong (W^*V, (V')^*W') \cong (\mathcal{L}(W', V),\ \mathcal{L}(V, W'))$.
(4) $\operatorname{Sub}_L B \cong (\mathrm{M}_{n\times m}(\mathbb{F}), \mathrm{M}_{m\times n}(\mathbb{F}))$,

where $n = \dim_{\mathbb{F}} W$ and $m = \dim_{\mathbb{F}} V$.

EXERCISE 11.80. Following Exercise 2.99, let $L = \mathfrak{fo}(\langle\,,\rangle)$, $B = [x, H^{\perp}]$ and $C = [y, H^{\perp}]$. Show:

(1) $\operatorname{Ker}_L B = B + \mathbb{F}[x, y] + [H^{\perp}, H^{\perp}]$.
(2) $\operatorname{Sub}_L B \cong (B, C)$.

CHAPTER 12

An Artinian Theory for Lie Algebras

The module-theoretic characterization of semiprime Artinian rings (R is unital and completely reducible as a left R-module) cannot be translated to Jordan systems by just replacing left ideals by inner ideals: if we take, for instance, the Jordan algebra $\mathrm{M}_2(\mathbb{F})^+$ of 2×2-matrices over a field F, any nontrivial inner ideal of $\mathrm{M}_2(\mathbb{F})^+$ has dimension 1, so it cannot be complemented as a $\mathbb{F}$-subspace by any other inner ideal. Nevertheless, O. Loos and E. Neher succeeded in getting the appropriate characterization: a Jordan pair $V = (V^+, V^-)$ (over an arbitrary ring of scalars) is a direct sum of simple Artinian Jordan pairs if and only if it is complemented in the following sense: for any inner ideal B of V^σ there exists an inner ideal C of $V^{-\sigma}$ such that each of them is complemented as a submodule by the kernel of the other. In particular, a simple Jordan pair is complemented if and only if it is Artinian. The notion of complemented inner ideal makes sense for Lie algebras, and examples of complemented inner ideals are the extremes of a finite $\mathbb{Z}$-grading of a nondegenerate Lie algebra. Complemented inner ideals are studied in Section 12.1. In Section 12.2 it is proved that any finite family of Peirce compatible idempotents of the subquotient of a Lie algebra with respect to an abelian inner ideal can be lifted to a finite family of Peirce compatible idempotents of the Lie algebra. Using this result, together with the structure theorem of nondegenerate Artinian Jordan pairs, it is shown in Section 12.3 that any abelian inner ideal B of finite length of a nondegenerate Lie algebra L yields a finite $\mathbb{Z}$-grading $L = L_{-n} \oplus \cdots \oplus L_n$ such that $B = L_n$, and hence B is a complemented inner ideal. In particular, simple nondegenerate Artinian Lie algebras are complemented. The converse is also true; even more, in Section 12.4 it is proved that a Lie algebra is complemented if and only if it is a direct sum of simple nondegenerate Artinian Lie algebras. Finally, in Section 12.5, a unified approach to inner ideals of Jordan pairs and abelian inner ideals of Lie algebras is outlined.

12.1. Complemented inner ideals

Throughout this section, L will be a Lie algebra over a ring of scalars Φ in which $2, 3$ and 5 are invertible. We follow the exposition given in [**FLGGL08**] which, as mentioned in the introduction of the chapter, is the Lie version of the work by O. Loos and E. Neher [**LN94**] for Jordan pairs.

DEFINITION 12.1. An inner ideal M of a Lie algebra L is said to be *complemented* if there exists an inner ideal N of L such that

$$L = M \oplus \operatorname{Ker}_L N = N \oplus \operatorname{Ker}_L M.$$

Then N (resp. M) is called a *complement* of M (resp. N).

LEMMA 12.2. *Let M, N be abelian inner ideals of a Lie algebra L which are complemented by each other. Then the Jordan pairs* $\operatorname{Sub}_L M$ *and* (M, N) *are isomorphic.*

PROOF. The pair of linear maps

$$(1_M, \varphi) : (M, L/\operatorname{Ker}_L M) \to (M, N),$$

where φ is the canonical linear isomorphism of $L/\operatorname{Ker}_L M$ onto N, is an isomorphism of Jordan pairs (see Section 11.1 for definition). □

LEMMA 12.3. *Let $L = L_{-n} \oplus \cdots \oplus L_n$ be a $(2n+1)$-grading of a Lie algebra L. If L is nondegenerate, then the abelian inner ideals L_n and L_{-n} are complemented by each other.*

PROOF. It follows from Proposition 11.39(ii). □

LEMMA 12.4. *Let L be nondegenerate and let I be an ideal of L. We have:*
(i) $\operatorname{Ker}_L I = \operatorname{Ann}(I)$.
(ii) *If I is complemented as an inner ideal, then $L = I \oplus \operatorname{Ann}(I)$ and I is the only complement of itself.*

PROOF. (i) Let $x \in \operatorname{Ker}_L I$. Then $[[x, I], I] = 0$ and hence $\operatorname{ad}^2_{[x,I]} L = 0$, which implies $[x, I] = 0$ by nondegeneracy of L. Therefore, $\operatorname{Ker}_L I \subset \operatorname{Ann}(I)$; the reverse inclusion is trivial.

(ii) Let M be an inner ideal of L such that

$$L = I \oplus \operatorname{Ker}_L M = M \oplus \operatorname{Ann}(I).$$

For any $x \in M$ we have $\operatorname{ad}^2_x \operatorname{Ann}(I) \subset M \cap \operatorname{Ann}(I) = 0$. Hence, by Proposition 4.10, $M \subset \operatorname{Ann}(\operatorname{Ann}(I))$. Using the Modular Law and the nondegeneracy of L we get that $L = M \oplus \operatorname{Ann}(I)$ implies $M = \operatorname{Ann}(\operatorname{Ann}(I))$, so M is an ideal and therefore we have by (i)

$$\operatorname{Ker}_L M = \operatorname{Ann}(M) = \operatorname{Ann}(\operatorname{Ann}(\operatorname{Ann}(I))) = \operatorname{Ann}(I).$$

Thus $L = I \oplus \operatorname{Ann}(I)$ and $I = \operatorname{Ann}(\operatorname{Ann}(I))$ is the only complement of I. □

LEMMA 12.5. *Let B be an inner ideal of a strongly prime Lie algebra L.*
(1) *If $\operatorname{Ker}_L B \neq 0$, then B is abelian.*
(2) *If B is complemented by an inner ideal, then the following conditions are equivalent:*
  (i) *B is abelian,*
  (ii) *B is proper,*
  (iii) *any complemented of B is abelian.*

PROOF. (1) Note first that for any inner ideal B of a Lie algebra L,

$$[[B, [B, B]], L] \subset [B, B].$$

Indeed, let $b_1, b_2, b_3 \in B$ and $a \in L$. Then

$$\begin{aligned} [[b_1, [b_2, b_3]], a] &= [[b_1, a], [b_2, b_3]] + [b_1, [[b_2, b_3], a]] = [[[b_1, a], b_2], b_3] \\ &+ [b_2, [[b_1, a], b_3]] + [b_1, [[b_2, b_3], a]] \subset [B, B]. \end{aligned}$$

Let $0 \neq x \in \operatorname{Ker}_L B$. Then $[x,[[B,[B,B]],L]] \subset [x,[B,B]] \subset [[x,B],B] = 0$ and hence $[B,[B,B]] = 0$ by Theorem 7.10. Now, by nondegeneracy of L,

$$[[B,B],[[B,B],L]] \subset [[B,B],B] = 0$$

implies $[B,B] = 0$.

(2) Note that (i) $\Rightarrow$ (ii) is trivial since a prime Lie algebra cannot be abelian. Suppose now that B is a complemented inner ideal of L.

(ii) $\Rightarrow$ (iii). Let C be a complement of B. If C is not abelian, then $\operatorname{Ker}_L C = 0$ by (1). Hence $L = B \oplus \operatorname{Ker}_L C = B$, a contradiction since B is proper.

(iii) $\Rightarrow$ (i). Again by (1), if B is not abelian, then $\operatorname{Ker}_L B = 0$ which implies that any complement C is equal to L, a contradiction since L is not abelian. □

By [**LN94**, Proposition 3.8], a nondegenerate Jordan pair V is von Neumann regular if and only if every principal inner ideal is complemented. A similar result holds for Lie algebras.

PROPOSITION 12.6. *Let e be a Jordan element of a nondegenerate Lie algebra L. Then e is von Neumann regular if and only if the abelian inner ideal $[e] = \operatorname{ad}_e^2 L$ has an* abelian *complement.*

PROOF. Let $e \in L$ be von Neumann regular. Using Lemma 5.8 and Proposition 5.11, extend e to an idempotent (e,f) with associated 5-grading

$$L = L_{-2} \oplus L_{-1} \oplus L_0 \oplus L_1 \oplus L_2,\ L_2 = [e],\ \text{and } L_{-2} = [f].$$

It follows from Lemma 12.3 that $[f]$ is an abelian complement of $[e]$.

Suppose conversely that $[e]$ is complemented by an abelian inner ideal B. Then $L = [e] \oplus \operatorname{Ker}_L B$ and hence $e = [e,[e,a]] + x$ for some $a \in L$ and $x \in \operatorname{Ker}_L B$. Since B is abelian, $[x,[x,B]] \subset \operatorname{Ker}_L B$ by Lemma 11.36(ii). Moreover,

$$x = e - [e,[e,a]] \in \Phi e + [e] = (e)$$

implies $[x,[x,L]] \subset [e]$. Thus

$$[x,[x,B]] \subset [e] \cap \operatorname{Ker}_L B = 0.$$

On the other hand, since by Proposition 11.41(ii) $\operatorname{Ker}_L[e] = \operatorname{Ker}_L(e)$,

$$[x,[x,B]] \subset [(e),[(e),\operatorname{Ker}_L(e)]] = 0.$$

Therefore, $[x,[x,L]] = [x,[x,B \oplus \operatorname{Ker}_L[e]]] = 0$, which implies $x = 0$ by nondegeneracy of L, so $e = [e,[e,a]]$ is von Neumann regular. □

PROPOSITION 12.7. *Let L be a Lie algebra such that any abelian inner ideal of L has a (not necessarily abelian) complement. Then L is nondegenerate.*

PROOF. Let x be an absolute zero divisor of L. Then $M = \Phi x$ is an abelian inner ideal of L with $\operatorname{Ker}_L M = L$. Let N be a complement of M. We have that $L = N \oplus \operatorname{Ker}_L M$ implies $N = 0$ and hence $L = M \oplus \operatorname{Ker}_L N$ implies $M = 0$, so $x = 0$, which proves that L is nondegenerate. □

12.2. Lifting idempotents

We develop in this section a useful technique for lifting any finite family of Peirce compatible idempotents of the subquotient of a Lie algebra with respect to an abelian inner ideal to a finite family of Peirce compatible idempotents of the Lie algebra (related results in the Jordan setting can be found in [**McC82**]). As in the previous section, Jordan pairs considered here are linear, i.e. they are Jordan pairs over a ring of scalars Φ in which 2 and 3 are invertible, while for Lie algebras we will also assume the invertibility of 5 (because we will use the Jacobson–Morozov decomposition associated to an idempotent of a Lie algebra, where this restriction on the scalars is required). By Example 11.1(2), any Jordan pair $V = (V^+, V^-)$ gives rise to a Jordan triple system $T = V^+ \oplus V^-$, with triple product defined by $\{x^+ \oplus x^-, y^+ \oplus y^-, z^+ \oplus z^-\} := \{x^+, y^-, z^+\} \oplus \{x^-, y^+, z^-\}$. This allows to transfer notions and results on triple systems to Jordan pairs. For instance, any idempotent (e^+, e^-) of a Jordan pair V yields the *tripotent* $e = e^+ \oplus e^-$ of the Jordan triple system T, i.e. $P(e)e = Q_{e^+}e^- \oplus Q_{e^-}e^+ = e^+ \oplus e^-$. We will use this fact in what follows without mention of it.

DEFINITION 12.8. A family $\{(e_i^+, e_i^-)\}_{i\in I}$ of idempotents of a Jordan pair V is called *Peirce compatible* if for any pair of indexes $(i, j) \in I \times I$, there exists $\beta_{ij} \in \{0, 1, 2\}$ such that

$$(e_i^+, e_i^-) \in V_{\beta_{ij}}(e_j^+, e_j^-),$$

where $V_{\beta_{ij}}(e_j^+, e_j^-)$ denotes the β_{ij}-Peirce space of V with respect to the idempotent (e_j^+, e_j^-) (Proposition 11.13).

Two particular cases of Peirce-compatibility are orthogonality and collinearity:

- (e_i^+, e_i^-) and (e_j^+, e_j^-) are *orthogonal*, written as

$$(e_i^+, e_i^-) \perp (e_j^+, e_j^-),$$

  if $(e_j^+, e_j^-) \in V_0(e_i^+, e_i^-)$, equivalently, [**Loo75**, 5.11], $(e_i^+, e_i^-) \in V_0(e_j^+, e_j^-)$. In this case, $(e_i^+ + e_j^+, e_i^- + e_j^-)$ is also an idempotent of V.

- (e_i^+, e_i^-) and (e_j^+, e_j^-) are *collinear*, written as

$$(e_i^+, e_i^-) \top (e_j^+, e_j^-),$$

  if $(e_j^+, e_j^-) \in V_1(e_i^+, e_i^-)$ and $(e_i^+, e_i^-) \in V_1(e_j^+, e_j^-)$.

DEFINITION 12.9. A family $\{(e_i, f_i)\}_{i\in I}$ of idempotents of a Lie algebra L is called *compatible* if for any $i, j \in I$, $[h_i, h_j] = 0$, where $h_i = [e_i, f_i]$ for any index $i \in I$; and it is called *Peirce compatible* if for any pair of indexes $(i, j) \in I \times I$, $(e_i, f_i) \in L^{h_j}_{\alpha_{ij}} \times L^{h_j}_{-\alpha_{ij}}$ for some $\alpha_{ij} \in \{0, \pm 1, \pm 2\}$, with $L^{h_j}_{\alpha_{ij}}$ denoting the α_{ij}-space of L relative to the idempotent (e_i, f_i) (see Theorem 5.11).

PROPOSITION 12.10. *Peirce compatibility implies compatibility.*

PROOF. If $(e_i, f_i) \in L^{h_j}_{\alpha_{ij}} \times L^{h_j}_{-\alpha_{ij}}$, then $h_i = [e_i, f_i] \in [L^{h_j}_{\alpha_{ij}}, L^{h_j}_{-\alpha_{ij}}] \subset L^{h_j}_0$, so $[h_j, h_i] = 0$. □

As for Jordan pairs, two special cases of Peirce compatibility of idempotents of a Lie algebra are orthogonality and collinearity:

- (e_i, f_i) and (e_j, f_j) are *orthogonal*, written as
$$(e_i, f_i) \perp (e_j, f_j)$$
if $(e_i, f_i) \in L_0^{h_j} \times L_0^{h_j}$ and $(e_j, f_j) \in L_0^{h_i} \times L_0^{h_i}$.
- (e_i, f_i) and (e_j, f_j) are *collinear*, written as
$$(e_i, f_i) \top (e_j, f_j)$$
if $(e_i, f_i) \in L_1^{h_j} \times L_{-1}^{h_j}$ and $(e_j, f_j) \in L_1^{h_i} \times L_{-1}^{h_i}$.

LEMMA 12.11. *Let $\mathcal{F} = \{(e_i, f_i)\}_{i=1}^n$ be a finite family of compatible idempotents of a Lie algebra L. We have:*

(i) *$\mathcal{F}$ gives rise to a finite $\mathbb{Z}^n$-grading*
$$L = \bigoplus_{\alpha \in \mathbb{Z}^n} L_\alpha, \quad \textit{where } \alpha = (\alpha_1, \dots, \alpha_n) \in \mathbb{Z}^n \textit{ and } L_\alpha = \bigcap_{i=1}^n L_{\alpha_i}^{h_i}.$$

(ii) *For each $(\lambda_1, \dots, \lambda_n) \in \mathbb{N}^n$, the grading in* (i) *induces a finite $\mathbb{Z}$-grading*
$$L = \bigoplus_r L_r, \quad \textit{where } L_r = \sum_{\lambda_1\alpha_1 + \cdots + \lambda_n\alpha_n = r} L_\alpha.$$

PROOF. Straightforward. □

LEMMA 12.12. *Let B be an abelian inner ideal of a Lie algebra L, and let (e_1, f_1) and (e_2, f_2) be compatible idempotents of L such that $e_1, e_2 \in B$ and $[e_1]$ is a minimal inner ideal. Then $e_1 \in L_k^{h_2}$ if and only if $[e_1] = L_2^{h_1} \subset L_k^{h_2}$, $k \in \{0, 1, 2\}$.*

PROOF. Set $V := \mathrm{Sub}_L\, B$. Suppose that $e_1 \in L_k^{h_2}$ (the reverse implication is trivial because $e_1 \in [e_1]$). Since both $L_2^{h_2}$ and $L_0^{h_2} \cap B$ are inner ideals of L (Proposition 11.56(iv)), only the case $k = 1$ needs to be considered.

For $k = 0, 1, 2$, let $E_k^+(e_i, \overline{f}_i)$ be the k-Peirce projection on V^+ relative to the *Jordan pair* idempotent $(e_i, \overline{f}_i)$, $i = 1, 2$. Since $E_k^+(e_i, \overline{f}_i)$ is the restriction of $p_{ik}(\mathrm{ad}_{h_i})$ to B for a certain polynomial $p_{ik}(\lambda)$ over the ring of integers (Remark 5.12), it follows from the compatibility of the idempotents (e_1, f_1) and (e_2, f_2) that for $j, k \in \{0, 1, 2\}$, the Peirce projections $E_k^+(e_1, \overline{f}_1)$ and $E_j^+(e_2, \overline{f}_2)$ commute pairwise. Hence
$$[e_1] = V_2^+(e_1, \overline{f}_1) = [e_1] \cap V_0^+(e_2, \overline{f}_2) \oplus [e_1] \cap V_1^+(e_2, \overline{f}_2) \oplus [e_1] \cap V_2^+(e_2, \overline{f}_2).$$
If $[e_1] \cap V_0^+(e_2, \overline{f}_f) + [e_1] \cap V_2^+(e_2, \overline{f}_2) = 0$, then $[e_1] \subset V_1^+(e_2, \overline{f}_2) \subset L_1^{h_2}$ and we are done. Suppose then that $[e_1] \cap V_k^+(e_2, \overline{f}_2) \neq 0$ for $k = 0$ or $k = 2$ (the two cases being similar). Since $[e_1]$ is a minimal inner ideal, $[e_1] \cap V_k^+(e_2, \overline{f}_2) = [e_1]$, so $e_1 \in [e_1] \subset V_k^+(e_2, \overline{f}_2) \subset L_k^{h_2}$ for $k = 0$ or $k = 2$, which is a contradiction. □

LEMMA 12.13. *Let M be an inner ideal of a Lie algebra L and let (e_1, f_1) and (e_2, f_2) be two orthogonal idempotents of L such that $e_1, e_2 \in M$. Then $L_1^{h_1} \cap L_1^{h_2}$ is contained in M.*

PROOF. Let $x \in L_1^{h_1} \cap L_1^{h_2}$. For $i = 1, 2$, we have $[e_i, [f_i, x]] = [h_i, x] = x$ since $[e_i, x] \in [L_2^{h_i}, L_1^{h_i}] \subset L_3^{h_i} = 0$. Hence

$$x = \operatorname{ad}_{e_1} \operatorname{ad}_{f_1} \operatorname{ad}_{e_2} \operatorname{ad}_{f_2} x = \operatorname{ad}_{e_1}[[f_1, e_2], [f_2, x]] \\ + \operatorname{ad}_{e_1} \operatorname{ad}_{e_2}[f_1, [f_2, x]] \in [M, [M, L]] \subset M$$

since $[f_1, e_2] \in [L_0^{h_2}, L_2^{h_2}] \subset L_2^{h_2} = [e_2, [e_2, L]] \subset M$. □

DEFINITION 12.14. Let B be an abelian inner ideal of a Lie algebra L and let $V = (B, L/\operatorname{Ker}_L B)$ be the subquotient of L with respect to B. We say that a family $\{(e_i, \overline{g}_i)\}_{i \in I}$ of idempotents of V *lifts to* a family of idempotents of L if for each index $i \in I$ there exists an element $f_i \in L$ such that (e_i, f_i) is an idempotent of L and $\overline{g}_i = \overline{f}_i$.

PROPOSITION 12.15. *Let B be an abelian inner ideal of a Lie algebra L and let $V = (B, L/\operatorname{Ker}_L B)$. Then any finite family $\{(e_i, \overline{g}_i)\}_{1=1}^n$ of Peirce compatible idempotents of V lifts to a finite family $\{(e_i, f_i)\}_{1=1}^n$ of Peirce compatible idempotents of L. Furthermore,*

(i) *$(e_i, \overline{g}_i)$ is a division idempotent if and only if so is (e_i, f_i),*
(ii) *if $(e_i, \overline{g}_i)$, $(e_j, \overline{g}_j)$ are orthogonal (resp. collinear), then (e_i, f_i), (e_j, f_j) are orthogonal (resp. collinear).*

PROOF. We will use induction on the number of Peirce compatible idempotents of V. Begin with the case of a single idempotent $(e, \overline{g})$. As $[[e, g], e] = \{e, \overline{g}, e\} = 2e$, we have by Lemma 5.8 and Theorem 5.11 that for any representative g of the coset $\overline{g}$ there exists $f \in L$ such that (e, f) is an idempotent of L and $[e, f] = [e, g]$. Furthermore,

$$(12.1)\quad \overline{g} = \{\overline{g}, e, \overline{g}\} = \overline{[g, [e, g]]} = \overline{[g, [e, f]]} = \overline{[[g, e], f]} + \overline{[e, [g, f]]} = \overline{[[f, e], f]} = 2\overline{f},$$

since $[e, [g, f]] \in [B, L] \subset \operatorname{Ker}_L B$ by Lemma 11.36(ii), which proves that any idempotent of V can be lifted to an idempotent of L. Set $h = [e, f]$ as usual. By Proposition 11.56(4) the Peirce subspaces of V^+ relative to the idempotent $(e, \overline{f})$ are given by

$$(12.2)\qquad V_k^+(e, \overline{f}) = B \cap L_k^h, \; k = 0, 1, 2.$$

Suppose that given k Peirce compatible idempotents $(e_1, \overline{g}_1), \ldots, (e_k, \overline{g}_k)$ of V there exist k Peirce compatible idempotents $(e_1, f_1), \ldots, (e_k, f_k)$ of L such that $\overline{f}_i = \overline{g}_i$ for $1 \leq i \leq k$. Since Peirce compatibility implies compatibility (Preposition 12.10), $[h_i, h_j] = 0$ for all $1 \leq i, j \leq k$, where $h_i = [e_i, f_i]$. By Lemma 12.11, these compatible idempotents (e_i, f_i) give rise to a finite $\mathbb{Z}^k$-grading

$$(12.3)\qquad L = \bigoplus_{\alpha \in \mathbb{Z}^k} L_\alpha, \text{ for } \alpha = (\alpha_1, \ldots, \alpha_k) \in \mathbb{Z}^k, \text{ with } L_\alpha = \bigcap_{i=1}^k L_{\alpha_i}^{h_i},$$

and where $L_\alpha = 0$ if $|\alpha| := \sum_{i=1}^k |\alpha_i| > 2k$. Write $x \in \operatorname{Ker}_L B$ as a sum $x = \sum x_\alpha$ relative to the grading (12.3). We claim

$$(12.4)\qquad x_\alpha \in \operatorname{Ker}_L B \text{ for any } \alpha \in \mathbb{Z}^k.$$

By Lemma 11.36(ii), $[h_i, \operatorname{Ker}_L B] \subset [[B, L], \operatorname{Ker}_L B] \subset \operatorname{Ker}_L B$ for each $1 \leq i \leq k$, and since $L_\alpha \subset L_{\alpha_i}^{h_i}$, we have by Lemma 5.12 that $x_\alpha = p_i(\operatorname{ad}_{h_i})$ for certain polynomials $p_i(\lambda) \in \mathbb{Z}[\lambda]$. Hence $x_\alpha \in \operatorname{Ker}_L B$, as claimed.

Let $(e_{k+1}, \overline{g}_{k+1})$ be the next idempotent of V to be lifted. By the Peirce compatibility of the idempotents $(e_i, \overline{g}_i)$, for each index $1 \leq i \leq k$ there exists $\beta_i \in \{0, 1, 2\}$ such that $(e_{k+1}, \overline{g}_{k+1}) \in V_{\beta_i}(e_i, \overline{f}_i)$, with $V^+_{\beta_i}(e_i, \overline{f}_i) \subset L^{h_i}_{\beta_i}$ by (12.2), so putting $\beta = (\beta_1, \ldots, \beta_k) \in \mathbb{Z}^k$, we have that $e_{k+1} \in L_\beta$. We claim that we can take a representative of the coset $\overline{g}_{k+1}$ in $L_{-\beta}$. Indeed, let $g_{k+1} = \sum_{\alpha\in\mathbb{Z}^k} g_\alpha$ be the decomposition in (12.3) of a representative of $\overline{g}_{k+1}$. Since $\overline{g}_{k+1} \in \bigcap_{i=1}^k V^-_{\beta_i}(e_i, \overline{f}_i)$, for all $1 \leq i \leq k$, we have

$$\overline{[h_i, g_{k+1}]} = \overline{[[e_i, f_i], g_{k+1}]} = -\{\overline{f}_i, e_i, \overline{g}_{k+1}\} = -\{\overline{g}_i, e_i, \overline{g}_{k+1}\} = -\beta_i \overline{g}_{k+1}.$$

So $\sum_\alpha(\alpha_i+\beta_i)g_\alpha = [h_i, g_{k+1}]+\beta_i g_{k+1} \in \operatorname{Ker}_L B$, and hence, by (12.4), $g_\alpha \in \operatorname{Ker}_L B$ whenever $\alpha_i + \beta_i$ is invertible in Φ (equivalently, when $\alpha_i + \beta_i \neq 0$, since we have $|\alpha_i + \beta_i| \leq |\alpha_i| + |\beta_i| \leq 4$), which proves that $\overline{g}_{k+1} = \overline{g}_{-\beta}$, as claimed.

Suppose then that $g_{k+1} \in L_{-\beta}$. By the case of a single idempotent, there exists $f_{k+1} \in L$ such that (e_{k+1}, f_{k+1}) is an idempotent of L, with

$$[e_{k+1}, f_{k+1}] = [e_{k+1}, g_{k+1}] \in [L_\beta, L_{-\beta}] \subset L_{(0,\ldots,0)}.$$

Thus $h_{k+1} = [e_{k+1}, f_{k+1}] \in L^{h_i}_0$ for all $1 \leq k$, so $[h_{k+1}, h_i] = 0$, which proves that any finite family $\{(e_i, \overline{g}_i)\}_{i=1}^n$ of Peirce compatible idempotents of V lifts to a family $\{(e_i, f_i)\}_{i=1}^n$ of compatible idempotents of L.

Let us now see that we can replace the family $\{(e_i, f_i)\}_{1=1}^n$ of compatible idempotents of L by a new family $\{(e_i', f_i')\}_{1=1}^n$ of *Peirce compatible* idempotents such that $\overline{f_i'} = \overline{f_i}$, for all $1 \leq i \leq n$. By Peirce compatibility of the idempotents $(e_i, \overline{f}_i)$ of V, for any couple of indexes $i, j \in \{1, \ldots, n\}$ there exists $\beta_{ij} \in \{0, 1, 2\}$ such that $[h_i, e_j] = \{e_i, \overline{f}_i, e_j\} = \beta_{ij} e_j$, i.e. $e_j \in L_\beta$ for $\beta = (\beta_{1j}, \ldots, \beta_{nj})$. By Lemma 11.46, we can change each f_i to f_i' satisfying: (i) $f_i' \in L_{-\beta}$, (ii) (e_i, f_i') is an idempotent of L, and (iii) $[e_i, g_i'] = \pi_{(0,\ldots,0)}([e_i, g_i]) = [e_i, g_i] = h_i$, by the compatibility of the idempotents (e_i, g_i). We have by (12.1) that $\overline{f}_i = \overline{f}'_i$.

(i) It follows from Proposition 11.56(iii). For (ii), let $\beta_{ij} \in \{0, 1, 2\}$ such that $(e_i, f_i) \in (L^{h_j}_{\beta_{ij}}, L^{h_j}_{-\beta_{ij}})$. Hence, if $(e_i, \overline{f}_i)$, $(e_j, \overline{f}_j)$ are orthogonal (the collinear case being similar), then $\{e_j, \overline{f}_j, e_i\} = [h_j, e_i] = \beta_{ij} e_i$ implies $\beta_{ij} = 0$, which proves that $(e_i, f_i) \perp (e_j, f_j)$. □

12.3. A construction of gradings of Lie algebras

In this section we present a method for constructing gradings of Lie algebras. It requires the existence of an abelian inner ideal B of a Lie algebra L such that the subquotient of L with respect to B is covered by a finite grid. Then one obtains a finite $\mathbb{Z}$-grading $L = L_{-n} \oplus \cdots \oplus L_n$ such that $B = L_n$.

Dealing with the general case [**FLGGLN07**] would involve the use of quite technical results of 3-graded root systems [**LN04**]. According with the spirit of this book, we will restrict ourselves to the particular case that L is nondegenerate and B has finite length, where only basic structure theory of Jordan pairs is needed.

THEOREM 12.16. *Let L be a nondegenerate Lie algebra over a ring of scalars Φ in which $2, 3, 5$ are invertible. Then for every nonzero abelian inner ideal B of L of finite length there exists a finite $\mathbb{Z}$-grading $L = L_{-n} \oplus \cdots \oplus L_n$ such that $L_n = B$, whence B is a complemented inner ideal.*

PROOF. *Suppose first that the ideal of L generated by B is simple.*

Set $V = (B, L/\operatorname{Ker}_L B)$. Then V is a simple Artinian Jordan pair (Proposition 11.47(iv,v)) and therefore isomorphic to one of the Jordan pairs listed in Corollary 11.49. By [**Loo75**, 12.13], Jordan pairs of Hermitian matrices, Clifford pairs and Albert pairs contain invertible elements, while a Jordan pair of skew-symmetric matrices $(\mathrm{A}_n(K), \mathrm{A}_n(K))$ contains invertible elements if and only if n is even.

(a) Suppose that V contains an invertible element $e \in B$. Then e is von Neumann regular and hence, by Theorem 5.11, can be extended to an idempotent (e, f) of L yielding the 5-grading $L^h_{-2} \oplus L^h_{-1} \oplus L^h_0 \oplus L^h_1 \oplus L^h_2$, where $h = [e, f]$ and $L^h_2 = [e] = B$, by the invertibility of e.

(b) Suppose now that V is a Jordan pair of rectangular matrices over a division ring, and let $\{(e_{ij}, \overline{f}_{ij})\}$ be a $n \times m$*-rectangular matrix system* for V, i.e.

(i) each $\{(e_{ij}, \overline{f}_{ij})\}$ is a division idempotent, $1 \le i \le n$, $1 \le j \le m$,

(ii) for $(i, j) \neq (r, s)$, the idempotents $(e_{ij}, \overline{f}_{ij})$, $(e_{rs}, \overline{f}_{rs})$ are orthogonal if $i \neq r$ and $j \neq s$, and collinear otherwise,

(iii) $V = (B, L/\operatorname{Ker}_L B) = \sum_{i,j} V_2(e_{ij}, \overline{f}_{ij})$.

By Preposition 12.15, the family $\{(e_{ij}, \overline{f}_{ij})\}$ can be lifted to a family $\mathcal{E} = \{(e_{ij}, f_{ij})\}$ of Peirce compatible division idempotents of L satisfying the Lie analogue of condition (ii) above and such that B is generated as an inner ideal by the e_{ij}'s. Let $L = \bigoplus_{\alpha \in \mathbb{Z}^{nm}} L_\alpha$ be the finite $\mathbb{Z}^{nm}$-grading given by the family $\mathcal{E}$, and let

$$L = \bigoplus L_r, \quad \text{where } L_r = \bigoplus_{\alpha_{11}+\alpha_{12}\cdots+\alpha_{nm}=r} L_\alpha, \quad \text{for } \alpha = (\alpha_{11}, \alpha_{12}, \dots, \alpha_{nm})$$

be the associated finite $\mathbb{Z}$-grading (Lemma 12.11(ii) for $\lambda_{11} = \cdots = \lambda_{nm} = 1$).

Given e_{ij}, for any pair of indexes $(r, s) \in \{1, \dots, n\} \times \{1, \dots, m\}$, we have:

- $e_{ij} \in V_2^+(e_{rs}, \overline{f}_{rs})$ if $(r, s) = (i, j)$,
- $e_{ij} \in V_0^+(e_{rs}, \overline{f}_{rs})$ if $i \neq r$ and $j \neq s$,
- $e_{ij} \in V_1^+(e_{rs}, \overline{f}_{rs})$ in the $n + m - 2$ remaining cases.

Since $e_{ij} \in V_1^+(e_{rs}, \overline{f}_{rs}) \subset L_1^{h_{rs}}$ occurs exactly in $n + m - 2$ cases, $e_{ij} \in L_{n+m}$ for every (i, j). Hence, by Lemma 12.12, $[e_{ij}] \subset L_{n+m}$, and therefore $B = V^+ \subset L_{2n-2}$ by (iii). Now we claim that $L_r \subset B \subset L_{n+m}$ for any $r \ge n + m$. Given $x \in L_r$, we can suppose that x is homogeneous with respect to each h_{ij}, i.e. $x \in L_\alpha$ with $\alpha = (\alpha_{11}, \alpha_{12}, \dots, \alpha_{nm})$ such that $\alpha_{11} + \alpha_{12} \cdots + \alpha_{nm} = r \ge n + m$. If any of these α_{ij} is equal to 2, then $x \in L_2^{h_{ij}} = [e_{ij}] \subset B \subset L_{n+m}$. Otherwise there exist at least $m + n$ idempotents (e_{ij}, f_{ij}) such that $[h_{ij}, x] = x$; but within any subset of $n + m$ idempotents of a $n \times m$-rectangular matrix system, there are at least two of them, say (e_{ij}, f_{ij}), (e_{rs}, f_{rs}), which are orthogonal. Then, by Lemma 12.13, $x \in L_1^{h_{ij}} \cap L_1^{h_{rs}} \subset B \subset L_{n+m}$. Therefore, $L_{n+m} = B$ and $L_r = 0$ for $r > n + m$. Since L is nondegenerate, L_r does actually vanish for any integer r such that $|r| > n + m$.

(c) Suppose that V is a Jordan pair of skew-symmetric matrices over a field $\mathbb{F}$, where n is odd and greater than 4, and let $\{(e_{ij}, \overline{f}_{ij})\}$ be a symplectic matrix system, $i < j \in \{1, 2, \dots n\}$, of V (see [**Neh87**, II.2.4]), that is

(i) each $(e_{ij}, \overline{f}_{ij})$ is a division idempotent of V,

(ii) two idempotents $(e_{ij}, \overline{f}_{ij})$ and $(e_{rs}, \overline{f}_{rs})$ are orthogonal if and only if $\{i,j\} \cap \{r,s\} = \emptyset$, and collinear if they share exactly one index,

(iii) $V = (B, L/\operatorname{Ker}_L B) = \sum_{i,j} V_2(e_{ij}, \overline{f}_{ij})$.

As in the previous case, the family $\{(e_{ij}, \overline{f}_{ij})\}$ is lifted to a family $\mathcal{E} = \{(e_{ij}, f_{ij})\}$ of Peirce compatible division idempotents of L satisfying the Lie analogue of condition (ii); moreover, B is generated as an inner ideal by the e_{ij}'s. Let $L = \bigoplus_k L_k$, with $L_k = \bigoplus L_\alpha$, be the finite $\mathbb{Z}$-grading defined by the family $\mathcal{E}$, where all $\lambda_{ij} = 1$ (Lemma 12.11(ii)), and fix $i, j \in \{1, \dots, n\}$, $i < j$. By (ii), for any pair (r,s), $r < s$,

- $e_{ij} \in L_2^{h_{rs}}$ if $(i,j) = (r,s)$,
- $e_{ij} \in L_1^{h_{rs}}$ if (i,j) and (r,s) share exactly one index,
- $e_{ij} \in L_0^{h_{rs}}$ otherwise.

Since $e_{ij} \in L_1^{h_{rs}}$ occurs exactly in $2n-4$ cases, $e_{ij} \in L_{2n-2}$ for every (i,j). Hence it follows from Lemma 12.12 that $[e_{ij}] \subset L_{2n-2}$ and by (iii) that $B = V^+ \subset L_{2n-2}$.

Let $x \in L_r$ for $r \geq 2n-2$, which can be supposed to be homogeneous with respect to each h_{ij}, i.e. $x \in L_\alpha$ with $\alpha = (\alpha_{ij})$ such that $\sum \alpha_{ij} = r$. If any of these α_{ij} equals 2, then $x \in L_2^{h_{ij}} \subset B \subset L_{2n-2}$. Otherwise there exist at least $2n-2$ idempotents with $[h_{ij}, x] = x$; but within any subset of $2n-2$ idempotents in a symplectic matrix system of size n there always exist two of them which are orthogonal, and therefore $x \in B \subset L_{2n-2}$ by Lemma 12.13. Then $L_{2n-2} = B$ and $L_r = 0$ for $r > 2n-2$. Since L is nondegenerate, $L_r = 0$ for any integer r such that $|r| > 2n-2$.

(d) Suppose finally that V is a Bi-Cayley pair. Then V contains two collinear idempotents (e, f_1) and (e_2, f_2) such that $B = [e_1, [e_1, L]] \oplus [e_2, [e_2, L]]$. Lift them to two collinear idempotents (e_1, f_1) and $(e_2, f_2)\}$ of L and consider the associated $\mathbb{Z}^2$-grading $L = \bigoplus_{\alpha \in \mathbb{Z}^2} L_\alpha$. This grading gives rise to a 7-grading (Lemma 12.11(ii) for $\lambda_1 = \lambda_2 = 1$) such that $B = L_{(1,2)} \oplus L_{(2,1)} = L_3$.

The general case. By Proposition 11.61, $B = \bigoplus_{i=1}^n B_i$, where each $B_i = B \cap S_i$ for some simple component S_i of $\operatorname{Soc}(L)$ and the subquotient $V_i = (B_i, L/\operatorname{Ker}_L B_i) \cong (B_i, S_i/\operatorname{Ker}_{S_i} B_i)$ is a simple Artinian Jordan pair. As was proved in each one of the particular cases, for every $i = 1, \dots, n$, there exists a finite family of Peirce compatible idempotents $\mathcal{E}_i = \{(e_k^i, f_k^i)\}_{k=1}^{m_i}$ with $e_k^i, f_k^i \in S_i$ and $h_k^i = [e_k^i, f_k^i]$. Let

$$L = \bigoplus L_{\overline{\alpha}^i}, \quad \overline{\alpha}^i = (\alpha_1^i, \dots, \alpha_{m_i}^i) \in \mathbb{Z}^{m_i}, \tag{12.5}$$

where

$$L_{\overline{\alpha}^i} = \bigcap_{k=1}^{m_i} L_{\alpha_k^i}^{h_k^i} = \bigcap_{k=1}^{m_i} \{x \in L \ : \ [h_k^i, x] = \alpha_k^i x\},$$

be the $\mathbb{Z}^{m_i}$-grading given by the family $\mathcal{E}_i$ (Lemma 12.11(i)). Then $B_i = L_{n_i}$ for the $(2n_i+1)$-grading of L determined by (12.5).

Let $\mathcal{E}$ be the family built by the ordered juxtaposition of the families $\mathcal{E}_1, \mathcal{E}_2, \dots, \mathcal{E}_s$. Then $\mathcal{E}$ becomes a family of $m := \sum m_i$ Peirce compatible idempotents of L. Keeping the notation of (12.5), grade L with respect to the family $\mathcal{E}$ as follows:

$$L = \bigoplus L_{\overline{\alpha}}, \quad \overline{\alpha} = (\overline{\alpha}^1, \dots, \overline{\alpha}^s) \in \mathbb{Z}^{m_1} \times \cdots \times \mathbb{Z}^{m_s}, \tag{12.6}$$

where

$$L_{\overline{\alpha}} = \bigcap_{i=1}^{s} L_{\overline{\alpha}^i}.$$

Let n be the least common multiple of $n_1, \ldots, n_s$, $r_i = n/n_i$, and take

$$\lambda_1^i = \cdots = \lambda_{m_i}^i = r_i$$

for each $i = 1, \ldots, s$ (Lemma 12.11(ii)). We have the new $\mathbb{Z}$-grading in L (which we distinguish from the previous ones putting a hat on its components L_i):

$$L = \bigoplus \hat{L}_t, \quad \hat{L}_t = \bigoplus_{(\overline{\alpha}^1,\ldots,\overline{\alpha}^s)\in\Delta_t} L_{(\overline{\alpha}^1,\ldots,\overline{\alpha}^s)}, \tag{12.7}$$

where $\Delta_t = \{(\overline{\alpha}^1, \ldots, \overline{\alpha}^s) : \sum_{i=1}^s r_i(\alpha_1^i + \cdots + \alpha_{m_i}^i) = t\}$.

We claim that $B = \hat{L}_n$. As was seen, for $1 \le i \le n$, $B_i = L_{n_i}$ (the $(2n_i+1)$-grading determined by (12.5)); but, by (12.7), $L_{n_i} \subset \hat{L}_n$. Then $B = \oplus B_i \subset \hat{L}_n$. Conversely, if $x \in \hat{L}_t$ with $t \ge n$, we have two possibilities:

- Every idempotent $(e_k^i, f_k^i) \in \mathcal{E}$ such that $[h_k^i, x] \neq 0$ belongs to the same $\mathcal{E}_i$. In this case $x \in L_{(0,\ldots,\overline{\alpha}^i,\ldots,0)}$ and $t = r_i(\alpha_1^i + \cdots + \alpha_{m_i}^i) \ge n$, so necessarily $\alpha_1^i + \cdots + \alpha_{m_i}^i = n_i$, i.e. $x \in B_i \subset \hat{L}_n$.
- There exist two idempotents in different families, say $(e_k^i, f_k^i) \in \mathcal{E}_i$ and $(e_{k'}^j, f_{k'}^j) \in \mathcal{E}_j$, with $[h_k^i, x] \neq 0 \neq [h_{k'}^j, x]$. In this case, either $[h_k^i, x] = 2x$, so $x \in B_i$ (or $[h_k^j, x] = 2x$, so $x \in B_j$) or $[h_k^i, x] = x = [h_{k'}^j, x]$, which implies by Lemma 12.13 that $x \in B \subset \hat{L}_n$.

Since L is nondegenerate, $\hat{L}_t = 0$ for all $|t| > n$, which completes the proof that L is $(2n+1)$-graded with $B = \hat{L}_n$. □

COROLLARY 12.17. *Let L be a simple Lie algebra over a field $\mathbb{F}$ of characteristic 0. If L contains an abelian inner ideal B whose subquotient is a simple exceptional Jordan pair, then $L \cong \mathrm{TKK}(\mathrm{Sub}_L B)$. Hence, L is of type E_6 or E_7 if the Jordan pair $\mathrm{Sub}_L B$ is Bi-Cayley or Albert.*

PROOF. Since simple exceptional Jordan pairs are Artinian, B has finite lenght. Then, by Theorem 12.16, $B = L_n$ is the extreme of a $(2n+1)$-grading of L for some $n \ge 1$, with associated Jordan pair $V = (L_n, L_{-1})$ isomorphic to $\mathrm{Sub}_L B$ (Proposition 11.39(iii)). Hence $L \cong \mathrm{TKK}(\mathrm{Sub}_L B)$ (Theorem 11.34). As mentioned in Remarks 11.50, L is of type E_6 or E_7 if $\mathrm{Sub}_L B$ is Bi-Cayley or Albert. □

REMARKS 12.18. (1) The above corollary seems to confirm the aphorism of K. McCrimmon [**McC04**]: *Nine times out of ten, when you open up a Lie algebra you find a Jordan algebra inside which makes it tick.*

(2) In a recent paper, E. García and M. Gómez Lozano have proved the following result closely related to the above corollary: Let L be a strongly prime Lie algebra over a ring of scalars Φ in which 2 and 3 are invertible, and let $a \in L$ be a Jordan element. If $\mathrm{Ker}\{a\}$ is not a subalgebra of L, then the Jordan algebra L_a is special. Note that if B is an Albert algebra with a as unit element, then $\mathrm{TKK}(\mathrm{Sub}_L B) \cong L = L_{-1} \oplus L_0 \oplus L_1$ implies (Proposition 11.39) that $\mathrm{Ker}(B) = \mathrm{Ker}\{a\} = L_0 + L_1$ is a subalgebra of L.

12.4. Complemented Lie algebras

A Lie algebra L is said to be complemented if every inner ideal of L is complemented. In this section it is proved that a Lie algebra L over a ring of scalars Φ in which $2, 3, 5$ are invertible is complemented if and only if it is a direct sum of simple nondegenerate Artinian Lie algebras. Thus, as conjectured by G. Benkart in the introduction of [**Ben77**], inner ideals in Lie algebras play a role analogous to Jordan inner ideals in the development of an Artinian theory for Lie algebras.

DEFINITION 12.19. A Lie algebra L is said to be *complemented* if any inner ideal of L is complemented, and it is called *abelian complemented* if every abelian inner ideal is complemented by an abelian inner ideal. As will be seen later, complemented Lie algebras are actually abelian complemented.

PROPOSITION 12.20. *Let L be a complemented Lie algebra and let I be an ideal of L. Then every inner ideal of I is an inner ideal of L and I is a complemented Lie algebra.*

PROOF. Since L is nondegenerate by Proposition 12.7, we have by Lemma 12.4 that $L = I \oplus \mathrm{Ann}(I)$ and $\mathrm{Ann}(I) = \mathrm{Ker}_L I$. Let M be an inner ideal of I. Then

$$[M, [M, L]] = [M, [M, I \oplus \mathrm{Ann}(I)]] = [M, [M, I]] \subset M,$$

which proves that M is an inner ideal of the Lie algebra L. Denote by π the projection of $L = I \oplus \mathrm{Ann}(I)$ onto I. It is easy to see that for any inner ideal P of L we have

$$\mathrm{Ker}_I \pi(P) = I \cap \mathrm{Ker}_L P = \pi(\mathrm{Ker}_L P). \tag{12.8}$$

Let N be a complement of M in L. Then $\pi(N)$ is an inner ideal of I. We claim that $\pi(N)$ is actually a complement of M in I. Indeed, by the Modular Law and (12.8), $L = M \oplus \mathrm{Ker}_L N$ implies

$$I = M \oplus I \cap \mathrm{Ker}_L N = M \oplus \mathrm{Ker}_I \pi(N).$$

And $L = N \oplus \mathrm{Ker}_L M$ implies, again by (12.8),

$$I = \pi(L) = \pi(N) + \pi(\mathrm{Ker}_L M) = \pi(N) + \mathrm{Ker}_I \mathrm{M}.$$

Since $\pi(n) \in \pi(N) \cap \mathrm{Ker}_I M \subset \pi(N) \cap \mathrm{Ker}_L M$, $n \in N \cap \mathrm{Ker}_L M = 0$, so the sum $\pi(N) + \mathrm{Ker}_I \mathrm{M}$ is actually direct. Therefore, $\pi(N)$ is a complement of M in I. □

• Let $V = (V^+, V^-)$ be a Jordan pair. Following [**LN94**], an inner ideal $B \subset V^\sigma$ is said to be *complemented* if there exists an inner ideal $C \subset V^{-\sigma}$ (called a *complement* of B) such that

$$V^\sigma = B \oplus \mathrm{Ker}_V C \text{ and } V^{-\sigma} = C \oplus \mathrm{Ker}_V B.$$

A Jordan pair V is *complemented* if any inner ideal of V is complemented.

PROPOSITION 12.21. *Suppose that L is abelian complemented. Then for every abelian inner ideal M of L the subquotient $V = (M, L/\mathrm{Ker}_L M)$ of L with respect to M is a complemented Jordan pair.*

PROOF. Let B be an inner ideal of V contained in M. Then B is an abelian inner ideal of L and therefore it is complemented by an abelian inner ideal C of L. Let $\pi : L \to L/\mathrm{Ker}_L M$ denote the canonical projection. We claim that $\pi(C) = (C + \mathrm{Ker}_L M)/\mathrm{Ker}_L M$ is a complement of B in V. By Proposition 11.47(ii),

$\pi(C)$ is an inner ideal of V. Furthermore, for any $m \in M$, $m \in \operatorname{Ker}_V \pi(C)$ if and only if

$$[[C, m], C] \subset C \cap \operatorname{Ker}_L M \subset C \cap \operatorname{Ker}_L B = 0.$$

Therefore, $\operatorname{Ker}_V \pi(C) = M \cap \operatorname{Ker}_L C$. Hence, by the Modular Law,

$$M = M \cap (B \oplus \operatorname{Ker}_L C) = B \oplus (M \cap \operatorname{Ker}_L C) = B \oplus \operatorname{Ker}_V \pi(C).$$

On the other hand, $L = C \oplus \operatorname{Ker}_L B$ implies

$$L/\operatorname{Ker}_L M = \pi(C) + \pi(\operatorname{Ker}_L B) \subset \pi(C) + \operatorname{Ker}_V B.$$

But the sum $L/\operatorname{Ker}_L M = \pi(C) + \operatorname{Ker}_V B$ is direct by [**LN94**, Lemma 3.1], since V is nondegenerate by Propositions 12.7 and 11.47(iii).

Let N be an abelian complement of M in L. By Lemma 12.2, we have the isomorphism $(M, L/\operatorname{Ker}_L M) \cong (M, N) \cong (L/\operatorname{Ker}_L N, N)$. Hence one proves as before that any inner ideal of V contained in $L/\operatorname{Ker}_L M$ has a complement. □

PROPOSITION 12.22. *Let L be a strongly prime Lie algebra. If L is complemented, then it satisfies both chain conditions on abelian inner ideals.*

PROOF. Let $\{M_i\}$ be an ascending or descending chain of abelian inner ideals of L. Then $M = \cup M_i$ is also an abelian inner ideal, so we can consider the subquotient $V = (M, L/\operatorname{Ker}_L M)$. Since L is abelian complemented by Lemma 12.5, we have by Propostion 12.21 that V is a complemented Jordan pair. Then V coincides with its socle by [**LN94**, Proposition 5.9]. As V is strongly prime (Proposition 11.47(iii)), V is actually simple. Then, by [**LN94**, Theorem 5.2], V satisfies both chain conditions on inner ideals. Since each M_i is an inner ideal of V, the chain $\{M_i\}$ is stationary. □

LEMMA 12.23. *Let $L = \bigoplus I_\alpha$ be a direct sum of ideals each of which is a simple nondegenerate Artinian Lie algebra. Then any inner ideal B of L is of the form $B = \bigoplus B_\alpha$, where for each index α, either $B_\alpha = I_\alpha$ or B_α is an abelian inner ideal of L contained in I_α.*

PROOF. For each index α, denote by π_α the projection of $L = \bigoplus I_\alpha$ onto I_α. Then $B_\alpha = \pi_\alpha(B)$ is an inner ideal of the Lie algebra I_α. It is clear that $B \subset \bigoplus B_\alpha$. Conversely, if $B_\alpha = I_\alpha$, we have by the simplicity of I_α that

$$B_\alpha = I_\alpha = [I_\alpha, [I_\alpha, I_\alpha]] = [B, [B, I_\alpha]] \subset B.$$

Suppose then that B_α is a proper inner ideal of I_α. Since I_α is a simple nondegenerate Artinian Lie algebra, it follows from Proposition 3.31 that B_α is abelian, so we can consider the Jordan pair $V_\alpha = (B_\alpha, I_\alpha/\operatorname{Ker}_{I_\alpha} B_\alpha)$. As L satisfies the dcc on principal inner ideals within B_α, it follows from Propositions 11.16 and 11.51 that any $b_\alpha \in B_\alpha$ is von Neumann regular in I_α, i.e. there exists $x_\alpha \in I_\alpha$ such that

$$b_\alpha = [b_\alpha, [b_\alpha, x_\alpha]] = [b, [b, x_\alpha]] \in [B, [B, I_\alpha]] \subset B,$$

which completes the proof. □

THEOREM 12.24. *For a Lie algebra L, the following conditions are equivalent:*

(i) *L is complemented,*
(ii) *L is a direct sum of ideals each of which is a simple nondegenerate Artinian Lie algebra.*

In this case, L is abelian complemented.

PROOF. (i) $\Rightarrow$ (ii). Assume that L is complemented and therefore nondegenerate by Proposition 12.7. Regarded L as a left module over its multiplication algebra, Lemma 12.4 implies that L is completely reducible, i.e. $L = \bigoplus I_\alpha$ is a direct sum of ideals each of which is a simple Lie algebra, which is also complemented by Proposition 12.20. Since any proper inner ideal of I_α is abelian by Lemma 12.5, it follows from Proposition 12.22 that I_α is Artinian. Note that this also proves that I_α is abelian complemented.

(ii) $\Rightarrow$ (i). Suppose first that L is a simple nondegenerate Artinian Lie algebra, and let B be an inner ideal of L. If $B = L$ then B is its own complement, so we may suppose that B is proper. Then B is abelian (Proposition 3.31) and has finite length (Proposition 11.47(vi)). So B is one of the extremes of a finite $\mathbb{Z}$-grading of L (Theorem 12.16) and hence complemented by Lemma 12.3.

Consider now the general case that $L = \bigoplus I_\alpha$ is a direct sum of ideals each of which is a simple nondegenerate Artinian Lie algebra, and let B be an inner ideal of L. By Lemma 12.23, $B = \bigoplus B_\alpha$, where each B_α is an inner ideal of I_α. By the simple case we have previously analyzed, each B_α is complemented by an inner ideal C_α of I_α. Then $C = \bigoplus C_\alpha$ is a complement of B.

Suppsoe that L is complemented. If the inner ideal $B = \bigoplus B_\alpha$ is abelian, then any complement C_α of B_α is abelian, and therefore $C = \bigoplus C_\alpha$ is an abelian complement of B, so L is abelian complemented. □

12.5. A unified approach to inner ideals

In this section we show that inner ideals of Jordan pairs and abelian inner ideals of Lie algebras are essentially the same mathematical object. This fact allows us to pass from one of these algebraic structures to the other without changing the intrinsic properties of the inner ideal, when required.

By an *inner ideal* we will mean here an abelian inner ideal of a Lie algebra, or an inner ideal of a Jordan pair.

- Let B and C be inner ideals over the same ring of scalars Φ, i.e, B is an inner ideal of $\mathfrak{B}$ and C is an inner ideal of $\mathfrak{C}$, where $\mathfrak{B}$ and $\mathfrak{C}$ are Lie algebras of Jordan pairs (even one of them can be a Lie algebra and the other a Jordan pair) over the same ring of scalars Φ. We will say that these inner ideals B, C are *Jordan isomorphic* if so are its respective subquotients, i.e. if $\mathrm{Sub}_{\mathfrak{B}}\ B \cong \mathrm{Sub}_{\mathfrak{C}}\ C$ as Jordan pairs. An interesting example of Jordan isomorphism which takes place in a kind of inner ideals called point spaces will be studied in Section 13.3.
- By the *geometry* of an inner ideal B we mean the *inner structure* of B, i.e. the lattice of the inner ideals contained in B plus *orthogonality relations*: $x \perp y$ means $\mathrm{ad}_x \mathrm{ad}_y = 0$ (equivalently, $\mathrm{ad}_y \mathrm{ad}_x = 0$) if Lie; $Q_{x,y} = 0$ if Jordan. $\mathrm{ad}_x \mathrm{ad}_y = 0$ if Lie, $Q_{x,y} = 0$ if Jordan.

FACT 12.25. *Any abelian inner ideal of a Lie algebra can be regarded as an inner ideal of a Jordan pair, and conversely, any inner ideal of a Jordan pair can be realized as an abelian inner ideal of a Lie algebra. Moreover, in both directions the inner ideal keeps up its geometry.*

PROOF. Let B be an abelian inner ideal of a Lie algebra L. Then $B = V^+$ where $V = \mathrm{Sub}_L\, B$, the inner ideals of L contained in B are the inner ideals of V

contained in V^+ (Proposition 11.47(i)), and for any couple of elements $x, y \in B$ and any a in L, $\{x, \overline{a}, y\} = [[x, a], y]$, so $x \perp_L y$ if and only if $x \perp_V y$.

Suppose now that M is an inner ideal of a Jordan pair V. Then M is an abelian inner ideal of the Lie algebra T contained in T_1, where $T = T_{-1} \oplus T_0 \oplus T_1$ is the TKK-algebra of the Jordan pair V. Since $T_1 = V^+$ we have by the grading properties that the inner ideals of V contained in V^+ are precisely the (abelian) inner ideals of T contained in T_1, and for $x, y \in M \subset T_1$ and $a \in T$, we have $[x, [y, a]] = [x, [y, a_{-1}]] = -\{y, a_{-1}, x\}$. Thus $x \perp_V y$ if and only if $x \perp_T y$. □

Recall (Proposition 11.47(iii)) that if L is nondegenerate, then the Jordan pair $\mathrm{Sub}_L B$ is nondegenerate, and that if V is nondegenerate, so is the Lie algebra $\mathrm{TKK}(V)$ (Propostion 11.25(4)).

FACT 12.26. *An abelian inner ideal B living in a large house L can move to a flat T*

$$B \to V = \mathrm{Sub}_L B \to T = \mathrm{TKK}(V) = T_{-1} \oplus T_0 \oplus T_1,$$

which fits better with its real necessities. And in its new home T, the abelian inner ideal $B = T_1$ will no longer be a single: it will find a partner, the left extreme T_{-1} of the grading.

PROOF. Let $B = T_1$, where $T = T_1 \oplus T_0 \oplus T_{-1}$ is the TKK-algebra of the Jordan pair $\mathrm{Sub}_L B$. By Lemma 12.3, B is complemented by the abelian inner ideal T_{-1}. □

WARNING 12.27. In spite of what we have just said, we must be careful when using the transference from Lie to Jordan or from Jordan to Lie, because not everything keeps up (something is gone with the wind!). Let us see an example: *Does the following argument show that any Jordan element of a Lie algebra is von Neumman regular?* (We remind the reader of the fact (Exercise 5.43) that this assertion is not true.) Let a be a Jordan element of a Lie algebra L and set $B = \mathrm{ad}_a^2 L$. Then B is a complemented abelian inner ideal of T. Hence, by Proposition 12.6, a is von Neumman regular in T and therefore also in L. *What is wrong in this argument it is the fact that while B is a principal inner ideal in L, it does not remain as a principal inner ideal in T: the element $a \in L$ does not necessarily belong to T.*

Inner ideals with principal chain condition.

LEMMA 12.28. *Let $\{B_\alpha\}$ be a family of inner ideals of H (where H denotes a Lie algebra or a Jordan pair). If $B_\alpha \perp B_\beta$ for any couple of indexes α, β with $\alpha \neq \beta$, then the sum $B = \sum_\alpha B_\alpha$ is an inner ideal of H. Moreover, if H is nondegenerate, then this sum is direct.*

PROOF. To unify notation, for $b, c \in B$ and $x \in H$, set $[[b, x], c] := \{b, x, c\}$. Since the inner ideals B_α are mutually orthogonal, we have

$$\{B, H, B\} = \sum_\alpha \{B_\alpha, H, B_\alpha\} \subset \sum_\alpha B_\alpha = B.$$

Suppose now that H is nondegenerate and let $b \in B_\alpha \cap \sum_{\beta \neq \alpha} B_\beta$. Then

$$\{b, H, b\} \subset \sum_{\beta \neq \alpha} \{B_\alpha, H, B_\beta\} = 0,$$

so $b = 0$ by nondegeneracy of H. □

THEOREM 12.29. *Let B be an abelian inner ideal of a nondegenerate Lie algebra L. If B satisfies principal dcc, then $B = \bigoplus B_\alpha$ is an orthogonal sum of inner ideals B_α, and for each index α, $B_\alpha = W_\alpha^+$, where W_α is a simple Jordan pair coinciding with its socle and therefore one of those described in Theorem 11.19. A similar result holds for inner ideals of a nondegenerate Jordan pair.*

PROOF. Let V be the subquotient of L with respect to B. Since $V^+ = B$ and B has dcc on principal inner ideals, $V^+ = \mathrm{Soc}(V^+)$ by Propostion 11.16. Hence $\mathrm{Soc}(V) = \bigoplus W_\alpha$, where each W_α is a simple Jordan pair coinciding with its socle and $B = \bigoplus B_\alpha = \bigoplus W_\alpha^+$. □

12.6. Exercises

EXERCISE 12.30. Prove Lemma 12.11.

EXERCISE 12.31. Let A be a ring. A left ideal L of A will be called *annihilator complemented* if there exists a right ideal R of A such that

$$A = L \oplus \mathrm{lann}(R) = R \oplus \mathrm{rann}(L).$$

Show that if $e \in A$ is an idempotent, then the left ideal Ae is annihilator complemented.

EXERCISE 12.32. Let A be a simple ring. Show that the following conditions are equivalent:

(i) Any left ideal L of A is annihilator complemented,
(ii) $A = \mathrm{Soc}(A)$ and $L = \mathrm{lann}(\mathrm{rann}(L))$ for any left ideal L of A,
(iii) $A = \mathcal{F}(X)$, where X is a left vector space over a division ring Δ,
(iv) Any left ideal L of A is of the form $L = Ae$ for some idempotent $e \in Q_s(A)$.

EXERCISE 12.33. Let $(X, Y, \langle\,,\rangle)$ be a pair of dual vector spaces over a division ring Δ and let $\{x_i\}_{i=1}^4 \subset X$, $\{y_i\}_{i=1}^4 \subset Y$ be such that $\langle x_i, y_j\rangle = \delta_{ij}$. Set

$$e_{12} = (y_1^* x_2, y_2^*, x_1),\ e_{13} = (y_1^* x_3, y_3^*, x_1),\ e_{34} = (y_3^* x_4, y_4^*, x_3).$$

Show:

(1) e_{12}, e_{13}, e_{34} are idempotents of the Lie algebra $\mathfrak{fsl}_Y(X)$,
(2) $e_{12} \perp e_{34}$ and $e_{12} \top e_{13}$.

EXERCISE 12.34. As in the preceding exercise, find examples of orthogonal and collinear idempotentes in the Lie algebras $\mathfrak{fo}(X, \langle\,,\rangle)$ and $\mathfrak{fsp}(X, \langle\,,\rangle)$.

EXERCISE 12.35. In the Lie algebra $L = \mathfrak{sl}_4(\mathbb{F})$, where $\mathbb{F}$ is a field of characteristic 0 or $p > 3$. Show that the subspace $B = \mathbb{F}[12] + \mathbb{F}[13]$ is an abelian inner ideal and defines a 7-grading on L such that $B = L_3$.

EXERCISE 12.36. Let e and f be orthogonal idempotents of a unital ring R. Show:

(i) eRf is an abelian inner ideal of the Lie algebra $L = R^-$ contained in $[R, R]$.
(ii) If $e + f = 1$, then L has a 3-grading such that $eRf = L_1$.
(iii) If $e + f \neq 1$, then L has a 5-grading such that $eRf = L_2$.

EXERCISE 12.37. Suppose that R is a unital ring with involution $*$ and that $e \in R$ is an idempotent such that e and e^* are orthogonal. Show:

(i) e^*Ke is an abelian inner ideal of the Lie algebra $K = \mathrm{Skew}(R, *)$.
(ii) If $e + e^* = 1$, then K admits a 3-grading such that $e^*Ke = K_1$.
(iii) If $e + e^* \neq 1$, then $e^*Ke = K_2$ for some 5-grading of K.

EXERCISE 12.38. Following the notation in Exercises 2.99 and 11.80, let e_0, e_1, and e_2 be the projections of $X = H \oplus H^\perp = \mathbb{F}x \oplus \mathbb{F}y \oplus H^\perp$ onto $\mathbb{F}x$, $H^\perp$, and $\mathbb{F}y$ respectively. Show:

(i) $e_0, e_1, e_2 \in \mathcal{L}_X(X)$ with $e_0^* = e_2$ and $e_1^* = e_1$.
(ii) These projections induce a 5-grading in the ring $R = \mathcal{F}_X(X)$, with

$$R_k = \sum_{i-j=k} e_i R e_j, \ k = \pm 2, \pm 1, 0.$$

(iii) Each R_k is $*$-invariant with $\mathrm{Skew}(R_2, *) = \mathrm{Skew}(R_{-2}, *) = 0$.
(iv) $K = \mathfrak{fo}(X, \langle\,,\rangle)$ admits a 3-grading, with $K_i = \mathrm{Skew}(R, *)$, $i = \pm 1, 0$.
(v) $K_1 = [x, H^\perp]$, $K_0 = \mathbb{F}[x, y] + [H^\perp, H^\perp]$, and $K_{-1} = [y, H^\perp]$.
(vi) $\mathfrak{fo}(X, \langle\,,\rangle) \cong \mathrm{TKK}(V)$.
(vii) The abelian inner ideals $[x, H^\perp]$ and $[y, H^\perp]$ of $\mathfrak{fo}(X, \langle\,,\rangle)$ are complemented by each other.

EXERCISE 12.39. Following Exercise 11.79, consider the pair of dual vector spaces $(V'^\perp, V^\perp)$. Noting that W is a finite-dimensional subspace of $V^\perp$, show:

(1) There exists $W' \leq V'^\perp$ such that (W', W) is a pair of dual vector spaces.
(2) The abelian inner ideals $B = W^*V$ and $C = V'^*W'$ of $\mathfrak{fo}(\langle\,,\rangle)$ are complemented by each other.

EXERCISE 12.40. Keeping on with the question of the above exercise, I propose taking an algebraic approach. Set $R = \mathcal{F}_Y(X)$. Show:

(1) $B = eRf$, where e and f are idempotent of R such that $fe = 0$.
(2) Replacing f by $g = f - ef$ we get two orthogonal idempotents such that $B = eRg$.
(3) Show that $C = gRe$ is an abelian inner ideal and that B and C are complemented by each other.

CHAPTER 13

Inner Ideal Structure of Lie Algebras

In this chapter we study the inner ideal structure of the following types of Lie algebras: Lie algebras coming from a centrally closed prime ring of characteristic 0 or greater than 3 (Section 13.1); Lie algebras of skew-symmetric elements of a centrally closed prime ring R with involution of characteristic 0 or greater than 5 ($\operatorname{char}(R) \neq 2, 3$ if R is simple) (Section 13.2); and exceptional simple Lie algebras over an algebraically closed field of characteristic 0 (Section 13.5). Since, as will be seen in the next chapter, every simple $(2n+1)$-graded Lie algebra of characteristic 0 or greater than $4n+1$ lies in one of the three types listed above, our results determine in fact the abelian inner ideals of the simple Lie algebras satisfying some kind of finiteness condition or containing a nonzero von Neumann regular element. Other topics we deal with in this chapter appear in Section 13.3, where an interesting sort of abelian inner ideals, called point spaces and consisting of extremal elements, are described in finitary orthogonal algebras; and in Section 13.4, where the study of the abelian inner ideals of Lie algebras of skew-symmetric elements is sharpened in the case of simple rings with involution and minimal one-sided ideals.

13.1. Lie inner ideals of prime rings

Following [**FL14**], we study in this section the abelian inner ideals of the Lie algebra R^-, where R is a centrally closed prime ring of characteristic 0 or greater than 3, thus extending the work of G. Benkart [**Ben76**] for simple Artinian rings, which is here revisited. New abelian inner ideals, which we call special, occur when making this extension. A necessary and sufficient condition for R^- to have a special inner ideal is the existence of an element $x \in R$ such that $x^2 = 0$ and x is not von Neumann regular. For this reason, special inner ideals do not appear in the context of simple Artinian rings.

Standard inner ideals. Throughout this subsection, R will be a semiprime associative algebra over a ring of scalars Φ in which 6 is invertible, Z the center of R, L a subalgebra of the Lie algebra R^-, and $\{a, b, c\} = abc + cba$ the Jordan triple product of $a, b, c \in R$.

LEMMA 13.1. *Let U be a Φ-submodule of R such that $U^2 = 0$. Then the following conditions are equivalent:*

(i) *U is an abelian inner ideal of L,*
(ii) *U is an inner ideal of L,*
(iii) $\{U, L, U\} \subset U$.

PROOF. $U^2 = 0$ implies $[U, U] = 0$, so (i) $\Leftrightarrow$ (ii), and

$$[[u, x], v] = uxv + vxu = \{u, x, v\},$$

for all $u, v \in U$ and $x \in L$, which proves the equivalence of (ii) and (iii). $\square$

DEFINITION 13.2. An inner ideal U of L such that $U^2 = 0$ will be called *isotropic*. As noted above, isotropic inner ideals are abelian. In fact, they are the keystone in the construction of the two types of abelian inner ideals of the Lie algebra R^-.

PROPOSITION 13.3. *Every principal inner ideal of R^- is isotropic.*

PROOF. Let $x \in R$ be a Jordan element of the Lie algebra R^-. It follows from Proposition 3.35(1) that $(\mathrm{ad}_x^2 R)^2 = 0$. □

DEFINITION 13.4. An abelian inner ideal B of L is said to be *standard* if

$$B = U + \Omega,$$

where U is an isotropic inner ideal of L and Ω is a Φ-submodule of $Z \cap L$.

NOTATION 13.5. Given an abelian inner ideal B of L, we set

$$\mathrm{U}(B) = \{x \in B + Z : \ x^2 = 0\}.$$

LEMMA 13.6. *Let B be an abelian inner ideal of L. We have:*

(i) *If $B \subset \mathrm{U}(B) + Z$, then $\mathrm{U}(B)$ is an isotropic inner ideal of L with*

$$\{\mathrm{U}(B), L, \mathrm{U}(B)\} = [[\mathrm{U}(B), L], \mathrm{U}(B)] \subset B.$$

(ii) *If in addition $\mathrm{U}(B) \subset B$, then $B = \mathrm{U}(B) \oplus (B \cap Z)$ is a standard inner ideal of L.*

PROOF. (i) It is clear that $B+Z$ is an abelian subalgebra of L. Set $U := \mathrm{U}(B)$. If $B \subset U + Z$, then $U + U \subset B + Z \subset U + Z$. Thus for any $u, v \in U$ there exists $w \in U$ and $z \in Z$ such that $u+v = w+z$, with $z = 0$ since $u+v-w$ is a nilpotent central element and R is semiprime. This proves that $U + U \subset U$, and since U is clearly invariant under Φ, U is a Φ-submodule of R. Then, for any $u, v \in U$, $2uv = (u+v)^2 = 0$ implies $uv = 0$. Hence, for any $x \in L$,

$$\{u, x, v\} = uxv + vxu = [[u, x], v] \in [[B + Z, L], B + Z] = [[B, L], B] \subset B,$$

with $[[u, x], v]^2 = (uxv + vxu)^2 = 0$. This proves that U is an isotropic inner ideal of L satisfying $\{U, L, U\} \subset B$.

(ii) Suppose in addition that $\mathrm{U}(B) \subset B$. Then the Modular Law applied to the inclusion $B \subset \mathrm{U}(B) + Z$ yields the equality $B = \mathrm{U}(B) + (Z \cap B)$, where the sum is direct since R is semiprime. □

THEOREM 13.7. *Let B be an abelian inner ideal of L. Then B is standard if and only if the following condition, denoted by* (ST), *holds:*

$$\mathrm{U}(B) \subset B \subset \mathrm{U}(B) + Z.$$

PROOF. By Lemma 13.6, condition (ST) is sufficient for B to be standard. Suppose then that $B = U + \Omega$, where U is an isotropic inner ideal of L and Ω a Φ-submodule of $Z \cap L$. It is clear that $U \subset \mathrm{U}(B)$, and by the Modular Law, $\mathrm{U}(B) \subset B + Z \subset U \oplus Z$ implies $\mathrm{U}(B) = U$, which proves that B satisfies (ST). □

PROPOSITION 13.8. *Let π be the canonical epimorphism of R^- onto $\overline{R} = R^-/Z$. If B is an abelian inner ideal of R^- such that $\pi(b)$ is von Neumann regular in $\overline{R}$ for every $b \in B$, then B is a standard inner ideal of R^-.*

PROOF. Set $U = \bigcup_{b\in B} \operatorname{ad}_b^2 R$. Since B is an inner ideal of R^-, we have that $U \subset B$. Moreover, by Proposition 3.35(1), $u^2 = 0$ for every $u \in U$, which proves that $U \subset \mathrm{U}(B)$. Now let $b \in B$. Von Neumann regularity of $\pi(b)$ in $\overline{R}$ implies that there exist $x \in R$ and $z \in Z$ such that $b = [[b,x],b] + z$. Therefore,

$$B \subset U + Z \subset \mathrm{U}(B) + Z.$$

Thus, by Theorem 13.7, we only need to verify that $\mathrm{U}(B) \subset B$. But, by Lemma 13.6(i), this is a direct consequence of the von Neumann regularity of every element of $\mathrm{U}(B)$ in R, which we will prove now.

Let $u = b + w \in \mathrm{U}(B)$, where $b \in B$, $w \in Z$ and $u^2 = 0$. Consider the isomorphisms $R_u^- \cong \overline{R}_{\pi(u)} = \overline{R}_{\pi(b)}$ given in Lemma 8.45. Since $\pi(b)$ is von Neumann regular in $\overline{R}$, the Jordan algebra $\overline{R}_{\pi(u)}$ is unital by Proposition 8.61(i), and therefore so is $R_u^+ \cong R_u^-$. This fact, together with the semiprimeness of R, proves that u is von Neumann regular in R: if $\overline{v}$ is the unit element of R_u^+, then for any $x \in R$ we have $(u - uvu)x(u - uvu) = u(x - vux - xuv + vuxuv)u = 0$, whence $u = uvu$ since R is semiprime. □

LEMMA 13.9. *Let R be a simple ring of characteristic 0 or $p > 3$ and let B be an abelian inner ideal of the Lie algebra $R' = [R,R]$. We have:*

(i) *$\mathrm{U}(B)$ is an isotropic inner ideal of R^-.*
(ii) *If R is von Neumann regular, then B is a standard inner ideal of R^-.*

PROOF. Since B is an abelian inner ideal of R', for any $b \in B$, we have

$$\operatorname{ad}_b^4 R \subset \operatorname{ad}_b^3 [R,R] = 0.$$

Then, by Lemma 4.37, there exists $z \in Z$ such that $(b-z)^2 = 0$, so $B \subset \mathrm{U}(B) + Z$. Hence, by Lemma 13.6(i), $\mathrm{U}(B)$ is an isotropic inner ideal of R'. Let u be a nonzero element in $\mathrm{U}(B)$. By simplicity of R, we have $R = RuR$ and hence

$$[[u,R],u] = uRu = u[Ru,R]u \subset uR'u = [[u,R'],u] \subset \mathrm{U}(B),$$

which proves that $\mathrm{U}(B)$ is actually an isotropic inner ideal of R^-.

Suppose now that R is von Neumann regular. Then if follows from Lemma 13.6(i) that $\mathrm{U}(B) = \{\mathrm{U}(B), R, \mathrm{U}(B)\} \subset B$, which implies by Lemma 13.6(ii) that $B = \mathrm{U}(B) \oplus (Z \cap B)$ is a standard inner ideal of R^-. □

Special inner ideals. Throughout this subsection, R will denote a unital semiprime associative algebra over a field $\mathbb{F}$ of characteristic not 2, Z the center of R, and L a subalgebra of the Lie algebra R^-. We will show how to construct abelian inner ideals of L which are not standard.

THEOREM 13.10. *Let U be an isotropic inner ideal of L and let $f : U \to \mathbb{F}1$ be a nonzero linear map such that $[[U,L],U] \subset \ker(f)$. Then the vector space*

$$\operatorname{Inn}(U,f) := \{u + f(u) : u \in U\}$$

is an abelian inner ideal of L which is not standard.

PROOF. Set $B = \operatorname{Inn}(U,f)$. We have:

(1) B is an abelian inner ideal of L.

Indeed, $[[B,L],B] = [[U,L],U] \subset \ker(f) \subset B$ since $u = u + f(u) \in U \cap B$ for any $u \in \ker(f)$, and $[B,B] \subset [U,U] \subset U^2 = 0$.

(2) $U \cap B = \ker(f)$.

As noted in (1), $\ker(f) \subset U \cap B$. Conversely, let $v \in U \cap B$, $v = u + f(u)$ for some $u \in U$. Then $v - u = f(u) \in U \cap Z = 0$, so $v = u \in \ker(f)$.

(3) $\mathrm{U}(B) = U$.

By definition, $U \subset B + Z$, and since $U^2 = 0$, we have $U \subset \mathrm{U}(B)$. Conversely, let $b + z \in \mathrm{U}(B)$, $b = u + f(u)$. Then

$$0 = (b+z)^2 = (u + (f(u)+z))^2 = 2(f(u)+z)u + (f(u)+z)^2$$

implies $f(u) + z = 0$ since the sum $Z + U$ is direct, so $b + z = u \in U$.

(4) B is not standard.

By Theorem 13.7 it is enough to see that $\mathrm{U}(B) = U$ is not contained in B. Suppose otherwise that $U \subset B$. Then we would have by (2) that $U = U \cap B = \ker(f)$, which is a contradiction since $f \neq 0$. □

DEFINITION 13.11. An abelian B of L will be called *special* if $B = \mathrm{Inn}(U, f)$ for some isotropic inner ideal U of L and a nonzero linear functional of U such that $[[U, L], U] \subset \ker(f)$.

EXAMPLE 13.12. Let x be an element of R which is not von Neumann regular and such that $x^2 = 0$. Set $U = \mathbb{F}x \oplus xRx$ and let $f : U \to \mathbb{F}1$ be the linear map defined by the conditions $f(xRx) = 0$ and $f(x) = 1$. Then $B = \{u + f(u) : u \in U\}$ is a special inner ideal of R^-.

PROOF. From $x^2 = 0$ it follows that U is an isotropic inner ideal of R^- with $[[U, R], U] \subset xRx = \ker(f)$. Hence, by Theorem 13.10, $B = \mathrm{Inn}(U, f)$ is a special inner ideal of R^-. □

There is no difference between standard and special inner ideals from the Jordan point of view.

PROPOSITION 13.13. *Let $B = \mathrm{Inn}(U, f)$ be a special inner ideal of R^-. Then the subquotients of the Lie algebra R^- with respect to the abelian inner ideals B and U are isomorphic.*

PROOF. Set $L = R^-$. It is clear that $\mathrm{Ker}_L U = \mathrm{Ker}_L B$. Denote this vector subspace of L by K. Then the pair of linear maps $(\varphi, 1_{L/K}) : (U, L/K) \to (B, L/K)$, where $\varphi : U \to B$ is defined by $\varphi(u) = u + f(u)$, $u \in U$, and $1_{L/K}$ is the identity map on the vector space L/K, is a Jordan pair isomorphism of $\mathrm{Sub}_L U$ onto $\mathrm{Sub}_L B$: It is clear that φ is onto, and $\varphi(u) = 0$ implies $u = -f(u)$, so $u \in U \cap Z = 0$. Moreover, as $[[U, R], U] \subset \ker(f)$, we have

$$\begin{aligned}\varphi([[u, \bar{a}], v]) &= [[u, a], v] + f([[u, a], v]) = [[u, a], v]\\ &= [[u + f(u), a], v + f(v)] = [[\varphi(u), \bar{a}], \varphi(v)]\end{aligned}$$

for $u, v \in U$, $a \in R$; and

$$\{\bar{a}, u, \bar{b}\} = \overline{[[a, u], v]} = \overline{[[a, u + f(u)], b]} = \{\bar{a}, \varphi(u), \bar{b}\}$$

for $a, b \in R$, $u \in U$. □

Centrally closed prime rings. Throughout this subsection, R will be a (not necessarily unital) centrally closed prime ring of characteristic 0 or greater than 3, with Z denoting its center, and $\mathbb{F}$ the centroid (equals the extended centroid) of R, so R can be regarded as an algebra over the field $\mathbb{F}$. Any simple ring is centrally closed. We describe the abelian inner ideals of R^-.

Abelian inner ideals of R^- are *semi-standard*, in the sense that they hold the second inclusion of the (ST)-condition.

LEMMA 13.14. *Let B be an abelian inner ideal of R^-. Then $B \subset \mathrm{U}(B) + Z$. Therefore B will be standard if and only $\mathrm{U}(B) \subset B$.*

PROOF. It follows from Theorems 4.35 and 13.7. □

PROPOSITION 13.15. *Let B be an abelian inner ideal of R^-. If any of the following conditions holds, then B is standard:*

(i) *R is nonunital,*
(ii) *every element of square 0 in R is von Neumann regular,*
(iii) *$Z \subset B$,*
(iv) *B is a maximal abelian inner ideal.*

In case (i), *$B = \mathrm{U}(B)$ is in fact an isotropic inner ideal.*

PROOF. (i) If R is nonunital, then $Z = 0$ by Lemma 1.22. Hence $\mathrm{U}(B) \subset B$ and B is in fact isotropic by Lemma 13.14.

(ii) By Lemma 13.14, $B \subset \mathrm{U}(B) + Z$. Hence, if every element of square 0 in R is von Neumann regular, it follows from Lemma 13.6(i):

$$\mathrm{U}(B) = \{\mathrm{U}(B), R, \mathrm{U}(B)\} \subset B,$$

which proves that B is standard by Lemma 13.6(ii).

(iii) The inclusion $Z \subset B$ implies $\mathrm{U}(B) \subset B$. Now Lemma 13.14 applies.

(iv) It follows from (iii) since $B + Z$ is an abelian inner ideal and therefore equal to B by maximality of B. □

COROLLARY 13.16. *Let X be a left vector space over a division ring Δ of characteristic not 2 or 3 and set $R = \mathrm{End}_\Delta(X)$. Then every abelian inner ideal of R^- (regarded as a Lie algebra over $Z(\Delta)$) is standard.*

PROOF. The ring R is prime (in fact, primitive) and von Neumann regular. Moreover, it is centrally closed over $Z(\Delta)$ by [**BMM96**, Theorem 4.3.7(ix)]. Thus we can apply Proposition 13.15(ii) to get that any abelian inner ideal of R^- is standard. □

THEOREM 13.17. *Let B be an abelian inner ideal of R^-. Then one of the following possibilities holds:*

(i) *B is standard: $B = U$ or $B = U \oplus Z$, where U is isotropic,*
(ii) *R is unital and $B = \mathrm{Inn}(V, f)$ is special.*

PROOF. If R is not unital, then every abelian inner ideal of R^- is isotropic by Proposition 13.15. Suppose then that R is unital (therefore $Z = \mathbb{F}1$). If $1 \in B$, then Z is contained in B and we have by Proposition 13.15(iii) that $B = U \oplus Z$ is standard. Consider finally the case that $1 \notin B$. Then, again by Proposition 13.15(iii), we have that $B \oplus Z = U \oplus Z$ for some isotropic inner ideal U of R. Given $u \in U$, suppose that $u + z$ and $u + z'$ are in B for $z, z' \in Z$. Then $z = z'$ (since otherwise 1 would belong to B, what has been discarded). This proves that there exists a linear functional f of the $\mathbb{F}$-vector space U such that

$$B = \{u + f(u) : u \in U\}.$$

If $f = 0$, then $B = U$ and we have case (i). Suppose then that f is nonzero and let $u \in U$ be such that $f(u) = 1$. Then $U = \mathbb{F}u \oplus \ker(f)$. Clearly, $U \cap B = \ker(f)$, and since $[[U, R], U] = [[B, R], B]$, we have $[[U, R], U] \subset U \cap B = \ker(f)$. Thus $B = \mathrm{Inn}(U, f)$ is special, which completes the proof. □

We end this subsection with an example of a primitive associative algebra R which is not centrally closed but with the property that all abelian inner ideals of R^- are standard.

EXAMPLE 13.18. Let X be an infinite-dimensional complex vector space, let $\mathcal{F}(X)$ be the simple complex associative algebra of all finite rank operators on X, and set $R = \mathcal{F}(X) \oplus \mathbb{R}1_X$. Then R is a primitive associative algebra over the real field $\mathbb{R}$ which is not centrally closed (its extended centroid is the complex field). However, every abelian inner ideal B of R^- is standard. In fact, either $B = U$, where U is an isotropic inner ideal, or $B = \mathbb{R}1_X \oplus U$.

Simple rings. Following a joint work with A. A. Baranov [**BFL16**], we characterize in this subsection the maximal isotropic inner ideals of the Lie algebra R^- of a simple ring R.

DEFINITION 13.19. By an *inner ideal* of a ring R we mean an additive subgroup M of R such that $MRM \subset M$. An inner ideal U of R will be called a *regular inner ideal* if $U^2 = 0$. Note that any inner ideal of R is an inner ideal of the Jordan algebra R^+, and that any regular inner ideal of R is an isotropic inner ideal of the Lie algebra R^-.

LEMMA 13.20. *For an additive subgroup U of a ring R the following conditions are equivalent:*

(i) *U is a regular inner ideal of R.*
(ii) *There exist a left ideal $\mathcal{L}$ and a right ideal $\mathcal{R}$ of R such that*

$$\mathcal{R}\mathcal{L} \subset U \subset \mathcal{R} \cap \mathcal{L} \quad \textit{and} \quad \mathcal{L}\mathcal{R} = 0.$$

PROOF. (i) $\Rightarrow$ (ii). Taking $\mathcal{L} = U + RU$ and $\mathcal{R} = U + UR$, it is easily seen that (ii) holds.

(ii) $\Rightarrow$ (i). Clearly, $U^2 \subset \mathcal{L}\mathcal{R} = 0$ and $URU \subset \mathcal{R}R\mathcal{L} \subset \mathcal{R}\mathcal{L} \subset U$. □

NOTATION 13.21. We have a Galois connection between the lattice $\mathcal{I}_r(A)$ of all right ideals of a ring R and the lattice $\mathcal{I}_l(A)$ of all left ideals of R given by

$$\mathcal{R} \mapsto \mathrm{lann}(\mathcal{R}), \quad \mathcal{L} \mapsto \mathrm{rann}(\mathcal{L}),$$

where $\mathrm{lann}(\mathcal{R})$ ($\mathrm{rann}(\mathcal{L})$) denotes the left annihilator (right annihilator) of $\mathcal{R}$ ($\mathcal{L}$). We set $\overline{\mathcal{L}} := \mathrm{lann}(\mathrm{rann}(\mathcal{L}))$ and $\overline{\mathcal{R}} := \mathrm{rann}(\mathrm{lann}(\mathcal{R}))$ to denote the corresponding closures of $\mathcal{L}$ and $\mathcal{R}$.

DEFINITION 13.22. By an *orthogonal pair* of a ring R we mean a pair $(\mathcal{R}, \mathcal{L})$, where $\mathcal{R}$ is a nonzero right ideal and $\mathcal{L}$ is a nonzero left ideal of R such that $\mathcal{L}\mathcal{R} = 0$.

LEMMA 13.23. *For an orthogonal pair* $(\mathcal{R},\ \mathcal{L})$ *the following conditions are equivalent:*

(i) $\mathcal{R} = \overline{\mathcal{R}}$ *and* $\mathcal{L} = \operatorname{lann}(\mathcal{R})$,
(ii) $\mathcal{R} = \operatorname{rann}(\mathcal{L})$ *and* $\mathcal{L} = \operatorname{lann}(\mathcal{R})$,
(iii) $\mathcal{L} = \overline{\mathcal{L}}$ *and* $\mathcal{R} = \operatorname{rann}(\mathcal{L})$.

PROOF. It suffices to prove that (i)⇔(ii). The proof of (ii)⇔(iii) is similar. Suppose that $\mathcal{L} = \operatorname{lann}(\mathcal{R})$. Then $\operatorname{rann}(\mathcal{L}) = \operatorname{rann}(\operatorname{lann}(\mathcal{R}) = \overline{\mathcal{R}}$. □

NOTATION 13.24. We say that $(\mathcal{R}_1,\ \mathcal{L}_1) \subset (\mathcal{R}_2,\ \mathcal{L}_2)$ if and only if $\mathcal{R}_1 \subset \mathcal{R}_2$ and $\mathcal{L}_1 \subset \mathcal{L}_2$. This gives a partial order on the set of the orthogonal pairs of the ring R.

PROPOSITION 13.25. *Let* $(\mathcal{R},\ \mathcal{L})$ *be an orthogonal pair of a ring* R. *We have:*

(i) $(\overline{\mathcal{R}},\ \operatorname{lann}(\mathcal{R}))$ *and* $(\operatorname{rann}(\mathcal{L}),\ \overline{\mathcal{L}})$ *are maximal orthogonal pairs.*
(ii) *Any orthogonal pair is contained in a maximal one.*
(iii) $(\mathcal{R},\ \mathcal{L})$ *is maximal if and only if it satisfies the equivalent conditions of* Lemma 13.23.

PROOF. (i) Since $\mathcal{R} \subset \overline{\mathcal{R}}$ and $\mathcal{L} \subset \operatorname{lann}(\mathcal{R})$, both $\overline{\mathcal{R}}$ and $\operatorname{lann}(\mathcal{R})$ are nonzero; and since $\operatorname{lann}(\mathcal{R}) = \operatorname{lann}(\overline{\mathcal{R}})$, we have that $(\overline{\mathcal{R}},\ \operatorname{lann}(\mathcal{R}))$ is an orthogonal pair. Suppose now that $(\overline{\mathcal{R}},\ \operatorname{lann}(\mathcal{R}))$ is contained in an orthogonal pair $(\mathcal{R}',\ \mathcal{L}')$. Then $\overline{\mathcal{R}} \subset \mathcal{R}'$ implies $\operatorname{lann}(\mathcal{R}') \subset \operatorname{lann}(\overline{\mathcal{R}}) = \operatorname{lann}(\mathcal{R})$, so $\mathcal{L}' \subset \operatorname{lann}(\mathcal{R}') \subset \operatorname{lann}(\mathcal{R}) \subset \mathcal{L}'$, which proves that $\operatorname{lann}(\mathcal{R}) = \mathcal{L}'$. Hence $\mathcal{R}' \subset \operatorname{rann}(\mathcal{L}') = \overline{\mathcal{R}} \subset \mathcal{R}'$. Then $\overline{\mathcal{R}} = \mathcal{R}'$. Thus the orthogonal pair $(\overline{\mathcal{R}},\ \operatorname{lann}(\mathcal{R}))$ is maximal. Similarly, one can prove that $(\operatorname{rann}(\mathcal{L}),\ \overline{\mathcal{L}})$ is a maximal orthogonal pair.

(ii) Let $(\mathcal{R},\ \mathcal{L})$ be an orthogonal pair. As noted in the proof of (i), $(\mathcal{R},\ \mathcal{L})$ is contained in the maximal orthogonal pair $(\overline{\mathcal{R}},\ \operatorname{lann}(\mathcal{R}))$. Similarly, $(\mathcal{R},\ \mathcal{L})$ is also contained in the maximal orthogonal pair $(\operatorname{rann}(\mathcal{L}),\ \overline{\mathcal{L}})$.

(iii) Suppose that $(\mathcal{R},\ \mathcal{L})$ is maximal. Then $(\mathcal{R},\ \mathcal{L}) \subset (\overline{\mathcal{R}},\ \operatorname{lann}(\mathcal{R}))$ implies $\mathcal{R} = \overline{\mathcal{R}}$ and $\mathcal{L} = \operatorname{lann}(\mathcal{R})$. □

The following lemma shows that the orthogonal pair $(\mathcal{R},\ \mathcal{L})$ associated to a nonzero regular inner ideal U of a simple ring R is defined by U almost uniquely.

LEMMA 13.26. *Let* U *be a nonzero regular inner ideal of a simple ring* R *with associated orthogonal pair* $(\mathcal{R},\ \mathcal{L})$. *Then* $UR = \mathcal{R}R$ *and* $RU = R\mathcal{L}$. *In particular, if* R *is unital, then* $UR = \mathcal{R}$ *and* $RU = \mathcal{L}$.

PROOF. Since U is nonzero, we have by simplicity of R that $\mathcal{L}R\mathcal{R} = R$. Hence

$$\mathcal{R}R = \mathcal{R}\mathcal{L}R\mathcal{R} \subset UR\mathcal{R} = UR \subset \mathcal{R}R,$$

which proves that $UR = \mathcal{R}R$. Similarly, one proves that $RU = R\mathcal{L}$. □

LEMMA 13.27. *Let* R *be a simple ring and let* $(\mathcal{R}_1,\ \mathcal{L}_1)$ *and* $(\mathcal{R}_2,\ \mathcal{L}_2)$ *be maximal orthogonal pairs of* R. *Then* $\mathcal{R}_1 \cap \mathcal{L}_1 \subset \mathcal{R}_2 \cap \mathcal{L}_2$ *implies* $(\mathcal{R}_1,\ \mathcal{L}_1) = (\mathcal{R}_2,\ \mathcal{L}_2)$.

PROOF. Set $\mathcal{L} := \mathcal{L}_1 + \mathcal{L}_2$ and $U_j := \mathcal{R}_j \cap \mathcal{L}_j$, $j = 1, 2$. By Lemma 13.26,

$$R\mathcal{L} = R\mathcal{L}_1 + R\mathcal{L}_2 = RU_1 + RU_2 \subset RU_2 = R\mathcal{L}_2 \subset \mathcal{L}_2.$$

We claim that $\mathcal{L}\mathcal{R}_2 = 0$. Otherwise $R = R\mathcal{L}\mathcal{R}_2$ and hence, by the formula displayed above, $R = R\mathcal{L}\mathcal{R}_2 \subset \mathcal{L}_2\mathcal{R}_2 = 0$, which is a contradiction. Thus $\mathcal{L}\mathcal{R}_2 = 0$ and hence

$$\mathcal{L}_1 \subset \mathcal{L} \subset \operatorname{lann}(\mathcal{R}_2) = \mathcal{L}_2.$$

Similarly, $\mathcal{R}_1 \subset \mathcal{R}_2$. But then $(\mathcal{R}_1, \mathcal{L}_1) = (\mathcal{R}_2, \mathcal{L}_2)$ by maximality of $(\mathcal{R}_1, \mathcal{L}_1)$. $\square$

THEOREM 13.28. *Let R be a simple ring. For an additive subgroup U of R the following conditions are equivalent:*

(i) *U is a maximal isotropic inner ideal of the Lie algebra R^-.*
(ii) *U is a maximal regular inner ideal of the ring R.*
(iii) *$U = \mathcal{R} \cap \mathcal{L}$, where $(\mathcal{R}, \mathcal{L})$ is a maximal orthogonal pair of R.*

PROOF. (i)$\Rightarrow$ (ii). Suppose that U is a maximal isotropic inner ideal of R^-. Put $V = URU + U$. Then $U \subset V$ and $V^2 = 0$. Since V is maximal, one has $U = V$. Therefore $VRV = URU \subset U$, so $V = U$ is a regular inner ideal of R. Clearly, V is maximal since U is maximal.

(ii) $\Rightarrow$(iii). By Lemma 13.20, there is an orthogonal pair $(\mathcal{R}, \mathcal{L})$ such that $U \subset \mathcal{R} \cap \mathcal{L}$. By taking closures if necessary, one can assume that the orthogonal pair $(\mathcal{R}, \mathcal{L})$ is maximal. By Lemma 13.20 again, $\mathcal{R} \cap \mathcal{L}$ is a regular inner ideal of R, so $U = \mathcal{R} \cap \mathcal{L}$ since U is maximal.

(iii) $\Rightarrow$(i). Let $U = \mathcal{R} \cap \mathcal{L}$, where $(\mathcal{R}, \mathcal{L})$ is a maximal orthogonal pair. Then $U^2 \subset \mathcal{L}\mathcal{R} = 0$, so U is an isotropic inner ideal of R^-. Let U' be a maximal isotropic inner ideal of R^- containing U. By the arguments above, $U' = \mathcal{R}' \cap \mathcal{L}'$, where $(\mathcal{R}', \mathcal{L}')$ is a maximal orthogonal pair. By Lemma 13.27, $(\mathcal{R}, \mathcal{L}) = (\mathcal{R}', \mathcal{L}')$, so $U = U'$ is a maximal isotropic ideal. $\square$

REMARK 13.29. Let R be a finite-dimensional (not necessarily semiprime) associative algebra over an algebraically closed field $\mathbb{F}$ of characteristic not 2 or 3. A. A. Baranov and H. Shlaka have studied in [**BS16**] a particular sort of isotropic inner ideals of the Lie algebra $L = R^{[k]}$, $k \geq 0$, where $R^{[0]} = R^-$ and $R^{[k]} = [R^{[k-1]}, R^{[k-1]}]$, $k \geq 1$, and proved that such inner ideals are regular.

Simple rings with minimal one-sided ideals. In this subsection we refine the description of the abelian inner ideals of a centrally closed prime ring R in the case that *R is a simple ring with minimal one-sided ideals*, equivalently, *satisfying the dcc on principal inner ideals.* (Notice that R is in particular Artinian if and only if it is unital.) We also describe the abelian inner ideals of its associated Lie algebras R' and $\overline{R}'$.

PROPOSITION 13.30. *Let U be an isotropic inner ideal of R^-. Then*

$$U = \mathcal{R} \cap \mathcal{L} = \mathcal{R}\mathcal{L},$$

where $(\mathcal{R}, \mathcal{L})$ is an orthogonal pair of the ring R.

PROOF. Note first that any isotropic inner ideal of R^- is an inner ideal of the Jordan algebra R^+. Regarded R as a ring of adjointable linear maps of finite rank $\mathcal{F}_Y(X)$ (see Section 2.2 for notation), it follows from [**FLGR99**, Theorem 3]:

$$U = W^*V = (W^*X) \cap (Y^*V) = \mathcal{R} \cap \mathcal{L},$$

where V and W are subspaces of X and Y respectively, $\mathcal{R} = W^*X$ a right ideal, and $\mathcal{L} = Y^*V$ a left ideal of R. Since R is von Neumann regular, $\mathcal{R} \cap \mathcal{L} = \mathcal{R}\mathcal{L}$, and $U^2 = 0$ implies $\langle V, W \rangle = 0$, equivalently, $\mathcal{L}\mathcal{R} = 0$, i.e. $(\mathcal{R}, \mathcal{L})$ is an orthogonal pair of the ring R. $\square$

COROLLARY 13.31. *If R is not unital, then every abelian inner ideal B of R' is a regular inner ideal of R. In fact, $B = \mathcal{R} \cap \mathcal{L}$, where $(\mathcal{R},\ \mathcal{L})$ is an orthogonal pair of R.*

PROOF. Since R is von Neumann regular, it follows from Lemma 13.9(ii) that B is a standard inner ideal of R^-; but since R is not unital, B is in fact isotropic. By Proposition 13.30, $B = \mathcal{R} \cap \mathcal{L}$, where $(\mathcal{R},\ \mathcal{L})$ is an orthogonal pair of R. □

THEOREM 13.32. *Let R be Artinian and let B be an abelian inner ideal of R^-. Then either $B = eRf$ or $B = eRf \oplus Z$, where e and f are idempotents of R such that $fe = 0$.*

PROOF. Since R is von Neumann regular, it follows from Proposition 13.15 that B is a standard inner ideal of R. Thus, by Theorem 13.17, either $B = U$ or $B = U \oplus Z$, where U is an isotropic inner ideal of R^-. Now it follows from the Jordan inner structure of simple Artinian rings (see [**FLGR99**, (16)] or [**McC71**, Theorem 1]) that $U = eRf$, where e, f are idempotents of R, with $fe = 0$ since $U^2 = 0$. □

COROLLARY 13.33. *Let R be Artinian and let B be an abelian inner ideal of R'. Then either $B = eRf$ or $B = eRf \oplus (Z \cap R')$, where e, f are idempotentes of R such that $fe = 0$.*

PROOF. Again by the von Neumann regularity of R, we have by Lemma 13.9 that B is a standard inner ideal of R^-. Now Theorem 13.32 applies. □

THEOREM 13.34. *Let R be a simple ring of characteristic not 2 or 3 satisfying the dcc on principal inner ideals and let $\overline{B}$ be a proper inner ideal of $\overline{R}' = R'/Z \cap R'$. Then $\overline{B} \cong \mathcal{R} \cap \mathcal{L}$, where $(\mathcal{R},\ \mathcal{L})$ is an orthogonal pair of R. If R is in particular Artinian, then $\overline{B} \cong eRf$, where e, f are idempotents of R with $fe = 0$.*

PROOF. Denote by B the inverse image of $\overline{B}$ by the canonical epimorphism of R' onto $\overline{R}'$. Then B is a proper, and hence abelian, inner ideal of R' [**Ben76**, Corollary 3.13]. Now Corollaries 13.31 and 13.33 apply. □

REMARK 13.35. Theorem 13.34 was proved by G. Benkart in [**Ben76**, Theorem 5.1] for the case of a simple Artinian ring.

Semiprime rings with dcc on principal one-sided ideals. We determine in this subsecttion the Lie inner ideal structure of a semiprime ring R with dcc on principal one-sided ideals, equivalently, coinciding with its socle, and in particular in the case that R is a semiprime Artinian ring.

PROPOSITION 13.36. *Let R be a semiprime ring such that 2, 3 are invertible in its centroid, and let B be an abelian inner ideal of R^-.*

(i) *If R coincides with its socle, then $B = \Omega \oplus \mathcal{R}\mathcal{L}$, where Ω is an additive subgroup of Z and $(\mathcal{R}, \mathcal{L})$ is an orthogonal pair of R.*
(ii) *If R is actually Artinian, then $B = \Omega \oplus eRf$, where Ω is as in (i) and e, f are orthogonal idempotents of R.*

PROOF. (i) By the structure of the socle, $R = \bigoplus M_\alpha$ is a direct sum of ideals each of which is a simple ring of characteristic not 2 or 3 coinciding with its socle. Using the von Neumann regularity of R, we get that any abelian inner ideal of R^- can be written as a sum $B = \bigoplus B_\alpha$, where B_α is an abelian inner ideal of the Lie algebra M_α^- for each index α. By we have just shown, each B_α is a standard inner ideal of R_α^-; in fact, $B_\alpha = \Omega_\alpha \oplus (\mathcal{R}_\alpha \cap \mathcal{L}_\alpha)$, with $(\mathcal{R}_\alpha, \mathcal{L}_\alpha)$ being an orthogonal pair of M_α, and where each Ω_α is either 0 or $Z(M_\alpha)$; but again by von Neumann regularity, $\mathcal{R}_\alpha \cap \mathcal{L}_\alpha = \mathcal{R}_\alpha \mathcal{L}_\alpha$. Setting $\Omega = \sum \Omega_\alpha$, $\mathcal{R} = \sum \mathcal{R}_\alpha$ and $\mathcal{L} = \sum \mathcal{L}_\alpha$, we get that $B = \Omega \oplus \mathcal{R}\mathcal{L}$, as desired.

(ii) If R is Artinian, then it coincides with its socle. Hence, by (i), $B = \Omega \oplus \mathcal{R}\mathcal{L}$. We can apply the structure of one-sided ideals of semiprime Artinian rings to show that $\mathcal{R} = eR$ and $\mathcal{L} = Rf$, where e, f are idempotents of R with $fe = 0$, or use [**McC71**, Theorem 1] to get that any isotropic inner ideal U of R^- is of the form eRf. By Exercise 12.40, we can change e to an idempotent orthogonal to f. □

Open question 13.37. As commented above, by [**McC71**, Theorem 1], any Jordan inner ideal M of a semiprime Artinian ring R is of the form eRf where e and f are idempotents of R, in particular, M is an inner ideal of the ring R. The same is true for Jordan inner ideals of a semiprime ring satisfying dcc on principal one-sided ideals, i.e. coinciding with its socle (see [**FLGR99**]). But it is unknown if this is true in general, or under what conditions on a ring R, Jordan inner ideals are actually inner ideals of R.

Minimal abelian inner ideals and minimal one-sided ideals. Let R be a semiprime ring. We prove in this subsection that $\overline{R}'$ contains minimal abelian inner ideals if and only if R has minimal one-sided ideals of square 0. If R is prime, then the second condition actually says that R is not a division ring.

Lemma 13.38. *Let R be a ring and let $a \in R$ be such that $a^2 = 0$. Then aRa is an abelian inner ideal of the Lie algebra R^- contained in $[R, R]$.*

Proof. As pointed out in Example 4.2, a is a Jordan element of the Lie algebra R^-, and it is clear that $aRa = [a, [a, R]] \subset [R, R]$. □

Definition 13.39. By a *Jordan inner ideal* of a ring R we mean an additive subgroup B of R such chat $xRx \subset B$ for every $x \in B$. Note that if R is actually an associative Φ-algebra with $1/2 \in \Phi$, then the Jordan inner ideals of R are the inner ideals of the Jordan algebra R^+.

The following auxiliary lemma is a *Jordan* characterization of the minimal one-sided ideals of a semiprime ring.

Lemma 13.40. *Let R be a semiprime ring and let a be a nonzero element of R. Then the following conditions are equivalent:*

(i) *aR is a minimal right ideal.*
(ii) *aRa is a minimal Jordan inner ideal.*
(iii) *Ra is a minimal left ideal.*

Proof. Suppose that aR is a minimal right ideal. By [**BMM96**, Proposition 4.3.3], $aR = eR$ where e is an idempotent such that eRe is a division ring and Re is a minimal left ideal. Then $a = ea = axa$ for some $x \in R$ and $aRa = eRea$. Let b be a nonzero element in aRa. Then $b = ca$ for some $c \in eRe$. Since aR is a minimal right ideal and Re is a minimal left ideal, we have that $bR = aR$ and

$$bRb = aRb = a(Rc)a = a(Re)a = aRa,$$

which proves that aRa is a minimal Jordan inner ideal of R.

Suppose conversely that $B = aRa$ is a minimal Jordan inner ideal of R. We must prove that aR is a minimal right ideal of R. Let ac be a nonzero element in aR. Since R is semiprime, $acxa \neq 0$ for some $x \in R$. By minimality of the Jordan inner ideal aRa, there exists $y \in R$ such that $(acxa)y(acxa) = a$. Then $a \in acR$, which proves that the right ideal aR is minimal. The equivalence of the conditions (ii) and (iii) follows by symmetry. □

LEMMA 13.41. *Let R be a semiprime ring such that 2 is invertible in its centroid, and let a be a nonzero element of R such that $a^2 = 0$. Then $[a, [a, R]] = aRa$, $aRa \cap Z = 0$, and the following conditions are equivalent:*

(i) *aR is a minimal right ideal of R.*
(ii) *aRa is a minimal abelian inner ideal of the Lie algebra R^-.*
(iii) *aRa is a minimal abelian inner ideal of the Lie algebra R'.*
(iv) *$aRa \cong \overline{[a, [a, R]]}$ is a minimal abelian inner ideal of the Lie algebra $\overline{R}$.*
(v) *aRa is a minimal abelian inner ideal of the Lie algebra $\overline{R'}$.*

PROOF. Let a be an element of R having square 0. Denote by B the abelian inner ideal aRa of the Lie algebra R^-. Since R is semiprime, Z does not contain nonzero nilpotent elements, so $B \cap Z = 0$ and therefore B can be regarded as an inner ideal of the Lie algebra $\overline{R}$ contained in $\overline{R'}$. Hence the equivalence of the conditions (ii), (iii), (iv), (v) follows easily. Thus we only need to verify that conditions (i) and (ii) are equivalent. But this is a consequence of Lemma 13.40 just observing that for any nonzero element $x \in aRa$, $xRx = \mathrm{ad}_x^2 R^-$, since $x^2 = 0$ and $\frac{1}{2} \in \Gamma(R)$. □

THEOREM 13.42. *Let R be a semiprime ring such that 6 is invertible in its centroid. Then the Lie algebra $\overline{R'}$ contains a minimal abelian inner ideal if and only if the ring R contains a minimal right ideal aR with $a^2 = 0$.*

PROOF. Let aR be a minimal right ideal of R such that $a^2 = 0$. It follows from Lemma 13.41 that $[a, [a, R]]$ can be regarded as a minimal abelian inner ideal of $\overline{R'}$.

Suppose conversely that $\overline{R'}$ contains a minimal abelian inner ideal $\overline{B}$. Without loss of generality we may suppose by Corollary 4.21 that R coincides with its unital central closure and $\overline{B}$ is a minimal abelian inner ideal of $\overline{R}$. Write $\overline{B} = B/Z$, where B is an inner ideal of R^- containing Z, and take $b \in B$ such that $\bar{b} \neq 0$. Since $\overline{B}$ is an abelian inner ideal, $\bar{b}$ is a Jordan element of $\overline{R}$, so $\mathrm{ad}_b^3 R \subset Z$, which implies by Proposition 3.35 that $\mathrm{ad}_b^3 R = 0$, i.e. b is a Jordan element of R^-. By Theorem 4.35, there exists $\lambda \in Z$ such that $(b - \lambda)^2 = 0$. Then $a = b - \lambda$ is a Jordan element of R^- with $a^2 = 0$. As $\overline{B}$ is a minimal abelian inner ideal, $\overline{B} = \mathrm{ad}_{\bar{a}}^2 \overline{R} = (aRa \oplus Z)/Z \cong aRa$, so aR is a minimal right ideal of R by Lemma 13.41. □

REMARK 13.43. As mentioned in Example 4.30, if R is a division ring of characteristic not 2 or 3, then the Lie algebra $\overline{R}'$ has no nontrivial inner ideal.

13.2. Lie inner ideals of prime rings with involution

Following [**BFLGL16**], we extend in this section the Lie inner ideal structure of simple Artinian rings with involution (initiated by G. Benkart [**Ben76**] and completed by her and the author [**BFL09**]) to centrally closed prime rings with involution.

Let R be a centrally closed prime ring of characteristic 0 or greater than 5 with involution $*$ and center Z, and let K be the Lie algebra of the skew-symmetric elements of R. We prove that there are at most three types of abelian inner ideals in K. The skew-symmetric versions of the standard and special inner ideals studied in the preceding section, and the so-called Clifford inner ideals, which only occur in prime rings with nonzero socle and orthogonal involution.

Standard inner ideals of skew-symmetric elements. Throughout this subsection, R will denote a semiprime ring (since we do not assume R to be unital, its center $Z = Z(R)$ may be equal to zero) with involution $*$. We also assume that 2 is invertible in its centroid $\Gamma = \Gamma(R)$. Then $*$ induces an involution, also denoted by $*$, in Γ, and $K = \mathrm{Skew}(R, *)$, $K' = [K, K]$, $\overline{K} = K/K \cap Z$, $\overline{K'} = K'/K' \cap Z$ are Lie algebras over the ring of scalars $\Phi := \mathrm{Sym}(\Gamma, *)$. By L we will mean any Φ-subalgebra of K, so an inner ideal of L will be always a Φ-submodule. As defined in the previous section, an inner ideal U of L is called isotropic if $U^2 = 0$. And an inner ideal B of L is called standard if $B = U + \Omega$, where U is an isotropic inner ideal of L and Ω is a Φ-submodule of $Z \cap L$. As a particular case of Theorem 13.7, we have that an abelian inner ideal B of L is standard if and only if

$$\mathrm{U}(B) \subset B \subset \mathrm{U}(B) + Z,$$

where $\mathrm{U}(B) = \{x \in B + Z : x^2 = 0\}$.

DEFINITION 13.44. Let R be a ring with involution $*$. An element $a \in R$ will be called *isotropic* if $aa^* = 0$, and we will say that the involution is *isotropic* if R contains nonzero isotropic elements.

LEMMA 13.45. *Let $R = \mathcal{L}_X(X)$, where X is a left vector space with a nondegenerate Hermitian or skew-Hermitian form $\langle\,,\rangle$ over a division algebra with involution Δ, and let $*$ be the adjoint involution. We have:*

(i) *$a \in R$ is isotropic if and only if Xa is a totally isotropic subspace.*
(ii) *The involution is isotropic if and only if $(X, \langle\,,\rangle)$ is isotropic.*

PROOF. (i) $\langle Xa, Xa\rangle = \langle X, Xaa^*\rangle = 0$ if and only if $aa^* = 0$.

(ii) By (i), if the involution is isotropic, then X contains nonzero isotropic vectors. Suppose conversely that $x \in X$ is a nonzero isotropic vector and set $a = x^*x$. Then $a \in R$ satisfies $aa^* = \pm x^*\langle x, x\rangle x = 0$. □

LEMMA 13.46. *Let $a \in R$ be isotropic. Then a^*Ka is an isotropic inner ideal of the Lie algebras K and $\overline{K}$.*

PROOF. Set $U = a^*Ka$. From $aa^* = 0$ we get $U^2 = 0$, and for any $b = a^*xa$, $x \in K$, and any $k \in K$, we have $[[b, k], b] = 2bkb = 2a^*(xaka^*x)a \in a^*Ka$. Since R is semiprime, $U \cap Z = 0$, so $U \cong (U \oplus K \cap Z)/K \cap Z$ is also an isotropic inner ideal of $\overline{K}$. □

Special inner ideals of skew-symmetric elements. Throughout this subsection, R will denote a unital semiprime ring with involution $*$ which does not act as the identity on its center Z. We will also assume that Z is a field of characteristic not 2. Then $K = \mathrm{Skew}(R, *)$ is both a Lie algebra and a Jordan triple system over the field $\mathbb{F} := \mathrm{Sym}(Z, *)$.

PROPOSITION 13.47. *Let L be an $\mathbb{F}$-subalgebra of K, let U be a nonzero isotropic inner ideal of L, and let $f : U \to \mathrm{Skew}(Z, *)$ be a nonzero $\mathbb{F}$-linear map such that $[[U, L], U] \subset \ker(f)$. Then $\mathrm{Inn}(U, f) := \{u + f(u) : u \in U\}$ is an abelian inner ideal of L which is not standard.*

PROOF. The proof is similar to that of Theorem 13.10. □

DEFINITION 13.48. An abelian inner ideal B of K is called *special* if it is of the form $\mathrm{Inn}(U, f)$, for some isotropic inner ideal U of K and some nonzero $\mathbb{F}$-linear map $f : U \to \mathrm{Skew}(Z, *)$ as in the above theorem.

PROPOSITION 13.49. *The Lie algebra K contains a special inner ideal if and only if there exists $x \in K$ such that $x^2 = 0$ and x is not von Neumann regular.*

PROOF. Let $B = \mathrm{Inn}(U, f)$ be a special inner ideal of K. Since $\{U, K, U\} = [[U, K], U] \subset \ker(f)$ and $\ker(f)$ is strictly contained in U, not every element of U can be von Neumann regular.

Conversely, suppose that K contains a nonzero element x of square 0 which is not von Neumann regular. Then $U = \mathbb{F}x \oplus xKx$ is an isotropic inner ideal of K. Define $f : U \to \mathrm{Skew}(Z, *)$ by taking $f(xKx) = 0$ and $f(x) = z$, where $0 \neq z \in \mathrm{Skew}(Z, *)$. It is clear that $[[U, K], U] = [[xKx, K], xKx] \subset xKx = \ker(f)$. So the pair (U, f) defines a special inner ideal of K. □

As in the case of inner ideals of a semiprime associative algebra whose center is a field, there is no difference between isotropic and special inner ideals from the Jordan point of view.

PROPOSITION 13.50. *Let U be a nonzero isotropic inner ideal of K and let $f : U \to \mathrm{Skew}(Z, *)$ be a nonzero $\mathbb{F}$-linear map such that $[[U, K], U] \subset \ker(f)$. Then the inner ideals U and $\mathrm{Inn}(U, f)$ are Jordan isomorphic.*

PROOF. It is similar to that of Proposition 13.13. □

Clifford inner ideals. Throughout this subsection, X will denote a vector space of dimension greater than 2 with a nondegenerate symmetric bilinear form $\langle\,,\rangle$ over a field $\mathbb{F}$ of characteristic not 2. The reader is referred to Section 2.2 for notation. Recall that given $x, y \in X$, we have the linear maps

(1) y^*x defined by $(x')y^*x = \langle x', y\rangle x$ for all $x' \in X$,

(2) $[x, y] = x^*y - y^*x$,

with $y^*x \in \mathcal{F}_X(X)$, $(y^*x)^* = x^*y$, and $[x, y] \in \mathfrak{fo}(X, \langle\,,\rangle)$.

The following composition rules hold.

(i) $a(x^*y)b = (xa^*)^*(yb)$, for all $x, y \in X$, $a \in \mathcal{L}_X(X)$, $b \in \mathrm{End}_{\mathbb{F}}(X)$,

(ii) $(x^*y)(z^*v) = x^*\langle y, z\rangle v$, for all $x, y, z, v \in X$.

PROPOSITION 13.51. *Let $B := [x, H^\perp] = \{[x, z] : z \in H^\perp\}$, where H is a hyperbolic plane of X and x is a nonzero isotropic vector of H. We have:*

(i) *B is an abelian inner ideal of $\mathfrak{o}(X, \langle\,,\rangle)$ contained in $\mathfrak{fo}(X, \langle\,,\rangle)$, with $b^3 = 0$ for every $b \in B$ and $c^2 \neq 0$ for some $c \in B$. Thus B is neither standard nor special.*

(ii) *B coincides with its centralizer in $\mathfrak{o}(X, \langle\,,\rangle)$, so it is a maximal abelian inner ideal of $\mathfrak{o}(X, \langle\,,\rangle)$.*

(iii) *If $H^\perp$ is anisotropic, then B is a minimal inner ideal. Otherwise B contains an extremal element and therefore an one-dimensional inner ideal.*

PROOF. (i) By Example 8.81, B is an abelian inner ideal of $\mathfrak{o}(X, \langle\,,\rangle)$ contained in $\mathfrak{fo}(X, \langle\,,\rangle)$ such that $b^3 = 0$ for every $b = [x, z] \in B$, and $b^2 = 0$ if and only if $z \in H^\perp$ is isotropic. Since $\dim_{\mathbb{F}} X > 2$, $H^\perp$ must contain some anisotropic vector, so there exists $c \in B$ such that $c^2 \neq 0$. This implies that B cannot be standard. Moreover, since the adjoint involution $*$ of $\mathcal{L}_X(X)$ is of the first kind, B cannot be special either.

(ii) Let $a \in \mathfrak{o}(X, \langle\,,\rangle)$ be such that

$$a(x^*z - z^*x) = (x^*z - z^*x)a, \; z \in H^\perp. \tag{13.1}$$

The proof will be complete if we prove that, for the isotropic vector $y \in H$ such that $\langle x, y\rangle = 1$, we have that $ya \in H^\perp$ and $a = [x, ya]$. Since $a^* = -a$, equation (13.1) can be written as

$$x^*(za) - z^*(xa) = (za)^*x - (xa)^*z, \; z \in H^\perp, \tag{13.2}$$

which evaluated in y yields

$$za = \langle y, za\rangle x - \langle y, xa\rangle z, \; z \in H^\perp. \tag{13.3}$$

Take an anisotropic vector $z \in H^\perp$ in (13.3). Then $0 = \langle za, z\rangle = -\langle y, xa\rangle\langle z, z\rangle$ which implies $\langle y, xa\rangle = 0$, and hence, again by (13.3),

$$za = \langle y, za\rangle x, \quad \text{for all} \;\; z \in H^\perp. \tag{13.4}$$

Evaluating (13.2) in z and applying (4), we get for any $z \in H^\perp$,

$$-\langle z, z\rangle xa = -\langle z, xa\rangle z = \langle za, x\rangle z = \langle\langle y, za\rangle x, x\rangle z = \langle y, za\rangle\langle x, x\rangle z = 0. \tag{13.5}$$

Taking an anisotropic vector $z \in H^\perp$ in (13.5), we get

$$xa = 0. \tag{13.6}$$

Thus $\langle ya, x\rangle = -\langle y, xa\rangle = 0$. Since $\langle y, ya\rangle = 0$, we also have $ya \in H^\perp$. Finally, using the decomposition $X = \mathbb{F}x \oplus \mathbb{F}y \oplus H^\perp$ we will see that $a = [x, ya]$, so completing the proof.

(i) $x[x, ya] = x(x^*ya - (ya)^*x) = -\langle x, ya\rangle x = 0$.
(ii) $y[x, ya] = y(x^*ya - (ya)^*x) = ya$. Now let $z \in H^\perp$. Then
(iii) $z[x, ya] = z(x^*ya - (ya)^*x) = -\langle z, ya\rangle x = \langle za, y\rangle x = za$ by (13.4).

This proves that $a = [x, ya] \in [x, H^\perp]$.

(iii) If $H^\perp$ is anisotropic, then $B = \mathrm{ad}_b^2\, \mathfrak{fo}(X, \langle\,,\rangle)$ for every nonzero element $b \in B$ (Example 8.81), so B is a minimal inner ideal. Suppose then that there exists a nonzero isotropic vector $z \in H^\perp$ and set $b = [x, z]$. Then it follows from Example 6.2 that $b = [x, z]$ is an extremal element, i.e.

$$\mathrm{ad}_b^2\, \mathfrak{fo}(X, \langle\,,\rangle) = b\mathfrak{fo}(X, \langle\,,\rangle)b$$

is an one-dimensional inner ideal. □

DEFINITION 13.52. Let L be a subalgebra of $\mathfrak{o}(X, \langle\,,\rangle)$ containing $\mathfrak{fo}(X, \langle\,,\rangle)$. An abelian inner ideal B of L is called *Clifford* if $B = [x, H^\perp]$, where H is a hyperbolic plane of X and $x \in H$ is a nonzero isotropic vector. This terminology is motivated by the fact that the subquotient of L with respect to B is the *Clifford Jordan pair* $(H^\perp, H^\perp)$ (see [**BFL09**, Proposition 4.4(1)]).

The following proposition is the converse of the statement (i) of Proposition 13.51. Recall (Definition 8.80) that a Jordan element b of the Lie algebra $\mathfrak{o}(X, \langle\,,\rangle)$ is called a Cilfford element if $b^3 = 0$ and $b^2 \neq 0$. As we show now, Clifford elements can only be found in CLifford inner ideals.

PROPOSITION 13.53. *Let L be a subalgebra of $\mathfrak{o}(X, \langle\,,\rangle)$ containing $\mathfrak{fo}(X, \langle\,,\rangle)$, and let B be an abelian inner ideal of L. If B contains a Clifford element, then B is a Clifford inner ideal.*

PROOF. Let $b \in B$ be a Clifford element of the centrally closed prime ring $\mathcal{L}_X(X)$. By Proposition 8.83(6), b^2 is symmetric with $\mathrm{rank}(b^2) = 1$, so $b^2 = \alpha x^* x$, where both $\alpha \in \mathbb{F}$ and $x \in X$ are nonzero. Now $b^3 = 0$ implies

$$\alpha^2 \langle x, x\rangle x^* x = b^2 b^2 = 0,$$

so the vector x is isotropic. Extend x to the *hyperbolic pair* (x, y), i.e. x, y are isotropic vectors and $\langle x, y\rangle = 1$ and set $H := \mathbb{F}x \oplus \mathbb{F}y$. We have the following relations:

(i) $yb^2 = y(\alpha x^* x) = \alpha\langle y, x\rangle x = \alpha x$,
(ii) $\langle yb, yb\rangle = -\langle y, yb^2\rangle = \langle y, \alpha x\rangle = -\alpha$, so yb is anisotropic,
(iii) $yb \in H^\perp$, since $\langle yb, y\rangle = 0$ and $\langle yb, x\rangle = \langle yb,\ \alpha^{-1} yb^2\rangle = \langle yb^3,\ \alpha^{-1} y\rangle = 0$.

Let $w \in H^\perp$ and set $a := [y, w]$. Then:

(iv) $xa = x(y^* w - w^* y) = \langle x, y\rangle w - \langle x, w\rangle y = w$,
(v) $b^2 a = \alpha(x^* x)a = \alpha x^* x a = \alpha x^* w$,
(vi) $ab^2 = \alpha a(x^* x) = -\alpha(xa)^* x = -\alpha w^* x$,
(vii) $bab = b(y^* w - w^* y)b = -(yb)^* wb + (wb)^* yb = [wb, yb^*]$,
(viii) $\mathrm{ad}_b^2 a = b^2 a + ab^2 - 2bab = \alpha[x, w] - 2[wb, yb^*]$.

Taking $w = yb$ in (viii), we get by (i)

$$\mathrm{ad}_b^2 [y, yb] = \alpha[x, yb] - 2[yb^2, yb] = \alpha[x, by] - 2[\alpha x, yb] = \alpha[x, yb],$$

so $[x, yb] \in B$. Since yb is anisotropic, it follows from Example 8.81(ii) that $[x, H^\perp] = \mathrm{ad}^2_{[x,yb]} \mathfrak{fo}(X, \langle\,,\rangle) \subset B$. Hence $B = [x, H^\perp]$ since $[x, H^\perp]$ is maximal by Proposition 13.51(ii). This proves that B is a Clifford inner ideal. $\square$

Now we describe Clifford inner ideals in algebraic terms. Recall that for any subset S of a ring R with involution, $\kappa(S) = \{a - a^* : a \in S\}$, and that an idempotent $e \in \mathcal{L}_X(X)$ is said to be $*$-orthogonal if $ee^* = 0 = e^* e$. Note that e is a rank-one $*$-orthogonal idempotent if and only if $e = x^* y$, where (x, y) is a hyperbolic pair.

PROPOSITION 13.54. *Let L be a subalgebra of $\mathfrak{o}(X, \langle\,,\rangle)$ containing $\mathfrak{fo}(X, \langle\,,\rangle)$. An abelian inner ideal B of L is Clifford if and only if $B = \kappa(eR(1-e))$, where R is any $*$-subalgebra of $\mathcal{L}_X(X)$ containing $L + \mathcal{F}_X(X)$, and e is a rank-one $*$-orthogonal idempotent.*

PROOF. Let $e = x^* y$, where (x, y) is a hyperbolic pair, let $H = \mathbb{F}x \oplus \mathbb{F}y$ be the associated hyperbolic plane, and set $f := e + e^*$. Then

$$eR = (x^* y)R = x^*(yR) = x^* X, \tag{13.7}$$

and given that $1 - f$ is the orthogonal projection on $H^\perp$, it follows from (13.7)

$$eR(1 - f) = x^* X(1 - f) = x^* H^\perp. \tag{13.8}$$

Since $\langle yb, y\rangle = 0$ for every $b \in \mathfrak{o}(X, \langle\,,\rangle)$, we have

$$ebe^* = (x^*y)b(x^*y)^* = (x^*yb)(y^*x) = x^*\langle yb, y\rangle x = 0. \tag{13.9}$$

Hence, for every $a \in R$,

$$\kappa(ea(1-f)) = \kappa(ea(1-e) - eae^*) = \kappa(ea(1-e)) - e\kappa(a)e^* = \kappa(ea(1-e)). \tag{13.10}$$

Then, by (13.8) and (13.10), $[x, H^\perp] = \kappa(x^*H^\perp) = \kappa(eR(1-f)) = \kappa(eR(1-e))$. $\square$

Centrally closed prime rings with involution. Let R be a prime ring with involution $*$ and center Z. Recall that any involution of R induces an involution in its extended centroid $\mathcal{C}$, and therefore it can be extended to its central closure $\mathcal{C}R$. And that R is said to be centrally closed if $\mathcal{C} = \Gamma$.

In this subsection we determine the abelian inner ideals of $K = \operatorname{Skew}(R, *)$, where R is a centrally closed prime ring of characteristic 0 or greater than 5 with involution, regarded as an algebra over the field $\mathbb{F} := \operatorname{Sym}(\mathcal{C}, *)$.

PROPOSITION 13.55. *Let R be a centrally closed prime ring of characteristic 0 or greater than 5 with involution $*$ of the first kind such that $[K, K] \neq 0$, and let B be an abelian inner ideal of K. We have:*

(i) *$b^3 = 0$ for every $b \in B$.*

If $c^2 \neq 0$ for some $c \in B$, then

(ii) *K is a subalgebra of $\mathfrak{o}(X, \langle\,,\rangle)$ containing $\mathfrak{fo}(X, \langle\,,\rangle)$, where X is a vector space of dimension greater than 2 with a nondegenerate symmetric bilinear form over the field $\mathbb{F}$, and B is a Clifford inner ideal of K.*

PROOF. (i) That $b^3 = 0$ for any $b \in B$ was proved in Theorem 4.42.

(ii) Suppose that there exists $c \in B$ such that $c^2 \neq 0$. Then c is a Clifford element of R and hence, by Proposition 8.83(5), R is a $\mathbb{F}$-subalgebra of $\mathcal{L}_X(X)$ containing $\mathcal{F}_X(X)$, where X is a vector space with a nondegenerate symmetric bilinear form over the field $\mathbb{F}$, so K is a subalgebra of $\mathfrak{o}(X, \langle\,,\rangle)$ containing $\mathfrak{fo}(X, \langle\,,\rangle)$, with $\dim_{\mathbb{F}} X > 2$ since $[K, K] \neq 0$. It follows from Proposition 13.53 that B is a Clifford inner ideal of K. $\square$

THEOREM 13.56. *Let R be a centrally closed prime ring of characteristic 0 or greater than 5 with involution $*$, let $\overline{\mathcal{C}}$ be the algebraic closure of the extended centroid $\mathcal{C}$ of R, and suppose that $\overline{\mathcal{C}} \otimes_{\mathcal{C}} R$ is not the full matrix algebra $\mathrm{M}_2(\overline{\mathcal{C}})$ with the transpose involution. If B is an abelian inner ideal of K, then one of the following possibilities holds:*

(i) *$B = V$ or $B = V \oplus \operatorname{Skew}(Z, *)$, where V is an isotropic inner ideal of K.*

(ii) *$B = \operatorname{Inn}(V, f)$ is special.*

(iii) *$B = \kappa(eR(1-e))$ is Clifford.*

Furthermore, in case (ii) *R is unital, $*$ is of the second kind, and $V = \operatorname{Skew}(U, *)$ where U is a $*$-invariant isotropic inner ideal of R^-, while in case* (iii) *R has nonzero socle and $*$ is orthogonal.*

PROOF. Suppose first that the involution $*$ is of the second kind and let ξ be a nonzero skew-symmetric element of $\mathcal{C}$. Then $R = K \oplus \xi K$. Set $C := B \oplus \xi B$. It is straightforward to see that C is an abelian inner ideal of R^-. By Theorem 13.17, we have three possibilities:

(1) $C = U$, where U is an isotropic inner ideal of R^-,

(2) R is unital and $C = U \oplus Z$, where U is as in (1),
(3) R is unital and $C = \{u+g(u) : u \in U\}$, where U is as in (1) and $g : U \to Z$ is a nonzero linear form such that $[[U, R], U] \subset \ker(g)$.

If $C = U$ as in (1), then $B = \mathrm{Skew}(U, *)$ is an isotropic inner ideal of K. Suppose then that C is as in (2) or (3). In both cases, U is $*$-invariant: $U^* \subset C^* = C \subset U \oplus Z$. Thus for any $u \in U$, $u^* = v + z$, where $u, v \in U$ and $z \in Z$. Since $u^* - v$ is nilpotent, $u^* - v = 0$, so $u^* = v \in U$ as claimed. If (2), then $B = \mathrm{Skew}(U, *) \oplus \mathrm{Skew}(Z, *)$, with $\mathrm{Skew}(U, *)$ being an isotropic inner ideal of K. If (3), then $B = \{v + f(v) : v \in V\}$, where $V = \mathrm{Skew}(U, *)$ is an isotropic inner ideal of K and $f : V \to \mathrm{Skew}(Z, *)$ is the restriction of g to V, satisfying $[[V, K], V] \subset \ker(f)$.

Suppose now that the involution $*$ is of the first kind. If $b^2 = 0$ for every $b \in B$, then B is an isotropic inner ideal. Thus we may assume that $c^2 \neq 0$ for some $c \in B$. Then we have by Proposition 13.55 that B is a Clifford inner ideal. □

REMARKS 13.57. (1) The Lie algebra $\mathrm{Skew}(\mathrm{M}_2(\mathbb{F}), *)$, where $\mathbb{F}$ is a field of characteristic not 2 and $*$ is the transpose involution, is an abelian inner ideal of itself which does not lie in any of the three cases of the theorem above, so the exception in the statement is not superfluous.

(2) Let B be an abelian inner ideal of $K = \mathrm{Skew}(R, *)$, where R is still prime but not necessarily centrally closed, and let $\mathcal{C}$ denote its extended centroid. Then $\mathrm{Sym}(\mathcal{C}, *)B$ is an abelian inner ideal of $\mathrm{Skew}(\mathcal{C}R, *)$ and therefore one of those described in Theorem 13.56.

Simple rings with involution. In this subsection R will be a simple ring of characteristic not 2 or 3 with involution $*$ and center Z. We will also assume that either $Z = 0$ or $\dim_Z R > 16$. As usual, we set $K = \mathrm{Skew}(R, *)$.

THEOREM 13.58. *Let B be an abelian inner ideal of K. Then B is either standard, special, or Clifford.*

PROOF. The proof of the above theorem works just replacing the reference to Theorem 4.42 in Proposition 13.55(i) by Theorem 4.47(ii). □

COROLLARY 13.59. *Let R and B be as above. We have:*

(i) *If the involution is of the first kind, then B is either isotropic or Clifford.*
(ii) *If R is von Neumann regular, then B is either standard or Clifford.*

PROPOSITION 13.60. *Suppose that the involution $*$ of R is of the first kind. If B is a proper inner ideal of $[K, K]$ such that $c^2 \neq 0$ for some $c \in B$, then B is a Clifford inner ideal of K.*

PROOF. By [**Ben76**, Theorem 4.21], B is abelian, and by Theorem 4.45, $b^3 = 0$ and $b^2Kb^2 = 0$ for any $b \in B$. Since items (2)-(6) in Proposition 8.83 follow from item (1), $c^2Kc^2 = 0$ implies that we can regard R as $\mathcal{F}_X(X)$, where X is a vector space with a nondegenerate bilinear form over the centroid Γ of R and c^2 has rank 1. If R is unital, then $\dim_Z R > 16$ and hence $\dim_\Gamma X > 4$. If R is not unital, then X is infinite-dimensional over Γ. In both cases, $[K, K] = K = \mathfrak{so}(X, \langle\, ,\rangle)$, so B is an abelian inner ideal of K. Since $c^2 \neq 0$, it follows from Corollary 13.59 that B is Clifford. □

13.3. Point spaces

Let L be a Lie algebra over a field $\mathbb{F}$. Following [**BFL09**], we study in this section a sort of abelian inner ideals B of L, called point spaces, characterized by the condition that every element of B is extremal. Point spaces of Lie algebras are the same as point spaces in Jordan pairs as we will see now.

Point spaces of Jordan pairs. In this subsection, V will denote a Jordan pair over a field $\mathbb{F}$ of characteristic not 2 or 3.

DEFINITION 13.61. A subspace $P \subset V^\sigma$, $\sigma = \pm$, is called a *point space* of V if $Q_x V^{-\sigma} = \mathbb{F}x$ for any nonzero $x \in P$. Note that any subspace of P is also a point space and $\mathbb{F}x$ is a minimal inner ideal of V for any nonzero element $x \in P$.

EXAMPLE 13.62. Following Exercise 11.64, let $V = (V^+, V^-)$ be the Jordan pair defined by a bilinear form $\langle\,,\rangle$ over an associative $\mathbb{F}$-algebra R. Then $V^\pm$ are point spaces if and only if $\langle\,,\rangle$ is nondegenerate and $R = \mathbb{F}$. In particular, the subspace $\mathrm{M}_{1\times r}(\mathbb{F})$ is a point space of the Jordan pair $(\mathrm{M}_{1\times r}(\mathbb{F}), \mathrm{M}_{r\times 1}(\mathbb{F}))$ of row and column matrices.

THEOREM 13.63. *Let $V = (V^+, V^-)$ be a nondegenerate Jordan pair. If V^+ is a point space, then V is the Jordan pair defined by a nondegenerate bilinear form. Moreover, if V^+ has finite dimension r, then V is isomorphic to $(\mathrm{M}_{1\times r}(\mathbb{F}), \mathrm{M}_{r\times 1}(\mathbb{F}))$.*

PROOF. For any nonzero $x \in V^+$, $Q_x V^- = \mathbb{F}x$ is a minimal inner ideal of V, so $\mathrm{Soc}(V^+) = V^+$. Let (e^+, e^-) be a nonzero idempotent of V and let $V = V_2 \oplus V_1 \oplus V_0$ be the corresponding Peirce decomposition. Then V_0 is a nondegenerate Jordan pair with V_0^+ being a point space. We claim that $V_0^+ = 0$, and hence that $V_0^- = 0$ too by the nondegeneracy of V_0. If V_0^+ were nonzero, then it would contain a nonzero idempotent $f = (f^+, f^-)$, and hence $Q_{e^+} V^- \cap Q_{f^+} V^- = 0$ by [**Loo75**, 5.4], which leads to the contradiction:

$$\mathbb{F}(e^+ + f^+) = Q_{e^+ + f^+} V^- = Q_{e^+} V^- \oplus Q_{f^+} V^- = \mathbb{F}e^+ \oplus \mathbb{F}f^+.$$

Therefore, $V = V_2 \oplus V_1$ is a nondegenerate Jordan pair coinciding with its socle. Further, V has capacity 1 (and hence it is simple) and two additional properties: (i) V has no invertible elements unless $V = (\mathbb{F}, \mathbb{F})$, and (ii) the coordinate system of V is the field $\mathbb{F}$ itself. By the classification of Jordan pairs of finite capacity [**Loo75**, 12.12] (see also [**Loo75**, 12.5] , V is the Jordan pair defined by a nondegenerate bilinear form $\langle,\rangle : V^+ \times V^- \to \mathbb{F}$ (see [**LN94**, 5.11] for a related result). Finally, if V^+ has finite dimension, say r, then V^- is canonically isomorphic to the dual space of V^+. Via the canonical isomorphism, we can identify the Jordan pair V with $(\mathrm{M}_{1\times r}(\mathbb{F}), \mathrm{M}_{r\times 1}(\mathbb{F}))$. □

Point spaces of Lie algebras. In this subsection, L will denote a Lie algebra over a field $\mathbb{F}$ of characteristic different from 2 and 3.

DEFINITION 13.64. A subspace P of L will be called a *point space* if $[P, P] = 0$ and $\mathrm{ad}_x^2 L = \mathbb{F}x$ for every nonzero element $x \in P$.

PROPOSITION 13.65. *Let P be a point space of a Lie algebra L. We have:*

(i) *P is an abelian inner ideal.*
(ii) *Any subspace of P is also a point space.*

(iii) *P is a point space of the Jordan pair* $\mathrm{Sub}_L P$.

PROOF. Straightforward. □

In [**DFLGGL12**] it is shown that the classical Lie algebras of types A_n, B_{n+1}, and D_{n+1} contain point spaces of dimension n, so, by (i), they also contain point spaces P_i of dimension $i = 1, \ldots, n$. For instance, $\bigoplus_{j=1}^n \mathbb{F}[j(n+1)]$ is a point space of $\mathfrak{sl}_{n+1}(\mathbb{F})$. On the contrary, any nonzero point space of a classical Lie algebras of type C_n is one-dimensional.

COROLLARY 13.66. *Assume that L is nondegenerate and let P be a point space of L. Then $\mathrm{Sub}_L P$ is the Jordan pair defined by a nondegenerate bilinear form over $\mathbb{F}$. In particular, if P has finite dimension r, then* $\mathrm{Sub}_L P \cong (\mathrm{M}_{1\times r}(\mathbb{F}), \mathrm{M}_{r\times 1}(\mathbb{F}))$.

PROOF. This follows from Theorem 13.63, since P is a point space of the nondegenerate Jordan pair $\mathrm{Sub}_L P = (P, L/\operatorname{Ker} P)$. □

Point spaces of finitary orthogonal algebras. In this subsection we describe the point spaces of the finitary orthogonal algebra $\mathfrak{fo}(X, \langle\,,\rangle)$ where X is a vector space of dimension (possibly infinite) greater than 4 over $\mathbb{F}$. This is by no means a restriction since, as will be seen later, in Lie algebras L coming from a simple ring with involution $*$ and minimal one-sided ideals, point spaces of dimension greater than 1 only occur when $*$ is the adjoint involution of a symmetric bilinear form, i.e. when L is a finitary orthogonal algebra.

LEMMA 13.67. *Any nonzero $b \in \mathfrak{fo}(X, \langle\,,\rangle)$ can be written as*

$$b = \sum_{k=1}^{n} [x_{2k-1}, x_{2k}],$$

where $\{x_1, ..., x_{2n}\}$ is linearly independent.

PROOF. Straightforward. □

LEMMA 13.68. *An $\mathbb{F}$-subspace P of $\mathfrak{fo}(X, \langle\,,\rangle)$ is a point space if and only if all of its elements have rank ≤ 2 and square 0; equivalently, for any nonzero element $a \in P$, $a = [x_1, x_2]$, where $\mathbb{F}x_1 + \mathbb{F}x_2$ is a totally isotropic two-dimensional subspace.*

PROOF. Let P be a point space of $K := \mathfrak{fo}(X, \langle\,,\rangle)$. We claim that $a^2 = 0$ for any $a \in P$. Otherwise P would be of the form $[x, H^\perp]$ by Proposition 13.53, which is not a point space. By Lemma 13.67, the rank of every element of K is even. Suppose that P contains an element a whose rank is ≥ 4, i.e. $a = \sum_{k=1}^n [x_{2k-1}, x_{2k}] \in P$ where $n \geq 2$ and the x_i are linearly independent. Taking $\{y_i\}$ dual to $\{x_i\}$, we obtain that $b = [y_2, y_1]$ is an element of K satisfying

$$1/2[[a, b], a] = aba = [x_1, x_2] \notin \mathbb{F}a,$$

which is a contradiction.

Conversely, if every nonzero element a in P has rank 2 and $a^2 = 0$, then P is a point space. Indeed, let $a = [x_1, x_2] \in P$ and $c \in K$. Then

$$\begin{aligned}
\frac{1}{2}[[a, c], a] &= aca = [x_1, x_2]c[x_1, x_2] = (x_1^* x_2 - x_2^* x_1)c(x_1^* x_2 - x_2^* x_1) \\
&= (x_1^* x_2 c - x_2^* x_1 c)(x_1^* x_2 - x_2^* x_1) \\
&= x_1^* \langle x_2 c, x_1 \rangle x_2 - x_2^* \langle x_1 c, x_1 \rangle x_2 \\
&- x_1^* \langle x_2 c, x_2 \rangle x_1 + x_2^* \langle x_1 c, x_2 \rangle x_1 = \langle x_2 c, x_1 \rangle a \in \mathbb{F}a,
\end{aligned}$$

since $\langle xc, x\rangle = 0$ for any $x \in X$ because c is skew-symmetric and $\langle\,,\rangle$ is symmetric. Therefore, we only need to show that P is abelian. First note that if $\{x_1, x_2\} \subset X$ is linearly independent, then

$$[x_1, x_2]^2 = x_1^*\langle x_2, x_1\rangle x_2 - x_2^*\langle x_1, x_1\rangle x_2 - x_1^*\langle x_2, x_2\rangle x_1 + x_2^*\langle x_1, x_2\rangle x_1 = 0$$

if and only if $\langle x_1, x_1\rangle = \langle x_1, x_2\rangle = \langle x_2, x_2\rangle = 0$. Now let $a = [x_1, x_2]$ and $b = [x_3, x_4]$ be two nonzero elements in P. Since $a + b$ has rank ≤ 2, $\{x_1, x_2, x_3, x_4\}$ is linearly dependent, say $x_4 = \alpha x_1 + \beta x_2 + \gamma x_3$, we have that $b = [x_3, \alpha x_1 + \beta x_2]$, where x_1, x_2 and x_3, satisfy the following orthogonal relations:

$$\langle x_i, x_i\rangle = 0, \text{ for } i = 1, 2, 3, \ \langle x_1, x_2\rangle = 0, \text{ and } \langle x_3, \alpha x_1 + \beta x_2\rangle = 0,$$

since $a^2 = b^2 = 0$. Now we have

$$\begin{aligned} [b, a] &= ba - ab = b[x_1, x_2] - [x_1, x_2]b \\ &= b(x_1^* x_2 - x_2^* x_1) - (x_1^* x_2 - x_2^* x_1)b \\ &= -(x_1 b)^* x_2 + (x_2 b)^* x_1 - x_1^*(x_2 b) + x_2^*(x_1 b) \\ &= [x_2, x_1 b] + [x_2 b, x_1]. \end{aligned}$$

But

$$\begin{aligned} b &= [x_3, \alpha x_1 + \beta x_2] = \alpha[x_3, x_1] + \beta[x_3, x_2] \\ &= \alpha(x_3^* x_1 - x_1^* x_3) + \beta(x_3^* x_2 - x_2^* x_3) \end{aligned}$$

implies by the orthogonal relations,

$$x_1 b = \alpha\langle x_1, x_3\rangle x_1 + \beta\langle x_1, x_3\rangle x_2 \text{ and } x_2 b = \alpha\langle x_2, x_3\rangle x_1 + \beta\langle x_2, x_3\rangle x_2.$$

Hence,

$$[b, a] = [x_2, x_1 b] + [x_2 b, x_1] = \alpha\langle x_1, x_2\rangle[x_2, x_1] + \beta\langle x_2, x_3\rangle[x_2, x_1] = 0$$

by the orthogonal relations again. □

The type of a point space of $\mathfrak{fo}(X, \langle\,,\rangle)$. A point space P of $\mathfrak{fo}(X, \langle\,,\rangle)$ (where X is assumed to be of dimension, possibly infinite, greater than 4 over $\mathbb{F}$) is said to be of *type 1* if there exists a nonzero vector u in the image of any nonzero $a \in P$. Point spaces which are not of type 1 are said to be of *type 2*. Let S be a totally isotropic subspace of X of dimension greater than 1. If u is a nonzero vector in S, then $P = [u, S]$ is a point space of $\mathfrak{fo}(X, \langle\,,\rangle)$ by Lemma 13.68, and P is of type 1 since u is in the image of any nonzero element of P. As will be seen next, every point space of type 1 of $\mathfrak{fo}(X, \langle\,,\rangle)$ has this form.

PROPOSITION 13.69. *Every point space P of type 1 of $\mathfrak{fo}(X, \langle\,,\rangle)$ is of the form $[u, S]$, where S is a totally isotropic subspace of X of dimension greater than one and u is a nonzero vector of S. Moreover, S is uniquely determined by P, and $[u, S] = [v, S]$ implies $v = \alpha u$ for some $\alpha \in \mathbb{F}$ if $\dim_{\mathbb{F}} S > 2$.*

PROOF. Let u be a nonzero vector which lies in the image of every nonzero element of P. Set $S := \{x \in X : [u, x] \in P\}$. Clearly, S is a subspace of X and $[u, S] \subset P$. The reverse inclusion also holds.

Let $a = [x_1, x_2] \in P$, where $\{x_1, x_2\}$ is linearly independent. Since $u \in Xa$, we have $u = \alpha_1 x_1 + \alpha_2 x_2$. Hence $[u, x_2] = [\alpha_1 x_1, x_2] = \alpha_1 a \in P$ implies $x_2 \in S$ (the same is true for x_1). Furthermore, one of the α_i, say α_1, is different from 0. Hence $[u, x_2] = [\alpha_1 x_1, x_2]$ implies $a = [x_1, x_2] = [u, \alpha_1^{-1} x_2] \in [u, S]$. Note now that $S = \{xa : a \in P,\ x \in X\}$ and therefore S is uniquely determined by P. Suppose finally that $P = [v, S]$ for another vector v of S. Then we have that v belongs to the image of $[u, x] \in P$ for any $x \in S$. Therefore, if the vectors u and v were linearly independent, we would have $\dim S = 2$, a contradiction. [1] $\square$

LEMMA 13.70. *Let P be a point space of $K = \mathfrak{fo}(X, \langle\,,\rangle)$. If P contains three elements $a = [x_1, x_2]$, $b = [x_1, x_3]$, and $c = [x_2, x_3]$ such that $\{x_1, x_2, x_3\}$ is linearly independent, then $P = \mathbb{F}a \oplus \mathbb{F}b \oplus \mathbb{F}c$; equivalently, $P = e^*Ke$ where $e \in \mathcal{F}_X(X)$ is an isotropic idempotent of rank 3. Moreover, P is of type 2 and a maximal point space.*

PROOF. Let N be a point space of K containing P, and let $[u, v]$ be any nonzero element of N. Since $[N, N] = 0$, it follows from Lemma 13.68 that each one of the sets $\{u, v, x_1, x_2\}$, $\{u, v, x_1, x_3\}$, and $\{u, v, x_2, x_3\}$ is linearly dependent. Hence $\{u, v\} \subset S$, where S denotes the linear span of $\{x_1, x_2, x_3\}$. Therefore, $N = \mathbb{F}a \oplus \mathbb{F}b \oplus \mathbb{F}c = P$. Moreover, P is of type 2 since

$$(\mathbb{F}x_1 \oplus \mathbb{F}x_2) \cap (\mathbb{F}x_1 \oplus \mathbb{F}x_3) \cap (\mathbb{F}x_2 \oplus \mathbb{F}x_3) = 0,$$

with $\mathbb{F}a \oplus \mathbb{F}b \oplus \mathbb{F}c = [S, S] = e^*Ke$ for any idempotent $e \in \mathcal{F}_X(X)$ such that $Xe = S$. $\square$

COROLLARY 13.71. *Suppose that X contains a totally isotropic subspace of dimension 5. Then $\mathfrak{fo}(X, \langle\,,\rangle)$ has two point spaces of dimension 3 which are not conjugate under any isomorphism of $\mathfrak{fo}(X, \langle\,,\rangle)$.*

PROOF. Let $\{x_1, x_2, x_3, x_4, x_5\}$ be a linearly independent subset of X whose linear span is totally isotropic, and let S denote the linear span of x_2, x_3, x_4. Then $P = [S, S]$ and $N = [x_1, \mathbb{F}x_1 + S]$ are point spaces of dimension 3. Furthermore, P is maximal by Lemma 13.70, but N is contained in the 4-dimensional point space $[x_1, \mathbb{F}x_1 + S + \mathbb{F}x_5]$. $\square$

THEOREM 13.72. *Let P be a point space of $K = \mathfrak{fo}(X, \langle\,,\rangle)$. Then either P is of type 1, or $P = e^*Ke$ for some isotropic idempotent e of rank 3 is a point space of type 2.*

PROOF. If $\dim P = 1$, then $P = \mathbb{F}[x_1, x_2]$ is of type 1. The same is true if $\dim P = 2$. Suppose that $P = \mathbb{F}a \oplus \mathbb{F}b$, where $a = [x_1, x_2]$ and $b = [x_3, x_4]$. Without loos of generality, we may assume by Lemma 13.68 that $x_4 = \alpha_1 x_1 + \alpha_2 x_2$, $\alpha_1, \alpha_2 \in \mathbb{F}$. Hence $P = [u, S]$, where $u = \alpha_1 x_1 + \alpha_2 x_2$ and $S = \mathbb{F}x_1 + \mathbb{F}x_2 + \mathbb{F}x_3$.

Suppose then that $\dim P \geq 3$ and let (a, b, c) be any triple of linearly independent elements of P. By Lemma 13.68, we may write $a = [x_1, x_2]$, $b = [x_3, y_1]$, and $c = [x_4, y_2]$, where $\{y_1, y_2\}$ is contained in the linear span of $\{x_1, x_2\}$, each of the sets $\{x_1, x_2, x_3\}$ and $\{x_1, x_2, x_4\}$ is linearly independent, and $\{x_3, x_4, y_1, y_2\}$ is linearly dependent.

If for some a, b, c as above we have that $\{y_1, y_2\}$ is linearly independent, then we can write $x_4 = \alpha x_3 + \beta y_1 + \gamma y_2$ with $\alpha \neq 0$, and hence $[x_3, y_2] \in P$. Since $[y_1, y_2]$ and $[x_3, y_1]$ also belong to P, we have by Lemma 13.70 that $P = e^*Ke$ for some isotropic idempotent e of rank 3.

If on the contrary, for fixed $a = [x_1, x_2]$ we have that $\{y_1, y_2\}$ is linearly dependent for any choice of b, c, then P is a point space of type 1. $\square$

[1] As observed in the footnote of Exercise 13.96, the grading induced by a point space $P = [u, S]$, where $\dim_{\mathbb{F}} S > 2$, of a finitary orthogonal algebra $\mathfrak{fo}(X, \langle\,,\rangle)$ is not special.

13.4. Inner ideals of rings with involution and minimal one-sided ideals

Let R be a simple ring of characteristic 0 or greater than 3 with involution $*$ and minimal one-sided ideals, and let Z be its center. Assume that either $Z = 0$ or $\dim_Z R > 16$. In this section we determine the abelian inner ideals of the Lie algebras $K = \mathrm{Skew}(R, *)$ and $K' = [K, K]$. In particular, we will consider the case where R is Artinian. It should be noted that, since R is von Neumann regular, one-sided ideals of R and isotropic inner ideals of R^- (resp. of K) are invariant under the centroid Γ of R (resp. under $\mathrm{Sym}(\Gamma, *)$). Furthermore, again by von Neumann regularity, only the isotropic inner ideals of $[K, K]$ need to be determined.

Let R be a simple ring of characteristic different from 2 with involution $*$ and minimal one-sided ideals. By Theorem 2.31, R can be regarded as the ring $\mathcal{F}_X(X)$ relative to a nondegenerate Hermitian or alternate form $\langle\,,\rangle$ over a division ring with involution $(\Delta, -)$, and with $*$ being the adjoint involution. In this context, $a \in \mathcal{F}_X(X)$ is isotropic if and only if its image Xa is a totally isotropic subspace (Lemma 13.45), $\mathcal{F}_X(X)$ is Artinian if and only if X is finite-dimensional over Δ, and as shown next, we may always suppose (without changing the involution) that the form $\langle\,,\rangle$ is either symmetric (then Δ is a field with the identity as involution), or skew-Hermitian.

REMARK 13.73. Suppose that $\langle\,,\rangle$ is Hermitian and that there exists a nonzero element $\xi \in \mathrm{Skew}(\Delta, -)$. Then the involution $-$ of Δ can be replaced by the new one defined by $\alpha \mapsto \tilde{\alpha} = \xi^{-1}\overline{\alpha}\xi$, $\alpha \in \Delta$, and the Hermitian form $\langle\,,\rangle$ over Δ can be replaced by $\langle\,,\rangle^\xi$, where $\langle x, y\rangle^\xi = \langle x, y\rangle\xi$, $x, y \in X$, without changing the adjoint involution of the ring $\mathcal{F}_X(X)$. This new form $\langle\,,\rangle^\xi$ is skew-Hermitian. Thus we may assume that the form $\langle\,,\rangle$ is symmetric (in this case Δ is a field with the identity as involution) or skew-Hermitian. If the former, we say that the involution $*$ is *orthogonal*, if the latter, we recall that it is called of symplectic type.

LEMMA 13.74. *Let R be a semiprime von Neumann regular ring of characteristic not 2 with involution (in particular, a simple ring containing minimal one-sided ideals), and let U be an isotropic inner ideal of K. Then $U \subset [K, K]$, so it can be regarded as an isotropic inner ideal of the Lie algebras K' and $\overline{K}'$.*

PROOF. By von Neumann regularity, given $u \in U$ there exists $v \in K$ such that $u = 2uvu$, which, together with the fact that $u^2 = 0$, implies

$$u = 2uvu = [[u, v], u] \in [K, K],$$

so U is an isotropic inner ideal of K'. That U can be regarded as an ideal of $\overline{K}'$ is follows as in the proof of Lemma 13.46. $\square$

Involution of the second kind. Recall that R is denoting a simple ring of characteristic 0 or greater than 3 with involution $*$ and minimal one-sided ideals such that either its center $Z = 0$ or $\dim_Z R > 16$, $K = \mathrm{Skew}(R, *)$, and $K' = [K, K]$.

PROPOSITION 13.75. *Suppose that the involution $*$ of R is of the second kind and let B be a proper inner ideal of K'. We have:*

(i) *B is a standard inner ideal of K.*
(ii) *If B is isotropic, then it can be described in algebraic terms:*

$$B = \mathrm{Skew}(\mathcal{L}^* \cap \mathcal{L}, *) = \mathrm{Skew}(\mathcal{L}^*\mathcal{L}, *),$$

where $\mathcal{L}$ is a left ideal of R such that $\mathcal{L}\mathcal{L}^* = 0$, and in geometric terms as

$$B = \mathrm{Skew}(V^*V, *) = [V, V],$$

where V is a totally isotropic subspace of X.

(iii) *If R is actually Artinian and B is isotropic, then $B = e^*Ke$, where e is an isotropic idempotent of R.*

PROOF. (i) By [**Ben76**, Theorem 4.26], B is abelian. Let ξ be a nonzero skew-symmetric element of $\Gamma(R)$. Then $C := B \oplus \xi B$ is an abelian inner ideal of R' and hence, by Lemma 13.9, a standard inner ideal of R^-, so, $C = U \oplus \Omega$, where U is an isotropic inner ideal of R^- and $\Omega = Z$ or $\Omega = 0$. We claim that U is invariant under the involution: Let $u \in U$. Since B is $*$-invariante, $u^* = v + \omega$, where $v \in U$ and $\omega \in \Omega$. Then u^* and v commute and they are nilpotent, so $\omega = u^* - v$ is a central nilpotent element and therefore equal to 0 since R is simple. Thus $U^* = U$ and $B = \mathrm{Skew}(C, *) = \mathrm{Skew}(U, *) \oplus \mathrm{Skew}(\Omega, *)$ is a standard inner ideal of K.

(ii) Suppose now that B is an isotropic inner ideal of K. It follows from (i) that $B = \mathrm{Skew}(U, *)$, where U is an isotropic inner ideal of R^- which is invariant under $*$. Hence, by Corollary 13.31, $U = \mathcal{R} \cap \mathcal{L}$, where $\mathcal{L}$ is a left ideal and $\mathcal{R}$ a right ideal of R such that $\mathcal{L}\mathcal{R} = 0$. Now $U = U^*$ implies $\mathcal{R} = \mathcal{L}^*$, and $\mathcal{L}\mathcal{R} = 0$ forces $\mathcal{L}\mathcal{L}^* = 0$. Regarded R as the ring $\mathcal{F}_X(X)$ with the adjoint involution, we have that $\mathcal{L} = X^*V$, where V is a subspace of X, $\mathcal{L}^* = V^*X$ and $\mathcal{L}^*\mathcal{L} = V^*V$. Now $0 = \mathcal{L}\mathcal{L}^* = (X^*V)(V^*X) = X^*\langle V, V\rangle X$ implies that V is totally isotropic.

(iii) If R is actually Artinian, then any left ideal of R is of the form $\mathcal{L} = Re$, where e is an idempotent of R. Now condition $\mathcal{L}\mathcal{L}^* = 0$ tells us that e is an isotropic idempotent. □

COROLLARY 13.76. *Let R and $*$ be as in the previous proposition. Then K and K' contain the same isotropic inner ideals.*

PROOF. It follows from Lemma 13.74 and Proposition 13.75. □

Involution of the first kind. Propositions 13.60 and 13.75 reduce the determination of the abelian inner ideals B of $K' = [K, K]$ to the case that the involution is of the first kind and B is isotropic. Since any abelian inner ideal of K' is contained in a maximal abelian inner ideal M of K', and under the present conditions (see Proposition 13.60) either M is a maximal isotropic inner ideal of $[K, K]$ or a Clifford inner ideal of $K' = K = \mathfrak{fo}(X, \langle\, ,\rangle)$, we may use this fact to determine all the abelian inner ideals of K'. We begin by considering those contained in a Clifford inner ideal.

PROPOSITION 13.77. *Let $M = [x, H^\perp]$ be a Clifford inner ideal of $\mathfrak{fo}(X, \langle\, ,\rangle)$, where $\dim_{\mathbb{F}} X > 2$ and $\mathrm{char}(\mathbb{F}) \neq 2$, and let B be an inner ideal of $\mathfrak{fo}(X, \langle\, ,\rangle)$ which is strictly contained in M. Then B is a point space of type 1.*

PROOF. Set $S = \{s \in H^\perp : [x, s] \in B\}$. It is clear that S is a subspace of $H^\perp$, and since B is strictly contained in M, it follows from Example 8.81(ii) that every vector $s \in S$ is isotropic, so S is totally isotropic. Then $B = [x, \mathbb{F}x + S]$ is a point space of type 1. □

Suppose now that M is a maximal isotropic inner ideal of $[K, K]$ and B an inner ideal of $[K, K]$ contained in M. We begin with the case that R is Artinian.

Isotropic inner ideals of $[K,K]$: The Artinian case. In this subsection R will denote a simple Artinian ring of characteristic $\neq 2,3$ with involution $*$ of the first kind and $\dim_Z R > 16$, realized as the complete ring $\mathrm{End}_\Delta(X)$ of linear transformations on a vector space X over a division ring with involution $(\Delta, -)$, where the involution $*$ is the adjoint relative to a nondegenerate symmetric or skew-Hermitian form $\langle\,,\rangle$ on X. As usual, we set $K = \mathrm{Skew}(R,*)$.

LEMMA 13.78. *Let $R = \mathcal{F}_X(X)$ with the adjoint involution, let S be a totally isotropic subspace of X, and let $e = \sum_{j=1}^n y_j^* x_j$ be an isotropic idempotent of R. We have:*

(i) *$[S,S]$ is an isotropic inner ideal of K,*
(ii) *$e^*Ke = [Xe, Xe]$ is an isotropic inner ideal of K, and*
(iii) *$e^*Ke = f^*Kf$, where $f \in R$ is a $*$-orthogonal.*

In fact, $f = \sum_{i=1}^r z_i^* x_i$, where the pairs $\{z_i, y_i\}$ span pairwise orthogonal hyperbolic planes. If R is Artinian, then $\operatorname{rank} f \leq r$, *with r being the Witt index of $\langle\,,\rangle$.*

PROOF. (i) For any $s,t \in S$, we have

$$[s,t]^2 = (s^*t - \epsilon t^*s)^2 = (s^*t)^2 - \epsilon(s^*t)(t^*s) - \epsilon(t^*s)(s^*t) + (t^*s)^2 = 0$$

since $\langle S,S\rangle = 0$. Set $u = [s,t]$. Then for any $a \in K$ we have

$$\begin{aligned}[][u,a],u] &= 2uau = 2(s^*t - \epsilon t^*s)a(s^*t - \epsilon t^*s)\\ &= 2(s^*ta - \epsilon t^*sa)(s^*t - \epsilon t^*s) \in [S,S],\end{aligned}$$

since

$$\begin{aligned}(s^*ta)^2 + (t^*sa)^2 &= s^*\langle ta,s\rangle t + t^*\langle sa,t\rangle s = s^*\langle ta,s\rangle t - \epsilon(\langle ta,s\rangle t)^*s\\ &= [s, \langle ta,s\rangle t] \in [S,S],\end{aligned}$$

and

$$2s^*\langle ta,t\rangle s = s^*\langle ta,t\rangle s - \epsilon(\langle ta,t\rangle s)^*s = [s,\langle ta,t\rangle s] \in [S,S].$$

Similarly, $2t^*\langle sa,s\rangle t \in [S,S]$.

(ii) By Lemma 2.28(3), $K = [X,X]$. Let $x,y \in X$. Then we have

$$e^*[x,y]e = e(x^*y - \epsilon y^*x)e = (xe)^*(ye) - \epsilon(ye)^*(xe) = [xe, ye].$$

Therefore, $e^*[X,X]e = [Xe, Xe]$.

(iii) Since e is isotropic, $e = \sum_{j=1}^n y_j^* x_j$, where the x_j span a totally isotropic subspace. Using [**Kap03**, 1-13(h)], which also works in the skew-Hermitian case, we can construct a sequence $z_1, \dots, z_n$ of vectors of X such that $(x_i, z_j) = \delta_{ij}$ and $(z_i, z_j) = 0$ for all i,j. Then $f = \sum_{j=1}^n z_j^* x_j$ is the required $*$-orthogonal idempotent. Indeed, by (i), $f^*Kf = [Xf, Xf]$ with $Xf = \sum_{j=1}^n \Delta x_j = Xe$. Since the pairs $\{x_j, z_j\}$ span pairwise hyperbolic planes, so $\operatorname{rank}(e) = \operatorname{rank}(f)$. □

PROPOSITION 13.79. *Let M be a maximal isotropic inner ideal of K'. Then $M = e^*Ke$ where e is a $*$-orthogonal idempotent of R.*

PROOF. Set $N := M + MRM$. Then N is an isotropic inner ideal of R^- and hence, Theorem by 13.32, $N = fRe$, where e and f are idempotents of R with $ef = 0$. It is clear that $N^* = N$, which implies $f = e^*$. Then e is isotropic and $N = e^*Re = e^*\,\mathrm{Sym}(R,*)e \oplus e^*Ke$. Hence $M \subset \mathrm{Skew}(N,*) = e^*Ke$. So $M = e^*Ke$ since M is a maximal isotropic inner ideal and e^*Ke is an isotropic inner ideal of $[K,K]$ (Lemma 13.74). Finally, by Lemma 13.78(iii), we may replace the isotropic idempotent e by a $*$-orthogonal one. □

PROPOSITION 13.80. *Suppose that* $M = e^*Ke$, *where* e *is a* $*$-*orthogonal idempotent of* R *and set* $L := \overline{K}'$. *Then* $\mathrm{Sub}_L M$ *is isomorphic to one of the following the Jordan pair:*

(1) $(\mathrm{Her}_r(\Delta,-),\mathrm{Her}_r(\Delta,-))$
(2) $(\mathrm{K}_r(\mathbb{F}),\mathrm{K}_r(\mathbb{F})))$,

where r *denotes the Witt index of* $\langle\,,\rangle$ *in both cases.*

PROOF. The $*$-orthogonal idempotent e induces a 5-grading

$$R = R_{-2}\oplus R_{-1}\oplus R_0\oplus R_1\oplus R_2$$

such that $R_2 = e^*Re$ and $R_{-2} = eRe^*$, and a 5-grading in L with $L_2 = e^*Ke$ and $L_{-2} = eKe^*$. Since L is nondegenerate, it follows from Proposition 11.39 that $\mathrm{Sub}_L M \cong (e^*Ke, eKe^*)$. By Lemma 13.78(iii), $e = \sum_{i=1}^r y_i^*x_i$, where the pairs (x_i, y_i) span pairwise orthogonal hyperbolic planes H_i and r is the Witt index of $\langle\,,\rangle$. Then $v := \sum_{i=1}^r y_i^*y_i \in eRe^*$ is an invertible element of the Jordan pair $(e^*Re,\, eRe^*)$ with inverse $u = \epsilon\sum_{i=1}^r x_i^*x_i \in e^*Re$, ($\epsilon = \pm$ according to whether $\langle\,,\rangle$ is symmetric or skew-Hermitian). Hence $(e^*Re,\, eRe^*)$ is isomorphic to the Jordan pair (S^+, S^+), where S is the unital ring with unit element u defined on the abelian group $(e^*Re, +)$ by the product $a\cdot b := avb$, for all $a, b\in e^*Re$. Since $v^* = \epsilon v$, the map $\star := \epsilon *$ defines an involution on the ring S such that

$$e^*Ke = \mathrm{Skew}(e^*Re, *) = \mathrm{Sym}(S,\star)$$

if $\langle\,,\rangle$ is skew-Hermitian, and

$$e^*Ke = \mathrm{Skew}(e^*Re, *) = \mathrm{Skew}(S,\star)$$

if $\langle\,,\rangle$ is symmetric. Then we have the Jordan pair isomorphisms

$$(e^*Ke, eKe^*) \cong (\mathrm{Sym}(S,\star),\mathrm{Sym}(S,\star)) \cong (\mathrm{Her}_r(\Delta,-),\mathrm{Her}_r(\Delta,-))$$

if $\langle\,,\rangle$ is skew-Hermitian, and

$$(eKe^*, e^*Ke) \cong (\mathrm{Skew}(S,\star),\mathrm{Skew}(S,\star)) \cong (\mathrm{K}_r(\mathbb{F}),\mathrm{K}_r(\mathbb{F})),$$

where $\mathbb{F}$ is a field if $\langle\,,\rangle$ is symmetric. □

PROPOSITION 13.81. *Let* B *be an isotropic inner ideal of* $[K,K]$.

(i) *If* $\langle\,,\rangle$ *is skew-Hermitian, then* $B = g^*Kg = [Xg, Xg]$, *where* g *is a* $*$-*orthogonal idempotent of* R.
(ii) *If* $\langle\,,\rangle$ *is symmetric, then either* B *is a point space of type* 1, *or* $B = g^*Kg$ *for some* $*$-*orthogonal idempotent of rank greater than* 2.

PROOF. By Proposition 13.79, $B\subset M = e^*Ke$ for a $*$-orthogonal idempotent e. Denote by L the Lie algebra $\overline{K}'$. If $\langle\,,\rangle$ is skew-Hermitian (resp. symmetric), we have by Proposition 13.80 that $\mathrm{Sub}_L M$ is isomorphic to the Jordan pair of Hermitian matrices $(\mathrm{Her}_r(\Delta,-),\, \mathrm{Her}_r(\Delta,-))$ (resp. to the Jordan pair of skew-symmetric matrices $(\mathrm{K}_r(\mathbb{F}),\, \mathrm{K}_r(\mathbb{F})))$, where r is the Witt index of $\langle\,,\rangle$ in both cases. Since by Proposition 11.47(i) the inner ideals of L contained in e^*Ke are precisely the inner ideals of $\mathrm{Sub}_L M$ contained in e^*Ke, all we must do it is to consider the inner ideal structures of the Jordan pairs $(\mathrm{Her}_r(\Delta,-),\, \mathrm{Her}_r(\Delta,-))$ and $(\mathrm{K}_r(\mathbb{F}),\, \mathrm{K}_r(\mathbb{F}))$.

Suppose first that $\langle\,,\rangle$ is skew-Hermitian. Following the notation of Proposition 13.80, we have the sequence of isomorphisms

$$\begin{aligned}\mathrm{Sub}_L\, M &\cong (e^*Ke, eKe^*) \cong (\mathrm{Sym}(S,\star), \mathrm{Sym}(S,\star))\\ &\cong (\mathrm{Her}_r(\Delta,-), \mathrm{Her}_r(\Delta,-)).\end{aligned}$$

By [**McC71**, §5, Theorem 2], any inner ideal of the Jordan algebra $J := \mathrm{Sym}(S,\star)$ is of the form $B = f^*\cdot J\cdot f = (f^*v)J(vf)$, where f is an idempotent of S and $v \in eRe^*$. Put $g := vf$. Then $g^2 = (vf)(vf) = v(fvf) = v(f\cdot f) = vf = g$, and $g = vf \in vS = v(e^*Re)$ implies $gg^* = (ge)(ge)^* = g(ee^*)g^* = 0$, so g is an isotropic idempotent of R. Moreover, since $v = ev$ and $v^* = -v$ (because we are assuming that $\langle\,,\rangle$ is skew-Hermitian), we have

$$g^*Kg = f^*v^*Kvf = f^*v^*e^*Kevf = f^*\cdot J\cdot f = B.$$

Assume now that $\langle\,,\rangle$ is symmetric. As in the previous case, we have the sequence of isomorphisms

$$\begin{aligned}\mathrm{Sub}_L\, M &\cong (e^*Ke, eKe^*) \cong (\mathrm{Skew}(S,\star), \mathrm{Skew}(S,\star))\\ &\cong (\mathrm{K}_r(\mathbb{F}), \mathrm{K}_r(\mathbb{F})).\end{aligned}$$

It follows from [**Neh91**, 3.2 (e)] that every inner ideal B of $(\mathrm{K}_r(\mathbb{F}), \mathrm{K}_r(\mathbb{F}))$ is either $\mathrm{K}_s(\mathbb{F})$ for some integer $s \le r$, or its subquotient is covered by a family of collinear idempotents. If the former, then $B = f^*\cdot J\cdot f = (f^*v)J(vf)$ for some idempotent f of S and some invertible v and hence, as in the previous skew-Hermitian case, $B = g^*Kg$ for an isotropic idempotent g of R. If the latter, B is a point space and therefore it is uniquely determined up to isomorphism by its dimension (Theorem 13.63). Since (Theorem 13.72) point spaces of type 2 are of the form e^*Ke for an isotropic idempotent of rank 3, we may assume that B is of type 1 in (ii). This completes the proof. □

We complete the study of the Artinian case by determining the inner ideals of the Lie algebra $\overline{K}'$. Denote by π the canonical epimorphism of K' onto $\overline{K}'$ and identify an inner ideal B of $\overline{K}'$ with its inverse image in K' module its center.

THEOREM 13.82. *Let $R = \mathrm{End}_\Delta(X)$ be a simple Artinian ring of characteristic different from 2 and 3 with involution $*$ such that $\dim_Z R > 16$, and let B be and abelian inner ideal of $\overline{K}'$. Then either*

(i) *$B = [x, H^\perp]$ is a Clifford inner ideal,*
(ii) *$B = e^*Ke = [Xe, Xe]$, where e is an isotropic idempotent of R, or*
(iii) *$B = [x, S]$ is a point space of type 1.*

PROOF. Let B' be the inverse image of B under π. Then (see [**Ben76**, Theorem 4.21 and 4.26]), B' is an abelian inner ideal of K'. If $*$ is of the second kind, it follows from (i) and (iii) of Proposition 13.75 that, module the center of K', $B' = e^*Ke$, where e is an isotropic idempotent of R. Suppose then that $*$ is of the first kind. If $b^2 \neq 0$ for some $b \in B'$, then B' is a Clifford inner ideal by Proposition 13.60. Otherwise B' is an isotropic inner ideal of K' and Proposition 13.81 applies. □

The non-Artinian case. Assume now that $R = \mathcal{F}_X(X)$ with the adjoint involution with respect to a nondegenerate Hermitian or skew-Hermitian form $\langle\,,\rangle$ over a division ring with involution $(\Delta,-)$. Even in the case that R is not Artinian, equivalently, X is infinite-dimensional over Δ, R can be still described as a direct limit of simple Artinian algebras R_α with the same type of involution as R. In fact, R is a *strongly local matrix ring* in the following sense: any finite subset of R is contained in an inner ideal of the form eRe for some self-adjoint finite rank idempotent e of R (see [**BMM96**, 4.6.15]). Note that in geometric terms, $eRe \cong \mathcal{F}_V(V) = \mathrm{End}_\Delta(V)$, where V is a nondegenerate finite-dimensional subspace of X; in fact, $V = Xe$; equivalently, e is the projection on V determined by the decomposition $X = V \oplus V^\perp$. Furthermore, $Z(eRe) = Z(\Delta)e$.

Suppose that R is not Artinian and set $K = \mathrm{Skew}(R,*)$ as usual. Since $Z = 0$, $L := [K,K]$ is a simple nondegenerate Lie algebra which is a direct limit of Lie algebras $L_\alpha = [K_\alpha, K_\alpha]$, with $K_\alpha = \mathrm{Skew}(e_\alpha R e_\alpha, *) = e_\alpha K e_\alpha$ for a self-adjoint finite rank idempotent e_α of R.

THEOREM 13.83. *Let $L = [K,K]$ be as above. If B is a proper inner ideal of L, then either*

(i) $B = \mathrm{Skew}(\mathcal{L}^*\mathcal{L}, *) = [V,V]$, *where $\mathcal{L}$ is a left ideal of R such that $\mathcal{L}\mathcal{L}^* = 0$ and V is a totally isotropic subspace of X,*
(ii) *B is a type 1 point space of dimension greater than* 1, *or*
(iii) $B = [x, H^\perp]$ *is a Clifford inner ideal of K.*

In cases (ii) *and* (iii), *Δ is necessarily a field and $L = \mathfrak{fo}(X, \langle\,,\rangle)$. Moreover, $B = [V,V]$ as in* (i) *is a nonzero point space of L (over $\mathbb{F} = \mathrm{Sym}(\Delta, -)$) if and only if $\langle\,,\rangle$ is symmetric and* $\dim_{\mathbb{F}} V = 2$ *or* 3, *or $\langle\,,\rangle$ is skew-symmetric and* $\dim_{\mathbb{F}} V = 1$.

PROOF. If the involution is of the second kind, then it follows from Proposition 13.75(ii) (since R is not unital) that B is as in (i). Suppose then that the involution is of the first kind. If $a^2 \neq 0$ for some $a \in B$, then we have by Proposition 13.60 that B is a Clifford inner ideal of K as in (iii), so we may assume that $a^2 = 0$ for every $a \in B$, and that B is abelian by [**Ben76**, Theorem 4.21]. Choose a *directed set* $\{e_\alpha\}$ of self-adjoint idempotents of $R = \mathcal{F}_X(X)$ of finite rank greater than 4 ($e_\alpha \leq e_\beta \Leftrightarrow e_\alpha R e_\alpha \subset e_\beta R e_\beta \Leftrightarrow X e_\alpha \subset X e_\beta$) and set $Z_\alpha = Z(R_\alpha) = Z(\Delta)e_\alpha$ and $L_\alpha = [K_\alpha, K_\alpha]$. Then $B_\alpha = B \cap L_\alpha$ is an isotropic inner ideal of L_α. Now it follows from Proposition 13.81 that for each index α either B_α is a point space of type 1 (in this case $\langle\,,\rangle$ is symmetric), or $B_\alpha = g_\alpha^* K_\alpha g_\alpha$ for some $*$-orthogonal idempotent g_α of R_α (of rank greater than 2 if $\langle\,,\rangle$ is symmetric by Theorem 13.72). If the former holds for all indices α, then B itself is a point space by Lemma 13.68. Suppose on the contrary that for some index α, $B_\alpha = g_\alpha^* K g_\alpha$, where g_α is a $*$-orthogonal idempotent (of rank greater than 2 if $\langle\,,\rangle$ is symmetric). Then for every $e_\beta \geq e_\alpha$, $B_\beta = g_\beta^* K g_\beta = [X g_\beta, X g_\beta]$, for some isotropic idempotent g_β of R, (Lemma 13.78(ii)). Since these idempotents e_β form a directed set, the same is true for the family of corresponding totally isotropic subspaces $X e_\beta$. Hence $V = \bigcup X e_\beta$ is a totally isotropic subspace of X and $B = [V,V]$ is as in (i). This completes the proof. □

Minimal abelian inner ideals of $\overline{K}'$. In this subsection it will be shown that minimal abelian inner ideals there exist in a Lie algebra $\overline{K}'$, where $K = \mathrm{Skew}(R,*)$ and R is a simple ring with involution, if and only if R contains minimal one-sided ideals and the involution is isotropic.

LEMMA 13.84. *Let $a \in R$ be isotropic, where only semiprimeness of R is required here.*

(i) *If $a^*Ka \neq 0$ and Ra is a minimal left ideal, then a^*Ka is a minimal abelian inner ideal of the Lie algebras K, K', $\overline{K}$, and $\overline{K'}$.*

(ii) *If $R = \mathcal{F}_X(X)$, where X is a left vector space of dimension greater than 2 with a nondegenerate skew-Hermitian form $\langle\,,\rangle$ over a division algebra with involution Δ of characteristic not 2, and $*$ is the adjoint involution, then the condition $a^*Ka \neq 0$ is automatic.*

PROOF. Let $0 \neq b \in a^*Ka \subset Ra$. Then $b^2 = 0$ and we have by minimality of Ra that given $x \in a^*Ka$, there exists $c \in R$ such that $x = cb$. As $x^* = -x$, we also have $x = -(cb)^* = -b^*c^* = bc^*$. Moreover, since the socle of a semiprime ring is a von Neumann regular ideal and 2 is invertible in the centroid of R, there exists $y \in K$ such that $x = xyx$. Therefore, $x = xyx = bc^*ycb = \frac{1}{2}[b, [c^*yc, b]] \in \operatorname{ad}_b^2 K$, because $b^2 = 0$ and $c^*yc \in K$, which proves that a^*Ka is a minimal inner ideal of K. Taking now $x = b$, we get $b = \operatorname{ad}_b^2 k$ for some $k \in K$. Thus for any nonzero element $b \in a^*Ka$, we have

$$a^*Ka = \operatorname{ad}_b^2 K = \operatorname{ad}_{\operatorname{ad}_b^2 k} K = \operatorname{ad}_b^2 \operatorname{ad}_k^2 \operatorname{ad}_b^2 K \subset \operatorname{ad}_b^2[K, K],$$

so a^*Ka is also a minimal inner ideal of K'. We have by Lemma 13.46 that a^*Ka can be regarded as a minimal abelian inner ideal of the Lie algebras $\overline{K}$ and $\overline{K'}$.

Suppose now that $R = \mathcal{F}_X(X)$, relative to a nondegenerate skew-Hermitian form. Then a nonzero rank-one operator $a = x^*y \in R$ is isotropic if and only if y is an isotropic vector. Then for any $b \in K$, we have

$$a^*ba = -(y^*x)b(x^*y) = -(y^*xb)(x^*y) = -y^*\langle xb, x\rangle y.$$

Let $z \in X$ be such that $\langle z, x\rangle = 1$ and take $b = z^*z \in K$. Then $xb = \langle x, z\rangle z = z$ and hence $a^*ba = y^*y \neq 0$, which proves that $a^*Ka \neq 0$. □

In the proof of the next proposition, a partial converse of the above lemma, Jordan results of Section 8.1 will be used.

PROPOSITION 13.85. *Let R be a prime ring with involution of characteristic 0 or greater than 3 which is not an order in a simple associative algebra of dimension at most 16 over its center, and let a be a nonzero skew-symmetric element with $a^2 = 0$. If $aKa = \operatorname{ad}_a^2 K$ is a minimal abelian inner ideal, then R has nonzero socle.*

PROOF. Note first that by Propositions 3.36 and 3.39 the Lie algebra $\overline{K}$ is strongly prime. Furthermore, since $(aKa)^2 = 0$, we have that $Z(R) \cap aKa = 0$. These facts imply that for any nonzero element $x \in aKa$, $xKx = aKa$. Then we may suppose (by replacing a by some aka if necessary) that there exists $b \in K$ such that $aba = a$. Denote by A the associative algebra defined in the abelian group aRa by the $\cdot_b$-product (Exercise 8.106). Then A is prime and unital, with $1_A = a$, and since a is skew-symmetric, the map $-*$ is an involution of A with $\operatorname{Sym}(A, -*) = aKa$. Using the minimality of the inner ideal aKa and the nondegeneracy of $\overline{K}$, we get by Proposition 8.20 that $\operatorname{Sym}(A, -*)$ is a division Jordan algebra, which implies by [**Her76**, Theorem 2.17] that $R_a \cong A$ is a simple Artinian ring, and hence, by the local characterization of the socle (Proposition 8.33), that $a \in \operatorname{Soc}(R)$. Therefore, R has nonzero socle. □

THEOREM 13.86. *Let R be a simple ring with involution of characteristic* 0 *or greater than* 5. *Suppose that either $Z(R) = 0$ or the dimension of R over $Z(R)$ is greater than* 16. *Then the simple and nondegenerate Lie algebra $\overline{K'}$ contains minimal abelian inner ideals if and only if the ring R contains minimal one-sided ideals and the involution $*$ is isotropic.*

PROOF. The Lie algebra $\overline{K'}$ is simple by Theorem 2.6, and nondegenerate by Propositions 3.36 and 3.39. Suppose that R has minimal one-sided ideals and $*$ is isotropic. As pointed out in Remark 13.73, we may assume that $R = \mathcal{F}_X(X)$ with respect to a nondegenerate, symmetric or skew-Hermitian, form $\langle\, , \rangle$, with $*$ being the adjoint involution. Moreover, X contains a nonzero isotropic vector. We deal separately with the two cases: (1) Suppose that $\langle\, , \rangle$ is skew-Hermitian, let x be a nonzero isotropic vector of X and set $a = x^*x$. Then a is isotropic and hence, by Lemma 13.84, a^*Ka is a minimal abelian inner ideal of $\overline{K'}$; (2) if $\langle\, , \rangle$ is symmetric, then (see Remark 2.32) $\overline{K'}$ is the finitary orthogonal algebra $\mathfrak{fo}(X, \langle\, , \rangle)$, so it contains minimal abelian inner ideals by Proposition 13.51.

Suppose conversely that $\overline{K}'$ contains a minimal abelian inner ideal $\overline{B}$. Since $\overline{K}' \cong [\overline{K}, \overline{K}]$, we have by Corollary 4.21 that $\overline{B}$ can be regarded as a minimal abelian inner ideal of $\overline{K}$. Then its pre-image B is a proper inner ideal of K and hence abelian by [**Ben76**, 4.21 and 4.26]. Let $b \in B$ be a noncentral element. Assume first that the involution is of the first kind. Then $b^3 = 0$ by Corollary 4.42 and there are two possibilities: (i) $b^2 \neq 0$, so b is a Clifford element and hence $R = \mathcal{F}_X(X)$ has nonzero socle; (ii) $b^2 = 0$. In this case, $\mathrm{ad}_b^2 K' = bK'b$ can be regarded as the minimal abelian inner ideal $\mathrm{ad}_{\overline{b}}^2 \overline{K}' = \overline{B}$. Hence, by Proposition 13.85, R has nonzero socle. If the involution is of the second kind, it follows from Theorem 4.47 that $(b - z)^2 \neq 0$ for some $z \in \mathrm{Skew}(Z(R), *)$. Taking $a = b - z$ we can argue as above (by replacing b by a) to get as in (ii) that R has nonzero socle. Finally, the involution is clearly isotropic in (ii) and when it is of the second kind, since in both cases R contains a nonzero skew-symmetric element of square 0, while in (i), a Clifford element is of the form $[x, z]$ with x being an isotropic vector, so taking $a = x^*x$ we get a nonzero isotropic element. This completes the proof of the theorem. □

13.5. Inner ideals of the exceptional Lie algebras

Following [**DFLGGL12**], we determine in this section the inner ideals of the simple exceptional Lie algebras E_6, E_7, E_8, F_4, and G_2 over an algebraically closed field $\mathbb{F}$ of characteristic 0. We adopt an approach different from that followed in the previous section based on the connection between abelian inner ideals and $\mathbb{Z}$-gradings obtained in Theorem 12.16. For any proper inner ideal B of such an algebra L there exists a finite $\mathbb{Z}$-grading $L = L_{-n} \oplus \cdots \oplus L_n$ with $B = L_n$. Since the $\mathbb{Z}$-gradings are always compatible with a root decomposition, B can be expressed as a sum of root spaces. As a general rule, we will use the same symbol to denote inner ideals of Jordan pairs and abelian inner ideals of Lie algebras which are Jordan isomorphic. Thus P_r will denote a point space (of a Jordan pair or a Lie algebra) of dimension r over $\mathbb{F}$ (recall that point spaces of the same dimension are Jordan isomorphic). When required, we will use accents to distinguish between inner ideals which are Jordan isomorphic but not conjugate.

Inner ideals of the simple finite-dimensional Jordan pairs. In this subsection we review the classification of the inner ideals of the simple finite-dimensional Jordan pairs over an algebraically closed field $\mathbb{F}$ of characteristic 0. By [**Loo75**, 17.4], such a Jordan pair is isomorphic to one of the following:

(I) *A Jordan pair* $\mathcal{M}_{p\times q} := (\mathrm{M}_{p\times q}(\mathbb{F}), \mathrm{M}_{q\times p}(\mathbb{F}))$, $Q_x y = xyx$, *of* $p \times q$ *matrices with entries in* $\mathbb{F}$.

The nonzero inner ideals of $\mathcal{M}_{p\times q}$ contained in $\mathrm{M}_{p\times q}(\mathbb{F})$ are (up to conjugation) of the form

$$M_{r\times s} := \sum_{1\leq i\leq r}^{1\leq j\leq s} \mathbb{F}[ij], \quad \text{with } r \leq p \text{ and } s \leq q,$$

where $[ij]$ denotes de (ij)-unit matrix. Moreover, the subquotient of $\mathcal{M}_{p\times q}$ with respect to $M_{r\times s}$ is isomorphic to $\mathcal{M}_{r\times s}$. This can be obtained from the classification of the inner ideals in Jordan pairs covered by grids [**Neh91**, 3.2], or from the geometric description of the inner ideals of Jordan pairs of finite rank adjointable operators [**FLGRGLSM98**, Prop. 2.4]. Note that for each positive integer r, $M_{1\times r}$ is a point space of dimension r.

(II) *A Jordan pair* $\mathcal{S}_n := (\mathrm{S}_n(\mathbb{F}), \mathrm{S}_n(\mathbb{F}))$, $Q_x y = xyx$, *of symmetric* $n \times n$ *matrices with entries in* $\mathbb{F}$ $(n \geq 2)$.

By [**Neh91**, 3.2(c)] or [**McC71**, Theorem 3], every nonzero inner ideal of $\mathcal{S}_n$ is (up to conjugation) of the form $S_r = e_r \mathrm{S}_n(\mathbb{F}) e_r$, for $1 \leq r \leq n$, where $e_r = [11] + \cdots + [rr]$, with subquotient isomorphic to $\mathcal{S}_r$.

(III) *A Jordan pair* $\mathcal{K}_n := (\mathrm{K}_n(\mathbb{F}), \mathrm{K}_n(\mathbb{F}))$, $Q_x y = -xyx$, *of skew-symmetric* $n \times n$ *matrices with entries in* $\mathbb{F}$ $(n \geq 4)$.

It follows from [**Neh91**, 3.2(e)] that $\mathcal{K}_n$ contains two types (up to conjugation) of nonzero inner ideals :

(i) $K_s := e_s \mathrm{K}_n(F) e_s$, for $2 \leq s \leq n$, where $e_s = [11] + \cdots + [ss]$, and
(ii) the point spaces $P_r = \sum_{j=2}^{r+1} \mathbb{F}([1j] - [j1])$ for $1 \leq r \leq n-1$.

Note that K_s is a point space if and only if $s \leq 3$ $(K_2 = P_1)$.

(IV) *A Clifford Jordan pair* $\mathcal{Q}_n := C(X, q)$, $Q_x y = q(x,y)x - q(x)y$, *defined by a nondegenerate quadratic form* q *on an* n*-dimensional vector space* X *over* F.

By [**McC71**, Theorem 6], the nonzero inner ideals of $\mathcal{Q}_n$ are $Q_n := X$ and the totally isotropic nonzero subspaces of X. Since every totally isotropic subspace is a point space, if $n = 2m$ or $n = 2m+1$, $\mathcal{Q}_n$ contains a (unique up to conjugation) maximal point space P_m of dimension m.

(V) *The Albert pair* $\mathcal{A} := (\mathrm{Her}_3(\mathcal{C}), \mathrm{Her}_3(\mathcal{C}))$, *defined by the exceptional Jordan algebra of* 3×3 *matrices over the Cayley algebra* $\mathcal{C}$ *over* $\mathbb{F}$.

By [**McC71**, Main Theorem], $\mathcal{A}$ exactly contains two maximal (proper) inner ideals up to conjugation: the 6-dimensional point space $P_6 = \mathbb{F}[11] + \epsilon\mathcal{C}[12] + \mathbb{F}\epsilon[13]$, where ϵ is a primitive idempotent of the Cayley algebra $\mathcal{C}$ (see also [**McC71**, p. 457]), and the Peirce-2-space determined by the Jordan algebra idempotent $e = [11] + [22]$, i.e. $Q_e(\mathrm{Her}_3(\mathcal{C}))$. Since $Q_e(\mathrm{Her}_3(\mathcal{C}))$ is a 10-dimensional simple Jordan algebra of capacity 2 over $\mathbb{F}$, it is the Jordan algebra defined by a nondegenerate quadratic form on a 10-dimensional vector space over $\mathbb{F}$ (see Example 8.4), so Sub $Q_{10} \cong \mathcal{Q}_{10}$. Furthermore, $\mathcal{A}$ contains two 5-dimensional point spaces which are not conjugate: $P_5 = \mathbb{F}[11]+\epsilon\mathcal{C}[12] \subset P_6 \cap Q_{10}$ and $P'_5 = \mathbb{F}[11]+\mathcal{C}[12]\epsilon \subset Q_{10}$, which is a maximal point space. Note that we had already met with this phenomenon in Corollary 13.71.

(VI) *The Bi-Cayley pair* $\mathcal{B} := (\mathrm{M}_{1\times 2}(C), \mathrm{M}_{2\times 1}(C))$, *where* C *is the Cayley-algebra over* $\mathbb{F}$.

The inner ideals of $\mathcal{B}$ are up to conjugation, $M_{1\times 2}(\mathcal{C})$, $\mathcal{C}[11]$, and the linear spans of the $+$-part of the families of collinear idempotents, following [**Neh91**, 3.2] and the notation therein. In fact, the subquotient of $\mathcal{B}$ with respect to $\mathcal{C}[11]$ is isomorphic to $\mathcal{Q}_8$, and the inner ideals determined by the families of collinear idempotents are the point spaces of $\mathcal{B}$ [**Neh91**, 3.3(1)]. By [**Loo75**, 12.10], $\mathcal{B}$ is isomorphic to the Peirce-1-space of the Albert pair $\mathcal{A}$ with respect to the idempotent $e_1 = [11]$, hence the families of collinear idempotents of $\mathcal{B}$ are those of $\mathcal{A}$ contained in the Peirce-1-space with respect to e_1, so we can apply the results obtained for the Albert pair to get the point spaces of the Bi-Cayley pair. Thus $\mathcal{B}$ contains a maximal point space of dimension 5 (the one obtained by eliminating the [11]-part of the inner ideal P_6 of the Albert pair), and two point spaces of dimension 4 which are not conjugate (those obtained by eliminating the [11]-part of the inner ideals P_5 and P'_5 of the Albert pair).

Inner ideals and $\mathbb{Z}$-gradings revisited. Throughout this subsection L will denote a simple finite-dimensional Lie algebra over an algebraically closed field $\mathbb{F}$ of characteristic 0. Given a $\mathbb{Z}$-grading $L = L_{-n} \oplus \cdots \oplus L_n$, the linear map $D : L \to L$ such that $D(x) = kx$ for any $x \in L_k$, $-n \le k \le n$, is a derivation of L and therefore $D = \mathrm{ad}_h$ for some semisimple element h belonging to a Cartan subalgebra H of L. Let $L = H \oplus (\bigoplus_{\alpha\in\Phi} L_\alpha)$ be root decomposition relative to H the root decomposition of L relative to H. Then $\alpha(h) \in \mathbb{Z}$ for any root α, because ad_h gives rise to the $\mathbb{Z}$-grading above. Moreover, it is always possible (and we will do it so) to take a basis $\Delta = \{\alpha_1, \ldots, \alpha_l\}$ of the root system such that for each $1 \le i \le l$, $p_i := \alpha_i(h)$ is a nonnegative integer. Let $\{h_1, \ldots, h_l\}$ a basis of H dual to Δ, i.e. $\alpha_j(h_i) = \delta_{ij}$. Then $h = \sum p_i h_i$. It is easy to see that the root space L_α, for $\alpha = \sum m_i\alpha_i$, is contained in the homogeneous component $L_{\sum m_i p_i}$ of the above $\mathbb{Z}$-grading. And the $\mathbb{Z}$-gradings of L are in correspondence with the labels $(p_1, \ldots, p_l)$ of nonnegative integers. Moreover, two $\mathbb{Z}$-gradings of L can be taken into one another by an outer automorphism if and only if the corresponding sets of labels can be taken into one another by an automorphism of the Dynkin diagram [**Vin75**].

Recall that $\mathbb{Z}$-gradings are closely related to abelian inner ideals: For any $\mathbb{Z}$-grading $L = L_{-n} \oplus \cdots \oplus L_n$, both L_n and L_{-n} (called the extremes of the grading) are abelian inner ideals of L. Conversely, it follows from Theorem 12.16 that every proper (equivalently, abelian) inner ideal B of L yields a finite $\mathbb{Z}$-grading $L = L_{-n} \oplus \cdots \oplus L_n$ such that $B = L_n$.

Suppose we have a $\mathbb{Z}$-grading $L = L_{-n} \oplus \cdots \oplus L_n$ given by the label of nonnegative integers $(p_1, \ldots, p_l)$. The extreme L_n of the grading is easy to determine. If we denote by $\tilde{\alpha} = \sum_{i=1}^{l} n_i\alpha_i$ the maximal root relative to the basis Δ, its corresponding root space $L_{\tilde{\alpha}}$ is contained in the extreme L_n of the $\mathbb{Z}$-grading, and $n = \sum n_i p_i$. Note that for any root $\alpha = \sum m_i\alpha_i \in \Phi$, the root space L_α is contained in L_n if and only if $\sum m_i p_i = \sum n_i p_i$; that is, if and only if $m_j = n_j$ for all j such that $p_j \ne 0$. Therefore, denoting by I the set of all $j \in \{1, \ldots, l\}$ such that $p_j \ne 0$, we have that $L_n = B_I$ for

$$B_I := \bigoplus_{\alpha\in\Phi}\{L_\alpha : \alpha = \sum m_i\alpha_i \text{ with } m_j = n_j \text{ for all } j \in I\}.$$

To summarize, for L, H and $\Delta = \{\alpha_1, \ldots, \alpha_l\}$ as above.

THEOREM 13.87. *Let B be a nonzero abelian inner ideal of L. Then there exists a subset $I \subset \{1, \ldots, l\}$ and an inner automorphism φ of L such that $\varphi(B) = B_I$.*

Note that for any subsets I, J of $\{1, \ldots, l\}$, $I \subset J$ implies $B_J \subset B_I$. In particular, the maximal abelian inner ideals of L are conjugate to $B_{\{i\}}$ for some $i \in \{1, \ldots, l\}$, although not conversely. Another interesting fact is that every chain of abelian inner ideals of L has length not greater than l. Furthermore, there is always a chain of abelian inner ideals with length just l. This is clear by recalling that for each $\alpha \in \Phi^+$, $\alpha \notin \Delta$, there exists $\alpha_i \in \Delta$ such that $\alpha - \alpha_i \in \Phi$.

The Lie and Jordan classifications of abelian inner ideals. Although the method described above to determine abelian inner ideals works for any simple finite-dimensional Lie algebra over an algebraically closed field of characteristic 0, we will restrict ourselves to the exceptional Lie algebras E_6, E_7, E_8, F_4, and G_2 (the classical ones: A_n, B_n, C_n, and D_n are particular cases those studied in the previous sections). For such an algebra L we will classify its abelian (equivalently, proper) inner ideals up to conjugation (Lie classification) and up to Jordan isomorphism (Jordan classification).

- *The Lie classification.* After choosing a Cartan subalgebra and a basis of the related root system, each abelian inner ideal will be conjugate to some B_I. Further conjugations will be obtained by means of diagram automorphisms and some special cases will be dealt separately using techniques of eigenvalues and traces of ad-semisimple elements.
- *The Jordan classification.* To determine the subquotient of an abelian inner ideal B of L, we compare the lattice of the inner ideals of L contained in B (provided by the Lie classification) with the lattice of the inner ideals of the classical Jordan pairs. While in most cases this information is enough to determine $\mathrm{Sub}_L\ B$, in others, as those of the Lie algebras of type E_6 or E_7, additional information about the inner ideal structure of L, as the eventual existencia of outer automorphisms, is required.

The inner ideal structure of E_6. Choose the system of positive roots of E_6 given in [**Bou81**, Planche V(II)] with maximal root $\alpha_1+2\alpha_2+2\alpha_3+3\alpha_4+2\alpha_5+2\alpha_6$. Following the process described above, every nonzero abelian inner ideal of E_6 is conjugate to one in the next diagram.

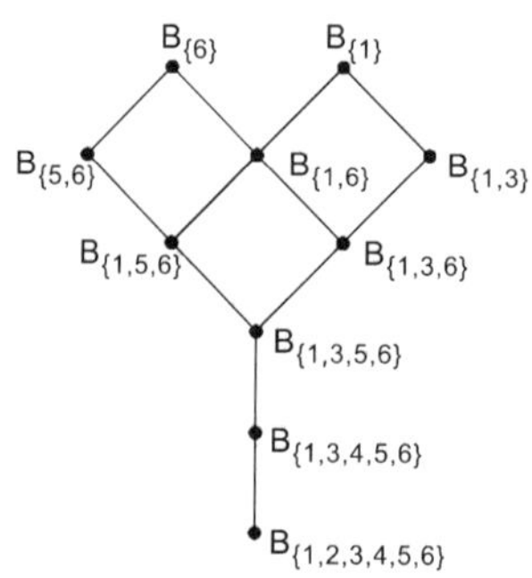

By folding the Dynkin diagram we get that $B_{\{6\}}$ is conjugate to $B_{\{1\}}$, as well as $B_{\{5,6\}}$ to $B_{\{1,3\}}$, and $B_{\{1,5,6\}}$ to $B_{\{1,3,6\}}$. The remaining cases correspond to not conjugate inner ideals because the dimensions are different. Therefore we conclude that the nonzero abelian inner ideals of E_6 up to conjugation and their corresponding subquotients are:

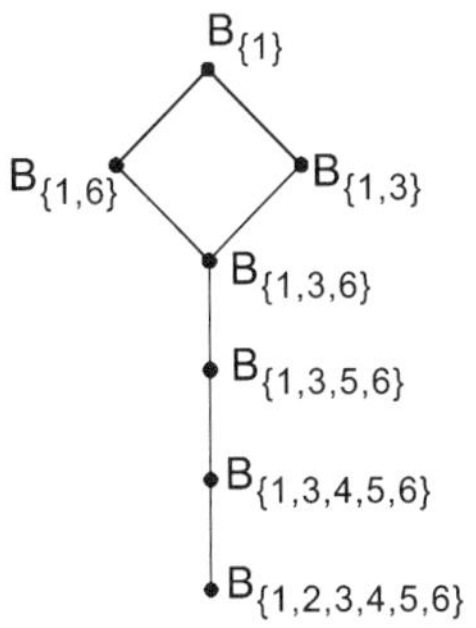

The Lie classification of E_6

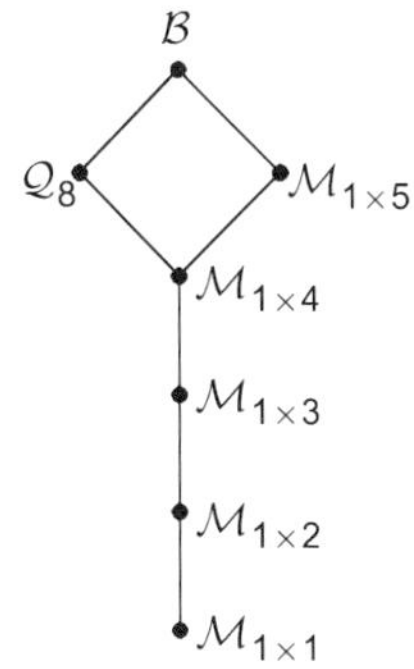

The Jordan classification of E_6

Since E_6 has a unique maximal abelian inner ideal up to conjugation, whose subquotient is isomorphic to the Bi-Cayley Jordan pair $\mathcal{B}$, the Lie algebra E_6 and the Jordan pair $\mathcal{B}$ have the same inner ideal structure un to Jordan isomorphism. However, the 2-order outer automorphism of E_6 connecting the inner ideals $B_{\{1\}}$ and $B_{\{6\}}$ yields two conjugate copies of the inner ideal structure of the Bi-Cayley Jordan pair $\mathcal{B}$ within E_6. This explains the apparently contradictory fact that while in $\mathcal{B}$ there are two 4-dimensional point spaces which are not conjugate, in E_6 there is a unique 4-dimensional inner ideal up to conjugation. This information about the inner ideals of E_6, their subquotients, and their corresponding TKK-algebras, is gathered in the next table:

Abelian inner ideals	dimensions	subquotients	TKK algebras
$B_{\{1\}}$	16	$\mathcal{B}$	E_6
$B_{\{1,6\}}$	8	$\mathcal{Q}_8$	D_5
$B_{\{1,3\}}$	5	$\mathcal{M}_{1\times 5}$	A_5
$B_{\{1,3,6\}}$	4	$\mathcal{M}_{1\times 4}$	A_4
$B_{\{1,3,5,6\}}$	3	$\mathcal{M}_{1\times 3}$	A_3
$B_{\{1,3,4,5,6\}}$	2	$\mathcal{M}_{1\times 2}$	A_2
$B_{\{1,2,3,4,5,6\}}$	1	$\mathcal{M}_{1\times 1}$	A_1

The inner ideal structure of E_7. Choose the system of positive roots of E_7 given in [**Bou81**, Planche V(II)] with maximal root

$$2\alpha_1 + +2\alpha_2 + 3\alpha_3 + 4\alpha_4 + 3\alpha_5 + 2\alpha_6 + \alpha_7.$$

Following the process described above, every nonzero abelian inner ideal of E_7 is conjugate to one in the next diagram.

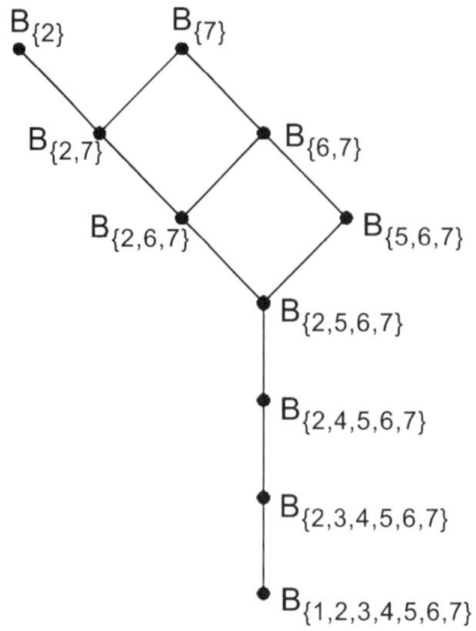

The Lie classification of E_7

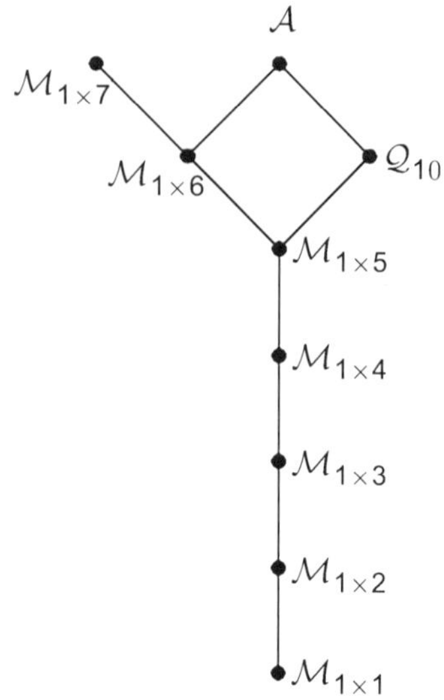

The Jordan classification of E_7

Note that E_7 contains two maximal abelian inner ideals up to conjugation: the 7-dimensional point space $B_{\{2\}}$, and $B_{\{7\}}$, whose subquotient is the Albert Jordan pair $\mathcal{A}$. So the lattice of the abelian inner ideals of E_7 contains the lattices of inner ideals of the Jordan pairs $\mathcal{M}_{1\times 7}$ and $\mathcal{A}$. Note also that the 5-dimensional inner ideals $B_{\{2,6,7\}}$ and $B_{\{5,6,7\}}$ are not conjugate, since $B_{\{5,6,7\}}$ is a maximal point space, while $B_{\{2,6,7\}}$ is contained in the 6-dimensional point space $B_{\{2,7\}}$; yet both $B_{\{2,6,7\}}$ and $B_{\{5,6,7\}}$ yield the same subquotient, $\mathcal{M}_{1\times 5}$, as point spaces of the same dimension. Additional information is gathered in the next table:

Abelian inner ideals	dimensions	subquotients	TKK algebras
$B_{\{7\}}$	27	$\mathcal{A}$	E_7
$B_{\{2\}}$	7	$\mathcal{M}_{1\times 7}$	A_7
$B_{\{6,7\}}$	10	$\mathcal{Q}_{10}$	D_6
$B_{\{2,7\}}$	6	$\mathcal{M}_{1\times 6}$	A_6
$B_{\{2,6,7\}}$ $B_{\{5,6,7\}}$	5	$\mathcal{M}_{1\times 5}$	A_5
$B_{\{2,5,6,7\}}$	4	$\mathcal{M}_{1\times 4}$	A_4
$B_{\{2,4,5,6,7\}}$	3	$\mathcal{M}_{1\times 3}$	A_3
$B_{\{2,3,4,5,6,7\}}$	2	$\mathcal{M}_{1\times 2}$	A_2
$B_{\{1,2,3,4,5,6,7\}}$	1	$\mathcal{M}_{1\times 1}$	A_1

The inner ideal structure of E_8. Choose the system of positive roots of E_8 given in [**Bou81**, Planche V(II)] with maximal root

$$2\alpha_1 + +3\alpha_2 + 4\alpha_3 + 6\alpha_4 + 5\alpha_5 + 42\alpha_6 + 3\alpha_7 + 2\alpha_8.$$

Following the process described above, every nonzero abelian inner ideal of E_8 is conjugate to one in the next diagram.

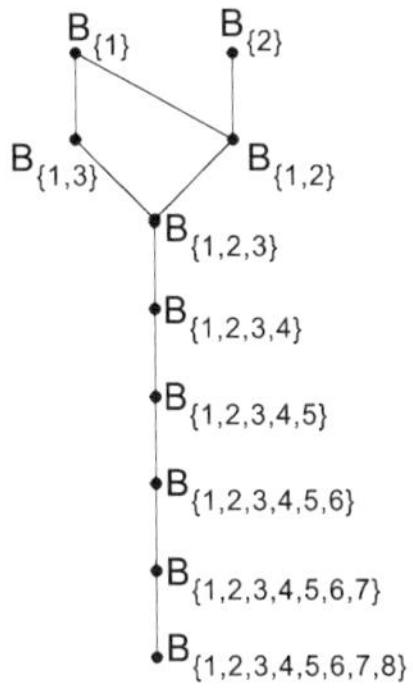

The Lie classification of E_8

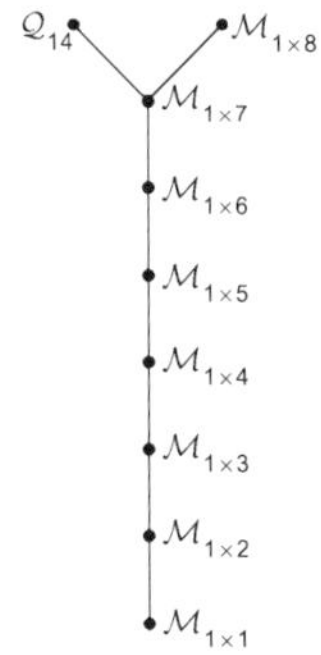

The Jordan classification of E_8

The inner ideals $B_{\{1,2\}}$ and $B_{\{1,3\}}$ are not conjugate. In fact, $B_{\{1,3\}}$ is a maximal point space, while $B_{\{1,2\}}$ is contained in the 8-dimensional point space $B_{\{2\}}$ (see the table below for dimensions). However, both $B_{\{1,2\}}$ and $B_{\{1,3\}}$ yield the same subquotient, $\mathcal{M}_{1\times 7}$. Further information about the inner ideal structure of E_8 is is given in the next table:

Abelian inner ideals	dimensions	subquotients	TKK algebras
$B_{\{1\}}$	14	$\mathcal{Q}_{14}$	D_8
$B_{\{2\}}$	8	$\mathcal{M}_{1\times 8}$	A_8
$B_{\{1,2\}}$ $B_{\{1,3\}}$	7	$\mathcal{M}_{1\times 7}$	A_7
$B_{\{1,2,3\}}$	6	$\mathcal{M}_{1\times 6}$	A_6
$B_{\{1,2,3,4\}}$	5	$\mathcal{M}_{1\times 5}$	A_5
$B_{\{1,2,3,4,5\}}$	4	$\mathcal{M}_{1\times 4}$	A_4
$B_{\{1,2,3,4,5,6\}}$	3	$\mathcal{M}_{1\times 3}$	A_3
$B_{\{1,2,3,4,5,6,7\}}$	2	$\mathcal{M}_{1\times 2}$	A_2
$B_{\{1,2,3,4,5,6,7,8\}}$	1	$\mathcal{M}_{1\times 1}$	A_1

The inner ideal structure of F_4. Choose the system of positive roots for F_4 described in [**Bou81**, Planche VIII (II)]. The nonzero abelian inner ideals of F_4 (up to conjugation), and their associated subquotients are:

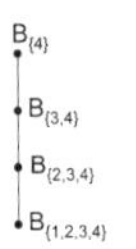

The Lie classification of F_4

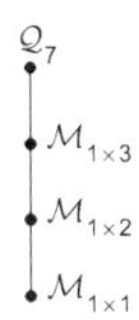

The Jordan classification of F_4

This information about inner ideals and subquotients, together with their corresponding TKK-algebras and Lie types, is gathered in the next table:

Abelian inner ideals	dimensions	subquotients	TKK algebras
$B_{\{4\}}$	7	$\mathcal{Q}_7$	B_4
$B_{\{3,4\}}$	3	$\mathcal{M}_{1\times 3}$	A_3
$B_{\{2,3,4\}}$	2	$\mathcal{M}_{1\times 2}$	A_2
$B_{\{1,2,3,4\}}$	1	$\mathcal{M}_{1\times 2}$	A_1

The inner ideal structure of G_2. Choose the system of positive roots for G_2 described in [**Bou81**, Planche IX (II)]. The nonzero abelian inner ideals of G_2 (up to conjugation), and their associated subquotients are:

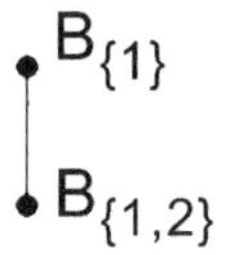

The Lie classification of G_2

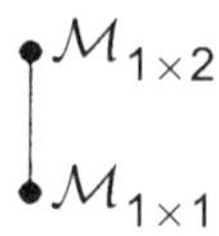

The Jordan classification of G_2

Further information about the inner ideal structure of G_2 is collected in the next table:

Abelian inner ideals	dimensions	subquotients	TKK algebras
$B_{\{1\}}$	2	$\mathcal{M}_{1\times 2}(F)$	A_2
$B_{\{1,2\}}$	1	$\mathcal{M}_{1\times 1}(F)$	A_1

13.6. Exercises

EXERCISE 13.88. Let U be an isotropic inner ideal of L and Ω a Φ-submodule of $Z\cap L$. Show that $U+\Omega$ is an abelian inner ideal of L.

EXERCISE 13.89. Let B be an inner ideal of a ring R. Show that $M=B+B^2$ is an inner ideal and a subring of R.

EXERCISE 13.90. Let M be an additive subgroup of a ring R. Show that the following conditions are equivalent:

(i) M is an inner ideal and a subring of R.
(ii) There exists a left ideal $\mathcal{L}$ and a right ideal $\mathcal{R}$ of the ring R such that

$$\mathcal{R}\mathcal{L}\subset M\subset \mathcal{R}\cap\mathcal{L}.$$

EXERCISE 13.91. Let M be an inner ideal of a ring R such that $M^2\subset M$. Show:

(i) If R is von Neumann regular, then $M=\mathcal{R}\cap\mathcal{L}=\mathcal{R}\mathcal{L}$, where $\mathcal{R}$ is a right ideal and $\mathcal{L}$ a left ideal of R.
(ii) If R is a semiprime Artinian ring, then $M=eRf$, where e and f are idempotents of R.

EXERCISE 13.92. Show that the pair of maps (φ^+,φ^-), where $\varphi^+(z)=[x,z]$ and $\varphi^-(v)=[v,y]$, $z,v\in H^\perp$, defines an isomorphism of the Clifford Jordan pair $C(H^\perp,\langle\,,\rangle)$ onto the Jordan pair $([x,H^\perp],[y,H^\perp])$.

EXERCISE 13.93. Via TKK-algebra, show that any inner ideal $M\subset\mathcal{F}(X_1,X_2)$ of the Jordan pair $(\mathcal{F}(X_1,X_2),\mathcal{F}(X_2,X_1))$ where $(X_1,Y_1,\langle\,,\rangle_1)$ and $(X_2,Y_2,\langle\,,\rangle_2)$ are pairs of dual vector spaces over a division ring Δ, is of the form $M=W_1^*V_2$, where $W_1\le Y_1$ and $V_2\le X_2$.

EXERCISE 13.94. Let X be a vector space of dimension greater than 2 with a nondegenerate symmetric bilinear form $\langle\,,\rangle$ over a field $\mathbb{F}$ of characteristic not 2. Show that any inner ideal M of the Jordan algebra $\mathrm{Sym}(\mathcal{F}_X(X),*)$, where $*$ denotes the adjoint involution, is of the form $M=\mathrm{Sym}(V^*V,*)$, where V is a subspace of X.

EXERCISE 13.95. Given X as above, show that any inner ideal of the Jordan triple system $\mathrm{Skew}(\mathcal{F}_X(X),*)$ is either $\mathrm{Skew}(V^*V,*)$, where V is a subspace of X, or a point space of $\mathfrak{fo}(X,\langle\,,\rangle)$.

EXERCISE 13.96. In the Lie algebra $K=\mathfrak{fo}(X,\langle\,,\rangle)$, consider the point space $[x,S]$. Our aim is to endow K with a 7-grading such that $K_3=[x,S]$. To this end, show the following assertions:

(i) Given a basis $\{x=x_1,x_2,\dots,x_n\}$ of S , there exist $y_1,y_2,\dots y_n\in X$ such that $\langle x_i,y_j\rangle=\delta_{ij}$ and $\langle y_i,y_j\rangle=0$ for all $1\le i,j\le n$.
(ii) Let $U=\mathbb{F}x_2\oplus\cdots\oplus\mathbb{F}x_n$, $V=\mathbb{F}y_2\oplus\cdots\oplus\mathbb{F}y_n$, $W=(\mathbb{F}x_1\oplus U\oplus V\oplus\mathbb{F}y_1)^\perp$. Then $X=\mathbb{F}y_1\oplus V\oplus W\oplus U\oplus\mathbb{F}x_1$.
(iii) Let e_0, e_1, e_2, e_3, e_4 be the projections of X onto $\mathbb{F}y_1$, V, W, U, $\mathbb{F}x_1$. Then e_0, e_1, e_2, e_3, e_4 are idempotents of the ring $\mathcal{L}_X(X)$ with $e_0^*=e_4$, $e_1^*=e_3$, and $e_2^*=e_2$.
(iv) These idempotents induce a 9-grading in the simple ring $R=\mathcal{F}_X(X)$ with $R_k=\sum_{i-j=k}e_iRe_j$.
(v) Each R_k is $*$-invariant with $\mathrm{Skew}(R_4,*)=\mathrm{Skew}(R_{-4},*)=0$.
(vi) $K=\mathfrak{fo}(X,\langle\,,\rangle)$ admits a 7-grading, with $K_i=\mathrm{Skew}(R_i,*)$.
(vii) $[x,S]=K_3$. [2]
(viii) The the inner ideals $[x,S]$ and $[y_1,V]$ are complemented by each other.

EXERCISE 13.97. Write a chain of inner ideals of maximal length for each one of the Lie algebras A_5,B_5,C_5, and D_5.

[2]This is an example of an *exceptional grading* (see [**Zel84b**] for definition). As proved by O. Smirnov in [**Smi99**, Lemma 6.4 and Theorem 6.5], the minimal possible dimension for the space X is 7.

CHAPTER 14

Classical Infinite-Dimensional Lie Algebras

As an application of Zelmanov's theorem for simple Lie algebras having a finite $\mathbb{Z}$-grading (Section 14.1) and the fact that any nonzero von Neumann regular element in a Lie algebra gives rise to a 5-grading, we determine in Section 14.2 the simple and nondegenerate Lie algebras over a field of characteristic 0 or greater than 7 containing minimal abelian inner ideals. Then, in Section 14.3, we give "Jordan-proof" of Baranov's structure theorem for simple finitary Lie algebras over an algebraically closed field of characteristic 0. This result is extended in Section 14.4 to strongly prime Lie algebras containing extremal elements. Finally, in Section 14.5, we give a new proof of the characterization, due to A. A. Baranov and J. Rowley, of simple locally finite Lie algebras of diagonal type as those containing a nonzero Jordan element, i.e. a nonzero abelian inner ideal. As observed in Section 14.6, over a field of characteristic 0, simple Lie algebras with a nontrivial finite $\mathbb{Z}$-grading are generated by nilpotent elements, in fact spanned by Jordan elements. I think it could be interesting to study this class of Lie algebras.

14.1. Simple Lie algebras with a finite $\mathbb{Z}$-grading

Recall that given an associative algebra R with center $Z = Z(R)$, we set $\overline{R}'$ to denote the Lie algebra $R'/R' \cap Z$, where $R' = [R, R]$, and that if R has an involution $*$, we write $\overline{K}'$ to denote the Lie algebra $K'/K' \cap Z$, where $K = \mathrm{Skew}(R, *)$ and $K' = [K, K]$. Zelmanov's theorem for simple Lie algebras with a finite grading [**Zel83a**, Theorem 1] can be rephrased as follows.

THEOREM 14.1. *Suppose that $L = L_{-n} \oplus \cdots \oplus L_n$, $L_n + L_{-n} \neq 0$, is a simple $(2n+1)$-graded Lie algebra over a field $\mathbb{F}$ of characteristic 0 or $p \geq 4n+1$. Then L is isomorphic to one of the following graded Lie algebras:*

(1) $\overline{R}'$, *where $R = R_{-n} \oplus \cdots \oplus R_n$ is a simple associative $\mathbb{Z}$-graded algebra.*

(2) $\overline{K}'$, $K = \mathrm{Skew}(R, *)$, *where $R = R_{-m} \oplus \cdots \oplus R_m$ for $m \geq n$, is a simple associative $\mathbb{Z}$-graded algebra with involution and $R_i^* = R_i$, $-n \leq i \leq n$.*

(3) *A Lie algebra of one of the types G_2, F_4, E_6, E_7, or E_8.*

REMARKS 14.2. (1) We remind the reader of the fact that any Lie algebra satisfying the conditions of Theorem 14.1 is nondegenerate (Propostion 3.27).

(2) O. Smirnov describes in [**Smi97**] the finite $\mathbb{Z}$-gradings of a simple associative algebra R. If R is unital, such a grading $R = \bigoplus_{i=-n}^{n} R_i$ is given by a complete system of orthogonal idempotents $\mathcal{E} = \{e_0, e_1, \ldots, e_n\}$, i.e. $R_i = \sum_{p-q=i} e_p R e_q$ for $i = -n, \ldots, n$. To extend his description of gradings to the case of a nonunital algebra, he introduces in [**Smi97**, Definition 4.4] the notion of a complete system of orthogonal submodules: $\{H_i : i = 0, \ldots, n\}$, $H = HRH$ for $H = \sum_{i=0}^{n} H_i$, and $H_i H_j = 0$ for $i \neq j$. Then $R = \sum_{p,q=0}^{n} R_{p,q}$, where $R_{p,q} = H_p R H_q$. In fact,

as observed by M. Siles in [**SM06**], one can find a complete system of orthogonal idempotents $\mathcal{E} = \{e_0, e_1, \ldots, e_n\}$ in the maximal left quotient algebra Q of R which induces the grading of R, i.e. $e_p R e_q = H_p R H_q$.

(3) If R has an involution $*$, it is proved in [**Smi99**, Corollary 4.5] that any finite $\mathbb{Z}$-pregrading of the Lie algebra $\overline{K}' = \sum_{i=-n}^{n} \overline{K}'_i$ is induced by a unique finite $\mathbb{Z}$-grading of $R = \sum_{i=-m}^{m} R_i$, $n \leq m$ and $R_i^* = R_i$. The case where $n < m$ is quite exceptional, since then R has minimal one-sided ideals and $*$ is orthogonal [**Smi99**, Theorem 5.4], equivalently, R contains a Clifford element (Proposition 8.83). This is the reason for the appearance in the original formulation of Zelmanov's theorem [**Zel83a**, Theorem 1] of Lie algebras which are the TKK-algebra of a Clifford Jordan pair (finitary orthogonal algebra, or type D_4 in the finite-dimensional case). The difference between n and m is nevertheless small: $m \leq 2n$ and $R_i = 0$ for $n < |i| < m$ [**Smi99**, Theorem 6.6].

COROLLARY 14.3. *Let L be a simple Lie algebra over a field $\mathbb{F}$ of characteristic* 0 *or greater than* 7 *containing a nonzero von Neumann regular element. Then $L = L_{-2}^h \oplus L_{-1}^h \oplus L_0^h \oplus L_1^h \oplus L_2^h$, $h = [e, f]$ for (e, f) an idempotent of L, so L is isomorphic to one of the Lie algebras listed in* Theorem 14.1.

PROOF. Given a nonzero von Neumann regular element $e \in L$, extend it to a $\mathfrak{sl}_2$-triple (e, f, h) (Lemma 5.8). Then we have by Theorem 5.11 that (e, f) is an idempotent of L with associated $\mathbb{Z}$-grading $L = L_{-2}^h \oplus L_{-1}^h \oplus L_0^h \oplus L_1^h \oplus L_2^h$, $h = [e, f]$, and Jordan pair $V(e, f) = (\mathrm{ad}_e^2 L, \mathrm{ad}_f^2 L)$. Now Theorem 14.1 applies. □

14.2. Simple Lie algebras with minimal abelian inner ideals

In this section we describe the simple Lie algebras containing minimal abelian inner ideals. As a consequence, we will get in the next section that any simple Lie algebra containing a nonzero extremal element over an algebraically closed field of characteristic 0 is finitary, thus proving the converse of Theorem 8.78.

THEOREM 14.4. [**DFFLGGL08**, Theorem 5.1] *Let L be a Lie algebra over a field of characteristic* 0 *or greater than* 7. *Then L is simple, nondegenerate, and contains minimal abelian inner ideals if and only if it is isomorphic to one of the following algebras:*

(1) $\overline{R}'$, *where R is a simple associative algebra with minimal one-sided ideals which is not a division algebra.*
(2) $\overline{K}'$, $K = \mathrm{Skew}(R, *)$, *where R is a simple associative algebra with isotropic involution and minimal one-sided ideals such that either $Z(R) = 0$ or* $\dim_{Z(R)} R > 16$.
(3) *A Lie algebra of one of the types G_2, F_4, E_6, E_7 or E_8 with nonzero Jordan elements.*

PROOF. We begin by checking case by case that each one of the Lie algebras listed above is simple and contains a minimal abelian inner ideal:

(1) Since R is not a division algebra, it contains a minimal right ideal aR with $a^2 = 0$; then $aRa = [aR, a] \subset [R, R]$ and hence the Lie algebra $\overline{R}'$ is simple (Theorem 2.5), nondegenerate (Propositions 3.35 and 4.7) and contains minimal abelian inner ideals (Theorem 13.42).

(2) It follows from Theorem 13.86.

(3) Since any simple exceptional Lie algebra is finite-dimensional over its centroid, the existence of a nonzero Jordan element plus the restriction on the characteristic imply that this algebra is nondegenerate and contains minimal abelian inner ideals.

Suppose conversely that L is a simple nondegenerate Lie algebra containing a minimal abelian inner ideal B. Then $B = \operatorname{ad}_b^2 L$ for any nonzero $b \in B$, so L contains a nonzero von Neumann regular element and Corollary 14.3 applies. Therefore L is one of the Lie algebras listed in Theorem 14.1. Using the same references as in the previous paragraph we get the coincidence of R with its socle in (1) and (2), and the fact that $*$ is isotropic in (2). □

Let R be a simple ring with minimal inner ideals realized as the ring $\mathcal{F}_Y(X)$ defined by a pair of dual vector spaces $(X, Y, \langle\, ,\rangle)$ over a division algebra Δ. Then the Lie algebra $\overline{R}'$ can be realized as $\mathfrak{fsl}_Y(X)/\mathfrak{fsl}_Y(X) \cap Z$, where $\mathfrak{fsl}_Y(X)$ is the special linear algebra (Section 2.2) and Z denotes the center of $\mathcal{F}_Y(X)$. If X is infinite-dimensional over Δ, then $Z = 0$; otherwise $Z = Z(\Delta)1_X$.

Recall that for any simple nondegenerate Lie algebra L, we denote by $\mathrm{DJP}(L)$ the isomorphic class of its division Jordan pairs (11.55).

PROPOSITION 14.5. *Let $\overline{R}' = \mathfrak{fsl}_Y(X)/\mathfrak{fsl}_Y(X) \cap Z$ be as above, where R is not a division ring, i.e. $\dim_\Delta X \geq 2$. We have:*

(i) $\mathrm{DJP}(\overline{R}') \cong (\Delta^+, \Delta^+)$.

(ii) $\Gamma(\overline{R}') = Z(\Delta)1_{\overline{R}'} \cong \Gamma(R)$.

PROOF. (i) Take $x_1, x_2 \in X$, $y_1, y_2 \in Y$ such that $\langle x_i, y_j\rangle = \delta_{ij}$, and set $e = y_1^* x_2$, $f = y_2^* x_1$. Then

$$e^2 = f^2 = 0, \ [[e, f], e] = 2efe = 2e, \ [[e, f], f] = -2fef = -2f,$$

so (e, f) is an idempotent of the Lie algebras R^- and R', and can be also regarded as an idempotent of $\overline{R}'$, with associated division Jordan pair

$$V(e, f) = (L_2^h, L_{-2}^h) = (eR'e, fR'f) = (eRe, fRf) = (y_1^*\Delta x_2, y_2^*\Delta x_1) \cong (\Delta^+, \Delta^+),$$

via the Jordan pair isomorphism $(\alpha, \beta) \mapsto (y_1^*\alpha x_2, y_2^*\beta x_1)$, for all $\alpha, \beta \in \Delta$.

(ii) By Lemma 11.33(ii), $\Gamma(\overline{R}') = \Gamma(V(e, f))$, and by [**McC99**, 3.5],

$$\Gamma(\Delta^+, \Delta^+) = \{(l_\alpha, l_\alpha) : \alpha \in Z(\Delta)\} \cong Z(\Delta) \cong \Gamma(R),$$

where l_α denotes the dilatation of ratio α. □

Suppose now that R is a simple ring with isotropic involution and minimal one-sided ideals such that either $Z = 0$ or $\dim_Z R > 16$. Then R can be realized as the algebra $\mathcal{F}_X(X)$, where X is a left vector space with a nondegenerate skew-Hermitian or symmetric bilinear form $\langle\, ,\rangle$ over a division ring with involution $(\Delta, -)$, and where $*$ is the adjoint involution. Since $*$ is isotropic, X contains nonzero isotropic vectors relative to the form $\langle\, ,\rangle$.

PROPOSITION 14.6. *Let R be as above and $K = \mathrm{Skew}(R, *)$. We have:*

(i) *If $\langle\, ,\rangle$ is skew-Hermitian, then* $\mathrm{DJP}(\overline{K}') \cong (\mathrm{Sym}(\Delta, -), \mathrm{Sym}(\Delta, -))$ *and* $\Gamma(\overline{K}') = \mathrm{Sym}(Z(\Delta), -)1_{\overline{K}'}$.

(ii) *If* $\langle\,,\rangle$ *is symmetric* (Δ *is then a field* $\mathbb{F}$ *with the identity as involution), then* $\mathrm{DJP}(\overline{K}')$ *is the Clifford Jordan pair defined by an anisotropic symmetric bilinear form on a vector space over* $\mathbb{F}$, *and* $\Gamma(K) = \mathbb{F}1_K$.

PROOF. (i) Let $x \in X$ be a nonzero isotropic vector. Then there exists an isotropic vector $y \in X$ such that $\langle x, y\rangle = 1$. Set $e := x^*x$ and $f := y^*y$. As in the proof of Proposition 14.5, (e, f) is an idempotent of the Lie algebras K, K' and $\overline{K}'$, with associated division Jordan pair

$$\begin{aligned} V(e,f) &= (eKe, fKf) = (x^* \operatorname{Sym}(\Delta,-)x, y^* \operatorname{Sym}(\Delta,-)y) \\ &\cong (\operatorname{Sym}(\Delta,-), \operatorname{Sym}(\Delta,-)) \end{aligned}$$

via the Jordan pair isomorphism $(\alpha,\beta) \mapsto (x^*\alpha x, y^*\beta y)$, for all $\alpha,\beta \in \operatorname{Sym}(\Delta,-)$. To prove that $\Gamma(\overline{K}') = \operatorname{Sym}(Z(\Delta),-)1_{\overline{K}'}$ we follow the path sketched in the proof of Proposition 14.5. By Lemma 11.33, $\Gamma(\overline{K}') = \Gamma(V(e,f))$, and by [**McC99**, 3.6],

$$\Gamma(\operatorname{Sym}(\Delta,-), \operatorname{Sym}(\Delta,-)) = \{(l_\alpha, l_\alpha) : \alpha \in \operatorname{Sym}(Z(\Delta),-)\}.$$

(ii) If $\langle\,,\rangle$ is symmetric (over a field $\mathbb{F}$), then $\overline{K}' = K' = K$ is the finitary orthogonal algebra $\mathfrak{fo}(X, \langle\,,\rangle)$. Since X contains nonzero isotropic vectors, we can decompose $X = H \oplus H^\perp$, where $H = \mathbb{F}x \oplus \mathbb{F}y$ is the hyperbolic plane defined by a hyperbolic pair (x, y). There are two possibilities:

(1) $H^\perp$ is anisotropic. In this case we have that $[x, H^\perp]$ is a minimal abelian inner ideal of $\mathfrak{fo}(X, \langle\,,\rangle)$ with associated division Jordan pair

$$\mathrm{DJP}(\mathfrak{fo}(X, \langle\,,\rangle) = ([x, H^\perp], [y, H^\perp]) \cong C(H^\perp, \langle\,,\rangle),$$

where, we recall, the isomorphism is given by $([x, z], [y, v]) \mapsto (z, -v)$, $z, v \in H^\perp$. Since $\Gamma(C(H^\perp, \langle\,,\rangle)) = \{(l_\alpha, l_\alpha) : \alpha \in \mathbb{F}\}$ by [**McC99**, Theorem 5.3], we obtain as in the previous cases that $\Gamma(\mathfrak{fo}(X, \langle\,,\rangle)) \cong \mathbb{F}$.

(2) $H^\perp$ is isotropic. Then $H^\perp$ contains a hyperbolic pair (v, z). It is easy to see that $([x, z], [y, v])$ is a division idempotent of $\mathfrak{fo}(X, \langle\,,\rangle)$, with associative division Jordan pair isomorphic to $(\mathbb{F}, \mathbb{F})$, which can be regarded as the Clifford Jordan pair defined by the one-dimensional vector space $\mathbb{F}$ with the quadratic form defined by the product. As before, $\Gamma(K) = \mathbb{F}1_K$. □

14.3. Simple finitary Lie algebras revisited

Recall that a Lie algebra over a field $\mathbb{F}$ is said to be finitary if it is isomorphic to a subalgebra of the Lie algebra $\mathfrak{fgl}(X)$ for some vector space X over $\mathbb{F}$. In this section we characterize the simple finitary Lie algebras over a field $\mathbb{F}$ of characteristic 0 or $p > 7$ containing minimal abelian inner ideals. In particular, simple finitary Lie algebras over a an algebraically closed field of characteristic 0, where the existence of minimal abelian inner ideals (in fact, of extremal elements) is guaranteed (Corollary 8.79).

THEOREM 14.7. *Let* L *be an infinite-dimensional central simple nondegenerate Lie algebra over a field* $\mathbb{F}$ *of characteristic* 0 *or* $p > 7$ *containing minimal abelian inner ideals. Then* L *is finitary if and only if for any division element* $a \in L$ *the division Jordan algebra* L_a *is PI. In this case,* L *is isomorphic to one of the following:*

(i) $\mathfrak{fsl}_Y(X)$, *where* $(X, Y; \langle\,,\rangle)$ *is an infinite-dimensional pair of dual vector spaces over a finite-dimensional division associative* $\mathbb{F}$*-algebra* Δ.

(ii) $[\mathrm{Skew}(\mathcal{F}_X(X), *), \mathrm{Skew}(\mathcal{F}_X(X), *)]$, *where* X *is an infinite-dimensional left vector space endowed with a nondegenerate isotropic skew-Hermitian form* $\langle\,,\rangle$ *over a finite-dimensional division* $\mathbb{F}$*-algebra with involution* Δ*, and where* $*$ *is the adjoint involution with respect to* $\langle\,,\rangle$.

(iii) $\mathfrak{fo}(X, \langle\,,\rangle)$, *where* X *is infinite-dimensional over* $\mathbb{F}$.

PROOF. Suppose that L is finitary and let $a \in L$ be a division element. By Proposition 8.77, L_a is an algebraic Jordan algebra of bounded degree and therefore a PI-algebra (Lemma 8.24).

Conversely, suppose that the class of isotopy $\mathrm{DJA}(L)$ of division Jordan algebras of L is PI (see Remarks 8.26(1)). Then, by Theorem 14.4, L is isomorphic to one of the following algebras:

(1) $\overline{R}'$, where R is a simple associative algebra with minimal one-sided ideals which is not a division algebra.

(2) $\overline{K}'$, $K = \mathrm{Skew}(R, *)$, where R is a simple associative algebra with isotropic involution and minimal one-sided ideals such that either $Z(R) = 0$ or $\dim_{Z(R)} R > 16$.

Let $L = \overline{R}'$ as in (1). By Proposition 14.5, $L = \mathfrak{fsl}_Y(X)/\mathfrak{fsl}_Y(X) \cap Z$, where $(X, Y, \langle\,,\rangle)$ is a pair of dual vector spaces over a central division $\mathbb{F}$-algebra Δ, and $DJA(L)$ is the Jordan algebra Δ^+. Since Δ^+ is PI by hypothesis, Δ is an associative PI-algebra and hence finite-dimensional over its center $\mathbb{F}$ by Kaplansky's theorem [**Coh03**, Theorem 8.3.6]. Then the vector space X is necessarily infinite-dimensional over Δ (because we have assumed that L is infinite-dimensional over $\mathbb{F}$) and hence $\mathfrak{fsl}_Y(X) \cap Z = 0$. So L is the special finitary Lie algebra $\mathfrak{fsl}_Y(X)$.

Suppose now that $L = \overline{K}'$ as in (2), ($R = \mathcal{F}_X(X)$, where X is a left vector space endowed with a nondegenerate isotropic symmetric or skew-Hermitian form $\langle\,,\rangle$ over a division algebra with involution $(\Delta, -)$, and where $*$ is the adjoint involution with respect to $\langle\,,\rangle$). If $\langle\,,\rangle$ is skew-Hermitian, we have by Proposition 14.6 that $\mathrm{DJA}(L) \cong \mathrm{Sym}(\Delta, -)$ and $\Gamma(L) = \mathrm{Sym}(\Delta, -)1_L$. Since the Jordan algebra $\mathrm{Sym}(\Delta, -)$ is PI by hypothesis, the associative algebra Δ is also PI (Amitsur's theorem [**Her76**, Theorem 6.5.2]) and hence finite-dimensional over $\mathbb{F}$. Then X is infinite-dimensional over Δ and L is the finitary Lie algebra

$$[\mathrm{Skew}(\mathcal{F}_X(X), *), \mathrm{Skew}(\mathcal{F}_X(X), *)].$$

If $\langle\,,\rangle$ is symmetric, then $L = K' = K$ is the finitary orthogonal Lie algebra $\mathfrak{fo}(X, \langle\,,\rangle)$. $\square$

COROLLARY 14.8. *Let* L *be a simple infinite-dimensional Lie algebra over a field of characteristic* 0 *or* $p > 7$ *containing an extremal element* e *which is not an absolute zero divisor. Then* L *is isomorphic to one of the simple finitary Lie* $\mathbb{F}$*-algebras:* $\mathfrak{fsl}_Y(X)$, $\mathfrak{fo}(X, \langle\,,\rangle)$, $\mathfrak{fsp}(X, \langle\,,\rangle)$.

PROOF. Let $e \in L$ be an extremal element which is not an absolute zero divisor, i.e. $\mathrm{ad}_e^2 L = \mathbb{F}e$ is a one-dimensional minimal inner ideal. Then L is nondegenerate (Corollary 14.3 and Propostion 3.27), central over $\mathbb{F}$ (Lemma 6.5), and $\mathrm{DJA}(L) = \mathbb{F}$. Now Theorem 14.7 applies. $\square$

Now the promised converse of Corollary 8.79.

COROLLARY 14.9. *Let L be a simple infinite-dimensional Lie algebra over an algebraically closed field of characteristic 0 containing a nonzero extremal element. Then L is finitary.*

PROOF. By Corollary 3.24, L is nondegenerate. Now Corollary 14.8 applies. □

THEOREM 14.10. *Let L be a central simple Lie PI-algebra over a field of characteristic 0. If L contains a nonzero von Neumann regular element. Then L is finite-dimensional.*

PROOF. As in the proof of Corollary 14.3, extend e to an idempotent (e, f) of L and consider the corresponding 5-grading $L = L^h_{-2} \oplus L^h_{-1} \oplus L^h_0 \oplus L^h_1 \oplus L^h_2$, $h = [e, f]$. By Theorem 11.32, the Jordan pair $V = (L^h_2, L^h_{-2}) = ([e], [f])$ is simple, and therefore so is the Jordan algebra $L_f \cong V_f = V^{(f)}$ (see Exercise 11.72). Since L_f is PI (Proposition 8.57), it follows from the structure theorem of simple Jordan PI-algebras (see Theorem 8.25) that L_f contains minimal inner ideals. So L contains division elements and for any division element $a \in L$, L_a is a division Jordan PI-algebra. By Theorem 14.7, L is finitary and therefore locally finite (Proposition 8.74). Then L is finite-dimensional by Lemma 10.1. □

Recall that a central simple Lie PI-algebra over a field of characteristic 0 is not necessarily finite-dimensional (Remark 8.58). However, it is not clear von Neumann regularity to be absolutely necessary. Any Jordan element $a \in L$ such that L_a is simple would work.

Classes of finitaryness. Let L be an infinite-dimensional central simple Lie algebra over a field $\mathbb{F}$ of characteristic 0. We can distinguish the following cases of finitaryness:

- $L = \mathfrak{fo}(X, \langle\, ,\rangle)$, where $\langle\, ,\rangle$ is anisotropic. This is precisely the case of a central simple finitary Lie algebra over $\mathbb{F}$ without nonzero abelian inner ideals.
- Let $L = \mathfrak{fo}(X, \langle\, ,\rangle)$ with $X = H \oplus H^\perp$, where $\dim_{\mathbb{F}} X > 2$, H is a hyperbolic plane, and $H^\perp$ is anisotropic, i.e. the Witt index of $\langle\, ,\rangle$ is one. In this case, DJP(L) is an infinite-dimensional Clifford Jordan pair over $\mathbb{F}$.
- L has a nonzero finite-dimensional abelian inner ideal. This last class of finitary Lie algebras have been studied by M. Brešar and the author proving the following.

THEOREM 14.11. [**BFL10**, Theorem 3.2] *Let L be an infinite-dimensional simple Lie algebra over a field $\mathbb{F}$ of characteristic 0. Then the following conditions are equivalent:*

(i) *There exists a derivation d of L such that d^2 is nonzero and has finite rank.*
(ii) *There exists a derivation d of L and $n \geq 2$ such that d^n is nonzero and has finite rank.*
(iii) *L contains a nonzero finite-dimensional abelian inner ideal.*
(iv) *L is finitary and there exists a nonzero linear map $a \in L \subset \mathcal{F}(X)$, where X is a vector space over $\mathbb{F}$, such that $a^2 = 0$.*

If $\mathbb{F}$ is algebraically closed, then the existence of a nonzero square zero $a \in L$ in (iv) *is automatic.*

Notice that in (iv), $\mathrm{ad}_a^2 L = aLa \subset a\mathcal{F}(X)a = a(X'^*X)a = W^*V$, where X' denotes here the dual of X, W and V are subspaces of X' and X respectively, and $\dim_{\mathbb{F}} W = \dim_{\mathbb{F}} V = \mathrm{rank}(a)$, so the abelian inner ideal $\mathrm{ad}_a^2 L$ has dimension less than or equal to the rank of a.

14.4. Strongly prime Lie algebras with extremal elements

All the vector spaces and Lie algebras considered in this section are over an algebraically closed field $\mathbb{F}$ of characteristic 0. Recall that a Lie algebra is called strongly prime if it is prime and nondegenerate.

LEMMA 14.12. *Let M be a simple Lie algebra realized as a subalgebra of its algebra of derivations* $\mathrm{Der}(M)$ *via the adjoint representation. Then any subalgebra L of* $\mathrm{Der}(M)$ *containing M is strongly prime.*

PROOF. By Proposition 2.14, L is prime and M is an essential ideal of L. And since M is nondegenerate (because $\mathrm{char}(\mathbb{F}) = 0$), it follows from Proposition 4.12 that L is nondegenerate as well. □

Suppose now that L is a strongly prime Lie algebra with minimal abelian inner ideals. Then $M := \mathrm{Soc}(L)$ is an ideal of L which is simple as a Lie algebra and, via the adjoint representation, $M \lhd L \leq \mathrm{Der}(M)$. We will use this fact to determine the strongly prime infinite-dimensional Lie algebras containing a nonzero extremal element. Since, by Corollary 14.8, M is isomorphic to one of the simple finitary Lie algebras: $\mathfrak{fsl}_Y(X)$, $\mathfrak{fo}(X, \langle\,,\rangle)$, or $\mathfrak{fsp}(X, \langle\,,\rangle)$, we just need to determine the algebras of derivations of each one of these algebras.

PROPOSITION 14.13. *Let X be infinite-dimensional in the three cases below. Then:*

(i) $\mathrm{Der}(\mathfrak{fsl}_Y(X)) \cong \mathfrak{gl}_Y(X)/\mathbb{F}1_X$.
(ii) $\mathrm{Der}(\mathfrak{fo}(X, \langle\,,\rangle)) \cong \mathfrak{o}(X, \langle\,,\rangle)$.
(iii) $\mathrm{Der}(\mathfrak{fsp}(X, \langle\,,\rangle)) \cong \mathfrak{sp}(X, \langle\,,\rangle)$.

PROOF. (i) Let $\mathcal{P} = (X, Y, \langle\,,\rangle)$ be an infinite-dimensional pair of dual vector spaces over $\mathbb{F}$. Consider the adjoint operator $a \mapsto \mathrm{ad}_a$ from $\mathfrak{gl}_Y(X)$ to $\mathrm{Der}(\mathfrak{fsl}_Y(X))$, whose kernel is $\mathbb{F}1_X$. We must show that this map is onto. Fix a pair of vectors $(x_0, y_0) \in X \times Y$ such that $\langle x_0, y_0\rangle = 1$ and consider the family $\{\mathcal{P}_\lambda\}_{\lambda\in\Lambda}$ of all subpairs of dual vector spaces $\mathcal{P}_\lambda = (X_\lambda, Y_\lambda, \langle\,,\rangle_\lambda)$ of $\mathcal{P}$ such that for each index λ, $\mathcal{P}_\lambda$ has finite dimension $n_\lambda \geq 2$ and $(x_0, y_0) \in X_\lambda \times Y_\lambda$. Then $\{\mathcal{P}_\lambda\}$ is a directed partially ordered set with respect to the inclusion and $\mathcal{P}$ is the direct limit of the $\mathcal{P}_\lambda$. Moreover, for each index λ, $\mathcal{P} = \mathcal{P}_\lambda \oplus \mathcal{P}_\lambda^\perp$, where $\mathcal{P}_\lambda^\perp = (Y_\lambda^\perp, X_\lambda^\perp, \langle\,,\rangle_\lambda^\perp)$ (see Proposition 2.24). Consequently, for each index λ, the Lie algebra $M := \mathfrak{fsl}_Y(X)$ contains a subalgebra $M_\lambda \cong \mathfrak{fsl}_{Y_\lambda}(X_\lambda) = \mathfrak{sl}_{n_\lambda}(\mathbb{F})$, which is simple since $n_\lambda \geq 2$, and M is the direct limit of the M_λ. Let $D \in \mathrm{Der}(M)$. Since $M = \bigcup M_\lambda$, for each index λ there exists $\mu_\lambda \in \Lambda$ such that both M_λ and $D(M_\lambda)$ are contained in M_{μ_λ}. Denote by D_λ the restriction of D to M_λ, regarded as a linear map of M_λ into M_{μ_λ}. By [**Jac79**, Theorem 9, page 80], there exists $d_\lambda \in M_{\mu_\lambda}$ such that $D_\lambda(a) = [d_\lambda, a]$ for all $a \in M_\lambda$. Furthermore, the restriction of d_λ to X_λ is uniquely determined up to a scalar multiple of the identity on X_λ, so *uniquely determined* by the additional condition $\langle x_0 d_\lambda, y_0\rangle = 0$. Define $d : X \to X$ by $xd = xd_\lambda$ whenever $x \in X_\lambda$ (notice that according to our notational convention, we keep writing linear maps of X acting on the right). Clearly, d is well defined, linear and satisfies $D(a) = [d, a]$ for all $a \in M = \mathfrak{f}sl_Y(X)$. Thus it remains to show that d has an adjoint relative to $\mathcal{P}$. We have that $[d, \mathfrak{f}sl_Y(X)] = D(\mathfrak{f}sl_Y(X)) \subset \mathfrak{f}sl_Y(X)$. Moreover, it follows from Lemma 2.25 that $\mathcal{F}_Y(X)d = (Y^*X)d = Y^*(Xd) \subset \mathcal{F}_Y(X)$. Hence $ad \in \mathcal{F}_Y(X)$ for all $a \in \mathfrak{f}sl_Y(X)$. Let us see that this fact implies that d has an adjoint, i.e. for any $y \in Y$ there exists a *unique* vector $y' \in Y$ such that $\langle xd, y\rangle = \langle x, y'\rangle$ for all $x \in X$. Indeed, given $y \in Y$ take a nonzero vector $x' \in X$ satisfying $\langle x', y\rangle = 0$. Then $y^*x' \in \mathfrak{f}sl_Y(X)$ and hence $d(y^*x') = [d, y^*x'] + (y^*x')d \in \mathcal{F}_Y(X)$, so $d(y^*x') = y'^*x'$ for a uniquely determined vector $y' \in Y$. Applying both terms of this equality to any $x \in X$, we get $\langle xd, y\rangle x' = \langle x, y'\rangle x'$, so $\langle xd, y\rangle = \langle x, y'\rangle$, since $x' \neq 0$.

The proofs of (ii) and (iii) are similar, and even easier than that of (i). To prove (ii), write $(X, \langle\, ,\rangle)$ as a direct limit of nondegenerate inner subspaces $(X_\lambda, \langle\, ,\rangle_\lambda)$ of finite dimensions $2n_\lambda \geq 8$ in order to assure that the corresponding subalgebras $M_\lambda \cong \mathfrak{f}o(X_\lambda)$ of $\mathfrak{f}o(X, \langle\, ,\rangle)$ are simple of type D_{n_λ} [**Jac79**, Theorem 9, page 141]. To prove (iii), take the nondegenerate inner subspaces $(X_\lambda, \langle\, ,\rangle_\lambda)$ of finite dimension $2n_\lambda \geq 6$ [**Jac79**, Theorem 8, page 140]. In both cases, the restrictions of d_λ to X_λ is uniquely determined and the linear operator $d : X \to X$, locally defined by the d_λ, satisfies $d^* = -d$. Let us finally show that the adjoint representation $a \to \mathrm{ad}_a$ is injective in both cases. Let $a \in \mathfrak{o}(X, \langle\, ,\rangle)$ be such that $[a, \mathfrak{f}o(X, \langle\, ,\rangle)] = 0$. Given $x \in X$, take $y \in X$ not belonging to the subspace spanned by $\{x, xa\}$, and let $z \in X$ be such that $\langle x, z\rangle = \langle xa, z\rangle = 0$ and $\langle y, z\rangle = 1$. The operator $y^*x - x^*y \in \mathfrak{f}o(X, \langle\, ,\rangle)$ and therefore, $[a, y^*x - x^*y] = -(y^*a)x + (x^*a)y - y^*(xa) + x^*(ya) = 0$. Evaluating this operator equality on z, we obtain $-\langle z, ya\rangle x - xa = 0$. Thus a is a scalar multiple of the identity, which is a contradiction since $a^* = -a$. Similarly, if $a \in \mathfrak{s}p(X, \langle\, ,\rangle)$ satisfies $[a, \mathfrak{f}sp(X, \langle\, ,\rangle)] = 0$, then for any nonzero $x \in X$ we have $[x^*x, a] = x^*(xa) + (xa)^*x = 0$. Let $y \in X$ be such that $\langle y, x\rangle = 1$. Evaluating the operator equality on y, we get $xa + \langle y, xa\rangle = 0$, so $a = 0$ as before. □

REMARK 14.14. The proof of the above proposition, taken from [**FLGGL04**, Theorem 6.2], is inspired in De La Harpe's methods for classical Banach–Lie algebras of compact operators on a Hilbert space [**DlH72**].

THEOREM 14.15. *Let L be an infinite-dimensional Lie algebra over $\mathbb{F}$. Then L is strongly prime and contains nonzero extremal elements if and only if it is, up to isomorphism, one of the following:*

(i) $(\mathfrak{f}sl_Y(X) + \mathbb{F}1_X)/\mathbb{F}1_X \leq L \leq \mathfrak{g}l_Y(X)/\mathbb{F}1_X$, *where (X, Y) is an infinite-dimensional pair of dual vector spaces over $\mathbb{F}$, with* $\mathrm{Soc}(L) = \mathfrak{f}sl_Y(X)$.

(ii) $\mathfrak{f}o(X) \leq L \leq \mathfrak{o}(X, \langle\, ,\rangle)$, *where X is an infinite-dimensional vector space over $\mathbb{F}$ with a nondegenerate symmetric bilinear form $\langle\, ,\rangle$. In this case* $\mathrm{Soc}(L) = \mathfrak{f}o(X)$.

(iii) $\mathfrak{f}sp(X) \leq L \leq \mathfrak{s}p(X)$, *where X is an infinite-dimensional vector space over $\mathbb{F}$ with a nondegenerate alternate bilinear form $\langle\, ,\rangle$, with* $\mathrm{Soc}(L) = \mathfrak{f}sp(X)$.

PROOF. By Lemma 14.12 and Proposition 14.13, the Lie algebras L listed above are strongly prime, and contain nonzero extremal elements by Examples 6.2. Suppose conversely that L is an infinite-dimensional strongly prime Lie algebra over $\mathbb{F}$ containing nonzero extremal elements. Then $\mathrm{Soc}(L)$ is a simple Lie algebra containing a nonzero extremal element and hence, by Corollary 14.8, isomorphic to one of the simple finitary Lie algebras $\mathfrak{f}sl_Y(X)$, $\mathfrak{f}o(X, \langle\, ,\rangle)$, $\mathfrak{f}sp(X, \langle\, ,\rangle)$. Using that

$L \subset \mathrm{Der}(\mathrm{Soc}(L))$, via the adjoint representation, and computing the algebra of derivations of each one of the finitary simple Lie algebras (Proposition 14.13), we complete the proof. □

14.5. Locally finite Lie algebras with abelian inner ideals

Recall that a (not necessarily associative) algebra A over a field $\mathbb{F}$ is said to be locally finite if every finitely generated subalgebra of A is finite-dimensional. As mentioned in the introduction of [**BR13**], although the full classification of simple locally finite algebras (over an algebraically closed field of characteristic 0) seems to be impossible to obtain, this is possible for finitary algebras (as we have just seen) and the so-called locally finite Lie algebras of diagonal type, which can be described as follows. Let R be a locally finite associative algebra over a an algebraically closed field of characteristic 0 and let L be nonzero subalgebra of R^-. It is clear that L is locally finite and algebraic, so it contains nonzero abelian inner ideals. A. Baranov and J. Rowley proved in [**BR13**, Theorem 1.1] that an infinite-dimensional locally finite simple Lie algebra over an algebraically closed field of characteristic 0 is of diagonal type if and only if has nonzero abelian inner ideals (in fact, either $L = R^-$ or $L = [\mathrm{Skew}(R,*), \mathrm{Skew}(R,*)$, where R is simple and locally finite [**BBZ04**]). An extension of this characterization to Lie algebras of positive characteristic $p > 7$ was carried out by H. Hennig in [**Hen14**] using Jordan methods to reduce the question to Lie algebras with a finite $\mathbb{Z}$-grading. We will adopt this Jordan point of view in our approach to Baranov–Rowley's characterization of simple locally finite Lie algebras of diagonal type.

Centroid and central polynomials. In this subsection, L will denote a Lie algebra over a field $\mathbb{F}$, Γ its centroid, and $\mathrm{Ad}(L)$ the multiplication ideal of L, i.e. the subalgebra of $\mathrm{End}_{\mathbb{F}}(L)$ generated by the adjoint operators ad_x, $x \in L$.

LEMMA 14.16. *Suppose that L is simple and locally finite and let $\gamma \in \Gamma \cap \mathrm{Ad}(L)$. Then γ is algebraic over $\mathbb{F}$.*

PROOF. Let $\gamma = \sum_{i=1}^{t} \mathrm{ad}_{a_{i,1}} \cdots \mathrm{ad}_{a_{i,n_i}}$. For any nonzero element $b \in L$, the subalgebra S of L generated by the set $\{b\} \cup \{a_{i,j}\}$ is finite-dimensional and invariant under γ, so the restriction of γ to S is algebraic, i.e. there exists a nonzero polinomial $p(x) \in \mathbb{F}[x]$ such that $p(\gamma)S = 0$. But any nonzero element in the centroid of a simple algebra is one-to-one, so $p(\gamma) = 0$, and therefore γ is algebraic over $\mathbb{F}$. □

PROPOSITION 14.17. *Suppose that L is simple and locally finite over $\mathbb{F}$. If L is finite-dimensional over Γ, then Γ is an algebraic extension of $\mathbb{F}$.*

PROOF. The proof takes part of the arguments of that of Lemma 10.10. Let $\dim_\Gamma L = d$. By the Jacobson Density Theorem (without unit), the associative Γ-subalgebra $\mathrm{Ad}(L)$ of $\mathrm{End}_\Gamma(L)$, which can be regarded as $\mathrm{M}_d(\Gamma)$, is the whole of $\mathrm{End}_\Gamma(L)$, so $\mathrm{End}_\Gamma(L)$ is generated by elements of the set $\{\mathrm{ad}_a : a \in L\}$. Let $c_d(x_1, \ldots, x_n)$ be a nonzero central polynomial of $\mathrm{M}_d(\Gamma)$. Then there exist $w_1, \ldots, w_n \in \mathrm{Ad}(L)$ such that $c := c_d(w_1, \cdots, w_n) \neq 0$, so $c \in \Gamma \cap \mathrm{Ad}(L)$. By Lemma 14.16, c is algebraic over $\mathbb{F}$. Let γ be an arbitrary element in Γ. Then $\gamma c \in \Gamma$, and since $c \in \mathrm{Ad}(L)$, $c = \sum_{i=1}^{t} \mathrm{ad}_{a_{i,1}} \cdots \mathrm{ad}_{a_{i,n_i}}$ and $\gamma c = \sum_{i=1}^{t} \mathrm{ad}_{\gamma(a_{i,1})} \cdots \mathrm{ad}_{a_{i,n_i}}$ belongs to $\mathrm{Ad}(L)$. Therefore, again by Lemma 14.16, γc is algebraic over $\mathbb{F}$, so $\gamma = (\gamma c)c^{-1}$ is also algebraic over $\mathbb{F}$. □

Locally finite associative algebras. In this subsection, R will denote an associative algebra over a field $\mathbb{F}$ of characteristic 0.

LEMMA 14.18. *Suppose $\mathbb{F}$ algebraically closed and R locally finite. Then every nonzero subalgebra of R^- is locally finite, algebraic, and theefore it contains a nonzero abelian inner ideal.*

PROOF. It is clear that L is locally finite and it follows from the fact that every element in R is algebraic that for any $a \in L$, ad_a is algebraic (see (i) $\Rightarrow$ (iii) of Lemma 8.76), which implies that L contains a nonzero Jordan element by Corollary 4.32, equivalently, a nonzero abelian inner ideal. $\square$

LEMMA 14.19. *Suppose $\mathbb{F}$ algebraically closed and R simple and locally finite. Then $[R, R] \cap Z(R) = 0$.*

PROOF. It follows by proving that any $a \in [R, R] \cap Z(R)$ generates a locally nilpotent ideal of R. Since R is locally finite, we may assume that the element a belongs to a finite-dimensional, although not necessarily simple, associative algebra R_1. Denote by N the (solvable) radical of R_1 and set $\overline{R_1} = R_1/N$. Since $\overline{R_1}$ is semisimple and $\mathbb{F}$ is an algebraically closed field of characteristic 0, we have $Z(\overline{R_1}) \cap [\overline{R_1}, \overline{R_1}] = \overline{0}$. Therefore $a \in N$. This proves that a generates a solvable ideal in R_1 and therefore a locally solvable ideal in R, so $a = 0$ by Proposition 1.25. $\square$

PROPOSITION 14.20. *Suppose $\mathbb{F}$ algebraically closed and R simple, nonabelian, and locally finite. Then $R' = [R, R]$ is a simple locally finite Lie algebra containing a nonzero abelian inner ideal.*

PROOF. By Lemma 14.19, $Z(R) \cap [R, R] = 0$ and hence R' is a simple Lie algebra (Theorem 2.5). That R' is locally finite and has nonzero abelian inner ideals follows from Lemma 14.18. $\square$

Given a subset X of R, denote by $\mathrm{Lie}(X)$ (resp. $\mathrm{Assoc}(X)$) the Lie (resp. associative) subalgebra of R generated by X.

LEMMA 14.21. [**Hen14**, Lemma 8] *Suppose that R has a nontrivial finite $\mathbb{Z}$-grading $R = R_{-m} \oplus \cdots \oplus R_m$ and that $a_1, \ldots, a_n$ are homogeneous elements of R of nonzero degree. If $\mathrm{Lie}(a_1, \cdots, a_n)$ is finite-dimensional, then $\mathrm{Assoc}(a_1, \cdots, a_n)$ is finite-dimensional as well.*

PROOF. Using a combinatorial argument, it is proved that there exists a positive integer N such that any associative product $a_1 \cdots a_{i_k}$ can be written as a sum of products of the form $\rho_1 \cdots \rho_h$, where each $\rho_i \in \mathrm{Lie}(a_1, \ldots, a_n)$ and $h \leq N$. Hence, if $\dim_{\mathbb{F}} \mathrm{Lie}(a_1, \cdots, a_n) = d$, then $\dim_{\mathbb{F}} \mathrm{Assoc}(a_1, \cdots, a_n) \leq d^{N+1}$. $\square$

In the next two results, R will be denote a simple locally finite associative algebra with involution over an algebraically closed field of characteristic 0.

LEMMA 14.22. $Z(R) \cap \mathrm{Skew}(R, *) = 0$.

PROOF. Since R is simple, either $Z(R) = 0$ or $Z(R)$ is a field extension of $\mathbb{F}$. If the latter, R has 1 and $Z(R) = \mathbb{F}1$ since $\mathbb{F}$ is algebraically closed. So $Z(R) \cap \mathrm{Skew}(R, *) = 0$ in any case. $\square$

PROPOSITION 14.23. *If R is infinite-dimensional and $K = \mathrm{Skew}(R, *)$ is not abelian. Then $K' = [K, K]$ is a simple locally finite Lie algebra containing a nonzero Jordan element.*

PROOF. By Lemma 14.22, $Z(R) \cap K' = 0$, so K' is a simple Lie algebra by Theorem 2.6. That K' is locally finite and has nonzero abelian inner ideals follows from Lemma 14.18. □

Locally finite Lie algebras of diagonal type. The next lemma is the key tool in the Jordan characterization of simple locally finite Lie algebras of diagonal type.

LEMMA 14.24. *Let L be a simple locally finite Lie algebra over a field of characteristic 0. Then any nonzero Jordan element in L yields a nonzero von Neumann regular element.*

PROOF. Let a be a nonzero Jordan element of L. It is cleat that the attached Jordan algebra L_a is a locally finite. We claim that L_a contains a nonzero idempotent. If it does not hold, it follows from Corollary 8.12 that L_a is locally nilpotent, so a belongs to the local Kostrikin radical of L by Proposition 9.31. Since L is locally nondegenerate by Proposition 9.30, this leads to a contradiction. Therefore L_a contains a nonzero idempotent and hence, by Proposition 8.59, L contains a nonzero von Neumann regular element. □

THEOREM 14.25. *Let L be a simple infinite-dimensional locally finite Lie algebra over an algebraically closed field $\mathbb{F}$ of characteristic 0. Then the following conditions are equivalent:*

(i) *L is a subalgebra of R^-, where R is locally finite associative $\mathbb{F}$-algebra.*
(ii) *L is algebraic.*
(iii) *L contains a nonzero Jordan element, equivalently, a nonzero abelian inner ideal.*
(iv) *L contains a nonzero von Neumann regular element.*
(v) *L has a nontrivial finite $\mathbb{Z}$-grading.*
(vi) *$L = [R, R]$ or $L = [\mathrm{Skew}(R, *), \mathrm{Skew}(R, *)]$, where R is a simple infinite-dimensional locally finite associative $\mathbb{F}$-algebra and $*$ is an involution.*

PROOF. (i) $\Rightarrow$ (ii). It follows from Lemma 14.18.
(ii) $\Rightarrow$ (iii). See Corollary 4.32.
(iii) $\Rightarrow$ (iv). It is proved in Lemma 14.24.
(iv) $\Rightarrow$ (v). It is Corollary 14.3.
(v) $\Rightarrow$ (vi). By Theorem 14.1, L is isomorphic to one of the three Lie algebras listed in there. Note first that case (3) must be discarded since by Proposition 14.17 L cannot be finite-dimensional over its centroid. In cases (1) and (2), it remains to show that the $\mathbb{Z}$-graded associative algebra R is locally finite.

Suppose that $L \cong \overline{R}' = [R, R]/[R, R] \cap Z(R)$, $R = R_{-2} \oplus \cdots \oplus R_2$. Since $[R, R] \cap Z(R)$ is a central ideal of $[R, R]$ and $\overline{R}'$ is locally finite, $R' = [R, R]$ is locally finite as well. Let $a_1, \ldots, a_n$ be elements of R. Since the $\mathbb{Z}$-grading is nontrivial and R is simple, R is generated by the set $\bigcup_{i\neq 0} R_i$, so we may assume that $a_1, \ldots, a_n$ are homogeneous elements of nonzero degree; moreover, it follows from Remarks 14.2(2) that $R_i \subset [R, R]$ for any $i \neq 0$. Since $\mathrm{Lie}(a_1, \ldots, a_n)$ is finite-dimensional, we have by Lemma 14.21 that $\mathrm{Assoc}(a_1, \ldots, a_n)$ is finite-dimensional as well, so R is locally finite, which implies by Lemma 14.19 that $[R, R] \cap Z(R) = 0$.

Suppose now that $L \cong \overline{K}'$, $K = K_{-2} \oplus \cdots \oplus K_2$. Since L is locally finite and $[K, K] \cap Z(R)$ is a central ideal of $[K, K]$, we have that $[K, K]$ is locally finite as well. Let $a_1, \ldots, a_n$ be elements of R. By [**Her69**, Theorem 2.13], R is generated as associative algebra by $[K, K]$. Therefore, $\mathrm{Assoc}(a_1, \ldots, a_n) \subset \mathrm{Assoc}(\{x_i\})$ for some finite subset $\{x_i\}$ of $[K, K]$. Hence, since, $\mathrm{Lie}(\{x_i\})$ is finite-dimensional, we can apply Lemma 14.21 to conclude that $\mathrm{Assoc}(\{x_i\})$, and in turn $\mathrm{Assoc}(a_1, \ldots, a_n)$, is finite-dimensional. Therefore, R is locally finite and hence, by Lemma 14.22, $[K, K] \cap Z(R) = 0$.

(vi) $\Rightarrow$ (i). It is trivial. □

For other characterizations of the simple infinite-dimensional locally finite Lie algebras of diagonal type over an algebraically closed field $\mathbb{F}$ of characteristic 0 the reader is referred to [**Bar15**, Theorem 5.1]. For the case of a field of positive characteristic we have

THEOREM 14.26. (Hennig) *Let L be a simple infinite-dimensional locally finite Lie algebras of diagonal type over an algebraically closed field $\mathbb{F}$ of characteristic $p > 7$. Then the following conditions are equivalent:*

(i) *L is locally nondegenerate and there is a nonzero element $a \in L$ such that* $\mathrm{ad}_a^{p-1} = 0$.
(ii) $L = \overline{R}'$ *or* $L = [\mathrm{Skew}(R, *), \mathrm{Skew}(R, *)]$, *where R is simple infinite-dimensional locally finite associative $\mathbb{F}$ and $*$ an involution.*

14.6. Simple Jordan algebras generated by ad-nilpotent elements

Let L be a simple Lie algebra over a field of characteristic 0 and suppose that L is generated by ad-nilpotent elements. By combining Kostrikin Descent Lemma with the Vandermonde argument, one proves that L is in fact spanned by Jordan elements. For brevity, such a Lie algebra will be called *isotropic*. Notice that any simple Lie algebra with a nontrivial finite $\mathbb{Z}$-grading is isotropic, so the class of the isotropic simple Lie algebras include those Lie algebras studied in this chapter. This fact is reflected in the following chain of implications for a simple Lie algebra L over a field of characteristic 0.

(1) L contains a nonzero extremal element.
(2) L has a nonzero finite-dimensional abelian inner ideal.
(3) L is locally finite and contains a nonzero Jordan element.
(4) L contains a nonzero von Neumann regular element.
(5) L has a nontrivial finite $\mathbb{Z}$-grading.
(6) L is isotropic.

OPEN QUESTION 14.27. Determine those infinite-dimensional central simple Lie algebra over a field of characteristic 0 which are isotropic.

14.7. Exercises

EXERCISE 14.28. An associative algebra R over a field $\mathbb{F}$ is called *finitary* if R is a subalgebra of $\mathcal{F}(X)$ for some vector space X over $\mathbb{F}$. Show that if R is finitary, then aRa is finite-dimensional for any $a \in R$.

EXERCISE 14.29. Let R be an associative $\mathbb{F}$-algebra. Show that R is simple and finitary if and only it is isomorphic to some $\mathcal{F}_Y(X)$, where $(X, Y, \langle \cdot, \cdot \rangle)$ is a pair of dual vector spaces over a finite-dimensional division $\mathbb{F}$-algebra Δ, equivalently, R is simple and contains an idempotent e such that eRe is a finite-dimensional division $\mathbb{F}$-algebra.

EXERCISE 14.30. Let R be a central simple (not necessarily unital) associative algebra over a field $\mathbb{F}$. Show that R if finitary if and only if it satisfies a generalized polynomial identity (see [**BMM96**] for definition of generalized polynomial identity).

EXERCISE 14.31. A Jordan algebra J over a field $\mathbb{F}$ of characteristic not 2 is called *finitary* if for any $a \in J$ the inner ideal $U_a J$ is finite-dimensional. Show:

(1) Any nondegenerate finitary Jordan algebra coincides with its socle.
(2) If R is a finitary associative $\mathbb{F}$-algebra, then the Jordan algebra R^+ is finitary.

EXERCISE 14.32. Let L be a simple finitary Lie algebra over an algebraically closed field $\mathbb{F}$ of characteristic 0. Show that for any nonzero Jordan element $a \in L$, the Jordan algebra L_a is simple and either finite-dimensional or an infnite-dimensional Clifford Jordan algebra.

EXERCISE 14.33. Give an example of a Jordan algebra which is simple and have finite capacity but which is not finitary.

EXERCISE 14.34. Let $\mathbb{F}$ be a field of characteristic 0. Show that the Lie algebra $\mathrm{Der}(\mathbb{F}[x])$ is not locally finite.

EXERCISE 14.35. Let R be a simple associative algebra over a field of characteristic 0 generated by nilpotent elements. Show that the Lie algebra $\overline{R}'$ is spanned by Jordan elements.

CHAPTER 15

Classical Banach–Lie algebras

In this chapter, the strongly prime Banach–Lie algebras with extremal elements are described. They turn out to be natural extensions of the classical Banach–Lie algebras of compact operators on Hilbert spaces. Other results gathered in this chapter are the uniqueness of the complete norm topology in primitive Banach–Lie algebras, the automatic continuity of derivations of nondegenerate Banach–Lie algebras with essential socle, and the fact that any nondegenerate Banach–Lie algebra spanned by extremal elements is finite-dimensional. A natural example of strongly prime Banach–Lie algebras with extremal elements occurs in the class of primitive compact Banach–Lie algebra, which are here introduced and studied.

15.1. Primitive Banach–Lie algebras and continuity of isomorphisms

Let A be a complex algebra. By an *algebra norm* of A we mean any norm $||\cdot||$ on the underlying complex vector space of A making continuous the product, i.e. there exists a positive number k such that $||xy|| \leq k||x||\,||y||$ for all $x, y \in A$. By a *normed algebra* we will mean a complex algebra A endowed with an algebra norm. If the norm is complete, then A is usually called a *complete normed algebra*. Nevertheless, we will use the terms Banach–Lie algebra and Banach–Jordan algebra instead to designate those Lie and Jordan complex algebras equipped with a complete algebra norm. As usual, by a *Banach algebra* we will mean a complete normed associative algebra.

LEMMA 15.1. *Let L be a Banach–Lie algebra and let $a \in L$ be a Jordan element of L. Then the Jordan algebra L_a becomes a Banach–Jordan algebra for the quotient norm.*

PROOF. Straightforward. □

We can measure the continuity of a linear map f acting between normed spaces X and Y by considering the so-called *separating subspace*

$$\mathrm{Sep}(f) = \{y \in Y : \text{ there exists } x_n \to 0 \text{ in } X \text{ with } f(x_n) \to y\}.$$

This subspace is in fact an ideal whenever f is an algebra epimorphism of normed algebras, and it follows from the closed graph theorem that if X and Y are Banach spaces, then f is continuous if and only if $\mathrm{Sep}(f) = 0$.

- Given a Jordan element a of a Lie algebra L, denote by $\pi_a : L \to L_a$ the linear map defined by $\pi_a(x) = \overline{x} = x + \mathrm{Ker}_L\{a\}$, $x \in L$. If L is a Banach–Lie algebra, then π_a is continuous.
- Following Remark 8.47, let $f : (L, a) \to (M, b)$ be a morphism of the category $(\mathcal{L}, \bullet)$. Then the induced homomorphism $\overline{f} : L_a \to M_b$ is given by the condition $\overline{f}\pi_a = \pi_b f$.

LEMMA 15.2. *Let L and M be Banach–Lie algebras, let $a \in L$ and $b \in M$ be Jordan elements, and let $f : L \to M$ be a homomorphism of Lie algebras with $f(a) = b$. Then $\pi_b(\mathrm{Sep}(f)) \subset \mathrm{Sep}(\overline{f})$.*

PROOF. Given $y \in \mathrm{Sep}(f)$, there exists $\{x_n\}$ in L such that $x_n \to 0$ and $f(x_n) \to y$. Since π_a and π_b are continuous maps,

$$\pi_a(x_n) \to 0 \text{ and } \overline{f}(\pi_a(x_n)) = \pi_b(f(x_n)) \to \pi_b(y) = \overline{y},$$

so $\overline{y} \in \mathrm{Sep}(\overline{f})$. Therefore, $\pi_b(\mathrm{Sep}(f)) \subset \mathrm{Sep}(\overline{f})$. □

PROPOSITION 15.3. *Let L and M be Banach–Lie algebras, and let $f : L \to M$ be an algebra epimorphism. Suppose that M is primitive at a nonzero Jordan element $b = f(a)$, where a is a Jordan element of L. Then f is continuous.*

PROOF. Let $\overline{f} : L_a \to M_b$ be the algebra homomorphism from the Banach–Jordan algebra L_a onto the primitive Banach–Jordan algebra M_b (Lemma 15.1) induced by f. By [**Aup82**, Theorem 2], $\overline{f}$ is continuous, which implies, by Lemma 15.2, that $\pi_b(\mathrm{Sep}(f)) = 0$, i.e. $\mathrm{ad}_b^2 \mathrm{Sep}(f) = 0$. Since $\mathrm{Sep}(f)$ is an ideal of M and M is strongly prime, we have by Proposition 4.10 that $\mathrm{Sep}(f) = 0$, so f is continuous. □

COROLLARY 15.4. *Every algebra isomorphism between primitive Banach–Lie algebras is continuous.*

PROOF. Let $f : L \to M$ be an algebra isomorphism between primitive Banach–Lie algebras. Take a nonzero Jordan element $a \in L$ such that the Jordan algebra L_a is primitive. Then $b = f(a)$ is a nonzero Jordan element of M and $M_b \cong L_a$ is primitive. By Proposition 15.3, f is continuous. □

COROLLARY 15.5. *Primitive Banach–Lie algebras have uniqueness of the complete norm topology.*

COROLLARY 15.6. *Let $f : L \to K$ be an algebra isomorphism between Banach–Lie algebras with essential socle. Then f is continuous.*

PROOF. By Theorem 5.22, $\mathrm{Soc}(L) = \bigoplus M_\alpha$ is a direct sum of minimal ideals, where each M_α is generated by a minimal abelian ideal (say B_α) of L. Since f is an algebra isomorphism of L onto K, $\mathrm{Soc}(K) = \bigoplus N_\alpha$, where $N_\alpha = f(M_\alpha)$ is a simple ideal of K generated by the minimal abelian inner ideal $C_\alpha = f(B_\alpha)$. For each index α, write $L_\alpha = L/\mathrm{Ann}_L(M_\alpha)$, $K_\alpha = K/\mathrm{Ann}_L(N_\alpha)$, and denote by $\pi_\alpha : L \to L_\alpha$, $\rho_\alpha : K \to K_\alpha$ the corresponding canonical algebra epimorphisms. Since both $\mathrm{Ann}_L(M_\alpha)$ and $\mathrm{Ann}_K(N_\alpha)$ are closed ideals, L_α and K_α become Banach–Lie algebras and the corresponding epimorphisms π_α and ρ_α are continuous. By Proposition 1.8 and Corollary 4.11, L_α and K_α are strongly prime Lie algebras with nonzero socle ($\mathrm{Soc}(L_\alpha) \cong M_\alpha$, $\mathrm{Soc}(K_\alpha) \cong N_\alpha$), so strongly primitive by Proposition 8.68. Hence it follows from Corollary 15.4 that the induced algebra isomorphism $f_\alpha : L_\alpha \to K_\alpha$, given by the condition $f_\alpha \pi_\alpha = \rho_\alpha f$, is continuous. so $\rho_\alpha(\mathrm{Sep}(f)) \subset \mathrm{Sep}(f_\alpha) = 0$, i.e. $\mathrm{Sep}(f) \subset \mathrm{Ann}_K(N_\alpha)$. Since by assumption $\mathrm{Soc}(K)$ is an essential ideal, $\bigcap \mathrm{Ann}_K(N_\alpha) = \mathrm{Ann}(\sum N_\alpha) = 0$, so $\mathrm{Sep}(f) = 0$ and therefore f is continuous. □

COROLLARY 15.7. *Strongly prime Banach–Lie algebras with nonzero socle and nondegenerate Banach–Lie algebras with dense socle have uniqueness of the complete norm topology.*

PROOF. It is clear that any strongly prime Lie algebras with nonzero socle has essential socle. Suppose now that L is a nondegenerate Banach–Lie algebras with dense socle. Then it follows by the continuity of the product that L has essential socle. □

REMARK 15.8. It follows from [**RP85**, Proposition 7.2] and [**RP91**, Theorem 3.3] that any prime complete normed algebra containing a minimal ideal has uniqueness of the complete norm topology.

Let R be a Banach algebra. By the continuity of the Lie product, R^- becomes a Banach–Lie algebra for the norm of R, $Z(R)$ is a closed ideal of R^-, the Lie algebra $\overline{R} = R^-/Z(R)$ is a Banach–Lie algebra for the quotient norm, and the canonical epimorphism $\pi : R^- \to \overline{R}$, $\pi(x) = \overline{x}$, is continuous.

Following [**Coh03**], an associative algebra is said to be semiprimitive if it has Jacobson radical equals zero (every semiprime associative algebra with essential socle is semiprimitive). M. I. Berenguer and A. R. Villena proved in [**BV99**] that if $f : R \to S$ is a Lie-algebra isomorphism between semiprimitive Banach algebras, then $\mathrm{Sep}(f) \subset Z(S)$. We finish the section giving a Jordan proof of two particular cases of this result.

PROPOSITION 15.9. *Let R and S be primitive Banach algebras with nonzero nilpotent elements and let $f : R^- \to S^-$ be an isomorphism of their associated Lie algebras. Then* $\mathrm{Sep}(f) \subset Z(S)$.

PROOF. By Proposition 8.71, $\overline{R}$ and $\overline{S}$ are strongly primitive Lie algebras. Then, by Corollary 15.4, the induced isomorphism $\overline{f} : \overline{R} \to \overline{S}$ is continuous. So $\mathrm{Sep}(\overline{f}) = 0$ and hence $\overline{\mathrm{Sep}(f)} \subset \mathrm{Sep}(\overline{f}) = 0$, i.e. $\mathrm{Sep}(f) \subset Z(S)$. □

LEMMA 15.10. *Let R be a semiprime associative normed complex algebra with essential socle, let $e \in R$ be a division idempotent, and let M be the ideal of R generated by e. We have:*

(i) *$eRe = \mathbb{C}e$.*
(ii) *Either $M = \mathbb{C}e$, or M is noncommutative and contains nonzero nilpotent elements. If the former, $M \subset Z(R)$. If the latter, $[M, M] \cap Z(M) = 0$ and the Lie algebra $\overline{M} = M^-/Z(M)$ is strongly prime with* $\mathrm{Soc}(\overline{M}) \cong [M, M]$.
(iii) *$\overline{R}$ is a nondegenerate Banach–Lie algebra with essential socle.*

PROOF. (i) Any division associative normed complex algebra is isomorphic to $\mathbb{C}$ (see [**CGRP14**, Corollary 1.1.43]).

(ii) It follows from (i) and Exercise 2.97 that $M \cong \mathcal{F}_Y(X)$, relative to a pair of dual vector spaces $(X, Y, \langle , \rangle)$ over $\mathbb{C}$. Then either $M = \mathbb{C}e$ (what occurs only in the case that the dual pair is one-dimensional), or M contains a subalgebra isomorphic to the full matrix ring $\mathrm{M}_n(\mathbb{C})$ for some $n > 1$ [**BMM96**, Theorem 4.3.11]. If the former is the case, then $[M, \mathrm{Soc}(R)] = 0$ and hence $M \subset Z(R)$ by Lemma 3.33, since $\mathrm{Soc}(R)$ is an essential ideal of R by assumption. If the latter, $[M, M] \cap Z(M) = 0$ and $\overline{M} = M^-/Z(M)$ is strongly prime (Proposition 3.35(3)), with minimal abelian inner ideals (Lemma 13.41), and $\mathrm{Soc}(\overline{M}) = [M, M]$ since $[M, M]$ is a simple ideal (Theorem 2.5).

(iii) As previously mentioned, $\overline{R}$ is a Banach–Lie algebra, which is nondegenerate by Proposition 3.35(2). Thus we only need to show that it has essential socle. Let $\{M_\alpha\}$ denote the collection of the simple components of $\mathrm{Soc}(R)$ which are not one-dimensional. It follows from (ii) and Lie-socle theory (Theorem 5.22), that for each index α, the derived ideal $[M_\alpha, M_\alpha]$ can be regarded as a simple component of $\mathrm{Soc}(\overline{R})$. Let S denote the pre-image of $\mathrm{Soc}(\overline{R})$ by π, and let I be an ideal of R^- such that $[\overline{I}, \mathrm{Soc}(\overline{R})] = 0$, i.e. $[I, S] \subset Z(R)$. For each index α we have $[I, [M_\alpha, M_\alpha]] \subset [M_\alpha, M_\alpha] \cap Z(R) = 0$, so $[[I, M_\alpha], [M_\alpha, M_\alpha]] \subset [I, [M_\alpha, M_\alpha]] = 0$, which implies $[I, M_\alpha] = 0$ by (ii). As the one-dimensional components of $\mathrm{Soc}(R)$ lie in $Z(R)$, $[I, \mathrm{Soc}(R)] = [I, \bigoplus M_\alpha] = \bigoplus [I, M_\alpha] = 0$. Again by Lemma 3.33, this implies $[I, R] = 0$, i.e. $\overline{I} = 0$. Therefore $\mathrm{Soc}(\overline{R})$ is an essential ideal of $\overline{R}$. □

COROLLARY 15.11. *Let R and S be semiprime Banach algebras with essential socle, and let $f : R^- \to S^-$ be an algebra isomorphism. Then* $\mathrm{Sep}(f) \subset Z(S)$.

PROOF. By Lemma 15.10(ii), $\overline{R}$ and $\overline{S}$ are nondegenerate Banach–Lie algebras with essential socle. Then, by Corollary 15.6, the induced isomorphism $\overline{f} : \overline{R} \to \overline{S}$ is continuous. As in the proof of Proposition 15.9, we have $\mathrm{Sep}(f) \subset Z(S)$. □

15.2. Banach–Lie algebras with extremal elements

In any nondegenerate Banach–Lie algebra, nonzero extremal elements there exist whenever the Lie algebra in question contains minimal abelian inner ideals.

LEMMA 15.12. *Let L be a nondegenerate Banach–Lie algebra. Then every minimal abelian inner ideal is one-dimensional, therefore it is generated by an extremal element.*

PROOF. Let B be a minimal abelian inner ideal of L. Take a nonzero element $b \in B$. Then $B = \mathrm{ad}_b^2 L$ and there exists $c \in L$ such that $[b, [c, b]] = b$. By Proposition 8.61, the Jordan algebra L_b is unital with $\overline{c}$ being the unit element, and by Proposition 8.63, L_b is a division Jordan algebra. Since L_b is also a Banach–Jordan algebra by Lemma 15.1, $L_b = \mathbb{C}\overline{c}$ by [**CGRP14**, Corollary 4.1.14]. Then $\mathbb{C}\overline{c} = L_b = U_{\overline{c}} L_b = \overline{\mathrm{ad}_c^2 \mathrm{ad}_b^2 L}$, and hence

$$\mathbb{C}b = \mathrm{ad}_b^2(\mathbb{C}c) = \mathrm{ad}_b^2 \mathrm{ad}_c^2 \mathrm{ad}_b^2 L = \mathrm{ad}_{\mathrm{ad}_b^2 c}^2 L = \mathrm{ad}_b^2 L = B,$$

as required. □

Automatic continuity of derivations. Making a digression in our path to a structure theorem for strongly prime Banach–Lie algebras with extremal elements, we show the automatic continuity of any derivation of such a Banach–Lie algebra. It is easy to see that if $D : A \to B$ is a derivation between complete normed algebras, then $\mathrm{Sep}(D)$ is a closed ideal of B.

LEMMA 15.13. *Let D be a derivation of a nondegenerate Banach–Lie algebra L. Then* $\mathrm{Soc}(L) \cap \mathrm{Sep}(D) = 0$.

PROOF. Since $\mathrm{Sep}(D)$ is an ideal of L, it follows from Theorem 5.22 that if $\mathrm{Sep}(D)$ had nonzero intersection with $\mathrm{Soc}(L)$, then $\mathrm{Sep}(D)$ would contain a minimal abelian inner ideal and therefore a nonzero extremal element e by Lemma 15.12. Let $y \in \mathrm{Sep}(D)$ be such that $\mathrm{ad}_e^2 y = e$ and let $\{x_n\} \subset L$ be such that $x_n \to 0$ and $D(x_n) \to y$. Then $\mathrm{ad}_e^2 L = \mathbb{C}e$ implies that for each positive integer n there is a complex number α_n such that $\mathrm{ad}_e^2 x_n = \alpha_n e$. By the continuity of the Lie product, $\alpha_n \to 0$. Using that D is a derivation, we get

$$D(\mathrm{ad}_e^2 x_n) = [D(e), [e, x_n]] + [e, [D(e), x_n]] + \mathrm{ad}_e^2 D(x_n).$$

Hence

$$0 = \lim \alpha_n D(e) = \lim D(\alpha_n e) = \lim D(\mathrm{ad}_e^2 x_n) = \lim \mathrm{ad}_e^2 D(x_n) = \mathrm{ad}_e^2 y = e,$$

a contradiction. Therefore, $\mathrm{Soc}(L) \cap \mathrm{Sep}(D) = 0$. □

COROLLARY 15.14. *Let L be a nondegenerate Banach–Lie algebra whose socle is essential. Then every derivation of L is automatically continuous. In particular, every derivation of a strongly prime Banach–Lie algebra containing an extremal element is continuous.*

REMARK 15.15. An argument similar to that of Lemma 15.13 was used in [**FLMZ01**, Theorem 3.6] as a part of the proof of the continuity of the two components of any derivation of a semiprimitive Banach–Jordan pair.

For the remainder of this section we follow the exposition given in [**FL11**].

Banach–Lie algebras of special type. Following [**BD73**], a *Banach pairing* is a pair of dual vector spaces $(X, Y, \langle\, ,\rangle)$ over $\mathbb{C}$ such that both X and Y are endowed with prefixed complete norms making the bilinear form $\langle\, ,\rangle$ continuous. A standard application of the closed graph theorem allows us to prove that the complete norms of X and Y making continuous the *nondegenerate* bilinear form $\langle\, ,\rangle$ are unique up to equivalence. By another application of the closed graph theorem we obtain that every $a \in \mathcal{L}_Y(X)$ is a norm-continuous operator on X, so $\mathcal{L}_Y(X)$ is a subalgebra of the Banach algebra $\mathrm{BL}(X)$ of all bounded linear operators on X. Although $\mathcal{L}_Y(X)$ needs not be complete for the *operator norm* $|\cdot|$, it has a natural structure of Banach algebra under the norm $|\cdot|'$ defined by $|a|' = \max\{|a|, |a^{\#}|\}$. As a consequence, we have the following

PROPOSITION 15.16. *Let $(X, Y, \langle\, ,\rangle, ||\cdot||)$ be a Banach pairing. We have:*

(i) *This is the unique Banach pairing structure on the pair of dual vector spaces $(X, Y, \langle\, ,\rangle)$ up to equivalence;*
(ii) *$\mathfrak{gl}_Y(X)$ and $\mathfrak{gl}_Y(X)/\mathbb{C}1_X$ are Banach–Lie algebras for the norm defined by $|a|' = \max\{|a|, |a^{\#}|\}$, where $|\cdot|$ denotes both the operator norm and its quotient norm.*

A result similar to the statement of the next proposition was proved for Jordan-Banach algebras in [**PGRRRPVMn92**, Proposition 2.1]. In fact, the proof given here takes inspiration from the arguments used there.

PROPOSITION 15.17. *Let $(X, Y, \langle\, ,\rangle)$ be an infinite-dimensional pair of dual vector spaces over the complex field, and let L be a Lie algebra such that*

$$(\mathfrak{fsl}_Y(X) + \mathbb{C}1_X)/\mathbb{C}1_X \leq L \leq \mathfrak{gl}_Y(X)/\mathbb{C}1_X.$$

If L is a Banach–Lie algebra for a complete norm $||\cdot||$, then

(i) *L has uniqueness of the complete norm topology,*
(ii) *$(X, Y, \langle\, ,\rangle)$ becomes a Banach pairing and the injection of $(L, ||\cdot||)$ into $(\mathfrak{gl}_Y(X)/\mathbb{C}1_X, |\cdot|')$ is continuous.*

Furthermore, any norm making $\mathfrak{gl}_Y(X)/\mathbb{C}1_X$ into a Banach–Lie algebra comes from the operator norm.

PROOF. Let $a \mapsto \overline{a}$ be the canonical epimorphism of $\mathfrak{gl}_Y(X)$ onto $\mathfrak{gl}_Y(X)/\mathbb{C}1_X$, and take $x \in X$ and $y \in Y$ such that $\langle x, y \rangle = 0$. Then $y^*x \in \mathfrak{fsl}_Y(X)$ with $(y^*x)^2 = \langle x, y\rangle y^*x = 0$. Hence, for any $a \in \mathfrak{gl}_Y(X)$, we have

$$[[\overline{y^*x}, \overline{a}], \overline{y^*x}] = 2(y^*x)a(y^*x) = 2\langle xa, y\rangle y^*x \tag{15.1}$$

(notice that we keep writing linear maps on X acting on the right). Furthermore, the scalar $\langle xa, y\rangle$ only depends on the coset $\overline{a}$: if $\overline{b} = \overline{a}$ then $xb - xa = \alpha x$ for some $\alpha \in \mathbb{C}$, so $\langle xb, y\rangle = \langle xa, y\rangle$ since $\langle x, y\rangle = 0$. Given $x_0 \in X$ and $y_0 \in Y$ such that $\langle x_0, y_0\rangle = 1$, we have (see Proposition 2.24)

$$X = \mathbb{C}x_0 \oplus \{y_0\}^{\perp},\ \{x_0\}^{\perp\perp} = \mathbb{C}x_0,\ Y = \mathbb{C}y_0 \oplus \{x_0\}^{\perp},\ \{y_0\}^{\perp\perp} = \mathbb{C}y_0. \tag{15.2}$$

From now on fix $x_0 \in X$ and $y_0 \in Y$ as above, and let $\pi : X \to \{y_0\}^{\perp}$ be the projection of X onto $\{y_0\}^{\perp}$ along $\mathbb{C}x_0$, i.e. $\pi(\alpha x_0 + x_1) = x_1$ for all $\alpha \in \mathbb{C}$ and $x_1 \in \{y_0\}^{\perp}$. Define $\varphi : L \to \{y_0\}^{\perp}$ by $\varphi(\overline{a}) = \pi(x_0 a)$ for all $\overline{a} \in L$ (again $\overline{b} = \overline{a}$ implies $x_0 b = x_0 a + \alpha x_0$ and hence $\pi(x_0 b) = \pi(x_0 a)$, so φ is well-defined). Then

$$\varphi \text{ is onto and } \ker(\varphi) = \bigcap_{y_1 \in \{x_0\}^{\perp}} \ker(\mathrm{ad}^2(\overline{y_1^* x_0})). \tag{15.3}$$

Indeed, given $x_1 \in \{y_0\}^{\perp}$, take $a = y_0^* x_1$. Since $< x_1, y_0 >= 0$, the coset $\overline{a}$ belongs to L and satisfies $\varphi(\overline{a}) = \pi(x_0 a) = \pi(< x_0, y_0 > x_1) = \pi(x_1) = x_1$ by the definition of π, which proves that φ is onto. The equality relative to $\ker(\varphi)$ follows from the following chain of equivalences for any $\overline{a} \in L$, obtained by using (15.1) and (15.2):

$$\varphi(\overline{a}) = 0 \Leftrightarrow x_0 a \in \mathbb{C}x_0 = \{x_0\}^{\perp\perp} \Leftrightarrow \overline{a} \in \bigcap_{y_1 \in \{x_0\}^{\perp}} \ker(\mathrm{ad}^2(\overline{y_1^* x_0})).$$

Since the operators $\mathrm{ad}^2(\overline{b})$ are continuous for any $\overline{b} \in L$, it follows from (15.3)) that $\ker(\varphi)$ is a closed subspace of L, so we can carry over the complete quotient norm of $L/\ker(\varphi)$ to $\{y_0\}^{\perp}$ and define a complete norm on the subspace $\{y_0\}^{\perp}$ as follows:

$$p(x_1) := \inf\{||\overline{a}|| : \overline{a} \in L \text{ and } \pi(x_0 a) = x_1\}, \tag{15.4}$$

for all $x_1 \in \{y_0\}^{\perp}$. Moreover, since $X = \mathbb{C}x_0 \oplus \{y_0\}^{\perp}$ by (15.2), we can extend the norm p to a complete norm, also denoted by p, on the whole X as follows:

$$p(\alpha x_0 + x_1) := |\alpha| + p(x_1), \text{ for all } \alpha \in \mathbb{C} \text{ and } x_1 \in \{y_0\}^{\perp}. \tag{15.5}$$

As the adjoint involution $a \mapsto a^{\#}$ defines an isomorphism of $\mathfrak{gl}_Y(X)$ onto $\mathfrak{gl}_X(Y)$ with $1_X^{\#} = 1_Y$, we can consider the Banach–Lie algebra $L^{\#}$ with norm $||\overline{a^{\#}}|| := ||\overline{a}||$ for any $\overline{a} \in L$. Then we obtain as above a complete norm q on Y given by

$$q(\beta y_0 + y_1) := |\beta| + q(y_1), \tag{15.6}$$

for all $\beta \in \mathbb{C}$ and $y_1 \in \{x_0\}^{\perp}$, where $q(y_1) = \inf\{||\overline{b}|| : \overline{b} \in L \text{ and } \rho(b^{\#} y_0) = y_1\}$, with $\rho(\beta y_0 + y_1) = y_1$ being the projection of $Y = \mathbb{C}y_0 \oplus \{x_0\}^{\perp}$ onto $\{x_0\}^{\perp}$.

The bilinear form $\langle , \rangle : X \times Y \to \mathbb{C}$ is continuous for the norms p and q defined above. Since both (X, p) and (Y, q) are Banach spaces, it suffices to show that $\langle , \rangle$ is separately continuous. Fix $y \in Y$ and let x be any vector of X. By (15.2), $y = \beta y_0 + y_1$ and $x = \alpha x_0 + x_1$, where $\alpha, \beta \in \mathbb{C}$, $y_1 \in \{x_0\}^{\perp}$ and $x_1 \in \{y_0\}^{\perp}$. By (15.3), there exists $\overline{a} \in L$ such that $\pi(x_0 a) = x_1$. Thus

$$\langle x, y\rangle = \alpha\beta + \langle x_1, y_1\rangle = \alpha\beta + \langle x_0 a, y_1\rangle.$$

Taking modules on both sides of this equation and using (15.1), we get

$$|\langle x,y\rangle| \le |\alpha||\beta| + |\langle x_0 a, y_1\rangle| \le |\alpha||\beta| + k||\overline{y_1^* x_0}||||\overline{a}||$$

for some positive constant k. Then taking the infimum over all $\overline{a}\in L$ such that $\pi(x_0 a) = x_1$, we obtain

$$|\langle x,y\rangle| \le |\alpha||\beta| + k||\overline{y_1^* x_0}||p(x_1) \le \max\{|\beta|,\, k||\overline{y_1^* x_0}||\}p(x),$$

which proves that fixed $y\in Y$, the linear form $x\mapsto\langle x,y\rangle$ is continuous. Dually, fixed $x\in X$, the linear form $y\mapsto\langle x,y\rangle$ is continuous. Note that, by Proposition 15.16, the norms p and q constructed above are, up to equivalence, independent of the choice of the pair (x_0, y_0).

Now we see that the injection of $(L, ||\cdot||)$ into $(\mathfrak{gl}_Y(X)/\mathbb{C}1_X, |\cdot|')$ is continuous. By the closed graph theorem, it suffices to show that if $\overline{a}_n \xrightarrow{||\cdot||} 0$ and $\overline{a}_n \xrightarrow{|\cdot|'} \overline{a}$, then $\overline{a} = 0$. Fix a nonzero vector $x\in X$ and let $y\in\{x\}^\perp$. Then the inequality $|\langle xa_n, y\rangle| \le k||\overline{y^* x}||||\overline{a}_n||$ implies $\langle xa_n, y\rangle \to 0$. On the other hand, since $\overline{a}_n \xrightarrow{|\cdot|'} \overline{a}$, by replacing $\{a_n\}$ by an equivalent sequence, if necessary, we may assume that $a_n \xrightarrow{|\cdot|} a$, so $xa_n \xrightarrow{p} xa$ and hence $\langle xa_n, y\rangle \to \langle xa, y\rangle$ by the continuity of the bilinear form just proved. Therefore $\langle xa, y\rangle = 0$ for all $y\in\{x\}^\perp$, so $xa\in\{x\}^{\perp\perp} = \mathbb{C}x$. Since x is any nonzero vector of X, this proves that a is a scalar multiple of the identity on X, so $\overline{a} = 0$. If L is the whole $\mathfrak{gl}_Y(X)/\mathbb{C}1_X$, then the two norms $||\cdot||$ and $|\cdot|'$ are equivalent. Finally, L is strongly prime with nonzero socle (Theorem 14.15) and hence, by Corollary 15.7, L has uniqueness of the complete norm topology. In particular, any norm making the algebra $\mathfrak{gl}_Y(X)/\mathbb{C}1_X$ into a Banach–Lie algebra comes from the operator norm. □

COROLLARY 15.18. *Let $(X, Y, \langle\,,\rangle)$ be an infinite-dimensional pair of dual vector spaces over $\mathbb{C}$, and let L be a subalgebra of $\mathfrak{fgl}_Y(X)$ containing $\mathfrak{fsl}_Y(X)$. Then L cannot be equipped with a complete algebra norm.*

PROOF. Suppose on the contrary that L admits a complete algebra norm $||\cdot||$. Then, by Proposition 15.17, $(X, Y, \langle\,,\rangle)$ becomes a Banach pairing and $||\cdot||$ majorizes the operator norm $|\cdot|$ of $\mathrm{BL}(X)$. Let $\{x_n\}\subset X$ and $\{y_n\}\subset Y$ be infinite sequences of vectors such that $\langle x_n, y_m\rangle = \delta_{nm}$ (Exercise 2.90), and set $a_n = y_{n+1}^* x_n$. Then $a_n\in\mathfrak{fsl}_Y(X)\subset L$ and the infinite series $\sum_{n=1}^\infty \frac{a_n}{2^n||a_n||}$ converges absolutely to an element $a\in L\subset\mathfrak{fgl}_Y(X)$. But since $|\cdot|\le k||\cdot||$, we have that $\sum_{n=1}^\infty \frac{a_n}{2^n||a_n||}$ also converges to a with respect to the operator norm. Hence $xa = \sum_{n=1}^\infty \frac{xa_n}{2^n||a_n||}$ for any $x\in X$. Taking $x = x_{m+1}$, we get that $x_{m+1}a = \frac{x_m}{2^m||a_m||}$ for any $m\ge 1$, which is a contradiction since a has finite rank. □

Banach–Lie algebras of orthogonal type. A Banach pairing $(X, Y, \langle\,,\rangle, ||\cdot||)$ with $X = Y$ will be called a *Banach inner product space* and will be denoted by $(X, \langle\,,\rangle, ||\cdot||)$. The following elemental lemma will be used in the proof of the next proposition.

LEMMA 15.19. *Let $a\in\mathfrak{o}(X,\langle\,,\rangle)$ be such that any isotropic vector $x\in X$ is an eigenvector for a. Then $a = 0$.*

PROOF. Let (x_1, y_1) and (x_2, y_2) be pairwise orthogonal hyperbolic pairs. Note that $x_i a = \lambda_i x_i$ implies $y_i a = -\lambda_i y_i$, and since both x_1+x_2 and x_1+y_2 are isotropic vectors, $(x_1+x_2)a = \lambda_1 x_1 + \lambda_2 x_2$ implies $\lambda_1 = \lambda_2$, and $(x_1+y_2)a = \lambda_1 x_1 - \lambda_2 y_2$ implies $\lambda_1 = -\lambda_2$. Thus $\lambda_1 = \lambda_2 = 0$. Hence $a = 0$, since X is spanned by isotropic vectors. □

PROPOSITION 15.20. *Let X be an infinite-dimensional complex vector space with a nondegenerate symmetric bilinear form $\langle\,,\rangle$, and let L be a subalgebra of $\mathfrak{o}(X, \langle\,,\rangle)$ containing $\mathfrak{fo}(X, \langle\,,\rangle)$. If L is a Banach–Lie algebra for a norm $\|\cdot\|$, then $(X, \langle\,,\rangle)$ becomes a Banach inner product space and the injection of $(L, \|\cdot\|)$ into $(\mathfrak{o}(X, \langle\,,\rangle), |\cdot|)$ is continuous. Furthermore, any norm making $\mathfrak{o}(X, \langle\,,\rangle)$ into a Banach–Lie algebra is equivalent to the operator norm.*

PROOF. Let $x \in X$ be a nonzero isotropic vector. Extend x to a hyperbolic plane $H = \mathbb{C}x + \mathbb{C}y$, where y is an isotropic vector and $\langle x, y\rangle = 1$. For any $a \in \mathfrak{o}(X, \langle\,,\rangle)$ and any isotropic vector $z \in H^\perp$, we have $[x, z]^2 = (x^* z - z^* x)^2 = 0$ and hence

$$\mathrm{ad}^2_{[x,z]}\, a = 2[x, z]a[x, z] = 2(x^* za - z^* xa)(x^* z - z^* x) = 2x^* \langle za, x\rangle z + z^* \langle xa, z\rangle x.$$

Since $\langle xa, z\rangle = \langle x, za^*\rangle = -\langle za, x\rangle$, the formula above implies

$$\mathrm{ad}^2_{[x,z]}\, a = 2\langle xa, z\rangle[x, z] \quad\text{and}\quad |\langle xa, z\rangle| \le k\|[x, z]\|\|a\| \tag{15.7}$$

for some positive constant k. Let x, a, and H be as above, and denote by $\mathrm{Isot}(H^\perp)$ the set of all the isotropic vectors of $H^\perp$. We have

$$a \in \bigcap_{z\in \mathrm{Isot}(H^\perp)} \ker(\mathrm{ad}^2_{[x,z]}) \Leftrightarrow xa \in H \Leftrightarrow xa \in \mathbb{C}x. \tag{15.8}$$

Indeed, it follows from (15.7) that $a \in \ker(\mathrm{ad}^2_{[x,z]})$ for all $z \in \mathrm{Isot}(H^\perp)$ if and only if xa is orthogonal to any $z \in \mathrm{Isot}(H^\perp)$, equivalently, if and only if $xa \in H^{\perp\perp} = H$, since any vector of $H^\perp$ is a sum of isotropic vectors. Now $xa = \alpha x + \beta y$, $\langle xa, x\rangle = 0$ and $\langle x, y\rangle = 1$ imply $xa \in \mathbb{C}x$; the reverse implication is clear.

Consider the orthogonal projection π of $X = H \oplus H^\perp$ onto $H^\perp$ and define the linear map $\varphi : L \to H^\perp$ by $\varphi(a) = \pi(xa)$ for all $a \in L$. By (15.8), we have $\ker(\varphi) = \bigcap_{z\in \mathrm{Isot}(H^\perp)} \ker(\mathrm{ad}^2_{[x,z]})$. Moreover,

$$\mathrm{im}(\varphi) = H^\perp. \tag{15.9}$$

Indeed, given $z \in H^\perp$, take $a = [y, z]$. Then $a \in \mathfrak{fo}(X, \langle\,,\rangle) \subset L$ satisfies

$$xa = x(y^* z) - x(z^* y) = \langle x, y\rangle z - \langle x, z\rangle y = z.$$

Since the operators ad^2_b are continuous for any $b \in L$, $\ker(\varphi)$ is a closed subspace of L, so we can carry over the complete quotient norm of $L/\ker(\varphi)$ to $H^\perp = \mathrm{im}(\varphi)$, i.e. we define a complete norm on the subspace $H^\perp$ by

$$p(z) := \inf\{\|a\| : a \in L \text{ and } \pi(xa) = z\}, \tag{15.10}$$

for all $z \in H^\perp$. We extend p to a complete norm, also denoted by p, on the whole $X = H \oplus H^\perp$ as follows:

$$p(h + z) := \sqrt{q(h)^2 + p(z)^2}, \tag{15.11}$$

for all $h = \alpha x + \beta y \in H$ and $z \in H^\perp$, and where $q(h) = \max\{|\alpha|, |\beta|\}$.

Now we show that the symmetric bilinear form $\langle\, ,\rangle : X \times X \to \mathbb{C}$ is continuous for the norm p. As the sum $X = H \oplus H^\perp$ is orthogonal and H is finite-dimensional, we need only see that the restriction of $\langle\, ,\rangle$ to $H^\perp \times H^\perp$ is continuous. Furthermore, since $H^\perp$ is a Banach space, it suffices to show that it is separately continuous. Fix a nonzero vector $v \in H^\perp$ and let $\tilde{v}$ denote the linear form $z \mapsto \langle z, v\rangle$, $z \in H^\perp$. Since φ is onto by (15.9), for any $z \in H^\perp$ there exists $a \in L$ such that $\pi(xa) = z$. Suppose first that v is isotropic. Then we have again by (15.9) that $|\langle xa, v\rangle| \leq k||[x,v]||||a||$. Taking the infimum over all $a \in L$ such that $\pi(xa) = z$ we obtain $|\langle z, v\rangle| \leq k||[x,v]||p(z)$, which proves the continuity of the map $\tilde{v}$ when v is isotropic. The general case of an arbitrary $v \in H^\perp$ reduces to the previous one taking into account that any $v \in H^\perp$ is a sum of isotropic vectors of $H^\perp$. Again we have by Proposition 15.16 that the norm p just constructed is, up to equivalence, independent of the choice of the hyperbolic basis (x, y).

Let us now see that the injection of $(L, ||\cdot||)$ into $\mathfrak{o}(X, \langle\, ,\rangle)$ is continuous. Suppose that $a_n \xrightarrow{||\cdot||} 0$ and $a_n \xrightarrow{|\cdot|} a$ for some $a \in \mathfrak{o}(X, \langle\, ,\rangle)$. By the closed graph theorem, we must show that $a = 0$, which in virtue of Lemma 15.19 is equivalent to verifying that any isotropic vector is an eigenvector for a. Fix a nonzero isotropic vector $u \in X$, let H be a hyperbolic plane containing u, and let z be an arbitrary isotropic vector of $H^\perp$. Then we have by (15.7),

$$|\langle ua, z\rangle| \leq k||[u,z]||||a||$$

for some positive constant k. Hence $a_n \xrightarrow{||\cdot||} 0$ implies $\langle ua_n, z\rangle \to 0$ and therefore $\langle ua, z\rangle = 0$ by the continuity of the bilinear form $\langle\cdot,\cdot\rangle$ and the convergence of $\{a_n\}$ to a with respect to the operator norm. As any vector of $H^\perp$ is a sum of isotropic vectors, we have actually proved that $ua \in H^{\perp\perp} = H$, equivalently, by (15.8), $ua \in \mathbb{C}u$, i.e. u is an eigenvector for a, as required.

Finally, as for Banach–Lie algebras of special type, L has uniqueness of the complete norm topology and hence any norm making the algebra $\mathfrak{o}(X, \langle\, ,\rangle)$ into a Banach–Lie algebra is equivalent to the operator norm. □

COROLLARY 15.21. *Let X be an infinite-dimensional complex vector space with a nondegenerate symmetric bilinear form $\langle\, ,\rangle$. Then $\mathfrak{fo}(X, \langle\, ,\rangle)$ cannot be equipped with a complete algebra norm.*

PROOF. It is similar to that of Corollary 15.18, by replacing $y^*_{n+1}x_n$ by $[y_n, x_n]$, where $\{(x_n, y_n)\}$ is an infinite sequence of pairwise orthogonal hyperbolic pairs. □

Banach–Lie algebras of symplectic type. We begin with an elemental lemma which will be used in the proof of the proposition below.

LEMMA 15.22. *Let $a \in \mathfrak{sp}(X, \langle\, ,\rangle)$ be such that $\langle xa, x\rangle = 0$ for every $x \in X$. Then $a = 0$.*

PROOF. For $x, y \in X$, we have

$$\begin{aligned} 2\langle xa, y\rangle &= \langle xa, y\rangle - \langle y, xa\rangle = \langle xa, y\rangle + \langle ya, x\rangle \\ &= \langle (x+y)a, x+y\rangle - \langle xa, x\rangle - \langle ya, y\rangle = 0. \end{aligned}$$

Hence $a = 0$ by nondegeneracy of $\langle\, ,\rangle$. □

PROPOSITION 15.23. *Let X be an infinite-dimensional complex vector space with a nondegenerate alternate bilinear form $\langle\,,\rangle$, and let L be a subalgebra of $\mathfrak{sp}(X,\langle\,,\rangle)$ containing $\mathfrak{fsp}(X,\langle\,,\rangle)$. If L is a Banach–Lie algebra for a norm $\|\cdot\|$, then $(X,\langle\,,\rangle)$ becomes a Banach inner product space and the injection of $(L,\|\cdot\|)$ into $(\mathfrak{sp}(X,\langle\,,\rangle),|\cdot|)$ is continuous. Moreover, any norm making the algebra $\mathfrak{sp}(X,\langle\,,\rangle)$ into a Banach–Lie algebra is equivalent to the operator norm.*

PROOF. Recall that for any vectors $x,y\in X$, the operator $[x,y]=x^*y+y^*x$ belongs to $\mathfrak{fsp}(X,\langle\,,\rangle)$, $(x^*y)^*=-y^*x$, In particular, $x^*x=\frac{1}{2}[x,x]\in\mathfrak{fsp}(X,\langle\,,\rangle)$ for all $x\in X$. Let $x,z\in X$ be orthogonal, let $a\in\mathfrak{sp}(X,\langle\,,\rangle)$ and set $b:=[x,z]$. As any vector of X is isotropic and x,z are orthogonal, $b^2=0$. Hence

$$-\frac{1}{2}\operatorname{ad}_b^2 a=bab=(x^*za+z^*xa)(x^*z+z^*x)=x^*\langle za,x\rangle z+x^*\langle za,z\rangle x+$$
$$z^*\langle xa,x\rangle z+z^*\langle xa,z\rangle x=\langle za,x\rangle[x,z]+\langle za,z\rangle x^*x+\langle xa,x\rangle z^*z,$$

since $\langle xa,z\rangle=-\langle x,za\rangle=\langle za,x\rangle$. We have thus proved

$$\frac{1}{2}\operatorname{ad}_{[x,z]}^2 a=\langle z,xa\rangle[x,z]+\langle z,za\rangle x^*x+\langle x,xa\rangle z^*z. \tag{15.12}$$

In particular, for any $x\in X$ and $a\in\mathfrak{sp}(X,\langle\cdot,\cdot\rangle)$, we have

$$\operatorname{ad}_{x^*x}^2 a=2\langle x,xa\rangle x^*x. \tag{15.13}$$

Let H be a hyperbolic plane of X and x a nonzero vector in H. For any $z\in H^\perp$, let ξ_z denote the linear operator $\operatorname{ad}_{[x,z]}^2:L\to L$. Then for $a\in L$, we have

$$xa\in\mathbb{C}x\Leftrightarrow a\in\bigcap_{z\in H^\perp}\xi_z^{-1}(\mathbb{C}x^*x). \tag{15.14}$$

By (15.12), for any $z\in H^\perp$,

$$a\in\xi_z^{-1}(\mathbb{C}x^*x)\ \Leftrightarrow\ xa\in H^{\perp\perp}\cap\{x\}^\perp=H\cap\{x\}^\perp=\mathbb{C}x.$$

Fix a hyperbolic plane H of X with hyperbolic basis (x,y). Then $X=\mathbb{C}x\oplus\{y\}^\perp$ by (15.2), so we can consider the corresponding projection $\pi:X\to\{y\}^\perp$. As in (15.3), define $\varphi:L\to\{y\}^\perp$ by $\varphi(a)=\pi(xa)$ for all $a\in L$. Then

$$\ker(\varphi)=\bigcap_{z\in H^\perp}\xi_z^{-1}(\mathbb{C}x^*x)\ \text{ and }\ \operatorname{im}(\varphi)=\{y\}^\perp=\mathbb{C}y\oplus H^\perp. \tag{15.15}$$

By (15.14), $\pi(xa)=0\Leftrightarrow xa\in\mathbb{C}x\Leftrightarrow a\in\bigcap_{z\in H^\perp}\xi_z^{-1}(\mathbb{C}x^*x)$, which proves the equality relative to the kernel. On the other hand, as $\{y\}^\perp=\mathbb{C}y\oplus H^\perp$, $y=(x)y^*y$, and $(x)[y,z]=z$ for any $z\in H^\perp$, we have that $\operatorname{im}(\varphi)=\{y\}^\perp$.

By (15.15) $\ker(\varphi)$ is a closed subspace of L, so we can carry over the complete quotient norm of $L/\ker(\varphi)$ to $\{y\}^\perp$ as follows:

$$p(u):=\inf\{\|a\|:a\in L\ \text{ and }\ \pi(xa)=u\}, \tag{15.16}$$

for all $u\in\{y\}^\perp$. Since $X=\mathbb{C}x\oplus\{y\}^\perp$ by (15.2), we can extend the norm p to a complete norm, also denoted by p, on the whole X as follows:

$$p(\alpha x+u):=|\alpha|+p(u),\ \text{ for all }\ \alpha\in\mathbb{C}\ \text{ and }\ u\in\{y\}^\perp. \tag{15.17}$$

Now we show that the alternate form $\langle\,,\rangle$ is continuous with respect to the norm p defined in (15.17). Again it is enough to prove that for any $w\in X$, the linear form defined by $(x')\tilde{w}=\langle x',w\rangle$ for all $x'\in X$ is continuous. Since $X=\mathbb{C}x\oplus\{y\}^\perp=\mathbb{C}x\oplus\mathbb{C}y\oplus H^\perp$, we can divide the proof into three cases.

Consider first the case that $w = x$. Then for any $\alpha, \beta \in \mathbb{C}$ and $z \in H^\perp$, we have $(\alpha x + \beta y + z)\tilde{x} = \langle \alpha x + \beta y + z, x\rangle = -\beta$. Now. by (15.15), there exist $\lambda \in \mathbb{C}$ and $a \in L$ such that $xa = \lambda x + \beta y + z$ and hence $\langle x, xa\rangle = \beta$. Then it follows from (15.13) that $|\beta| = |\langle x, xa\rangle| \leq k||x^*x||||a||$ for some positive constant k. Hence

$$|(\alpha x + \beta y + z)\tilde{x}| = |\beta| \leq k||x^*x||p(\beta y + z) \leq \max\{k||x^*x||, 1\}p(\alpha x + \beta y + z),$$

which proves the continuity of $\tilde{x}$. The case that $w = y$ is straightforward: for any $\alpha \in \mathbb{C}$ and $u \in \{y\}^\perp$, we have $|(\alpha x + u)\tilde{y}| = |\alpha| \leq |\alpha| + p(u) = p(\alpha x + u)$. The remaining case that $w \in H^\perp$ goes as follows. Since $X = \mathbb{C}x + \{y\}^\perp$ and $\tilde{w}|_{\mathbb{C}x} = 0$, it suffices to show that the restriction of $\tilde{w}$ to $\{y\}^\perp$ is continuous. Let $v \in \{y\}^\perp$ and take, by (15.15), $a \in L$ such that $\pi(xa) = v$. Now we have by (15.12)

$$\frac{1}{2}\operatorname{ad}^2_{[x,w]} a = \langle w, xa\rangle[x, w] + \langle w, wa\rangle x^*x + \langle x, xa\rangle w^*w.$$

Taking norms in the above formula we obtain

$$|\langle w, xa\rangle|||[x, w]|| \leq ||\frac{1}{2}\operatorname{ad}^2_{[x,w]} a|| + |\langle w, wa\rangle|||x^*x|| + |\langle x, xa\rangle|||w^*w||,$$

where by the continuity of the Lie product,

$$||\frac{1}{2}\operatorname{ad}^2_{[x,w]} a|| \leq k||[x, w]||^2||a||$$

for some constant integer k, and where by (15.13) and the above formula,

$$|\langle w, wa\rangle| \leq \frac{1}{||w^*w||}||\frac{1}{2}\operatorname{ad}^2_{w^*w} a|| \leq \frac{k}{2}||w^*w||||a||.$$

Similarly, $|\langle x, xa\rangle| \leq \frac{k}{2}||x^*x||||a||$. Hence

$$|\tilde{w}(v)| = |\langle w, \pi(xa)\rangle| = |\langle w, xa\rangle| \leq k(x, w)||a||$$

for some positive number $k(x, w)$ depending on x and w but independent of a. Taking now the infimum over all $a \in L$ such that $\pi(xa) = v$ we get $|\tilde{w}(v)| \leq k(x, w)p(v)$, which proves the continuity of $\tilde{w}$. Note that by Proposition 15.16(i) the complete norm p just constructed is independent of the choice of the hyperbolic pair (x, y).

Let us finally see that the injection of $(L, ||\cdot||)$ into $(\mathfrak{sp}(X, \langle\,,\rangle), |\cdot|)$ is continuous. Let $a_n \xrightarrow{||\cdot||} 0$ and $a_n \xrightarrow{|\cdot|} a$ for some $a \in \mathfrak{sp}(X, \langle\,,\rangle)$, and fix $x \in X$. It follows from (15.13) that $\langle x, xa_n\rangle \to 0$, so $\langle x, xa\rangle = 0$ by the continuity of the bilinear form and the convergence of $\{a_n\}$ to a with respect to the operator norm $|\cdot|$. Thus xa is orthogonal to x for every $x \in X$, which implies $a = 0$ by Lemma 15.22. □

COROLLARY 15.24. *Let X be an infinite-dimensional complex vector space with a nondegenerate alternate bilinear form $\langle\,,\rangle$. Then $\mathfrak{fsp}(X, \langle\,,\rangle)$ cannot be equipped with a complete algebra norm.*

PROOF. It is similar to that of Corollary 15.18. Take an infinite sequence $\{(x_n, y_n)\}$ of pairwise orthogonal hyperbolic planes and set $a_n = y_n^* y_n$. □

Strongly prime Banach–Lie algebras with extremal elements. As could be expected of what we proved in Theorem 14.15, infinite-dimensional strongly prime Banach–Lie algebras with extremal elements are classical.

THEOREM 15.25. *Let $(L, ||\cdot||)$ be an infinite-dimensional Banach–Lie algebra. Then L is strongly prime and contains extremal elements if and only if any one of the following statements holds:*

(i) *There exists an infinite-dimensional Banach pairing $(X, Y, \langle\,,\rangle, ||\cdot||)$ such that*
$$(\mathfrak{fsl}_Y(X) + \mathbb{C}1_X)/\mathbb{C}1_X \leq L \leq \mathfrak{gl}_Y(X)/\mathbb{C}1_X$$
and the injection $(L, ||\cdot||)$ into $(\mathfrak{gl}_Y(X)/\mathbb{C}1_X, |\cdot|')$ is continuous.

(ii) *There exists an infinite-dimensional Banach inner product space $(X, \langle\,,\rangle)$, where $\langle\,,\rangle$ is symmetric, such that*
$$\mathfrak{fo}(X, \langle\,,\rangle) \leq L \leq \mathfrak{o}(X, \langle\,,\rangle)$$
and the injection of $(L, ||\cdot||)$ into $(\mathfrak{o}(X, \langle\,,\rangle), |\cdot|)$ is continuous.

(iii) *There exists an infinite-dimensional Banach inner product space $(X, \langle\,,\rangle)$, where $\langle\,,\rangle$ is alternate, such that*
$$\mathfrak{fsp}(X, \langle\,,\rangle) \leq L \leq \mathfrak{sp}(X, \langle\,,\rangle)$$
and the injection of $(L, ||\cdot||)$ into $(\mathfrak{sp}(X, \langle\,,\rangle), |\cdot|)$ is continuous.

PROOF. By Theorem 14.15, each one of the Banach–Lie algebras L listed above is strongly prime and contains extremal elements. Suppose conversely that L is an infinite-dimensional strongly prime Banach–Lie algebra with extremal elements. Then, again by Theorem 14.15, L can be represented as a Lie algebra of one of the types special, orthogonal), or symplectic. Then Propositions 15.17, 15.20, and 15.23 apply in each one of the cases. □

A Banach–Lie algebra is said to be *topologically simple* if it is not abelian and it does not contain proper closed ideals. Topologically simple nondegenerate Lie algebras are strongly prime. Natural examples of topologically simple Banach–Lie algebras containing extremal elements are $(\overline{\mathfrak{fsl}_Y(X)}, |\cdot|')$, $(\overline{\mathfrak{fo}(X, \langle\,,\rangle)}, |\cdot|)$ and $(\overline{\mathfrak{fsp}(X, \langle\,,\rangle)}, |\cdot|)$, where in all the cases the bar denotes the norm-operator closure. From Theorem 15.25, we obtain the following partial converse.

COROLLARY 15.26. *Any infinite-dimensional Banach–Lie algebra $(L, ||\cdot||)$ which is nondegenerate, topologically simple, and contains extremal elements admits one of the following representations:*

(i) $\mathfrak{fsl}_Y(X) \leq (L, ||\cdot||) \leq (\overline{\mathfrak{fsl}_Y(X)}, |\cdot|')$,
(ii) $\mathfrak{fo}(X, \langle\,,\rangle) \leq (L, ||\cdot||) \leq (\overline{\mathfrak{fo}(X, \langle\,,\rangle)}, |\cdot|)$
(iii) $\mathfrak{fsp}(X, \langle\,,\rangle) \leq (L, ||\cdot||) \leq (\overline{\mathfrak{fsp}(X, \langle\,,\rangle)}, |\cdot|)$

where in all the cases the injections are continuous.

Banach–Lie algebras of compact operators on Hilbert spaces. Let $\mathfrak{H}$ be a complex Hilbert space with inner product denoted by $(\,,)$. A *conjugation* of H is a conjugate linear isometry which is also involutive, i.e. a map $\theta : H \to H$ satisfying $(x, y\theta) = (y, x\theta)$ and $x\theta^2 = x$ for all $x, y \in H$. An *anti-conjugation* is a map $\zeta : H \to H$ satisfying $(x, y\zeta) = -(y, x\zeta)$ and $x\zeta^2 = -x$ for all $x, y \in H$. The existence and uniqueness (up to linear isometry) of conjugations and anti-conjugations is guaranteed in infinite-dimensional Hilbert spaces [**HOSr84**, (7.5.6)].

NOTATION 15.27. Let $\mathfrak{H}$ be an infinite-dimensional complex Hilbert space and fix a conjugation θ and anti-conjugation ζ in $\mathfrak{H}$. Following [**DlH72**], we denote by

$\mathfrak{gl}(\mathfrak{H}, \mathcal{C}_\infty)$ the Banach–Lie algebra of all compact operators on $\mathfrak{H}$, by $\mathfrak{o}(\mathfrak{H}, \theta, \mathcal{C}_\infty)$ the orthogonal Banach–Lie algebra of compact operators on $\mathfrak{H}$,

$$\mathfrak{o}(\mathfrak{H}, \theta, \mathcal{C}_\infty) = \{a \in \mathfrak{gl}(\mathfrak{H}, \mathcal{C}_\infty) : \theta a^* \theta = -a\},$$

and by $\mathfrak{sp}(\mathfrak{H}, \zeta, \mathcal{C}_\infty)$ the symplectic Banach–Lie algebra of compact operators on $\mathfrak{H}$,

$$\mathfrak{sp}(\mathfrak{H}, \zeta, \mathcal{C}_\infty) = \{a \in \mathfrak{gl}(\mathfrak{H}, \mathcal{C}_\infty) : \zeta a^* \zeta = a\}.$$

PROPOSITION 15.28. *Let $(\mathfrak{H}, (\,,))$ be an infinite-dimensional complex Hilbert space and fix a conjugation θ in $\mathfrak{H}$. We have:*

(i) *$\langle x, y\rangle := (x, y\theta)$, for all $x, y \in \mathfrak{H}$, defines a nondegenerate symmetric bilinear form which makes $\mathfrak{H}$ into a Banach inner product space with respect to the Hilbert norm.*

(ii) *$\mathfrak{gl}(\mathfrak{H})$ coincides with the Banach–Lie algebra of bounded linear operators on $\mathfrak{H}$. In fact, $a^\# = \theta a^* \theta$ for all $a \in \mathrm{BL}(\mathfrak{H})$, where a^* denotes the Hilbert adjoint of a.*

(iii) *$\mathfrak{gl}(\mathfrak{H}, \mathcal{C}_\infty) = \overline{\mathfrak{fgl}(\mathfrak{H})} = \overline{\mathfrak{fsl}(\mathfrak{H})}$ and it is a topologically simple nondegenerate Banach–Lie algebra with extremal elements.*

(iv) *$\mathfrak{o}(\mathfrak{H}, \theta, \mathcal{C}_\infty) = \overline{\mathfrak{fo}(\mathfrak{H})}$ is a topologically simple nondegenerate Banach–Lie algebra with extremal elements.*

PROOF. (i) For any $x, y \in \mathfrak{H}$, we have $\langle x, y\rangle = (x, y\theta) = (y, x\theta) = \langle y, x\rangle$, which proves that the bilinear form $\langle\,,\rangle$ is symmetric; the nondegeneracy is clear and the continuity follows from the Cauchy–Schwartz inequality and the fact that θ is an isometry.

(ii) By the closed graph theorem, for a linear operator a in $\mathfrak{H}$ the following are equivalent: (1) $a \in \mathfrak{gl}(\mathfrak{H})$, (2) $a \in \mathrm{BL}(\mathfrak{H})$, (3) a has an adjoint with respect to the Hilbert product. Furthermore, in this case,

$$\langle xa, y\rangle = (xa, y\theta) = (x, y\theta a^*) = (x, y\theta a^* \theta\theta) = \langle x, y\theta a^* \theta\rangle = \langle x, ya^\#\rangle.$$

(iii) and (iv). By a well known result of P. R. Halmos [**Hal52**], $\mathfrak{gl}(\mathfrak{H}, \mathcal{C}_\infty)$ and $\mathfrak{o}(\mathfrak{H}, \theta, \mathcal{C}_\infty)$ coincide with their derived ideals and hence with their closures, with respect to the operator norm of the special Lie algebra $\mathfrak{fsl}(\mathfrak{H})$ and the finitary orthogonal algebra $\mathfrak{fo}(\mathfrak{H})$ respectively, since compact operators on Hilbert spaces are limits of sequences of finite rank operators with respect to the norm topology. Then both Banach–Lie algebras $\mathfrak{gl}(\mathfrak{H}, \mathcal{C}_\infty)$ and $\mathfrak{o}(\mathfrak{H}, \theta, \mathcal{C}_\infty)$ are topologically simple, nondegenerate, and contain extremal elements. □

PROPOSITION 15.29. *Let $(\mathfrak{H}, (\,,))$ be an infinite-dimensional complex Hilbert space and fix an anti-conjugation ζ of $\mathfrak{H}$. We have:*

(i) *$\langle x, y\rangle := (x, y\zeta)$, for all $x, y \in \mathfrak{H}$, defines a nondegenerate alternate bilinear form which makes $\mathfrak{H}$ into a Banach inner product space with respect to the Hilbert norm.*

(ii) *$\mathfrak{gl}(\mathfrak{H})$ coincides with the Banach–Lie algebra of bounded linear operators on $\mathfrak{H}$. Moreover, $a^\# = -\zeta a^* \zeta$ for all $a \in \mathrm{BL}(\mathfrak{H})$.*

(iii) *$\mathfrak{sp}(\mathfrak{H}, \zeta, \mathcal{C}_\infty) = \overline{\mathfrak{fsp}(\mathfrak{H})}$ is a topologically simple nondegenerate Banach–Lie algebra with extremal elements.*

PROOF. It is similar to that of Proposition 15.28. □

Banach–Lie algebras spanned by extremal elements. As should be expected of Corollaries 15.18, 15.21, and 15.24, nondegenerate Banach–Lie algebras spanned by extremal elements are necessarily finite-dimensional.

THEOREM 15.30. *Let* $(L, ||\cdot||)$ *be a nondegenerate Banach–Lie algebra. Then the following conditions are equivalent:* (i) $L = \mathrm{Soc}(L)$, (ii) L *is spanned by extremal elements, and* (iii) L *is finite-dimensional.*

PROOF. By Lemma 15.12, $\mathrm{Soc}(L)$ is spanned by extremal elements, so we only need to prove that (i) $\Rightarrow$ (iii). Note first that $L = \mathrm{Soc}(L) = \bigoplus_{\alpha\in\Lambda} M_\alpha$ is a direct sum of minimal ideals, and for each index α, $M_\alpha = \mathrm{Ann}(\sum_{\beta\neq\alpha} M_\beta)$, which implies that M_α is closed and therefore a simple Banach–Lie algebra spanned by extremal elements. And second, there are only finitely many M_α. If the family $\{M_{\alpha_n}\}_{\alpha\in\Lambda}$ were infinite, we could construct an infinite sequence of nonzero elements $a_n \in M_{\alpha_n}$ and take $a = \sum_{n=1}^{\infty} \frac{a_n}{2^n||a_n||}$. Since $a \in \sum_{\alpha\in\Lambda} M_\alpha$, there exists a finite subset of indexes $\{\beta_1, \ldots, \beta_r\} \subset \Lambda$ such that $a \in M_{\beta_1} \oplus \cdots \oplus M_{\beta_r}$. Let α_m be distinct from all the β_i. Then, for each β_i, $[M_{\beta_i}, M_{\alpha_m}] \subset M_{\beta_i} \cap M_{\alpha_m} = 0$, but $[a, M_{\alpha_m}] = [a_m, M_{\alpha_m}] \neq 0$, a contradiction. This allows us to reduce the question to the case that L is simple. Now, if L were infinite-dimensional, it would be either $\mathfrak{f}sl_Y(X)$, $\mathfrak{f}o(X, \langle\,,\rangle)$ or $\mathfrak{f}sp(X, \langle\,,\rangle)$, with X being infinite-dimensional in all the cases (Theorem 15.25). But as proved in Corollaries 15.18, 15.21, and 15.24, no one of these algebras admits a complete algebra norm. □

15.3. Compact elements in Banach–Lie algebras

Following [**Ale68**], a Banach algebra R is said to be *compact* if for any element $a \in R$, the linear map given by $b \mapsto aba$ $(b \in R)$ is compact. Since this is a "Jordan notion", it makes sense for any Banach–Jordan algebra. A Banach–Jordan algebra J is said to be *compact* if for any element $a \in J$, the linear map $U_a : J \to J$ is compact.

DEFINITION 15.31. By a *compact element* of a Banach–Lie algebra L we mean a Jordan element $a \in L$ such that the linear map ad_a^2 acting on the Banach space L is compact. A Banach–Lie algebra L is said to be *compact* if every Jordan element of L is compact.

EXAMPLE 15.32. Let R be a compact Banach algebra and let $a \in R$ be an element of square 0. Then a is a compact element of the Banach–Lie algebra R^-: for any $b \in R$, $\mathrm{ad}_a^2 b = -2aba$.

LEMMA 15.33. *Let* L *be a Banach–Lie algebra and let* $e \in L$ *be von Neumann regular. Then* e *is a compact element of* L *if and only if the inner ideal* $(e) = \mathrm{ad}_e^2 L$ *is finite-dimensional.*

PROOF. By Theorem 5.11, there exists $f \in L$ such that $(e, f, [e, f])$ is a $\mathfrak{sl}_2$-triple and $\frac{1}{4}\mathrm{ad}_e^2\mathrm{ad}_f^2 = 1_{(e)}$. Hence, if e is compact, then the identity operator on the Banach space (e) is compact. Thus (e) is finite-dimensional. The converse is obvious. □

LEMMA 15.34. *Let* L *be a Banach–Lie algebra and let* $a \in L$ *be a compact element. Then the Banach–Jordan algebra* L_a *is compact.*

PROOF. For any $b \in L$, the corresponding $U_{\bar{b}}$-operator of the Jordan algebra L_a, where $\bar{b} = \pi_a(b)$, is given by $U_{\bar{b}}\bar{x} = \pi_a \operatorname{ad}_b^2 \operatorname{ad}_a^2 x$, $x \in L$. Since ad_a^2 is compact and the linear map $\pi_a : L \to L_a$ is continuous, the composition $\pi_a \operatorname{ad}_b^2 \operatorname{ad}_a^2$ is a compact operator, so L_a is compact. □

PROPOSITION 15.35. *Let L be a Banach–Lie algebra which is primitive at a nonzero compact element. Then L contains nonzero extremal elements.*

PROOF. By Lemma 15.34, the primitive Banach–Jordan algebra L_a is compact. Then, by [**FL85**, Lemma 6.2], it contains a minimal inner ideal, which implies, by Proposition 8.51(3), that L contains a minimal abelian inner ideal, and hence an extremal element by Lemma 15.12. □

It follows from the above result that primitive compact Banach–Lie algebras belong to the class of the strongly prime Banach–Lie algebras with extremal elements we have studied in the previous section. In each one of the three types, special, orthogonal and symplectic, we will consider those consisting of compact operators and check which of them are compact Banach–Lie algebras, according with our definition.

NOTATION 15.36. Given a complex Banach space X, we write $\mathrm{KL}(X)$ to the denote the Banach algebra of compact linear operators on X.

LEMMA 15.37. *Let R be an infinite-dimensional primitive compact Banach algebra. Then the Banach–Lie algebra R^- is primitive and compact.*

PROOF. By [**Ale68**, Theorem 5.1], R has nonzero socle and therefore it is a centrally closed prime ring. Moreover, since R is infinite-dimensional, it has nonzero nilpotent elements and trivial center. Then, by Corollary 8.70, R^- is primitive, and by Theorem 4.35, every Jordan element of R^- is of square 0 and hence a compact element by Example 15.32. Thus R^- is primitive and compact. □

COROLLARY 15.38. *Let X be an infinite-dimensional complex Banach space. Then the primitive Banach–Lie algebra $\mathrm{KL}(X)^-$ is compact.*

PROOF. $\mathrm{KL}(X)$ is a primitive Banach algebra [**Ale68**, Theorem 5.4]. Now Lemma 15.37 applies. □

The above corollary says to us that primitive compact Banach–Lie algebras can be found in the class of Banach–Lie algebras of special type. However, we cannot expect to find them among Banach–Lie algebras of orthogonal type.

PROPOSITION 15.39. *No infinite-dimensional Banach–Lie algebra of orthogonal type is compact.*

PROOF. Let L be an infinite-dimensional Banach–Lie algebra of orthogonal type, i.e there exists an infinite-dimensional Banach inner product space $(X, \langle\, , \rangle)$, where $\langle\, , \rangle$ is symmetric, such that $\mathfrak{fo}(X, \langle\, , \rangle) \leq L \leq \mathfrak{o}(X, \langle\, , \rangle)$ and the injection of $(L, || \cdot ||)$ into $(\mathfrak{o}(X, \langle\, , \rangle), | \cdot |)$ is continuous. Taking $c = [x, z]$ as in Example 8.81, we get a von Neumann regular element (in fact, a Clifford element) c in L such that $\operatorname{ad}_c^2 L$ is infinite-dimensional. By Lemma 15.33, c is not compact, so L is not compact. □

15.4. Exercises

EXERCISE 15.40. Let X and Y be normed vector spaces and let $f : X \to Y$ be a linear map. Show that $\mathrm{Sep}(f)$ is a closed subspace of Y.

EXERCISE 15.41. Show that every semiprime associative algebra with essential socle is semiprimitive.

EXERCISE 15.42. Complete the proofs of Corollaries 15.21 and 15.24.

EXERCISE 15.43. Let $(X, \langle\,,\rangle)$ be an infinite-dimensional alternate Banach inner product space $(X, \langle\,,\rangle)$. Show that the Banach–Lie algebra $\mathfrak{sp}(X, \langle\,,\rangle) \cap \mathrm{KL}(X)$ is compact.

EXERCISE 15.44. [**Kap48a**, Theorem] Let A be a nontrivial von Neumann regular Banach algebra. Show:

(1) A is semiprimitive and contains *reduced* idempotents, $eAe = \mathbb{C}e$.
(2) A is idempotent-finite, i.e. it dos not contain infinite sequence of nonzero idempotents.
(3) A is unital and $1 = e_1 + \cdots + e_n$ is a sum of orthogonal reduced idempotents.
(4) $A \cong \bigoplus_{k=1}^{k=m} \mathrm{M}_k(\mathbb{C})$ for some $1 \le m \le n$.

EXERCISE 15.45. Let A be a semiprime Banach algebra. Show that the socle of A coincides with the largest von Neumann regular ideal of A.

EXERCISE 15.46. [**BK88**, Principal theorem] Let J be a nontrivial von Neumann regular Banach–Jordan algebra. Show that J is reduced.

EXERCISE 15.47. [**FLRP86a**, Theorem 2] Let J be a nondegenerate Banach–Jordan algebra. Show that the socle of J coincides with the largest von Neumann regular ideal of J.

EXERCISE 15.48. Let L be a simple Banach–Lie algebra such that

(i) L is generated by ad-nilpotent elements,
(ii) every Jordan element of L is von Neumann regular.

Show that L finite-dimensional.

Bibliography

[AAJZ15] Adel Alahmadi, Hamed Alsulami, S. K. Jain, and Efim Zelmanov, *Finite generation of Lie algebras associated with associative algebras*, J. Algebra **426** (2015), 69–78, DOI 10.1016/j.jalgebra.2014.10.056. MR3301901

[AC96] José A. Anquela and Teresa Cortés, *Primitive Jordan pairs and triple systems*, J. Algebra **184** (1996), no. 2, 632–678, DOI 10.1006/jabr.1996.0280. MR1409234

[AC98] José A. Anquela and Teresa Cortés, *Local-to-global inheritance of primitivity in Jordan algebras*, Arch. Math. (Basel) **70** (1998), no. 3, 219–227, DOI 10.1007/s000130050187. MR1604064

[ACGGL05] José A. Anquela, Teresa Cortés, Esther García, and Miguel Gómez-Lozano, *Polynomial identities and speciality of Martindale-like covers of Jordan algebras*, J. Pure Appl. Algebra **202** (2005), no. 1-3, 1–10, DOI 10.1016/j.jpaa.2005.01.004. MR2163397

[ACLM96] José A. Anquela, Teresa Cortés, Ottmar Loos, and Kevin McCrimmon, *An elemental characterization of strong primeness in Jordan systems*, J. Pure Appl. Algebra **109** (1996), no. 1, 23–36, DOI 10.1016/0022-4049(95)00086-0. MR1386811

[ACM95] José A. Anquela, Teresa Cortés, and Fernando Montaner, *Local inheritance in Jordan algebras*, Arch. Math. (Basel) **64** (1995), no. 5, 393–401, DOI 10.1007/BF01197216. MR1323915

[ACZ15] José A. Anquela, Teresa Cortés, and Efim Zelmanov, *The Kurosh problem for Jordan nil systems over arbitrary rings of scalars*, J. Algebra **444** (2015), 313–327, DOI 10.1016/j.jalgebra.2015.08.004. MR3406180

[ACZ16] José A. Anquela, Teresa Cortés, and Efim Zelmanov, *Local nilpotency of the McCrimmon radical of a Jordan system*, Tr. Mat. Inst. Steklova **292** (2016), no. Algebra, Geometriya i Teoriya Chisel, 7–15, DOI 10.1134/S0371968516010015. Reprinted in Proc. Steklov Inst. Math. **292** (2016), no. 1, 1–9. MR3665922

[AFS17] Bruce Allison, John Faulkner, and Oleg Smirnov, *Weyl images of Kantor pairs*, Canad. J. Math. **69** (2017), no. 4, 721–766, DOI 10.4153/CJM-2016-047-1. MR3679693

[Ale68] J. C. Alexander, *Compact Banach algebras*, Proc. London Math. Soc. (3) **18** (1968), 1–18, DOI 10.1112/plms/s3-18.1.1. MR0229040

[AMC94] José A. Anquela, Fernando Montaner, and Teresa Cortés, *On primitive Jordan algebras*, J. Algebra **163** (1994), no. 3, 663–674, DOI 10.1006/jabr.1994.1036. MR1265856

[Aup82] Bernard Aupetit, *The uniqueness of the complete norm topology in Banach algebras and Banach Jordan algebras*, J. Funct. Anal. **47** (1982), no. 1, 1–6, DOI 10.1016/0022-1236(82)90097-0. MR663830

[Bah76] Juri A. Bahturin, *Simple Lie algebras satisfying a nontrivial identity*, Serdica **2** (1976), no. 3, 241–246. MR0447358

[Bar99] A. A. Baranov, *Finitary simple Lie algebras*, J. Algebra **219** (1999), no. 1, 299–329, DOI 10.1006/jabr.1999.7856. MR1707673

[Bar15] Alexander Baranov, *Simple locally finite Lie algebras of diagonal type*, Lie algebras and related topics, Contemp. Math., vol. 652, Amer. Math. Soc., Providence, RI, 2015, pp. 47–60, DOI 10.1090/conm/652/12970. MR3453048

[BBZ04] Y. A. Bahturin, A. A. Baranov, and A. E. Zalesski, *Simple Lie subalgebras of locally finite associative algebras*, J. Algebra **281** (2004), no. 1, 225–246, DOI 10.1016/j.jalgebra.2004.06.006. MR2091969

[BD73] Frank F. Bonsall and John Duncan, *Complete normed algebras*, Springer-Verlag, New York-Heidelberg, 1973. Ergebnisse der Mathematik und ihrer Grenzgebiete, Band 80. MR0423029

[Ben76] Georgia Benkart, *The Lie inner ideal structure of associative rings*, J. Algebra **43** (1976), no. 2, 561–584, DOI 10.1016/0021-8693(76)90127-7. MR0435149

[Ben77] Georgia Benkart, *On inner ideals and ad-nilpotent elements of Lie algebras*, Trans. Amer. Math. Soc. **232** (1977), 61–81, DOI 10.2307/1998926. MR0466242

[BFL94] Khalid Bouhya and Antonio Fernández López, *Jordan-∗-triples with minimal inner ideals and compact* JB∗*-triples*, Proc. London Math. Soc. (3) **68** (1994), no. 2, 380–398, DOI 10.1112/plms/s3-68.2.380. MR1253508

[BFL09] Georgia Benkart and Antonio Fernández López, *The Lie inner ideal structure of associative rings revisited*, Comm. Algebra **37** (2009), no. 11, 3833–3850, DOI 10.1080/00927870802545729. MR2573222

[BFL10] Matej Brešar and Antonio Fernández López, *On a class of finitary Lie algebras characterized through derivations*, Proc. Amer. Math. Soc. **138** (2010), no. 12, 4161–4166, DOI 10.1090/S0002-9939-2010-10644-5. MR2680042

[BFL16] A. A. Baranov and Antonio Fernández López, *Maximal zero product subrings and inner ideals of simple rings*, Preprint(ArXiv:1609.04609) (2016).

[BFLGL16] Jose Brox, Antonio Fernández López, and Miguel Gómez Lozano, *Inner ideals of Lie algebras of skew elements of prime rings with involution*, Proc. Amer. Math. Soc. **144** (2016), no. 7, 2741–2751, DOI 10.1090/proc/12903. MR3487211

[BFLL17] Jose Brox, Antonio Fernández López, and Miguel Gómez Lozano, *Clifford elements in Lie algebras*, J. Lie Theory **27** (2017), no. 1, 283–294. MR3536548

[BGGL14] Jose Brox, Esther García, and Miguel Gómez Lozano, *An elemental characterization of orthogonal ideals in Lie algebras*, Mediterr. J. Math. **11** (2014), no. 4, 1061–1067, DOI 10.1007/s00009-013-0344-3. MR3268807

[BGGL16] Jose Brox, Esther García, and Miguel Gómez Lozano, *Jordan algebras at Jordan elements of semiprime rings with involution*, J. Algebra **468** (2016), 155–181, DOI 10.1016/j.jalgebra.2016.06.036. MR3550861

[BI77] G. M. Benkart and I. M. Isaacs, *On the existence of ad-nilpotent elements*, Proc. Amer. Math. Soc. **63** (1977), no. 1, 39–40, DOI 10.2307/2041060. MR0432721

[BK88] Mohamed Benslimane and El Amin Mokhtar Kaïdi, *Structure des algèbres de Jordan-Banach non commutatives complexes régulières ou semi-simples à spectre fini* (French), J. Algebra **113** (1988), no. 1, 201–206, DOI 10.1016/0021-8693(88)90189-5. MR928061

[BM79] W. E. Baxter and W. S. Martindale III, *Central closure of semiprime nonassociative rings*, Comm. Algebra **7** (1979), no. 11, 1103–1132, DOI 10.1080/00927877908822394. MR535925

[BMM96] K. I. Beidar, W. S. Martindale III, and A. V. Mikhalev, *Rings with generalized identities*, Monographs and Textbooks in Pure and Applied Mathematics, vol. 196, Marcel Dekker, Inc., New York, 1996. MR1368853

[BMS87] K. I. Beĭdar, A. V. Mikhalëv, and A. M. Slin′ko, *A primality criterion for nondegenerate alternative and Jordan algebras* (Russian), Trudy Moskov. Mat. Obshch. **50** (1987), 130–137, 261; English transl., Trans. Moscow Math. Soc. (1988), 129–137. MR912055

[Bou81] Nicolas Bourbaki, *Éléments de mathématique* (French), Masson, Paris, 1981. Groupes et algèbres de Lie. Chapitres 4, 5 et 6. [Lie groups and Lie algebras. Chapters 4, 5 and 6]. MR647314

[BR13] A. A. Baranov and J. Rowley, *Inner ideals of simple locally finite Lie algebras*, J. Algebra **379** (2013), 11–30, DOI 10.1016/j.jalgebra.2012.12.021. MR3019243

[BS94] Yu. A. Bakhturin and Kh. Shtrade, *Locally finite-dimensional simple Lie algebras* (Russian, with Russian summary), Mat. Sb. **185** (1994), no. 2, 3–32, DOI 10.1070/SM1995v081n01ABEH003618; English transl., Russian Acad. Sci. Sb. Math. **81** (1995), no. 1, 137–161. MR1264772

[BS16] A. A. Baranov and H. Shlaka, *Jordan-Lie inner ideals of finite dimensional associative algebras*, Preprint(ArXiv:1609.05196) (2016).

[BV99] M. I. Berenguer and A. R. Villena, *Continuity of Lie isomorphisms of Banach algebras*, Bull. London Math. Soc. **31** (1999), no. 1, 6–10, DOI 10.1112/S0024609398005128. MR1651037

[CCFL12] Juan Carlos Cabello, Miguel Cabrera, and Antonio Fernández López, *π-complemented algebras through pseudocomplemented lattices*, Order **29** (2012), no. 3, 463–479, DOI 10.1007/s11083-011-9214-4. MR2979643

[CCR12] J. C. Cabello, M. Cabrera, and R. Roura, *π-complementation in the unitisation and multiplication algebras of a semiprime algebra*, Comm. Algebra **40** (2012), no. 9, 3507–3531, DOI 10.1080/00927872.2011.590954. MR2981149

[CGGL12] D. Ceretto, E. García, and M. Gómez Lozano, *A radical for graded Lie algebras*, Acta Math. Hungar. **136** (2012), no. 1-2, 16–29, DOI 10.1007/s10474-012-0233-4. MR2925753

[CGRP14] Miguel Cabrera García and Ángel Rodríguez Palacios, *Non-associative normed algebras. Vol. 1*, Encyclopedia of Mathematics and its Applications, vol. 154, Cambridge University Press, Cambridge, 2014. The Vidav-Palmer and Gelfand-Naimark theorems. MR3242640

[Cha90] A. D. Chanyshev, *Local finite-dimensionality of algebraic Lie algebras* (Russian), Uspekhi Mat. Nauk **45** (1990), no. 6(276), 153–154, DOI 10.1070/RM1990v045n06ABEH002694; English transl., Russian Math. Surveys **45** (1990), no. 6, 143–144. MR1101350

[Che89] V. I. Chernousov, *The Hasse principle for groups of type E_8* (Russian), Dokl. Akad. Nauk SSSR **306** (1989), no. 5, 1059–1063; English transl., Soviet Math. Dokl. **39** (1989), no. 3, 592–596. MR1014762

[CIR08] Arjeh M. Cohen, Gábor Ivanyos, and Dan Roozemond, *Simple Lie algebras having extremal elements*, Indag. Math. (N.S.) **19** (2008), no. 2, 177–188, DOI 10.1016/S0019-3577(09)00003-2. MR2489325

[Coh03] P. M. Cohn, *Further algebra and applications*, Springer-Verlag London, Ltd., London, 2003. MR1953987

[CSUW01] Arjeh M. Cohen, Anja Steinbach, Rosane Ushirobira, and David Wales, *Lie algebras generated by extremal elements*, J. Algebra **236** (2001), no. 1, 122–154, DOI 10.1006/jabr.2000.8508. MR1808349

[D'A92] Alain D'Amour, *Quadratic Jordan systems of Hermitian type*, J. Algebra **149** (1992), no. 1, 197–233, DOI 10.1016/0021-8693(92)90013-C. MR1165208

[DFFLGGL08] Cristina Draper, Antonio Fernández López, Esther García, and Miguel Gómez Lozano, *The socle of a nondegenerate Lie algebra*, J. Algebra **319** (2008), no. 6, 2372–2394, DOI 10.1016/j.jalgebra.2007.10.042. MR2388311

[DFLGGL12] C. Draper, A. Fernández López, E. García, and M. Gómez Lozano, *The inner ideals of the simple finite dimensional Lie algebras*, J. Lie Theory **22** (2012), no. 4, 907–929. MR3052677

[DlH72] Pierre de la Harpe, *Classical Banach-Lie algebras and Banach-Lie groups of operators in Hilbert space*, Lecture Notes in Mathematics, Vol. 285, Springer-Verlag, Berlin-New York, 1972. MR0476820

[DM95] Alain D'Amour and Kevin McCrimmon, *The local algebras of Jordan systems*, J. Algebra **177** (1995), no. 1, 199–239, DOI 10.1006/jabr.1995.1294. MR1356368

[DM00] Alain D'Amour and Kevin McCrimmon, *The structure of quadratic Jordan systems of Clifford type*, J. Algebra **234** (2000), no. 1, 31–89, DOI 10.1006/jabr.2000.8452. MR1799478

[EK13] Alberto Elduque and Mikhail Kochetov, *Gradings on simple Lie algebras*, Mathematical Surveys and Monographs, vol. 189, American Mathematical Society, Providence, RI; Atlantic Association for Research in the Mathematical Sciences (AARMS), Halifax, NS, 2013. MR3087174

[Eri72] Theodore S. Erickson, *The Lie structure in prime rings with involution*, J. Algebra **21** (1972), 523–534, DOI 10.1016/0021-8693(72)90011-7. MR0308203

[Fau73] John R. Faulkner, *On the geometry of inner ideals*, J. Algebra **26** (1973), 1–9, DOI 10.1016/0021-8693(73)90032-X. MR0367002

[FL85] Antonio Fernández López, *Modular annihilator Jordan algebras*, Comm. Algebra **13** (1985), no. 12, 2597–2613, DOI 10.1080/00927878508823292. MR811525

[FL11] Antonio Fernández López, *Banach-Lie algebras with extremal elements*, Q. J. Math. **62** (2011), no. 1, 115–129, DOI 10.1093/qmath/hap032. MR2774357

[FL14] Antonio Fernández López, *Lie inner ideals are nearly Jordan inner ideals*, Proc. Amer. Math. Soc. **142** (2014), no. 3, 795–804, DOI 10.1090/S0002-9939-2013-11809-5. MR3148514

[FL16] Antonio Fernández López, *A Jordan approach to finitary Lie algebras*, Q. J. Math. **67** (2016), no. 4, 565–571. MR3609846

[FLG13] Antonio Fernández López and Artem Yu. Golubkov, *Lie algebras with an algebraic adjoint representation revisited*, Manuscripta Math. **140** (2013), no. 3-4, 363–376, DOI 10.1007/s00229-012-0546-1. MR3019131

[FLGGL04] Antonio Fernández López, Esther García, and Miguel Gómez Lozano, *The Jordan socle and finitary Lie algebras*, J. Algebra **280** (2004), no. 2, 635–654, DOI 10.1016/j.jalgebra.2004.06.013. MR2089256

[FLGGL06] Antonio Fernández López, Esther García, and Miguel Gómez Lozano, *Inner ideals of finitary simple Lie algebras*, J. Lie Theory **16** (2006), no. 1, 97–114. MR2196417

[FLGGL07] Antonio Fernández López, Esther García, and Miguel Gómez Lozano, *The Jordan algebras of a Lie algebra*, J. Algebra **308** (2007), no. 1, 164–177, DOI 10.1016/j.jalgebra.2006.02.035. MR2290916

[FLGGL08] Antonio Fernández López, Esther García, and Miguel Gómez Lozano, *An Artinian theory for Lie algebras*, J. Algebra **319** (2008), no. 3, 938–951, DOI 10.1016/j.jalgebra.2007.10.038. MR2379087

[FLGGL09] Gang Yu, *Upper bounds for finite additive 2-bases*, Proc. Amer. Math. Soc. **137** (2009), no. 1, 11–18, DOI 10.1090/S0002-9939-08-09430-6. MR2439419

[FLGGLN07] Antonio Fernández López, Esther García, Miguel Gómez Lozano, and Erhard Neher, *A construction of gradings of Lie algebras*, Int. Math. Res. Not. IMRN **16** (2007), Art. ID rnm051, 34, DOI 10.1093/imrn/rnm051. MR2353091

[FLGR98] Antonio Fernández López and Eulalia García Rus, *Algebraic lattices and nonassociative structures*, Proc. Amer. Math. Soc. **126** (1998), no. 11, 3211–3221, DOI 10.1090/S0002-9939-98-04443-8. MR1459121

[FLGR99] Antonio Fernández López and Eulalia García Rus, *Inner ideals in quadratic Jordan algebras of infinite capacity*, Int. J. Math. Game Theory Algebra **9** (1999), no. 1, 35–54. MR1697472

[FLGRGLSM98] A. Fernández López, E. García Rus, M. Gómez Lozano, and M. Siles Molina, *Goldie theorems for associative pairs*, Comm. Algebra **26** (1998), no. 9, 2987–3020, DOI 10.1080/00927879808826324. MR1635898

[FLGRSCSM94] A. Fernandez Lopez, E. García Rus, E. Sánchez Campos, and M. Siles Molina, *Strong regularity and Hermitian Hilbert triples*, Quart. J. Math. Oxford Ser. (2) **45** (1994), no. 177, 43–55, DOI 10.1093/qmath/45.1.43. MR1269289

[FLMZ01] A. Fernández López, H. Marhnine, and C. Zarhouti, *Derivations on Banach-Jordan pairs*, Q. J. Math. **52** (2001), no. 3, 269–283, DOI 10.1093/qjmath/52.3.269. MR1865901

[FLRP86a] Antonio Fernández López and Ángel Rodríguez-Palacios, *On the socle of a noncommutative Jordan algebra*, Manuscripta Math. **56** (1986), no. 3, 269–278, DOI 10.1007/BF01180768. MR856365

[FLRP86b] Antonio Fernandez Lopez and Angel Rodriguez Palacios, *Primitive noncommutative Jordan algebras with nonzero socle*, Proc. Amer. Math. Soc. **96** (1986), no. 2, 199–206, DOI 10.2307/2046152. MR818443

[FLTB03] Antonio Fernández López and Maribel Tocón Barroso, *Strongly prime Jordan pairs with nonzero socle*, Manuscripta Math. **111** (2003), no. 3, 321–340, DOI 10.1007/s00229-003-0372-6. MR1993498

[Gar03] Esther García, *Inheritance of primeness by ideals in Lie algebras*, Int. J. Math. Game Theory Algebra **13** (2003), no. 6, 481–484. MR2115392

[Gar04] Esther García, *Tits-Kantor-Koecher algebras of strongly prime Hermitian Jordan pairs*, J. Algebra **277** (2004), no. 2, 559–571, DOI 10.1016/j.jalgebra.2003.10.016. MR2067619

[GGL07] Esther García and Miguel Gómez Lozano, *An elemental characterization of strong primeness in Lie algebras*, J. Algebra **312** (2007), no. 1, 132–141, DOI 10.1016/j.jalgebra.2006.11.003. MR2320450

[GGL09] Esther García and Miguel Gómez Lozano, *A note on a result of Kostrikin*, Comm. Algebra **37** (2009), no. 7, 2405–2409, DOI 10.1080/00927870802255683. MR2536928

[GGL11] Esther García and Miguel Gómez Lozano, *A characterization of the Kostrikin radical of a Lie algebra*, J. Algebra **346** (2011), 266–283, DOI 10.1016/j.jalgebra.2011.08.013. MR2842081

[GGL12] Dancheng Lu and Tongsuo Wu, *Regularity, a^*-invariant and the associated graded module to a filtration*, Comm. Algebra **40** (2012), no. 10, 3745–3752, DOI 10.1080/00927872.2011.594471. MR2982894

[GGLN11] Esther García, Miguel Gómez Lozano, and Erhard Neher, *Nondegeneracy for Lie triple systems and Kantor pairs*, Canad. Math. Bull. **54** (2011), no. 3, 442–455, DOI 10.4153/CMB-2011-023-9. MR2857919

[Grz92] Piotr Grzeszczuk, *On nilpotent derivations of semiprime rings*, J. Algebra **149** (1992), no. 2, 313–321, DOI 10.1016/0021-8693(92)90018-H. MR1172431

[GS05] A. N. Grishkov and I. P. Shestakov, *A characterization of Lie algebras of skew-symmetric elements*, Acta Appl. Math. **85** (2005), no. 1-3, 157–159, DOI 10.1007/s10440-004-5598-0. MR2128909

[Hal52] Paul R. Halmos, *Commutators of operators*, Amer. J. Math. **74** (1952), 237–240, DOI 10.2307/2372081. MR0045310

[Hen14] Johanna Hennig, *Simple, locally finite dimensional Lie algebras in positive characteristic*, J. Algebra **413** (2014), 270–288, DOI 10.1016/j.jalgebra.2014.04.027. MR3216608

[Her69] I. N. Herstein, *Topics in ring theory*, The University of Chicago Press, Chicago, Ill.-London, 1969. MR0271135

[Her70] I. N. Herstein, *On the Lie structure of an associative ring*, J. Algebra **14** (1970), 561–571, DOI 10.1016/0021-8693(70)90103-1. MR0255610

[Her76] I. N. Herstein, *Rings with involution*, The University of Chicago Press, Chicago, Ill.-London, 1976, Chicago Lectures in Mathematics. MR0442017

[Her94] I. N. Herstein, *Noncommutative rings*, Carus Mathematical Monographs, vol. 15, Mathematical Association of America, Washington, DC, 1994. Reprint of the 1968 original; With an afterword by Lance W. Small. MR1449137

[HM81] Leslie Hogben and Kevin McCrimmon, *Maximal modular inner ideals and the Jacobson radical of a Jordan algebra*, J. Algebra **68** (1981), no. 1, 155–169, DOI 10.1016/0021-8693(81)90291-X. MR604300

[HOSr84] Harald Hanche-Olsen and Erling Størmer, *Jordan operator algebras*, Monographs and Studies in Mathematics, vol. 21, Pitman (Advanced Publishing Program), Boston, MA, 1984. MR755003

[Hum72] James E. Humphreys, *Introduction to Lie algebras and representation theory*, Springer-Verlag, New York-Berlin, 1972. Graduate Texts in Mathematics, Vol. 9. MR0323842

[Jac58] N. Jacobson, *A note on three dimensional simple Lie algebras*, J. Math. Mech. **7** (1958), 823–831, DOI 10.1512/iumj.1958.7.57047. MR0097432

[Jac64] Nathan Jacobson, *Structure of rings*, American Mathematical Society Colloquium Publications, Vol. 37. Revised edition, American Mathematical Society, Providence, R.I., 1964. MR0222106

[Jac68] Nathan Jacobson, *Structure and representations of Jordan algebras*, American Mathematical Society Colloquium Publications, Vol. XXXIX, American Mathematical Society, Providence, R.I., 1968. MR0251099

[Jac79] Nathan Jacobson, *Lie algebras*, Dover Publications Inc., New York, 1979, Republication of the 1962 original. MR559927 (80k:17001)

[Kap48a] Irving Kaplansky, *Regular Banach algebras*, J. Indian Math. Soc. (N.S.) **12** (1948), 57–62. MR0029106

[Kap48b] Irving Kaplansky, *Rings with a polynomial identity*, Bull. Amer. Math. Soc. **54** (1948), 575–580, DOI 10.1090/S0002-9904-1948-09049-8. MR0025451

[Kap03] Irving Kaplansky, *Linear algebra and geometry*, revised ed., Dover Publications, Inc., Mineola, NY, 2003, A second course. MR2001037

[Kos90] A. I. Kostrikin, *Around Burnside*, Ergebnisse der Mathematik und ihrer Grenzgebiete (3) [Results in Mathematics and Related Areas (3)], vol. 20, Springer-Verlag, Berlin, 1990. Translated from the Russian and with a preface by James Wiegold. MR1075416

[Lam99] T. Y. Lam, *Lectures on modules and rings*, Graduate Texts in Mathematics, vol. 189, Springer-Verlag, New York, 1999. MR1653294

[Lev53] Jakob Levitzki, *On the structure of algebraic algebras and related rings*, Trans. Amer. Math. Soc. **74** (1953), 384–409, DOI 10.2307/1990809. MR0053089

[LN94] Ottmar Loos and Erhard Neher, *Complementation of inner ideals in Jordan pairs*, J. Algebra **166** (1994), no. 2, 255–295, DOI 10.1006/jabr.1994.1151. MR1279257

[LN04] Ottmar Loos and Erhard Neher, *Locally finite root systems*, Mem. Amer. Math. Soc. **171** (2004), no. 811, x+214. MR2073220

[Loo75] Ottmar Loos, *Jordan pairs*, Lecture Notes in Mathematics, Vol. 460, Springer-Verlag, Berlin-New York, 1975. MR0444721

[Loo89] Ottmar Loos, *On the socle of a Jordan pair*, Collect. Math. **40** (1989), no. 2, 109–125 (1990). MR1094683

[Loo90] Ottmar Loos, *Filtered Jordan systems*, Comm. Algebra **18** (1990), no. 6, 1899–1924, DOI 10.1080/00927879008824000. MR1071080

[Loo91] Ottmar Loos, *Finiteness conditions in Jordan pairs*, Math. Z. **206** (1991), no. 4, 577–587, DOI 10.1007/BF02571365. MR1100842

[Mar01] C. Martínez, *The ring of fractions of a Jordan algebra*, J. Algebra **237** (2001), no. 2, 798–812, DOI 10.1006/jabr.2000.8617. MR1816717

[McC69] Kevin McCrimmon, *The radical of a Jordan algebra*, Proc. Nat. Acad. Sci. U.S.A. **62** (1969), 671–678, DOI 10.1073/pnas.62.3.671. MR0268238

[McC71] Kevin McCrimmon, *Inner ideals in quadratic Jordan algebras*, Trans. Amer. Math. Soc. **159** (1971), 445–468, DOI 10.2307/1996022. MR0279145

[McC82] Kevin McCrimmon, *Compatible Peirce decompositions of Jordan triple systems*, Pacific J. Math. **103** (1982), no. 1, 57–102. MR687964

[McC85] Kevin McCrimmon, *Reduced elements in Jordan triple systems*, J. Algebra **97** (1985), no. 2, 540–564, DOI 10.1016/0021-8693(85)90063-8. MR813001

[McC99] Kevin McCrimmon, *Jordan centroids*, Comm. Algebra **27** (1999), no. 2, 933–954, DOI 10.1080/00927879908826470. MR1672019

[McC04] Kevin McCrimmon, *A taste of Jordan algebras*, Universitext, Springer-Verlag, New York, 2004. MR2014924 (2004i:17001)

[Mey72] Kurt Meyberg, *Lectures on algebras and triple systems*, The University of Virginia, Charlottesville, Va., 1972. Notes on a course of lectures given during the academic year 1971–1972. MR0340353

[MM83] W. S. Martindale III and C. Robert Miers, *On the iterates of derivations of prime rings*, Pacific J. Math. **104** (1983), no. 1, 179–190. MR683736

[MM89] W. S. Martindale III and C. Robert Miers, *Nilpotency and generalized Lie ideals*, J. Algebra **127** (1989), no. 1, 244–254, DOI 10.1016/0021-8693(89)90287-1. MR1029416

[Mon99] Fernando Montaner, *Local PI theory of Jordan systems*, J. Algebra **216** (1999), no. 1, 302–327, DOI 10.1006/jabr.1998.7755. MR1694558

[MZ99] Consuelo Martínez and Efim Zelmanov, *Nil algebras and unipotent groups of finite width*, Adv. Math. **147** (1999), no. 2, 328–344, DOI 10.1006/aima.1999.1848. MR1734527

[Neh87] Erhard Neher, *Jordan triple systems by the grid approach*, Lecture Notes in Mathematics, vol. 1280, Springer-Verlag, Berlin, 1987. MR911879

[Neh91] Erhard Neher, *Jordan pairs with finite grids*, Comm. Algebra **19** (1991), no. 2, 455–478, DOI 10.1080/00927879108824149. MR1100357

[Neh96] Erhard Neher, *Lie algebras graded by 3-graded root systems and Jordan pairs covered by grids*, Amer. J. Math. **118** (1996), no. 2, 439–491. MR1385288

[OR79] J. Marshall Osborn and M. L. Racine, *Jordan rings with nonzero socle*, Trans. Amer. Math. Soc. **251** (1979), 375–387, DOI 10.2307/1998700. MR531985

[Pch86] S. V. Pchelintsev, *Prime algebras and absolute divisors of zero* (Russian), Izv. Akad. Nauk SSSR Ser. Mat. **50** (1986), no. 1, 79–100, 207. MR835567

[PGRRRPVMn92] J. Pérez González, L. Rico Romero, Á. Rodríguez-Palacios, and A. R. Villena Muñoz, *Prime Jordan-Banach algebras with nonzero socle*, Comm. Algebra **20** (1992), no. 1, 17–53, DOI 10.1080/00927879208824330. MR1145324

[Pre86a] A. A. Premet, *Inner ideals in modular Lie algebras* (Russian, with English summary), Vestsī Akad. Navuk BSSR Ser. Fīz.-Mat. Navuk **5** (1986), 11–15, 123. MR876665

[Pre86b] A. A. Premet, *Lie algebras without strong degeneration*, Mat. Sb. (N.S.) **129(171)** (1986), no. 1, 140–153. MR830100 (87g:17018)

[RP85] Ángel Rodríguez-Palacios, *The uniqueness of the complete norm topology in complete normed nonassociative algebras*, J. Funct. Anal. **60** (1985), no. 1, 1–15, DOI 10.1016/0022-1236(85)90055-2. MR780101

[RP91] Ángel Rodríguez-Palacios, *An approach to Jordan-Banach algebras from the theory of nonassociative complete normed algebras*, Ann. Sci. Univ. Clermont-Ferrand II Math. (1991), no. 27, 1–57 (1992). MR1164710

[Sch66] Richard D. Schafer, *An introduction to nonassociative algebras*, Pure and Applied Mathematics, Vol. 22, Academic Press, New York-London, 1966. MR0210757

[Sel67] G. B. Seligman, *Modular Lie algebras*, Ergebnisse der Mathematik und ihrer Grenzgebiete, Band 40, Springer-Verlag New York, Inc., New York, 1967. MR0245627

[Shi57] A. I. Širšov, *On rings with identity relations* (Russian), Mat. Sb. N.S. **43(85)** (1957), 277–283. MR0095192

[SM04] Mercedes Siles Molina, *Algebras of quotients of Lie algebras*, J. Pure Appl. Algebra **188** (2004), no. 1-3, 175–188, DOI 10.1016/j.jpaa.2003.10.005. MR2030813

[SM06] Mercedes Siles Molina, *Finite $\mathbb{Z}$-gradings of simple associative algebras*, J. Pure Appl. Algebra **207** (2006), no. 3, 619–630, DOI 10.1016/j.jpaa.2005.10.005. MR2265543

[Smi97] Oleg N. Smirnov, *Simple associative algebras with finite* **Z***-grading*, J. Algebra **196** (1997), no. 1, 171–184, DOI 10.1006/jabr.1997.7087. MR1474168

[Smi99] Oleg N. Smirnov, *Finite* **Z***-gradings of Lie algebras and symplectic involutions*, J. Algebra **218** (1999), no. 1, 246–275, DOI 10.1006/jabr.1999.7880. MR1704686

[SZ19] Ivan Shestakov and Efim Zelmanov, *A finite presentation of Jordan algebras*, Internat. J. Algebra Comput. **28** (2018), no. 8, 1705–1716, DOI 10.1142/S0218196718400155. MR3899231

[The85] Armin Thedy, *z-closed ideals of quadratic Jordan algebras*, Comm. Algebra **13** (1985), no. 12, 2537–2565, DOI 10.1080/00927878508823290. MR811523

[Vin75] È. B. Vinberg, *The classification of nilpotent elements of graded Lie algebras* (Russian), Dokl. Akad. Nauk SSSR **225** (1975), no. 4, 745–748. MR0506488

[Zel79a] E. I. Zel′manov, *Jordan division algebras* (Russian), Algebra i Logika **18** (1979), no. 3, 286–310, 385. MR566787

[Zel79b] E. I. Zel′manov, *Primary Jordan algebras* (Russian), Algebra i Logika **18** (1979), no. 2, 162–175, 253. MR566779

[Zel82] E. I. Zel′manov, *Absolute zero divisors and algebraic Jordan algebras* (Russian), Sibirsk. Mat. Zh. **23** (1982), no. 6, 100–116, 206. MR682911

[Zel83a] E. I. Zel′manov, *Lie algebras with algebraic associated representation* (Russian), Mat. Sb. (N.S.) **121(163)** (1983), no. 4, 545–561. MR716113

[Zel83b] E. I. Zel′manov, *Prime Jordan algebras. II* (Russian), Sibirsk. Mat. Zh. **24** (1983), no. 1, 89–104, 192. MR688595

[Zel84a] E. I. Zel′manov, *Characterization of the McCrimmon radical* (Russian), Sibirsk. Mat. Zh. **25** (1984), no. 5, 190–192. MR762254

[Zel84b] E. I. Zel′manov, *Lie algebras with finite gradation* (Russian), Mat. Sb. (N.S.) **124(166)** (1984), no. 3, 353–392. MR752226

[Zel90] E. I. Zel′manov, *Solution of the restricted Burnside problem for groups of odd exponent*, Izv. Akad. Nauk SSSR Ser. Mat. **54** (1990), no. 1, 42–59, 221. MR1044047

[Zel91] E. I. Zel′manov, *Solution of the restricted Burnside problem for 2-groups*, Mat. Sb. **182** (1991), no. 4, 568–592. MR1119009

[Zel92a] E. I. Zel′manov, *On periodic compact groups*, Israel J. Math. **77** (1992), no. 1-2, 83–95, DOI 10.1007/BF02808012. MR1194787

[Zel92b] Efim Zelmanov, *Nil rings and periodic groups*, KMS Lecture Notes in Mathematics, Korean Mathematical Society, Seoul, 1992. With a preface by Jongsik Kim. MR1199575

[Zel17] Efim Zelmanov, *Lie algebras and torsion groups with identity*, J. Comb. Algebra **1** (2017), no. 3, 289–340, DOI 10.4171/JCA/1-3-2. MR3681577

[ZK90] E. I. Zel′manov and A. I. Kostrikin, *A theorem on sandwich algebras* (Russian), Trudy Mat. Inst. Steklov. **183** (1990), 106–111, 225. Translated in Proc. Steklov Inst. Math. **1991**, no. 4, 121–126; Galois theory, rings, algebraic groups and their applications (Russian). MR1092020

[Zm78] E. I. Zel′manov, *Jordan algebras with finiteness conditions* (Russian), Algebra i Logika **17** (1978), no. 6, 693–704, 747. MR555097

[ZS73] K. A. Ževlakov and I. P. Šestakov, *Local finiteness in the sense of Širšov* (Russian), Algebra i Logika **12** (1973), 41–73, 120. MR0399188

[ZSSS82] K. A. Zhevlakov, A. M. Slin′ko, I. P. Shestakov, and A. I. Shirshov, *Rings that are nearly associative*, Pure and Applied Mathematics, vol. 104, Academic Press, Inc. [Harcourt Brace Jovanovich, Publishers], New York-London, 1982. Translated from the Russian by Harry F. Smith. MR668355

Index of Notations

Index

SELECTED PUBLISHED TITLES IN THIS SERIES

240 **Antonio Fernández López,** Jordan Structures in Lie Algebras, 2019

237 **Dusa McDuff, Mohammad Tehrani, Kenji Fukaya, and Dominic Joyce,** Virtual Fundamental Cycles in Symplectic Topology, 2019

236 **Bernard Host and Bryna Kra,** Nilpotent Structures in Ergodic Theory, 2018

235 **Habib Ammari, Brian Fitzpatrick, Hyeonbae Kang, Matias Ruiz, Sanghyeon Yu, and Hai Zhang,** Mathematical and Computational Methods in Photonics and Phononics, 2018

234 **Vladimir I. Bogachev,** Weak Convergence of Measures, 2018

233 **N. V. Krylov,** Sobolev and Viscosity Solutions for Fully Nonlinear Elliptic and Parabolic Equations, 2018

232 **Dmitry Khavinson and Erik Lundberg,** Linear Holomorphic Partial Differential Equations and Classical Potential Theory, 2018

231 **Eberhard Kaniuth and Anthony To-Ming Lau,** Fourier and Fourier-Stieltjes Algebras on Locally Compact Groups, 2018

230 **Stephen D. Smith,** Applying the Classification of Finite Simple Groups, 2018

229 **Alexander Molev,** Sugawara Operators for Classical Lie Algebras, 2018

228 **Zhenbo Qin,** Hilbert Schemes of Points and Infinite Dimensional Lie Algebras, 2018

227 **Roberto Frigerio,** Bounded Cohomology of Discrete Groups, 2017

226 **Marcelo Aguiar and Swapneel Mahajan,** Topics in Hyperplane Arrangements, 2017

225 **Mario Bonk and Daniel Meyer,** Expanding Thurston Maps, 2017

224 **Ruy Exel,** Partial Dynamical Systems, Fell Bundles and Applications, 2017

223 **Guillaume Aubrun and Stanisław J. Szarek,** Alice and Bob Meet Banach, 2017

222 **Alexandru Buium,** Foundations of Arithmetic Differential Geometry, 2017

221 **Dennis Gaitsgory and Nick Rozenblyum,** A Study in Derived Algebraic Geometry, 2017

220 **A. Shen, V. A. Uspensky, and N. Vereshchagin,** Kolmogorov Complexity and Algorithmic Randomness, 2017

219 **Richard Evan Schwartz,** The Projective Heat Map, 2017

218 **Tushar Das, David Simmons, and Mariusz Urbański,** Geometry and Dynamics in Gromov Hyperbolic Metric Spaces, 2017

217 **Benoit Fresse,** Homotopy of Operads and Grothendieck–Teichmüller Groups, 2017

216 **Frederick W. Gehring, Gaven J. Martin, and Bruce P. Palka,** An Introduction to the Theory of Higher-Dimensional Quasiconformal Mappings, 2017

215 **Robert Bieri and Ralph Strebel,** On Groups of PL-homeomorphisms of the Real Line, 2016

214 **Jared Speck,** Shock Formation in Small-Data Solutions to 3D Quasilinear Wave Equations, 2016

213 **Harold G. Diamond and Wen-Bin Zhang (Cheung Man Ping),** Beurling Generalized Numbers, 2016

212 **Pandelis Dodos and Vassilis Kanellopoulos,** Ramsey Theory for Product Spaces, 2016

211 **Charlotte Hardouin, Jacques Sauloy, and Michael F. Singer,** Galois Theories of Linear Difference Equations: An Introduction, 2016

210 **Jason P. Bell, Dragos Ghioca, and Thomas J. Tucker,** The Dynamical Mordell–Lang Conjecture, 2016

209 **Steve Y. Oudot,** Persistence Theory: From Quiver Representations to Data Analysis, 2015

For a complete list of titles in this series, visit the AMS Bookstore at **www.ams.org/bookstore/survseries/**.